Renewable Energy - Resources, Challenges and Applications

*Edited by Mansour Al Qubeissi,
Ahmad El-kharouf and Hakan Serhad Soyhan*

Published in London, United Kingdom

IntechOpen

Supporting open minds since 2005

Renewable Energy - Resources, Challenges and Applications
http://dx.doi.org/10.5772/intechopen.81765
Edited by Mansour Al Qubeissi, Ahmad El-kharouf and Hakan Serhad Soyhan

Contributors
Boumediene Allaoua, Brahim Mebarki, Salah Tamalouzt, Nabil Benyahia, Abdelmounaim Tounzi, Amar Bousbaine, Martin Khzouz, Evangelos Gkanas, Ahmed Elwardany, Mahmoud Omar Amer, Samuel Asumadu-Sarkodie, Margaret Adobea Oduro, Samuel Gyamfi, Francis Kemausuor, Tuğçe Demirdelen, Emel Bakmaz, Kemal Aygul, Burak Esenboga, Mehmet Tumay, Khalid O. Moh. Yahya, Mohamed Salem, Nassim Iqteit, Sajjad Ahmad Khan, Huseyin Salvarli, Mustafa Seckin Salvarli, Miroslav Stanković, Pavlović Stefan, Dalibor Marinković, Marina Tišma, Margarita Gabrovska, Dimitrinka Nikolova, Shezan Arefin, İzzet Yüksek, İlker Karadağ, Remember Samu, Murat Fahrioglu, Festus Victor Bekun, Pankaj Kumar, Nitai Pal, Kumar Avinash Chandra, Himanshu Sharma, Sanjay Patel, Mohit Kumar, Carmen Luisa Barbosa Guedes, Jonathan Baumi, Caroline Milani Bertosse, Giovanni Rinaldi, Solomon Uhunamure, Nthaduleni Samuel Nethengwe, David Tinarwo, Lyubomyr Nykyruy, Grzegorz Wisz, Valentyna Yakubiv, Iryna Hryhoruk, Rostyslav Yavorskyi, Zhanna Zapukhlyak, John Paravantis, Nikoletta Kontoulis, Felix Farret, Emanuel Vieira, Faik Hamad, Muzaffar Ali, Rubina Kamal, Nadeem Ahmed Sheikh, Murat Kunelbayev, Chenzhen Ji, Kai Wang, Zhen Qin, Wei Tong, T. Cetin Akinci, Omer Akgun, Nimrah Khalid, Sarmad Ahmad Qamar, Muhammad Asgher, Belkacem Draoui, Abdelghani Draoui, Rubeena Kousar, Yedilkhan Amirgaliyev, Kalizhanova Aliya, Ainur Kozbakova, Omirlan Auelbekov, Nazbek Katayev

First published in London, United Kingdom, 2020 by IntechOpen
IntechOpen is the global imprint of INTECHOPEN LIMITED, registered in England and Wales, registration number: 11086078, 7th floor, 10 Lower Thames Street, London, EC3R 6AF, United Kingdom
Printed in Croatia

British Library Cataloguing-in-Publication Data
A catalogue record for this book is available from the British Library

Additional hard and PDF copies can be obtained from orders@intechopen.com

Renewable Energy - Resources, Challenges and Applications
Edited by Mansour Al Qubeissi, Ahmad El-kharouf and Hakan Serhad Soyhan
p. cm.
Print ISBN 978-1-78984-283-8
Online ISBN 978-1-78984-284-5
eBook (PDF) ISBN 978-1-83962-155-0

An electronic version of this book is freely available, thanks to the support of libraries working with Knowledge Unlatched. KU is a collaborative initiative designed to make high quality books Open Access for the public good. More information about the initiative and links to the Open Access version can be found at www.knowledgeunlatched.org

For EU product safety concerns:
IN TECH d.o.o., Prolaz Marije Krucifikse Kozulić 3, 51000 Rijeka, Croatia, info@intechopen.com or visit our website at intechopen.com.

Meet the editors

Dr. Al Qubeissi is a Chartered Engineer, Fellow of the Higher Education Academy, member of several engineering organisations (including UnICEG, IAENG–ISME, IMechE and the Institute for Future Transport and Cities), and Senior Lecturer and Course Director for Master of Engineering programmes at CU. He is experienced in computational thermo-fluids, biofuels and energy systems. Other relevant expertise includes turbine combustion, PV/T and battery thermal management. His research efforts have been disseminated via 10s of publications in high impact refereed journals, conference proceedings and books. Since joining CU in 2015, Dr. Al Qubeissi has been involved in leading 10s of research projects and PhD theses. Prior to that role, he was a Lecturer at the University of Brighton, UK.

Dr. El-kharouf is a Research Fellow at the Centre for Fuel Cell and Hydrogen Research, School of Chemical Engineering, UoB, UK. He is the manager of the EPSRC Centre for Doctoral Training in Fuel Cells and their Fuels; a collaborative consortium of UoB, UCL, Imperial College London, University of Nottingham and Loughborough University. His primary research is focussed on the development of next generation automotive Polymer Electrolyte Fuel Cells (PEFCs), aiming for an operation temperature of up to 120°C and increasing the stack volumetric and gravimetric power density. Dr El-kharouf's other research activities include the development and testing of materials for PEM electrolysers, solid oxide fuel cell development and the integration of fuel cell and hydrogen technologies in transport and stationary applications.

Professor Soyhan has been a member of the Department of Mechanical Engineering, SU since 1992. He received his BEng (1992), MSc (1995) and PhD (2000) from Istanbul Technical University and undertook post-doctoral research in chemical kinetics at the Combustion Physics Division, Lund University, Sweden and on HCCI engines and chemical kinetics at Shell Global Solutions, Chester, UK. Currently, he is working on fuels and combustion studies in transport. He is the Head of Local Energy Research Association and Head of the Combustion Institute, Turkey. Professor Soyhan is the Founder and Director of TeamSan Co, a member of the Turkish Society of Mechanical Engineers, and an associate member of TUBITAK USETEG Committee on R&D projects of the Transportation, Defence and Energy Technologies Group.

Contents

Preface

Nowadays, most human activities, from stationary applications of power generation, construction and industrial production to transport in all its forms of automotive, railway, marine and aviation, are reliant on fossil fuels. In addition, there is a constant increase in demand for energy worldwide that exceeds the currently predicted conventional energy resources available, i.e., fossil fuels. However, the consumption of fossil fuels causes significant damage to global and regional environments due to emission of harmful greenhouse gases. The increasing pressures to protect the environment from greenhouse gas emissions while maintaining energy security calls on scientists and engineers to create and develop new technologies to meet the growing energy demands. Hence, the motivation for collating relevant findings and solutions in this book.

The importance of renewable energy resources to any society comes from the potential to address the need for clean, sustainable and affordable energy for all. In this book, an overview of emerging energy technologies is provided to address the energy system challenges. Selective research inputs were reviewed and organised with the aim to inform a broad range of beneficiaries from public readers to subject experts of the recent developments, and to provide critical analysis of the latest developments. The book highlights the global nature of the energy challenge by drawing on activities and case studies from all over the world. The book demonstrates many emerging approaches and technologies adopted for different countries based on their available resources, the current energy system and the economic situation. In the last few decades, several new technologies and novel proposals have been made for replacing the conventional depleting energy systems with sustainable clean energy alternatives. These are collated in the book and a synopsis of the topics addressed is provided as follows.

Provisions of renewable and efficient energy systems are key parts of the solution, but these are complicated with several political, economic and resource associated challenges. At present, the contribution of renewable energy to the world primary energy is insufficient to meet the required primary energy supplies. As described in Chapter 1, both developed and developing countries will continue to rely on the depleting fossil fuels in the coming decades. Many developing countries have been trying to restructure their energy sector. However, this is proving to be a very challenging task. New technology costs, market shares and politics are the main barriers for new energy developments to penetrate the market. Countries' investment into clean and sustainable energy systems is balanced against the need for rapid economic growth and the impact on social and industrial factors.

Many challenges regarding the feasibility of these solutions have been identified in Chapter 8. For example, the recent global trend of promoting electric vehicles as 'a silver bullet solution' to achieving clean transport is simplistic and optimistic. Transforming the transport sector into an electric based system makes it highly dependent and interlinked with the stationary energy system through the electric grid, and therefore a decarbonisation of transport requires a decarbonisation of the whole energy system. Alternative energy resources, such as biomass, wind and solar energy, are used to define an environmentally friendly and cost-effective approach. Such an approach can be adopted as a standalone energy provider or used in an existing power system with minimal modification; for example, some biofuels can be mixed with conventional fossil fuels without modifications to the internal combustion engines. If this is achieved, such an approach will be of great benefit to industries and societies, with direct impacts on the environment and economy. For instance, different types of bio-/fossil-fuel blends can lead to important products and applications.

It is important to understand the role of renewable energy in shaping energy security against the backdrop of global geopolitical, socioeconomic and technological uncertainties. The evolving definition of energy security during the 20th and early 21st century is discussed initially. The dimensions, components and metrics of energy security are reviewed in Chapter 2, including a novel definition of energy security that comprises physical availability, economic affordability, accessibility from a socio-political standpoint and environmental acceptability. The role of wind and solar energy are highlighted, with emphasis on the social acceptance of renewable energy in an energy security context. Other energy security indexes are discussed, focusing on sustainability and renewable energy. Denmark, Germany, China, Russia and the US are examined as case studies that help us to understand the transition to renewable energy in the context of co-operation among states.

Advancement in energy policies has stimulated the adoption of instruments used in the renewable energy sector and climate change mitigation. Renewable energy policies play a crucial role in the abatement of greenhouse gas emissions, by providing access to modern energy and energy security by diversifying energy supply. For example, and as described in Chapter 4, there are numerous policies developed in Ghana to improve the uptake of renewable energy for electricity production and to ensure efficient use of electrical energy. Some of the specific government policy objectives include reducing technical and commercial losses in power supply, supporting the modernization and expansion of the energy infrastructure to meet the growing demands, ensuring reliability and accelerating the development and utilization of renewable energy and energy-efficient technologies. These policies have defined targets and a defined period in which to be implemented. In the current book, a comparative analysis is made with South Africa's and Morocco's renewable energy sectorial policies on the basis of various adopted strategies to their achievements and what policy makers in Ghana can learn from that.

Despite the enormous advantages associated with biogas technology, the level of awareness and perception remains low. It is therefore important to increase awareness of the technology at household level in critical energy regions, such as South Africa. In Chapter 3, a case study is made for Limpopo Province of South Africa, where 200 households are sampled of which 72 households use biogas digesters and 128 are without digesters. Primary data collection is obtained with the use of open and closed-ended questionnaires. The study recommends interventions through more elaborate awareness and promotion programmes in disseminating the technology as well as provision of technical assistance, loans, credits and subsidies to households willing to adopt the technology. Also, an examination of the nexus between energy consumption, total population, greenhouse gas emissions, and per capita GDP is carried out in Chapter 5 to forecast another example of a policy challenged region (Zimbabwe) for its energy consumption by 2030. A time series data from 1980 to 2012 is utilised alongside econometric techniques to explore the causal relationship amongst the variables under review. The stationary test reveals the integration of all the data series of interest of order one. The autoregressive integrated moving average model forecasts Zimbabwe's 2030 energy demand around 0.183 Quadrillion BTU against the current 0.174 Quadrillion Btu. The empirical finding is indicative for policy and decision makers who design the energy policy framework geared towards achieving the universal access to modern energy technologies in Zimbabwe.

Fundamental and applied research on renewable energy is actively supported both in terms of making significant contributions to the scientific development, addressing the issue of energy independence and security of different countries. Detailed analyses of research findings in the fields of thermoelectricity, photoelectricity and bioenergy are made in Chapter 6 for two typical East European countries: Poland and Ukraine. This is to find regularities in the development of these areas in each state and to determine the prospects for joint research. Comparison of the state of research in these countries is an example of the analysis of the situation at the EU borders and may answer questions related to sustainable development, the mass transition to renewable energy, the refusal to use fossil fuels and nuclear power plants. Particular research is considered in each state and the possibility of organising joint scientific research in order to mutually enhance the scientific potential is demonstrated.

The feasibility analysis on the different challenges in deployment of renewable energy is made to understand the global sustainability factors and envisage their future. The emphasis on cost and efficiency for technological advancement is the basic element for mass adaptation of renewable energy. At the same time, huge available resources, favourable economies and large social-economic benefits attract major parts of the globe to the transition from conventional to renewable energy. A typical example of such a scenario is noticeable in the case of India. In Chapter 7, the authors have explored the major options and barriers towards the deployment of different renewable energies in India, which will act as a catalyst to achieve India's dream renewable energy target of 175 GW by 2022.

Economic development experienced by the world during the last two decades has led to a strong growth in demand for energy in the transport sector, with major developments in the different modes of transport. Currently, with facing the global energy challenges, the transport sector is still heavily dependent on petroleum products – a major strategic global issue. For instance, the aviation industry consumes about 177 billion litres of kerosene, moving more than 25,000 aircraft and 6 billion passengers. To achieve that, in 2015, the civil aviation industry generated about 781 million tons of CO_2 corresponding to 2% all anthropogenic emissions of this greenhouse gas and all required energy is derived from fossil sources. To reduce the environmental impact, we must create alternative energy sources to bring energy security and to do this, it is paramount to increase research and development of viable production of bio-kerosene. The authors aim to present some varieties of biomass and its derivatives being studied as raw material for new aviation fuels such as ethanol, butanol, fatty acid methyl esters and fossil oils in Chapter 12.

Air pollution, global warming and the rapid depletion of oil resources of the planet are now paramount concerning problems. Hence, the authors have provided a brief overview on some alternatives in the automotive engineering applications, such as electric vehicles, hybrid electric vehicles and technology of fuel cell electric vehicles in Chapter 9. There will also be a focus on hydrogen technologies for both stationary and mobility/transportation applications. Hydrogen production from sustainable resources for the generation of pure and low-cost hydrogen is described in Chapter 13.

Coal fly ash (CFA) is generated during combustion of coal for energy production. Many studies are based on its utilization as the most abundant, cheap aluminosilicate industrial residue, which is recognized as a risk for the environment and human health. The authors have given a focus on the origin of CFA, its chemical properties and its catalytic application for biodiesel production. The aluminosilicate nature and the presence of rare earth elements make CFA suitable for different adsorption, catalytic and extraction processes for obtaining valuable products including alternative fuels and pure elements. However, the presence of toxic elements is a potential environmental problem, which should be solved to avoid soil, water and air pollution. The proposed modification methods and properties are discussed in Chapter 10. Also, carbonization is the art of reinventing waste biomass into a carbon/energy rich charcoal. It therefore adopts the principles of renewable energy. Biochar production is not a new process, but the renewed interest in it with biomass is a great opportunity for commercial and scientific applications. The carbon can be extracted from the produced char to form the precious graphite and graphene. In Chapter 11, the authors provide a general overview about slow pyrolysis processes including carbonization and the torrefaction process, which is a mild carbonization process. Different carbonization processes and methodologies that vary in the process parameters will be addressed and the most promising ones are highlighted.

Bioplastics and bioinspired nanocomposites with nanoscale reinforcements are used in a broad range of applications, such as biomedical, electronics, durable goods and packaging materials. Several drawbacks of conventional materials such as hydrophilicity, low-heat deflection, poor conductivity and barrier properties can be efficiently overcome using biohybrid nanomaterials. Nano-reinforcements in composite materials deliver remarkably improved properties such as a decrease in hydrophilicity and increase in mechanical properties as compared with a neat biopolymer, which fails to exhibit these properties on its own. In Chapter 8 the authors present the recent trends towards nano-functional materials, renewable materials that are being applied to produce nano-bio-composites and their applications, especially in the biomedical and healthcare sector. This emerging concept will enhance the scope of nanohybrid materials for sustainable product development with improved properties than previously seen in synthetic polymer-based or natural polymer-based materials.

The nexus between the intermittent renewable energy supply and the current demand requires advanced control strategies. Chapter 14 deals with power control of a PV/wind system for power generation with dynamic input dataset. The main contribution of this chapter is that it is the first time using real data from a PV/wind system and observing the system reliability with real-time simulation results. The proposed system consists of a doubly-fed induction based wind generator, rotor-side converter, grid-side converter, solar arrays, dc-dc converter, grid and dynamic loads. The aim of the proposed strategy is to use wind and solar energy with maximum efficiency by simulating the real condition of wind and insolation with input datasets. The modelling and the validation of the operation of the system and its controllers is done using PSCAD/EMTDC.

The growing demand for renewable energy sources, in particular for solar technologies, requires more detailed studies to increase power and efficiency. Among them, thermoelectric energy conversion is a well-known technology used for decades including: solar thermal generators (STEG), radioisotope thermoelectric generators (RTG), automotive thermoelectric generators (ATG) and thermoelectric generators (TEG). Chapter 15 demonstrates that the thermoelectric effect (Seebeck effect) can be used to harness the thermal energy retained in photovoltaic panels to increase their overall efficiency with its direct conversion into electrical energy and vice versa. It is also observed that solar radiation can be converted directly into electric energy, as in photovoltaic modules, or can be converted directly into electricity, as in thermoelectric modules. It is emphasised that although the energy conversion by thermoelectric effect still has low electrical efficiency, this source is characterised by a high degree of reliability, low maintenance, appreciable durability, absence of moving parts and it allows the generation of electric energy through recovery of the thermal energy from several industrial processes. Various solar air conditioning technologies such as solar PV, absorption, desiccant and adsorption cooling systems are overviewed in Chapter 16. It includes feasibility and comparative analysis of numerous standalone and hybrid configurations of solar cooling systems that have been investigated in the past. In addition, recent developments in the use of solar energy as a regeneration source to dehumidify desiccant wheels in different applications are also discussed. Details of system technologies and climate-based performance comparison in terms of various performance factors e.g., COP_{th}, Q_{latent}, $Q_{sensible}$, COP_{solar}, SF, PES and $\eta_{collector}$ for solar assisted configurations are highlighted. It is observed that hybridization of solar solid desiccant systems results in a more efficient and cost-effective cooling system as latent and sensible loads are treated independently, especially when the regeneration process of the desiccant wheel is integrated with solar energy. In Chapter 17, the authors have considered the thermal and hydraulic analysis of transfer media motion mode in the flat solar collector. The authors have substantiated the thermal and hydraulic parameters of the flat plate solar collector. Solar flat plate collector's heat absorbing tubes hydraulic analysis has shown that using the heat transfer standard size might be located in a pipeline length of 2.5 times more than of the collector's body.

Offshore renewable technologies hold the potential to satisfy a considerable amount of the global energy demand in the coming years. In Chapter 18, the main sources of renewable energy related to the oceans (waves, tides and offshore winds) will be characterized and discussed, with reference to the challenges related in their use. Thus, the main devices capable of exploiting these resources will be presented. Their working principle, together with operational and technological requirements will be described, highlighting strengths and weaknesses of each technology and providing examples of past and current experiences. Elements of project management, as well as environmental impact and public perception, will be included. In Chapter 19, a wind speed analysis is made using the Signal Processing Technique. The formation of wind energy and the necessary stages to produce electrical energy are discussed. The hourly wind speed data of the last ten years of Istanbul was studied and analysed using the Fourier method. The data obtained from the study can be considered as a background for the wind farms in the region. Also, tall buildings have the potential of generating wind energy. However, there are many challenges to incorporating wind generation into urban areas. These include planning issues besides visual impacts. As to integration, there is a need for a combined approach that considers wind energy harvesting besides these issues. In Chapter 20, a multidisciplinary approach is made to fill the gap between the architectural design and the wind engineering processes. The chapter presents design strategies from the literature to integrate wind energy to tall buildings using computational fluid dynamics simulation.

Solar energy increases its popularity in many fields, from buildings, food productions to power plants and other industries, due to its clean and renewable properties. To eliminate its intermittence feature, thermal energy storage is vital for efficient and stable operation of solar energy utilization systems. It is an effective way of decoupling the energy demand and generation, while playing an important role on smoothing their fluctuations. In Chapter 21, various types of thermal energy storage technologies are summarized and compared, including the latest studies on thermal energy storage materials and heat transfer enhancements. Then, the most up-to-date developments and applications of various thermal energy storage options in solar energy systems are summarized, with an emphasis on the material selections, system integrations, operational characteristics, performance assessments and technological comparisons. This chapter will be a useful resource for relevant researchers, engineers, policy-makers, technology users and engineering students in the field.

More contemporary technology must offer more exceptional energy-efficient applications at a lower cost. New technology must also have an ability to generate electric power from the conversion of wasted heat. Thermoelectric generators (TEGs) and their applications have gained momentum for their ability to use waste thermal energy. As shown in Chapter 22, TEG has demonstrated its efficiency and how it can offer increased potential by adding an MPPT algorithm to increase the power flow while decreasing the cost of operation. The limitations can be offset using lower-cost manufacturing materials and automated systems in the TEG units.

The rapid increase in demand for energy utilization is an unavoidable fact. Such large demand cannot be satisfied by the conventional power sources alone. Sustainable power sources (for example, solar and wind turbine-based energy systems) are the most effective and both economically and environmentally feasible. The hybrid renewable energy system is a recent concept in the field of sustainable development, which joins at least two renewable power sources like a wind turbine, solar module and other inexhaustible sources such as ocean energy, fuel cell etc. In Chapter 23, a survey is made on efficient hybridization of two sustainable sources such as solar modules and a wind farm into a perfect joined stage that can make this system more dependable and monetarily possible to operate. Simulation of a micro-grid connected renewable energy system is presented in Chapter 24, which comprises of a wind turbine based on doubly fed induction, photovoltaic and fuel cell generators. A hydrogen tank and water electrolyser are facilitated as long-term storage, and battery bank is utilized as a short-term storage. A global control strategy and an energy management strategy are proposed for the overall system. The effectiveness of this contribution is verified through computer simulations and satisfactory results are obtained.

This book aims to present the available energy resources, their emerging applications, and the recent advances in computational and experimental solutions. It is envisaged that this book will facilitate the formulation of new energy technologies, such as alternative fuels, effective solar energy harnessing techniques and approaches to facilitating renewable energy on a broader range of applications. The development and implementation of such technologies will support green economic growth, protect the environment, ensure energy security and create new jobs.

Dr. Mansour Al Qubeissi
Coventry University,
United Kingdom

Dr. Ahmad El-kharouf
University of Birmingham,
United Kingdom

Hakan Serhad Soyhan
Professor,
Sakarya University,
Turkey

Section 1

Economic and Regional Challenges

Chapter 1

For Sustainable Development: Future Trends in Renewable Energy and Enabling Technologies

Mustafa Seckin Salvarli and Huseyin Salvarli

Abstract

Energy demand in the world is nowadays growing further out of limits of installable generation capacity. Therefore, future energy demands should be met and improved efficiently and securely. Energy solutions should be supported by utilizing renewable energy sources. At present, the contribution of renewable energy to the world primary energy is not high to meet the primary energy and electricity supplies. Both developed and developing nations will necessarily continue to rely on fossil fuels in the coming decades. In developing countries, the situation is more inconvenient than that for developed countries. Many developing countries have been apparently trying to restructure their energy sectors. It seems that it is difficult to realize innovations. Cost, market share and policy are the main barriers for the development of renewable energy. In the strategy plans of many countries, the sustainable development in relation to the parameters such as economic, social and industrial is supported by their energy policies. New enabling technologies related to renewable energies will also help to reduce environmental costs, and thus the energy systems will be operated as both securely and economically without environmental problems. New renewable energy markets are surely required in both the wholesale and retail markets.

Keywords: renewable energy, sustainable development, environment, energy mix, trends, strategies, enabling technologies

1. Introduction

The demand for energy increases enormously. As indicated in [1], the industrial countries have 28% of the world's population, and they consume 77% of the world energy production. It is expected that today's world population will increase 1.26 times to reach 9.7 billion in 2050. Most of the world's population which include 90% of the population growth belong to the developing countries. By 2050, although the developed countries will be adopting more effective energy conservation policies, their energy consumption will not increase. However, in the developing countries, people generally have an aim to construct their own electricity-generating facilities.

According to the data given in Ref. [2], about 75% of the final energy demand and 67% of the electricity supply in 2016 will be met by the fossil fuels. As a basic energy resource in the world, coal is very important, and it is expected that its usage will be increased by 27% over 20 years.

 IntechOpen

It is expected that the reserves of fossil fuels will naturally come to an end. Thus, the alternative and renewable energies will be the most significant energy resources in the near future. This situation will be a reason to generate new jobs and to develop future industries.

The environment is being increasingly polluted because of rapid industrialization and human work. Sustainable development mainly covers the use of renewable energy, energy security, energy pricing, energy policy, renewable energy applications and smart grid technologies.

Two trends are currently related to the consumption of fossil resources and the global climate change. Renewable energy is fast emerging to both these problems. For the level development and life quality in a country, energy consumption is one of the most reliable indicators.

The data of parameters—such as economical, political, and partly environment and human life—are related to the present energy systems. According to the most of energy policies, the fundamental parameters are to save energy and use domestic energy sources. However, there will be a close relationship between the energy use and environment in the future.

While planning and building all industrial plants, their effects on the environment should be taken into consideration for improving the economy, supporting ecology and saving energy. Energy investments related to the environmental protection are to necessitate large financial resources. The success of any new technology will be measured by the parameter of cost-effectiveness that improves the environment. Thus, the growing energy demand in the world will be met by means of the clean power generation. It is a fact that clean and affordable energy will power progress toward achieving the sustainable development goals.

The emerging trends and new insights open up significant new business opportunities for the energy leaders and organizations to inform better decisions and enabling new technologies [3]. According to the review carried out on technology trends, the results may be identified and grouped as renewable energy, advanced materials and nanotechnology, advanced manufacturing technologies, information society technologies, life sciences, aerospace technologies and biotechnology, global change, green energy and ecosystem. In order to have high market growth and solve social problems, these technologies support strategic sectors, too [4].

2. Overall distribution of energy resources

At present, the contribution of renewable energy is not high to meet the primary energy and electricity supplies. Appropriate cost reductions, increase of the renewable energy industry and technology improvements are firstly related to government policy precision, private sector inventiveness and investment. In **Table 1**, the share of oil in total primary energy supply is the first level, and the fossil fuels include around 81% of total. It is aimed that the renewable energy will be used to displace fossil fuels as both environmentally safe and economically sustainable.

In the developing countries, the largest renewable energy source of global renewable supply, which includes solid, biofuels and charcoal, is 60.7% due to its use for residential heating and cooking. In **Table 2**, it is seen that the second largest source is hydropower, which provides 18.5% of renewable. The rest of renewable makes up a smaller share.

On the other hand, as will be seen in **Table 3**, the majority of renewables are consumed in the residential, commercial and public services sectors.

Renewables, which account for 24.5% of world electricity production, are the second largest contributor to global electricity production (**Table 4**).

Resources	Share (%)
Oil	31.8
Coal	27.1
Natural gas	22.2
Biofuels and waste	9.2
Nuclear	4.9
Hydro	2.5
Solar, wind, geothermal and tidal	1.8
Other	0.3

Table 1.
Fuel shares in world total primary energy supply (2017) [5].

Resources	Share (%)
Solid biofuels and charcoal	60.7
Hydro	18.5
Wind	5.1
Liquid biofuels	4.6
Geothermal	4.5
Solar, tidal	3.9
Biogases	1.7
Renewable municipal waste	0.9

Table 2.
Product shares in world renewable energy supply (2017) [5].

Sector	Share (%)
Residential, commercial and public	41.7
Electricity plants	35.1
Industry	10.5
Transport	4.4
Combined heat and power plants	3.0
Heat plants	0.5
Other	4.8

Table 3.
World sectoral consumption of renewables (2017) [5].

The average values for the world's total final consumption by sectors in 2017 are given in **Table 5**. The sectors of industrial, transport and residential energy use account for 37, 29 and 22%, respectively.

According to the data given in **Table 6**, renewable will have the fastest growth in the electricity sector, providing 29.4% of power demand in 2023, up from 23.9% in 2017.

Bio-energy (as solid, liquid or gaseous fuels) is, however, the largest source of growth in renewable consumption over the period 2018–2023 and will account for 30% of the growth in renewable consumption in this period due to the use of

Resources	Share (%)
Coal	38.5
Natural gas	23.0
Hydro	16.0
Nuclear	10.3
Solar, wind, geothermal and tidal	6.5
Oil	3.3
Biofuels and waste	2.0
Other	0.4

Table 4.
Fuel shares in world electricity production in 2017 [5].

Sector	Share (%)
Industry	37
Transport	29
Residential	22
Commerce and public services	8
Agriculture	2
Forestry	2
Other	2

Table 5.
World total final consumption by sector (2017) [6].

	Year	
	2017	**2023**
	Share (%)	
Renewable electricity	23.9	29.4
Renewable heat	10.3	11.8
Biofuels in road transport	3.4	3.8

Table 6.
Shares of renewables in 2017 and 2023 [7].

bio-energy in heat and transport. On the other hand, the rest of renewable, which include the 80% of the total final energy consumption, have less influence in the sectors of heat and transport. As predicted, although solar PV and wind energies continue to grow in the electricity sector, bio-energy will keep its place in the first level. During the period (2018–2023), renewables such as solar PV, wind, hydropower and bio-energy are expected to meet about 70% of global electricity generation growth. By 2023, the global electricity demand will be met by hydropower (16%), wind (6%), solar PV (4%) and bio-energy (3%). Biofuels in road transport have the lowest share of renewable, which is 3.4% in 2017 and 3.8% in 2023. Renewable heat consumption is also expected to reach a share of 11.8% by 2023. Due to the weaker policy support and additional barriers to deployment, the growth of renewable use in the transport and heat sectors is slower [7].

3. Energy and sustainable development

The energy systems can be an important reason of environmental impact for both developing and developed countries. Thus, a sustainable global energy system should provide to optimize efficiency and limit emissions. The technology and the global economy must also develop in harmony with a sustainable and steady development.

As the consumption of energy, especially from fossil fuels, increases, the global environmental problems are inevitable. Both developed and developing countries plan to enable the most appropriate energy systems and improve human, economic, social and environmental conditions for sustainable development. At present, there can be several challenges such as demographic, social, economic and technological trends for the long-term sustainability of the global energy systems.

As concluded in Ref. [2], to obtain sustainable energy systems, vigorous action should be mostly taken in the areas of energy diversity and efficiency, supply reliability, public trust, market-sensitive interventions, market-based climate change responses, cost reflective prices, technological innovation and development and regional integration of energy systems.

Government policies should be carefully planned for the production, replacement, transportation, distribution and usage of energy. Due to the energy-related environmental problems and challenges, countries should aim to protect the climate system, improve their policies and implement related preventions. Thus, the standards on reducing local air pollution should be also strengthened and implemented effectively and efficiently.

Dependence on conventional fossil resources, which is mostly produced in politically unstable countries, the current energy supply and use are highly unsustainable. To meet the present and future demands for improving conditions—such as human, economic, social and environmental—fundamental changes in technologies will be required everywhere. Some topics such as innovation, investment, work, organization and leadership should be taken into consideration.

There are three groups of critical factors shaping the energy future which are the global politics and economic situation, technology and energy policy and market development [8]. To ensure the energy need of a country, the environment, cultural heritage and rich natural sources should be applied. On the other hand, energy generation, transmission, distribution and trade should be also supported by using standardized equipment and materials.

Although the use of coal creates risks in local environmental pollution and greenhouse gas emissions, it somehow increases energy security. Carbon dioxide emissions per unit power at the point of use are high for coal. However, resources, such as coal and gas, will remain important [6, 8].

Diversification and utilization of the country's resources are always the key components that ensure sustainability and low-cost energy supply. The next investments on industry should be made for the clean technologies. Depending upon the technological developments, the quality of the cleaner environment will be also affected by the economic and political factors. To provide the resource diversity, the use of domestic renewable energy resources such as hydro, wind, solar, geothermal and biomass should generate more electricity.

By 2040, the world's energy supply mix will, however, consist of oil, gas, coal and low-carbon sources. As expected, to tackle pollution and reduce CO_2 emissions, the use of coal should be constrained. Renewable energies are both environmentally safe and economically sustainable when compared to fossil sources of energy.

The use of hydropower can ensure many profits for water supply and for irrigation in agriculture, but it has consequences for the aquatic ecosystems.

Geothermal power plant is sustainable and emits low emissions when compared to the conventional fossil fuel plants. If the pollutants are released from the power plant, an environmental damage could occur. Therefore cooled geothermal fluids are injected back into the earth, and the environmental risk is reduced.

The environmental impact of wind power when compared to the environmental impacts of fossil fuels is relatively minor. Depending on specific circumstances, the siting and operation of wind turbines may cause negative health effects on people who live in the vicinity of wind turbines.

The use of solar energy is rapidly increasing all over the world. There are, however, many arrangements on solar thermal and PV installed power, and it is also expected to be the same for the concentrating solar power systems.

Bio-energy is produced from biomass, which is a clean energy resource in relation to the type of biomass and conversion technology used.

4. Energy security, sustainability challenges and expectations

Energy, briefly, is principal to the challenge of sustainability with regard to the social, economic and environmental parameters. Thus, various environmental, economic, and development needs are associated with the transition to sustainable energy resources and systems. The local renewable resources, installation costs and policy structure will be principal factors.

Although the environmental impacts from energy production and use are local, the significant impacts related to the transport of pollutants in the atmosphere can occur on regional, continental and even transcontinental scales.

While electricity demand and sustainable development are rapidly growing worldwide, the goals of energy policy considering energy mix, efficiency, market and environmental standards should also be created to provide several rehabilitations on unlicensed electricity generation and renewable energy resources. Several main elements of the policies can be as follows:

- To ensure better free market prices than feed-in tariff
- To give extra encouraging sales tariff or domestically produced parts of renewable energy power plants
- To give priority to renewable energy when connecting to grid

Developing countries face energy challenges that are significant and increasing. However, many developing countries have some advantages in attempting to restructure their energy sectors and can have an opportunity to build cleaner and more efficient technologies. It is obvious that the situation in many ways for developing countries is more difficult than that for developed countries. A significant part of the population can have some difficulties in reaching to basic energy services due to the resource constraints. Many conventional technologies are likely to remain cheaper than sustainable energy technologies.

For the sustainable development of any country, renewable energy sources should be provided due to depleting fossil fuel level, climbing fossil fuel prices across the world and reducing environmental impacts. In order to meet future energy challenges, major types of renewable energy sources which have huge potential are solar, wind, hydro and biomass [9]. To have sustainable energy supply, as detailed in Ref. [10], there are several requirements such as climate compatibility, sparing use of resources, low risks, social equity and public acceptance that should be satisfied.

5. Barriers to renewable energy

While developing renewable energy, a lot of problems and barriers are faced. Some technologies have commercialized and industrialized to some extent, and with regard to the technologies, industries' scale and developing speed compared to those in developed countries are inevitable and have large gaps. The following barriers for the development of renewable energy may be classified in three groups:

1. Cost barriers. Traditional energy sources have a lower cost and price than renewable energies. As the production cost of renewable energy is higher than that of fossil fuels with the same technology, there are serious barriers for the commercialization and distribution in relation to renewable energy. The main reasons in high production cost of renewable energy are small scale and low production technology.

2. Market share barriers. The current development of renewable energy includes the cost barriers. However, a system operation reliability and decreasing production cost can be obtained by means of a developed market.

3. Policy barriers. Policy enactment and implementation are distinctive elements of the policy process. In the future, renewable energy should be developed to an industrial scale. Thus, depending on the support of policies, market share of renewable energy has to be increased.

In order to have more sustainable lifestyles, barriers related to the societal and cultural patterns must be prevented, and thus desirable and more sustainable alternatives and various incentives will be required. The current economic system still remains a barrier to change due to the existing belief in unlimited natural resources and in continuous economic growth.

On the other hand, the existing construction sector is a rather conservative industry. It is well known that new and more sustainable designs, building materials and construction methods are only emerging and being implemented slowly. The other challenge for the energy efficiency of buildings is related to high costs and long payback period for renovations.

6. Strategies, policies and measures of renewable energy developments

In implementing sustainable strategies, renewable energy has become a critical choice for countries. It is an inevitable situation that energy is the key element to drive social and economic development. However, as the fossil energy is commonly used, sustainability of the economy and environment suffers.

Renewable energies are clean and nonpolluting. They support and impel the goal of sustainable development. Therefore, the development of renewable energies is accelerated by forming policies and legislation with the fundamental incentives. The strategic goals realized for the renewable energies are basically increasing energy competitiveness, securing supply and protecting the environment.

Renewable energy resources are also chosen to substitute fossil fuels for organizing the energy structure and improving the energy supply safety. As renewable resources are local resources, they can be transformed directly or indirectly into electricity or liquid fuels.

In rural areas, the development of renewable energy resources can solve the problem of energy consumption and combine with the agriculture production procedure that increases the income of farmers. It is estimated that in 2050 renewable energy will approximately account for 30% of energy structure in the world.

The development of renewable energy relies on technology innovation and improvement of new high technology level that belong to industrialization and commercialization. It is a fact that the cost of renewable energy development is in some degree high. If the government's support and policy presentation cannot ensure a large-scale development, countries will not help to reduce cost, increase profit, maintain reliability and improve value of renewable energy.

Renewable energy is a basis for energy system in the future and supplies urgent needs for its environmental impacts, sustainable development and usage. Due to the current problems of energy and environment problems, it is necessary to impel the development and trends of renewable energies [11].

7. Global future trends in renewable energy

In Ref. [12], it is detailed that the global energy trends and their possible impacts are related to the issues such as supply and demand, energy access and environment and air pollution. Current policies have been planned to meet long-term climate goals under the Paris Agreement, and they will serve to reduce air pollution and ensure universal energy access. Due to the falling costs and supportive government policies, renewable technologies are being the first choice in power markets. It is possible that an enterprising utility will prefer to provide renewable energy at a low and fixed price to attract investment.

By 2040, it is expected that the global power mix will be sharing renewables in generation rising to over 40%. However, coal and gas will maintain to be the largest resources of energy.

Future electricity markets depending on the variability in supplies and power systems will have flexibility and adaptability. Market reforms, grid investments and new enabling technologies are required to be active in sharing of renewable [12].

As mentioned in Ref. [13], energy technologies have focused on the spreading of clean energy technologies with regard to the opportunities and challenges. The competitiveness and future development of industries are under the impression of global technology trends. Industrial dependency on foreign technology should be reduced by identifying problems for innovation and technology. At the present, to improve development of technologies, basic international trends can be as follows:

- Technology union
- Information and communications technology
- Digitisation
- Emphasis on high technology industries
- Recognition of importance of transnational corporations

While preferring an appropriate energy source in the energy mix, factors such as technological innovation, cost efficiencies, energy storage technology and increasing consumer demand are important for the handling of renewables and alternative resources. On the other hand, the growing influence of offshore wind will also attract new investors and could see more onshore wind suppliers and developers pivot into the sector.

8. Enabling technologies and applications

While the global energy demand is growing and the installation of new power plants is required, energy security and reliability should be improved, and alternative energy sources should be also investigated.

As concluded in [14], the elements such as high research and development intensity, rapid innovation cycles, high capital expenditure and highly skilled employment are provided to develop enabling technologies. The processes for goods and service innovation are met by the enabling technologies that are also multi-disciplinary and supporter of technology leaders on research attempts. Enabling technologies are essentially selected as follows:

- To address global challenges such as low-carbon energy or resource efficiency
- To support the development of new products
- To stimulate economic growth and provide jobs

In order to realize global renewable energy trends, a requirement for the combination of enabling trends and demand trends is provided to decrease costs and improve integration. Current enabling technologies are as follows:

- Advanced materials
- Advanced manufacturing systems
- Micro and nano-electronics
- Nanotechnology
- Industrial biotechnology
- Photonics

As also outlined in [14], to address societal challenges and accelerate the development of economy and the energy transition, advanced materials, advanced manufacturing systems and industrial biotechnology are essential. Due to the current digital evolution and the enormous benefits, the digital technologies are integrated into the process technologies, materials development and business model creation. Enabling technologies will also accelerate the creation of new markets, growth and jobs. There are primary technology developments and initiatives that are needed as follows:

- Creating advance materials for the use of energy efficiency (e.g. light weight), renewable energy generation and energy storage (e.g. battery components) or smart functionalities responding to stimulants (e.g. self-repair). The advance materials also create materials for construction, energy, mobility, food, health and electronics. 3D-printing polymer materials are used in automotive sector, lightweight design, medical sector and 3D printing.
- Developing advance process technologies and industrial biotechnology for more sustainable generation and alternative energy resources.

- Leveraging digital technologies for the use of advanced process control, enabling business models and creating new customer experiences. Digital technologies enable the transition from batch to flexible continuous intensified processes.

On the other hand, the development of technologies to turn CO_2 into a valuable resource and for its implementation in making polymers can help to reduce the use of petroleum. Process technologies enable the transformation of raw materials into materials which have a different chemical composition structure and properties than the input raw materials. Advanced process technologies are a specific type of enabling technologies that enable the chemical industry to provide all industrial value chains (e.g. construction, automotive, medical, electronics, energy) with the materials (solid, gas and liquid) and novel properties required to produce a vast range of user products.

As explained in [15, 16], solar energy can be converted to both electricity and thermal energy simultaneously by a hybrid photovoltaic/thermal (PV/T) system which can also maintain the energy demand of buildings. While designing PV/T systems to match with the operating conditions, performance analysis of such systems is important. The energy flow analysis should be used to consider economic limitations and applications. As expected, solar power has several advantages and has more competitive levels than other energy sources. For installation of onshore wind turbine, there may be a problem in land adequacy. Therefore, if the location is appropriate, offshore wind turbine that has higher economic costs is, however, an alternative development in these days.

Main challenges for wind energy are technical, social and environmental. However, for both developed and developing countries, wind energy also becomes an effective choice in maintaining green environment [17].

Basic emerging visions for future sustainable lifestyles may be summarized as follows:

- To shift the focus of design, planning and action from the individual to the community that enables communities to take responsibility. For instance, if a community is based on equity, mutual support and stakeholder involvement, a more connected communities and sustainable neighborhoods can be formed.

- To share goods and services supports collaborative infrastructure. Thus, the high impacts of individual consumption have been reduced by the community-based consumption. For instance, the applications in relation to the smart renewable energy support for distributed renewable energy generation and consumption.

- Without restricting the people's freedom of choice, sustainable options are required to be normalized. Several options can make sustainable choices easy and desirable and change the need for individual behaviour. For instance, according to the consumption levels and resource use, people will coordinate their behaviour.

There is a major emphasis on innovation in enabling technologies that can help integrate variable renewable resources into electricity systems.

9. Some aspects on marketing renewable energy

As the share of renewable energy sources is steadily increasing, it seems that much more flexibility for energy markets are required [18]. Large end users of electricity—such as retailers, manufacturers and technology companies—are important customers to purchase renewable power directly. There is an interaction between independent power producers, utilities and commercial and industrial users which is varying. By developing new models, the role of utilities changes in each transaction [19].

It is concluded that sustainable business models, which are more and more popular among different sectors, dedicate solutions such as designing a market model to ensure earnings for the stakeholders [20]. The application of these models can be classified in many different groups such as energy, innovation, marketing, entrepreneurship, developing countries, engineering, construction, mobility and transportation [21].

To encourage renewable energy market, the most frequently used policy tool is the feed-in tariff. Thus, a fixed price per unit of electricity sold is guaranteed during a period of agreement. While the feed-in tariffs provide rapid cost reductions of renewable, there is a risk that the renewable have to be subsidized by the governments for a long time. On the other hand, a transition from government-administered feed-in tariffs to auction systems has taken place during the past few years. By identifying the price for renewable energy contracts, it is aimed to obtain substantial cost reductions in renewable energy. Financing is an important factor in the deployment of renewable energy technologies. The main financing sources are public finance institutions, private investors and institutional investors [22].

Business-to-business marketing is known as industrial marketing, and its products are based on the functional consumption values such as price and quality. Business-to-business companies sell, rent and supply goods to other companies. Under the conditions of the globalized market, local customers do not simply purchase products from local suppliers. Business-to-business companies need to find new ways to stay relevant on the market as they are facing increased global competition. Companies must also approach their customers as humans with values to fulfil their personal needs. In contrast, the purpose of business-to-consumer marketing is to sell products or services directly to the consumers.

For sustainable solutions, business-to-business renewable energy companies can have a marketing advantage. However, marketing renewable energy is complex. Buying a renewable energy product is an investment. To help finance the investment, customers are frequently depended upon the support schemes which are not fixed and can vary in different countries.

It is expected that companies investing in renewable energy can use their environmental credentials for marketing purposes by supporting their use of renewable energy. Governments, having an interest in renewable energy and its benefits, secure different support schemes such as tax credits and subsidies. On the other hand, to make renewable energy more attractive, subsidies for fossil fuels are reduced [23].

In some energy markets, the needs of power systems with higher shares of variable renewable energy are reflected and responded to the trends of digitalization, decentralization and electrification. Gradual improvement of energy market pricing is imported. Package can generally include the real value of electricity in time, new dispatch rules, flexibility, economic energy resources, self-consumption and market link. Appropriate electricity market designs for changing models in power systems are necessary to speed up the energy transition. At the present, end users of electricity have more suppliers and innovative deals to choose and can easily switch tariffs and suppliers. The desired consequences for all end users cannot be, however, delivered by the retail market [24].

10. Conclusions

Fossil fuels are still maintaining the largest portion of energy consumption and keep on their increasing trend all over the world. In this situation, environmental pollution is somehow inevitable, whereas the renewable energy plants do not directly contribute any.

In the future, it is aimed that the main energy sources will become new and renewable energies. While the fossil fuels are inevitably running out, renewables are to be more important. They are effective in many areas such as continuous cost reductions, generating jobs, developing future industries and meeting energy and environmental targets.

The development and use of renewable energy will improve the energy security, environment, economy, mechanical manufacturing, construction, transportation and industry and also help to create new jobs. Energies of solar, wind and biomass can meet local energy demands and assist to improve the environmental protection. Current situation related to the energy demand encourages an enormous market for renewable energy. As predicted, the share of renewable in meeting global energy demand will grow to reach 12.4% in 2023.

In the longer term, if the investments in the renewable technologies continue, renewable will have the potential to make significant contributions to energy needs. Further, there are several technologies that include biofuels, and fuel cells also can contribute to heat, transport and electricity markets.

The share of fossil fuels in total primary energy supply is expected to include around 81% of total in 2023. By 2050, renewable energy will approximately account for 30% of energy structure in the world.

By providing a balanced resource diversification of countries for the primary energy resources, the share of domestic and renewable energy resources in the generation system can be increased to the maximum extent. As also aimed in the current strategy plans of many countries, targets should be obtained in time for supporting, developing and encouraging new environment-friendly practices in generation and services. The largest market share and the most of advanced renewable energy technologies belong to the leading developed countries such as the USA, Japan and the Europe.

In order to use less and cleaner energy in power plants, buildings, industrial facilities and transport systems, many energy-efficient enabling technologies are applied. These technologies could slash costs by up to 80 per cent, ensure energy savings by up to 30 per cent and help to slow global warming in the future. Thus, the countries could stay cost-effective and make sustainable progress. Marketing renewable energy can be also defined as the art of understanding consumers and their needs.

Author details

Mustafa Seckin Salvarli[1] and Huseyin Salvarli[2*]

1 Izmir Vocational School, Dokuz Eylul University, Buca, Izmir, Turkey

2 Aziz Nesin Bulvari, Karsiyaka, Izmir, Turkey

*Address all correspondence to: huseyin.salvarli@gmail.com

References

[1] UN. World Population Prospects 2019 Highlights [Internet]. 2019. Available from: https://population.un.org › Publications › Files [Accessed: 19 September 2019]

[2] WEC. World Energy Resources 2016 Summary [Internet]. 2016. Available from: https://www.worldenergy.org › wp-content › uploads › 2016/10 › World [Accessed: 19 September 2019]

[3] WEC. World Energy Scenarios [Internet]. 2019. Available from: https://www.worldenergy.org › assets › Scenarios_Report_FINAL_for_website [Accessed: 19 September 2019]

[4] Anastassios P. Technology Trends: A Review of Technologies and Policies Study on Technology Trends DTI [Internet]. 2012. p. 97. Available from: www.dti.gov.za › industrial_development › docs [Accessed: 19 September 2019]

[5] IEA. Renewables Information: Overview [Internet]. 2019 Edition. p. 12. Available from: https://webstore.iea.org/renewables-information-2019-overview [Accessed: 19 September 2019]

[6] IEA. World Energy Balances: Overview [Internet]. 2019 Edition. p. 23. Available from: https://webstore.iea.org › ... › Statistics & Data [Accessed: 19 September 2019]

[7] IEA. Renewables 2018 Analysis and Forecasts to 2O23 Executive Summary [Internet]. p. 10 Available from: https://webstore.iea.org › download › summary [Accessed: 19 September 2019]

[8] UN. Pathways to Sustainable Energy Exploring Alternative Outcomes. United Nations Publication. eISBN: 978-92-1-057736-6 ISSN 1014-7225 Copyright © United Nations; 2015

[9] Majid J, Sheeraz K, Mohammad R. Techno-economic feasibility analysis of solar photovoltaic power generation: A review. The Smart Grid and Renewable Energy. 2012;**3**:266-274

[10] Renewables Energies Innovation for the Future [Internet]. p. 129. Available from: https://www.dlr.de › system › publications › broschuere_ee_innov_zukunft_en [Accessed: 19 September 2019]

[11] Renewable Energy and Energy Efficiency in China. Current Status and Prospects for 2020. Worldwatch Institute. 2010. p. 50. Available from: www.worldwatch.org [Accessed: 19 September 2019]

[12] IEA. World Energy Outlook [Internet]. 2018. Available from: https://www.iea.org › weo [Accessed: 19 September 2019]

[13] IEA. Energy Technology Perspectives [Internet]. 2017. Available from: https://www.iea.org › etp [Accessed: 19 September 2019]

[14] SUSCHEM. Key Enabling Technologies in Horizon Europe Paper. 20 June 2018, p. 22. Available from: www.suschem.org › files › library › IMPACT_K... [Accessed: 19 September 2019]

[15] Athukoralaa AUCD, Jayasuriyaa WJA, Ragulageethana S, Pereraa ATD, Sirimannaa MPG, Attalagea RA. A techno-economic analysis for an integrated solar PV/T system with thermal and electrical storage – case study. Moratuwa Engineering Research Conference (MERCon); 2015. pp. 182-187

[16] Depuru S, Green RC, Nims D, Near C, Devabhaktuni V, Alam M. Solar energy: Trends and

enabling technologies. Renewable and Sustainable Energy Reviews. 2013;**19**:555-564

[17] Devabhaktuni V, Alam M, Boyapati P, Chandna P, Kumar A, Lack L, et al. Wind energy: Trends and enabling technologies. Renewable and Sustainable Energy Reviews. 2015;**53**:209-224

[18] Leutgöb K, Amann C, Tzovaras D, Ioannidis D. New business models enabling higher flexibility on energy markets. ECEEE Summer Study Proceedings. 2-040-19. 2019:235-245

[19] Herbes C, Friege C, editors. Marketing Renewable Energy: Concepts, Business Models and Cases. Cham, Switzerland: Springer International Publishing AG; April 2017. DOI: 10.1007/978-3-319-46427-5

[20] Nosratabadi S, Mosavi A, Shamshirband S, Kazimieras ZE, Rakotonirainy A, Chau KW. Sustainable business models: A review. Sustainability. 2019;**11**:1663. DOI: 10.3390/su11061663

[21] ACORE. Beyond Renewable Integration: The Energy Storage Value Proposition. American Council On Renewable Energy; 2016. pp. 1-35

[22] Facchinetti E, Eid C, Bollinger A, Sulzer S. Business model innovation for local energy management: A perspective from Swiss utilities. Frontiers in Energy Research. 2016;**4**:31. DOI: 10.3389/fenrg.2016.00031

[23] Nielsen LB. Marketing renewable energy. Culture, Communication & Globalization [thesis]. Aalborg University; 2018

[24] IRENA. Innovation Landscape for a Renewable-Powered Future: Solutions to Integrate Variable Renewable. Abu Dhabi: International Renewable Energy Agency; 2019. pp. 1-164

Chapter 2

Energy Security and Renewable Energy: A Geopolitical Perspective

John A. Paravantis and Nikoletta Kontoulis

Abstract

This chapter examines the role of renewable energy in shaping energy security against the backdrop of global geopolitical, socioeconomic, and technological uncertainties. The evolving definition of energy security during the twentieth and early twenty-first centuries is discussed initially. The dimensions, components, and metrics of energy security are reviewed, including the 4A definition of energy security that comprises physical availability; economic affordability; accessibility from a sociopolitical standpoint; and environmental acceptability. A novel energy security index is proposed, with the following components: physical availability; technology development; economic affordability; social accessibility; governance; unconventional threats; and natural environment. Of these, physical availability followed by technology development, economic affordability, and governance was rated as the most important, and the environment was rated as the least important by a small panel of experts. The roles of wind and solar energy are highlighted, with an emphasis on the social acceptance of renewable energy in an energy security context. Other energy security indexes are discussed, focusing on sustainability and renewable energy. Denmark, Germany, China, Russia, and the United States are examined as case studies that help understand the transition to renewable energy in the context of coopetition among states. As these countries face different political concerns, geopolitical realities, and energy security issues, they consider different policy approaches to address them.

Keywords: energy security, renewable energy, geopolitics, energy transition, case studies

1. Introduction

Energy has been crucial for economic growth throughout human history, the "precondition of all commodities, a basic factor equal with air, water, and earth" (E. F. Schumacher, Nobel laureate economist, 1977). The International Renewable Energy Agency (IRENA) [1] underlines that fossil fuels have been the foundation of the global energy system, and their role is deeply embedded in the relations among countries and regions.

The ability of a country to access the energy resources needed to maintain its national power (without compromising foreign policy, economic, social, and environmental objectives) is referred to as energy security. Energy security is paramount to human security [2] and has become an increasingly popular concept.

There is no universal definition of energy security [3, 4], which [5] has been aptly described as "slippery" and "polysemic." Consequently, energy security has become an umbrella term for different policy goals [6]. This ambiguity is affirmed by many existing definitions of energy security. A few years back, Sovacool [7] reported that there were at least 45 different definitions, and Ang et al. [8] identified 83 definitions in the existing literature. In a most recent study, Matsumoto et al. [9] confirmed that there are no uniform definitions or evaluation methods for assessing energy security.

Recently, a strong interest has emerged in favor of the integration of renewable energy in the energy mix as a priority measure of addressing energy security and climate change [10]. Energy security planning is increasingly geared toward establishing a low-carbon economy and achieving climate mitigation goals [11]. Article 2 of the Paris Agreement [12] requires countries to implement their nationally determined contributions and to increase their ambitions over time, necessary for keeping the rise in global mean temperature below 2°C [13].

At a time when society is increasing its demands for an accelerated transition to a low carbon energy system, the energy data for 2018 paint a worrying picture, with both energy demand and carbon emissions growing at the fastest rates seen for years [14]. Renewables and energy efficiency offer a safe, reliable, and affordable way to achieve massive decarbonization, in line with keeping the rise in global temperatures below 2°C [15].

This is a particularly exciting time to study the renewable energy dimension of energy security: the global energy landscape is in the middle of a game-changing revolution in source rock resources; consumer countries have turned into producers; producer countries have turned into consumers; and transit countries have turned into new players. The main goal of this chapter is to shed light on the role and the impact of renewables in energy security by (a) examining how renewable energy sources enter the dimensions, components, and metrics used in measuring energy security and (b) offering case studies of countries that help shed light on the link between renewables and energy security.

The remainder of this chapter is organized as follows: Section 2 provides a backdrop by linking the concept of energy security to renewable energy; Section 3 reviews the dimensions, components, and metrics of energy security and proposes a new energy security index, having experts rate their importance, discussing the role of renewable energy, and reviewing other indexes; and finally, Section 4 presents thoughts on the perspective of selected countries on energy security. The chapter is rounded up with conclusions.

This chapter is a review that presents the results of analytical research. It includes a descriptive part (expert interviews). Numerical computations were done with Microsoft Excel, and statistical analysis and graphing were done with Minitab Version 18.

2. Energy security and renewable energy

Since the turn of the twenty-first century, climate change and its link to fossil fuels have moved to the forefront of the political discourse, engaging political and industrial actors, academic researchers, and the society [16]. This has happened against a backdrop of geopolitical turmoil caused by a series of events of energy significance including:

- the second Gulf War (March 20, 2003 to December 18, 2011);
- Hurricane Katrina that caused catastrophic damage in Louisiana and Florida and underscored the importance of black-swan type of natural disasters (August 23, 2005);

- Germany's planned energy transition (*Energiewende*), a low carbon program relying heavily on renewable energy (targets set in September 2010);
- the natural gas crises between Ukraine and Russia that lasted from 2005 to 2009 and caused significant concern in the European Union (EU);
- the Arab Spring upheaval that changed the strategic balance in the Middle East and North Africa, with indirect implications for energy and geopolitics (started on December 18, 2010);
- the Fukushima Daiichi nuclear accident in Japan, showing the importance of securing energy installations from "unthinkable" natural threats (like tsunamis and earthquakes; March 11, 2011);
- the cyberattack on 35,000 computers of Aramco (the Saudi Arabia oil company that supplied 10% of global demand for oil; August 15, 2012), 1 year after the major cyberattack on Sony that compromised the personal details of 77 million PlayStation accounts (April 17–19, 2011);
- the Tigantourine gas plant hostage crisis in Algeria (January 16–19, 2013) with 39 fatalities of expatriates, the first major terrorist strike on a big energy facility;
- the adoption of the European Union Energy Union (February 25, 2015), a project of great geopolitical scope and significance;
- the signing (July 14, 2015) and later the unraveling (May 8, 2018) of the nuclear deal between Iran, the five permanent members of the United Nations (the United States, the United Kingdom, China, Russia, and France), Germany, and the EU, aiming to limit Iran's enrichment of uranium (2015);
- the lifting of the 1975 US oil export ban (September 10, 2015); and
- the Yemeni rebel attacks on two Saudi pumping stations with armed drones, the first such terrorist attack on energy installations (May 2019).

These events led to the realization that global geopolitics may threaten energy security. In this vein, renewable energy is considered a potential game changer in energy security. **Figure 1** plots the Google trends search interest for the terms "energy importance" (blue points) and "renewable energy" (green points) with dotted (pink) vertical lines corresponding to the start date of the events of the previous list. The presence of energy security and renewable energy in Google searches is a proxy of their relative importance in the global (online) discourse. **Figure 1** helps position this importance against the backdrop of global geopolitical activity. No Google trends data are available before 2004.

Energy security reached a peak in Google search interest from 2005 to 2009, corresponding to the Russia-Ukraine gas crises; its interest has been falling since then. Renewable energy Google search interest kept rising until 2009 and then fell until 2013 and has been rising since then. Its interest was also at a peak during the Russia-Ukraine gas crises and appears to have received a boost concurrent with the discussions of the EU Energy Union; the institution and then the unraveling of the Iran nuclear deal; and the lifting of the US oil export ban. Perhaps the cyberthreats of the events of Sony and Aramco helped precipitate this increased interest in renewable energy. As a result of these trends, renewable energy has received more Google search interest than energy security since 2015.

With the turn of the twenty-first century, the price of oil (arguably the world's most important commodity) climbed to record high and exhibited fluctuations that were difficult to predict. As depicted in **Figure 2**, a barrel of crude oil, which in average annual OPEC prices cost $2.70 in 1973, jumped to over $10 by 1978; further jumped and fluctuated around $30 until 1985; fell to around $20 or less until 1999; climbed to an unprecedented high of $94.10 in 2008, shocking the global economy; fell to just over $60 and climbed to a new high of $109.45 in 2012; and fell to just over $40 and back up to $69.52 in 2018. In 2019, only a preliminary average annual oil price per barrel was set at the time of writing ($64.05).

These fluctuations in the price of oil, especially the unpredictability after 2008, underscore that the world has entered a prolonged era of peak oil no matter for many decades shale oil and gas will continue to supply the world.

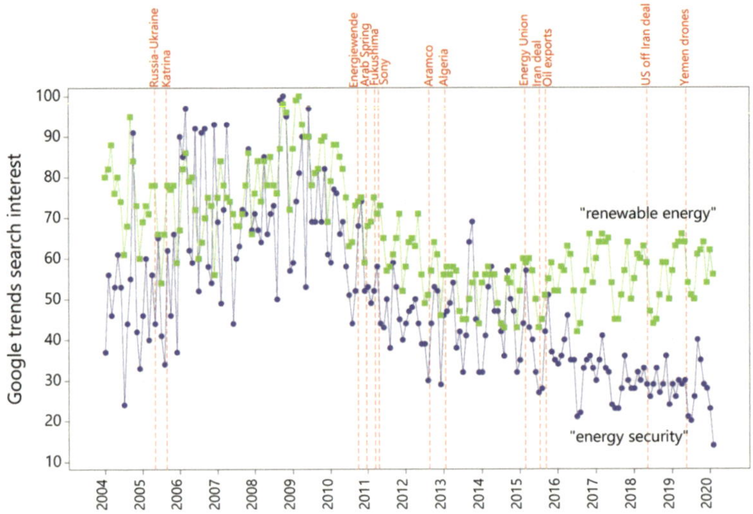

Figure 1.
Google trends (https://trends.google.com/trends/?geo=US) search interest of energy security and renewable energy compared to global geopolitical events.

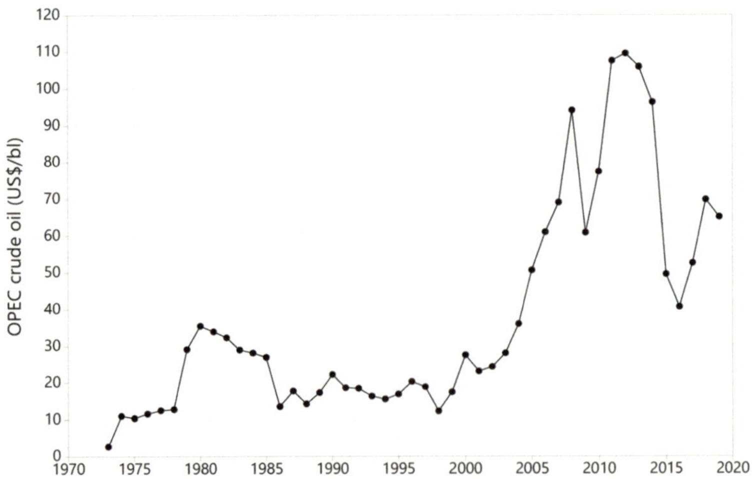

Figure 2.
Average annual OPEC price of a barrel of crude oil.

The climate is a chaotic system, which, considering the Ice Ages (the last of which ended just 12,000 years ago), has exhibited wide variability in the past. Nevertheless, all evidence at this point shows that global warming and its effects will intensify during the rest of the twenty-first century, so climate change will become a more concrete and tangible target [16]. Intensifying geopolitical tensions are also likely, mainly between the United States and China (with its expanding territorial claims in southeast Asia), with other emerging powers such as India (with its own territorial issues with neighboring countries), Indonesia (with its expanding population), Brazil (host of the world's most important ecosystem), Japan, and Germany, all of which the Economist Intelligence Unit has predicted will be among the countries with the highest nominal GDP per capita by 2050.

The need to address climate change and emerging geopolitical tensions will make energy policy a field of paramount importance in international relations, with energy security its most important aspect. The need to address climate change is perceived as urgent: 2018 carbon emissions grew by 2%, which is the fastest growth for 7 years [14]. Signs of a shift away from fossil fuels have become clearer [16], for example, since the 2011 Fukushima nuclear accident, Japan and Germany have moved toward ending their dependence on nuclear energy, while the Danish parliament has decided that Denmark will be fossil fuel free by 2050.

As new energy alternatives need to be less polluting than the sources they replace, renewable energy moves naturally into center stage. Renewable energy, at least four decades old, has gained increasing currency as a conceptual alternative to centralized energy sources (e.g., coal, nuclear power), which are considered environmentally destructive and dehumanizing [17]. Yet, renewable energy is not free of impacts on natural ecosystems, economy, society, and politics:

- Wind and solar energy are dilute fuels, requiring large expanses of land. The construction of onshore wind farms in particular necessitates clearing land areas with impacts on species such as tortoises, birds, and bats.

- Wind turbines have a lifespan of around 20–26 years, after which steel, cement, and other materials used in their construction must be recycled or properly disposed of as solid waste.

- Large renewable energy projects oftentimes are opposed by society on the grounds of visual esthetics and other intrusions into the way of life.

- The intermittent nature of wind and sunlight means that the energy they capture must be stored if they are to serve as the main energy source.

- Finally, although renewable energy is becoming more inexpensive, it continues to rely on state budgets.

For the transition to renewable energy to succeed, environmental impacts must be minimized, and the profile of renewable energy projects must be made more attractive to society. Most importantly, concerns about energy security [18] and the political economy of renewables must be addressed.

The International Energy Agency (IEA) was founded in 1974, with the intention of helping countries coordinate a collective response to major disruptions in the supply of oil by the establishment of mandatory strategic petroleum reserves (https://www.iea.org/about/ourmission). IEA defines renewable energy as "energy derived from natural processes that are replenished at a faster rate than they are consumed." The main renewable energy sources are wind, solar, biomass, hydropower, geothermal, and wave. Wind and solar energy have grown at an unprecedented rate and are arguably the most significant sources, with electric utilities buying into wind power almost without hesitation. Yet, they are called variable renewable energy sources because they share a unique problematic characteristic: the amount of power they generate varies with the weather and the time of day [1].

Considering that renewable technologies have the potential to contribute to energy security while meeting the environmental objectives at the regional, national, and global levels [19], IRENA (an official United Nations observer) was founded in 2009 to serve as a platform for international cooperation on the technologies, policies, and financial know-how on renewable energy (https://www.irena.org). At that time, renewable energy was only a marginal contributor to global primary energy and electricity supply [20]. Now renewable energy is growing rapidly in installed capacity and investments [20]. The global awareness about renewables is shown by numbers: by the end of 2018, renewable energy targets had been adopted in 169 countries at the national, state, or provincial level. Moreover, 135 countries have power regulatory policies [21].

Decreasing the dependence on fossil fuels and increasing the amount and percentage of renewables (and nuclear energy) will help mitigate climate change [22]. Domestic renewable energy may reduce the need of countries for energy imports [20] and consequently their dependence on exporter countries [18]. So, renewable energy is considered the most secure way to minimize energy supply risks by exploiting domestically controllable energy supplies [23].

3. Defining energy security

Many researchers have investigated energy security, defining it from a variety of angles. Energy security is a contested and complex term [24], which encapsulates concepts such as security of supply, reliability of infrastructures, affordability, and environmental friendliness [25]. Energy security means different things to different countries, depending on their geographical location; their natural resource endowment; their economic disposition [26]; their status as producer/exporter, consumer/importer, or transit [27]; their vulnerability to energy supply disruptions; their political system; their ideological views and perceptions [28]; and the status of their international relations, for example, reliance on Russian gas depends on historical experiences during the Cold War [29], as cited in [27].

Examined over different historical time frames, the concept of energy security is dynamic and fluid, with evolving energy policy challenges [6, 30]. The oil crises of 1973 and 1979 transformed oil supply from a military to a socio-political and economic issue for importing countries. The gas crises of 2006 and 2009 between Russia and Ukraine raised concerns about transit countries and brought back the use of energy as a geopolitical weapon [31]. As pointed out by Cherp and Jewell [32], a classic definition of energy security has been provided by Yergin [33], who visualized energy security as the assurance of "adequate, reliable supplies of energy at reasonable prices," adding a geopolitical component by qualifying that this assurance must be provided "in ways that do not jeopardize national values or objectives."

The IEA, a pioneer institution in energy security and the most important multinational energy platform, defines energy security as the "uninterrupted availability of energy sources at an affordable price" and considers it to have long- and short-term aspects. The IEA has restated the definition through the years to characterize energy security as the adequate, affordable, and reliable supply of energy.

Long-term energy security relates to "timely investments to supply energy in line with economic developments and environmental needs." Short-term energy security relates to "the ability of the energy system to react promptly to sudden changes in the supply-demand balance" [10, 34–36].

Countries have different energy security objectives depending on their role in the energy market: producer/exporter countries aim to ensure reliable demand for their commodities; consumer countries commonly aim toward diversity of energy supply, so as to minimize their dependence and maximize their security; and transit states try to make the best of their role as bridges connecting producers/exporters with their markets [26]. For consumer and transit countries, security of supply is important; for producer/exporter countries, security of demand is possibly as important as security of supply [27].

3.1 Dimensions and components of energy security

To highlight the role of renewable energy in energy security, one has to conceptualize and formulate the latter. Energy security is considered to be composed of a small number of dimensions, for example, technical, social, environmental, political, geological, and economic [37]; each dimension contains components, and each component may be measured by metrics, that is, quantitative or qualitative indicators. When all metrics, components, and dimensions are aggregated, an energy security index may be calculated based on available data.

In an extension to the original IEA definition of energy security, the Asia Pacific Energy Research Centre [4] highlighted the so-called four As of energy security: (1) availability of the supply of energy resources; (2) affordability of the price of energy resources, so that economic performance is not affected adversely; (3) accessibility to all social actors; and (4) acceptability from a sustainability standpoint. The first two As (availability and affordability) constitute the classic approach to energy security (the twentieth century), while the latter two (accessibility and acceptability) reflect contemporary environmental concerns, such as climate change, and sociopolitical issues, such as fuel poverty.

Ang et al. [8] argued that the most important dimension of energy security is availability, as this is taken into account in 99% of related studies. The term availability is also used to imply stable and uninterrupted supply of energy [38–40], while some authors use the term reliability to refer to the role of energy infrastructure [41, 42] and the production of electricity and heat [43]. As for accessibility, it has been at the center of energy security debates and policy approaches into the twenty-first century [44]. Goldthau and Sovacool [45] talked about the following three key energy challenges: energy security, energy justice, and a low carbon transition. They highlighted the need to consider energy security as a democracy issue; equity as an important aspect of accessibility; and global climate change as an important aspect of acceptability.

A similar set of four dimensions of energy security has been proposed by Sovacool and Rafey [46]: (1) availability, that is, diversifying the fuels, preparing the disruption recovery, and minimizing the dependence on foreign supplies; (2) affordability, that is, providing the affordable energy services and minimizing the price volatility; (3) efficiency and development, that is, improving the energy efficiency, altering the consumer attitudes, and developing the energy infrastructure; and (4) environmental and social stewardship, that is, protecting the natural environment, communities, and future generations.

Alhajji [47], a global energy expert, differentiated among six dimensions of energy security: economic, environmental, social, foreign policy, technical, and security. Vivoda [48] listed seven salient energy security dimensions: environment, technology, demand side management, sociocultural or political factors, human security, international elements like geopolitics, and the formulation of energy security policy and 44 attributes of energy security. Knox-Hayes et al. [49] extracted the following dimensions of energy security: (1) availability, indicating security of supply and affordability; (2) welfare, indicating equity and environmental quality; (3) efficiency, representing various factors including low energy intensity and small-scale energy (with some overlap with welfare); (4) affordability, indicating (among other factors) price affordability and small-scale energy; (5) environment, appearing to be very similar to welfare; (6) transparency, standing for equity, transparency, and education; (7) climate, connected to global climate change and having significant overlap with welfare and environment; and (8) equity, overlapping with other dimensions.

Sovacool and Mukherjee [2] presented the following dimensions with corresponding components: (1) availability, that is, security of supply and production, dependency, and diversification; (2) affordability, that is, price stability, access and equity, decentralization, and affordability; (3) technology development and efficiency, that is, innovation and research, safety and reliability, resilience and adaptive capacity, efficiency and energy intensity, and investment and employment; (4) environmental and social sustainability, that is, land use, water, climate change, and pollution; and (5) regulation and governance, that is, governance, trade and regional interconnectivity, competition and markets, and knowledge and access to information as well. Regarding energy independence, self-sufficiency may be a more pragmatic target since even a producer/exporter country cannot really extricate itself from the global energy markets and their vulnerabilities [50].

3.2 Metrics of energy security

There is a multitude of energy security indicators: Sovacool and Mukherjee [2] assembled 320 simple indicators and 52 complex indexes of energy security. Kruyt et al. [3] differentiated among simple indicators, such as reserves-to-production ratios, import dependence, energy prices, political stability and demand-side requirements, and aggregated indices. Sovacool and Brown [51] considered energy security to be defined according to the following criteria (i.e., dimensions), which may be measured with corresponding metrics: *availability*, measured by oil and natural gas import dependence and availability of alternative fuels; *affordability*, measured by retail electricity, gasoline, and petrol prices; *energy and economic efficiency*, measured by energy intensity, electricity use per capita, and average fuel economy of passenger vehicles; and *environmental stewardship*, measured by sulfur dioxide (SO_2) and carbon dioxide (CO_2) emissions.

In a paper evaluating the energy security performance of 18 countries from 1990 to 2010, Sovacool et al. [52] presented a more detailed list of dimensions, components, and corresponding metrics, adding the dimensions of regulation and governance, measured by energy exports; competition, measured by energy subsidies per capita; and information, measured by the completeness of energy data. Ang et al. [8] confirmed governance and added other dimensions such as infrastructure and energy efficiency.

An even more detailed definition of energy security involved the following dimensions, components, and corresponding metrics [53]:

1. *Availability*, measured by security of supply, equal to $\frac{\text{total production energy}}{\text{total consumed energy}}$; self-sufficiency, equal to $\frac{\text{imported energy}}{\text{total consumed energy}}$; diversification, measured by a diversity

index such as the Shannon-Wiener; renewable energy, equal to $\frac{\text{renewable energy}}{\text{total consumed energy}}$; and technological maturity, a qualitative metric.

2. *Affordability*, measured by price stability, equal to the deviations of price about a global mean value; dependency, equal to $\frac{\text{total imported energy}}{\text{population}}$; market liquidity, a qualitative metric; decentralization, equal to $\frac{\text{total energy by distributed and small-scale generation}}{\text{total energy production}}$; electrification, equal to the percentage of population with reliable access to grid; and equity, equal to the percentage of households depending on wood, straw, and so on for cooking and heating.

3. *Accessibility*, measured by the following qualitative metrics: import stability, trade, political stability, military power, and safety and reliability, all qualitative metrics.

4. *Acceptability*, measured by the following qualitative metrics: environment, a composite of several "micro aspects" that are "measured individually"; social satisfaction, national governance, international governance, transparency, and investment and employment.

It may be concluded that renewable energy is a factor in much of the research that aims to conceptualize, define, and measure energy security.

3.3 Synthesizing an energy security index

Having reviewed the dimensions, components, metrics, and methods of energy security, a novel energy security index will now be synthesized. An effort was made to combine all the dimensions and components meaningfully, without overlap. The following seven dimensions and components are proposed:

1. *Physical availability*, the historical bedrock of energy security [3, 26, 38, 54], accounting for security of supply; self-sufficiency (affected by oil and gas import dependence and accounting for the boosting effect of renewable energy); Strategic Petroleum Reserves (SPR, acting as a buffer and a deterrent); and energy diversification (accounting for the contribution of small-scale distributed renewable energy installations).

2. *Technology development*, accounting for (state and maturity of) infrastructure, for example, matching of available oil to refinery infrastructure; energy (grid) efficiency (the "fifth fuel"); onshore and offshore wind farms; energy consumption and conservation in the building sector, transportation systems, and the industry; decentralization, that is, diffusion of small scale and prosumer systems (mostly renewable energy); and research (intensity), development, and innovation (with a sizeable portion expected to be related to renewable energy).

3. *Economic affordability*, perhaps the second most important energy security dimension historically, accounting for affordability of electricity and gasoline prices (expressed in Purchasing Power Parity); stability (i.e., lack of volatility) and predictability of prices; competition, subsidization (per capita), and profitability; energy intensity (i.e., electricity use per capita and monetary unit of GDP); and fuel economy of passenger vehicles (also related to technology).

4. *Social accessibility*, that is, social stewardship, accounting for dependency (expressed as imported energy per capita); electrification, that is, percent of the population with (reliable) access to the electricity grid (potentially improved by resorting to renewable sources); energy democracy, for example, percent of households that are fuel poor (also likely to be improved by renewables); social equity, for example, percent of households relying on traditional energy sources (such as wood) for cooking and heating; and consumer awareness, knowledge, and attitudes, for example, toward renewable energy [55, 56].

5. *Governance*, taking into account: quality of governance, measured by, for example, the Worldwide Governance Indicators (WGI) of the World Bank (https://info.worldbank.org/governance/wgi) that rate "voice" (i.e., citizen participation) and accountability, political stability (which may be measured by the number of years since the previous regime change) and absence of violence, government effectiveness, regulatory quality, rule of law, and control of corruption (i.e., transparency and accountability, no crony capitalism); type of polity (democracy or otherwise); military power (possibly a qualitative variable); data quality and intelligence; good regulatory policies (e.g., avoiding over-regulation, setting reasonable and objective performance criteria, and avoiding picking winners and losers); and adoption of "fit" energy policies, that is, catering to all societal energy groups referred to as "tribes" [57, 58].

6. *Unconventional threats*, including asymmetric, paramilitary, or nonconventional threats to energy infrastructure, such as revolutions (e.g., Iranian revolution, Arab Spring); accidents caused by human error (likely to be less severe with renewable energy installations); durability and safety (of infrastructure, also related to technology); and terrorism incidents, including cyberwarfare (also likely to be of a less severe nature with renewables).

7. *Natural environment*, accounting for (existence of) tragedy of the commons (i.e., overexploitation of resources that are public goods, something that is irrelevant with renewable energy sources) and resource curse (i.e., the presence of abundant energy and natural resources in poor countries); (mitigation of) environmental pollution, for example, SO_2 emissions (per capita); (mitigation of) global climate change, for example, CO_2 emissions (per capita), affected very favorable with more use of renewable energy; forest cover; land use (management), probably the most important negative impact of onshore wind farms; water availability, that is, quality and quantity, (lack) of water stress and scarcity, and access to improved water quality; environmental (sustainability) management; health problems caused by environmental threats, for example, high concentration of toxic substances; and (impacts of) black-swan type of natural disasters.

How are the different dimensions of energy security perceived by different economic actors? In a paper examining seven suppositions about energy security in the United States, Sovacool [59] presented the following expert suppositions pertaining to energy security issues: (1) security of supply and trade; (2) energy democracy; (3) energy research; (4) energy efficiency; (5) affordability; (6) environmental pollution; and (7) climate change. It was concluded that the different dimensions of energy security are perceived differently by those working in different sectors of the economy:

- The private sector considered the following four energy security dimensions to be the most important (with a rating over 4.5 of 5): (1) conducting research and development on new and innovative energy technologies; (2) providing available and clean water; (3) minimizing the destruction of forests and the degradation of land and soil; and (4) minimizing air pollution.

- Among government occupations, more (i.e., eight) dimensions were rated over 4.5, including the four of the private sector plus the following: (5) reducing greenhouse gas emissions; (6) minimizing the impact of climate change; (7) assuring equitable access to energy services to all of its citizens; and (8) informing consumers and promoting social and community education about energy issues.

- With universities, even more dimensions were rated over 4.5, including the four of the private sector plus the following: (5) reducing greenhouse gas emissions; (6) minimizing the impact of climate change; (7) informing consumers and promoting social and community education about energy issues; (8) assuring equitable access to energy services to all citizens; (9) ensuring transparency and participation in energy permitting, siting, and decision making; and (10) having low energy intensity.

- The nonprofit sector rated the following dimensions over 4.5: (1) providing available and clean water; (2) minimizing air pollution; (3) conducting research and development on new and innovative energy technologies; (4) minimizing the destruction of forests and the degradation of land and soil; (5) reducing greenhouse gas emissions; (6) minimizing the impact of climate change; (7) informing consumers and promoting social and community education about energy issues; (8) assuring equitable access to energy services to all citizens; (9) ensuring transparency and participation in energy permitting, siting, and decision making; and (10) having a secure supply of coal, gas, oil, and/or uranium.

- Finally, those working in intergovernmental occupations rated the first two dimensions of the private sector and the following dimensions with a score over 4.5: (3) minimizing air pollution; (4) having a secure supply of coal, gas, oil, and/or uranium; (5) promoting trade in energy products, technologies, and exports; (6) reducing greenhouse gas emissions; (7) informing consumers and promoting social and community education about energy issues; (8) assuring equitable access to energy services to all citizens; and (9) having low energy intensity.

How are the dimensions of energy security covered by the research literature? In a paper examining 40 years of energy security trends, Brown et al. [60] found that 91 peer-reviewed academic articles covered the dimensions of energy security differently. In particular, availability was covered by 82% of the examined articles; affordability by 51% of the articles; energy and economic efficiency by 34% of the articles; and environmental stewardship by 26% of the articles. As to the precise nature of these dimensions of energy security, a Factor Analysis carried out by the authors concluded that: availability was mostly a function of oil import dependence, road fuel intensity, and natural gas import dependence (in decreasing order of importance); affordability was a function of electricity and gasoline retail prices; energy and economic efficiency were a function of electricity use per capita and energy per GDP intensity; and environmental stewardship was a function of CO_2 and SO_2 emissions.

To get an idea about the relative importance of the proposed seven dimensions of energy security, expert interviews were used as in other sources [7, 61]. A small panel of engineering, economic, and geopolitical energy experts was selected, including junior and senior academic faculties (with experience in energy, environment, transportation, and geopolitics) and senior professionals (with experience in environment and water management). One of the authors was included in the panel of experts interviewed [62]. The interviews contained a brief semi-structured part (the results of which are reported in this chapter) and a longer structured part (which is not reported here). During the semi-structured part, the experts were asked to (1) rate the importance of the seven dimensions of energy security and (2) give their opinion on the way the dimensions were defined. Input received during this phase was used to improve the scope of the dimensions and clarify the definitions. Although these expert ratings reflect the perspective of Greece at the time of writing, they are interesting.

Ratings were on a scale from 1 to 10. The experts' average ratings of the importance of each energy security dimension are shown in **Figure 3**. Physical availability was deemed to be the most important dimension (in accordance with its extensive coverage in the research literature), receiving an average rating of 8.8 (of 10). Technology development, economic affordability, and governance were next, with an average importance of 8. Social accessibility and unconventional threats received an average rating of 6.8. Finally, the natural environment was considered the least important dimension (the panel did not include experts working in the nonprofit sector), with an average rating of 5.8.

Some further interview findings were as follows:

- Most experts tended to rate dimensions nearer their discipline as more important, reflecting a form of cognitive bias.

- A couple of experts thought that there was a little overlap among some of the dimensions but could not suggest ways of overcoming it.

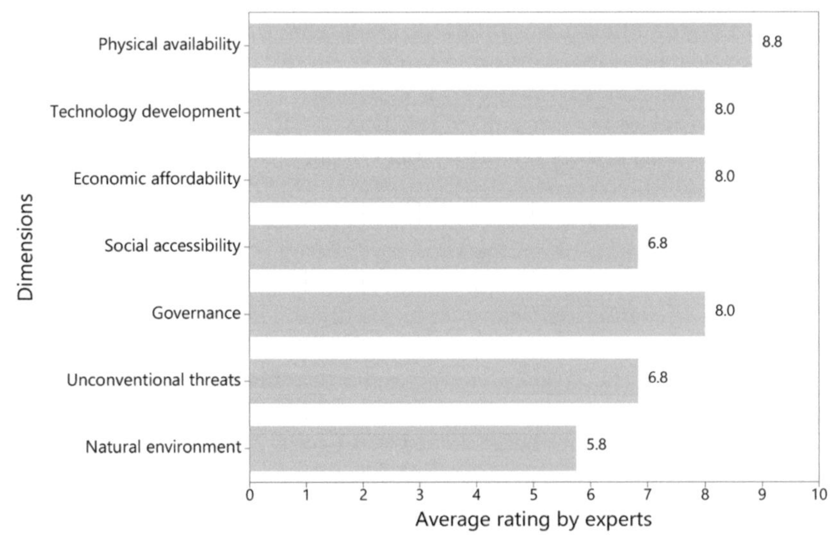

Figure 3.
Average expert rating of the importance of the dimensions of energy security.

- One expert (with intelligence background) argued that data quality and intelligence should be a separate dimension.

- There was uncertainty as to whether the impact of conventional warfare should be included in one of the existing dimensions or create an additional war dimension; the authors decided that conflict (especially war) is accounted indirectly through its impacts on almost all dimensions.

The experts were also asked to rate the importance of the dimensions of energy security at historical milestones that correspond to major geopolitical events. The more recent events were mentioned in the beginning of Section 2 of this chapter. Based on these ratings, weighted averages for the overall importance of energy security at each historical milestone were calculated. These are plotted in **Figure 4**.

In the eyes of energy experts reflecting the perspective of Greece, the importance of energy security has been increasing since the beginning of the twentieth century. Its highest ratings have appeared since 2010. This increase in the ratings of the importance of energy security is concurrent with the increased presence of renewable energy in the global online discourse (**Figure 1**). The potential role of renewable energy in energy security has never appeared more important than now.

3.3.1 The role of renewable energy in the dimensions of energy security

Although renewable energy has a much better greenhouse gas emission profile, it has environmental impacts like any other technology. Wind and solar energy are dilute fuels, requiring large expanses of land. The construction of onshore wind

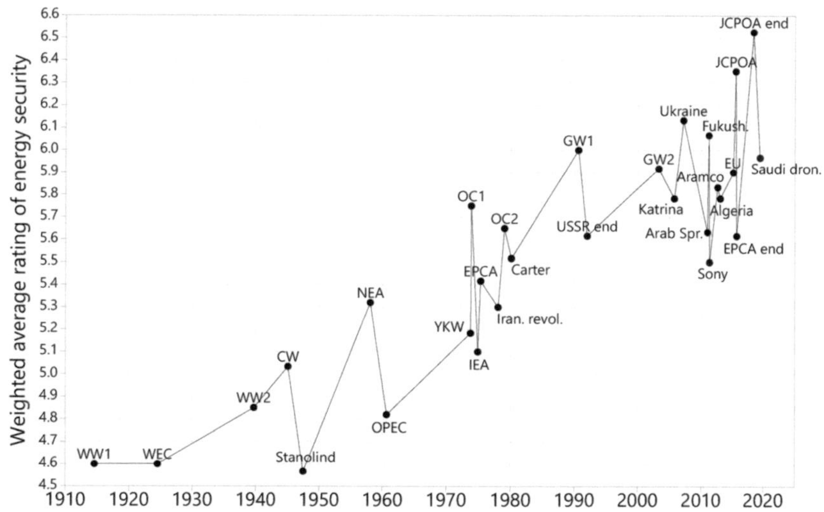

Figure 4.
Weighted average expert rating of the importance of energy security (WW1: First World War; WEC: Foundation of World Energy Council; WW2: Second World War; CW: Start of Cold War; Stanolind: first fracturing experiment; NEA: Foundation of European Nuclear Energy Agency; OPEC: Foundation of Organization of Petroleum Exporting Countries; YKW: Yom Kippur Arab-Israeli War; OC1: first oil crisis; IEA: Establishment of International Energy Agency; EPCA: Institution of US Oil Export Ban; Iran revol: Iranian revolution; OC2: second oil crisis; Carter: Carter Doctrine (on protection of Persian gulf by the United States); GW1: First Gulf War; USSR end: End of Cold War: GW2: Second Gulf War; Katrina: Hurricane Katrina; Ukraine: Start of Russia-Ukraine gas crises; Arab Spr.: Arab spring; Fukush.: Fukushima nuclear accident; Sony: cyberattack on Sony; Aramco: Cyberattack on Aramco; Algeria: gas plant hostage crisis in Algeria; EU: EU energy union; JCPOA: Institution of Iran Nuclear Deal; EPCA end: Lifting of US Oil Export Ban; JCPOA end: US withdrawal from Iran nuclear deal; Saudi dron.: Yemen drone attack).

farms necessitates clearing land areas with impacts on species such as tortoises, birds, and bats. The intermittent nature of wind and sunlight means that the energy they capture must be stored if they are to serve as the main energy source. Wind turbines have a lifespan of around 20–26 years, after which steel, cement, and other materials used in their construction must be recycled or properly disposed of. These impacts mean that the participation of renewable energy in the energy mix of a country creates negative environmental impacts. These may be partially offset by the positive environmental impact of the reduction in fossil fuel usage.

Looking back at the dimensions of energy security proposed by Sovacool and Rafey [46], the components of fuel diversification, disruption recovery, minimization of dependence on foreign supplies, minimization of price volatility, and support of sustainability (although with the introduction of aforementioned environmental impacts) are all served by the use of renewable energy for electricity production. Renewable energy improves the outlook of at least three of the four dimensions of energy security defined in that work.

Recalling the definition of energy security by Knox-Hayes et al. [49], it is argued that (further to the obvious connection of renewable energy to availability) the components of environmental quality (especially climate change) and small-scale energy production are improved by the use of renewable energy.

Turning to Sovacool and Mukherjee's work [2], the following components of security of supply and production should be favorably affected by renewable energy: dependency and diversification; price stability (regardless of the level of prices); decentralization and affordability (achieved with distributed small-scale installations); innovation and research (inherent in renewable energy); investment and employment (as new jobs are created in the renewable energy industry); environmental quality, especially climate change (with the aforementioned negative impacts of renewable energy); and trade and regional interconnectivity (e.g., with onshore wind farms and distributed small-scale systems). Renewable energy probably provides the best opportunity for a country to become more independent of the vulnerabilities of global energy markets and approach the goal of energy self-sufficiency [50] irrespective of its endowment in fossil fuel resources or its access to expensive nuclear energy technology.

Considering Ren and Sovacool's detailed presentation of an energy security index [53], renewable energy entered the dimensions of availability, as the percentage it represents of the total consumed energy; affordability, influencing the total energy produced by distributed and small-scale generation (a characteristic of renewable installations); accessibility, by improving the outlook of safety and reliability (as a secondary source); and acceptability, by helping with investment and employment.

The social acceptability of renewable energy has been reviewed by Stigka et al. [55] with empirical research carried out in a later work [56]. The socioeconomic and environmental disadvantages of renewable energy were discussed, and the 2014 renewable energy performance was presented for the EU countries, with Norway, Sweden, Latvia, Finland, Austria, Portugal, and Denmark having high renewable energy usage and being near their targets [55]. The same source also points out that social actors including local communities, local agencies, investors, nongovernmental organizations (NGOs), and local information networks are involved in renewable energy projects. Opposition to projects is not uncommon, per the NIMBY (not in my back yard) phenomenon, which led the authors to review the following barriers to renewable energy projects:

- economic and institutional factors, such as economic conditions in a region, issues with public or private ownership, lack of financial incentives, high investment costs (compared to fossil fuel alternatives), inefficiencies in the existing legal framework, complex licensing procedures, and bureaucratic problems;

- technical and planning factors, such as local geography and geomorphology, issues with the process of selecting an appropriate site (especially related to its previous usage) and planning problems;

- environmental and quality of life issues, such as landscape deterioration, visual intrusion, noise pollution and vibrations (related to the distance of residents from the renewable energy installations), disruption of nearby ecosystems, and impacts on the quality of life in the area; and

- factors related to public perceptions, such as lack of information or knowledge of renewable energy technologies, mistrust (which anxiety intensifying with ignorance), lack of impartiality, and suspicion toward investors.

The latter empirical research [56] found out that income and awareness of renewables are strong determinants of the willingness to accept renewable energy. Although esthetics could be more of a problem near tourist destinations, where economic, social, and cultural factors become involved [55], it was found that considerations related to tourism were low in the list of factors affecting the willingness to pay for renewable energy projects [56]. Renewable energy instigates the fear of uncontrolled development profits at the expense of the public good. So, steps must be taken for renewable energy to be accepted by local communities [55]. Education and the realization of positive impacts on the local economy may help in this direction.

3.4 Other energy security indexes

The literature on energy security indexes offers an insight into how the effects of energy security on the economy, society, and the environment are mediated directly or indirectly by renewable energy [38, 52, 63–65]. This section helps put the proposed energy security index in perspective.

Various studies [2, 34, 52, 66] have proposed a wide variety of energy security indexes, either to compare performance among countries or to track changes in a country's performance over time. There are indicators based on the perspective of the user [54] and others who link the concept of energy security with model-based scenario analyses in the context of addressing policy issues related to affordable energy and climate change [3].

The following are some well cited energy security index studies.

- Radovanović et al. [66] applied principal component analysis to assess the impact of individual indicators on an energy security index. They found energy intensity, GDP per capita, and carbon intensity to have the greatest impact on energy security.

- The *Vulnerability Index* [67] is a composite indicator, which considers five indicators: energy intensity; energy import dependency; ratio of energy-related carbon emissions to the total primary energy supply (TPES); electricity supply vulnerability; and lack of diversity in transport fuels [66].

- The six-factor *Risky External Energy Supply* [64] is entirely supply oriented and considers solely the level of diversification, with an emphasis on the assessment of transport safety of energy generating products [66].

- The *Aggregated Energy Security Performance Indicator* (AESPI) [68] has been developed by considering 25 individual indicators representing social, economic, and environmental dimensions. The indicator ranges from 0 to 10 and requires time series data for its estimation. AESPI helps assess the past energy security status of a country; it also helps evaluate the impacts of energy policies and plans on future energy security.

- The *Socioeconomic Energy Risk* is a composite index that considers the following indicators: energy source diversification, energy resource availability and feasibility, energy intensity, energy transport, energy dependence, political stability, market liquidity, and the GDP [66].

- The *US Energy Security Risk Index* [69] is a complex composite indicator obtained based on 83 individual indicators assessing geopolitical indicators, economic development, environmental concerns, and reliability [66].

- The concept of the "energy trilemma" is an attempt to balance the trade-offs among three major energy goals: energy security, economic competitiveness, and environmental sustainability [8]. The dimensions of energy trilemma are defined by the World Energy Council (WEC) [42] as: (a) energy security, that is, effective management of primary energy supply from domestic and external sources, reliability of energy infrastructure, and ability of energy providers to meet current and future demands; (b) energy equity, that is, accessibility and affordability of energy supply across the population; and (c) environmental sustainability, which encompasses the achievement of supply and demand-side energy efficiency, along with development of energy supply from renewable and other low-carbon sources. Related index efforts include the *Energy Architecture Performance Index* (EAPI), which was proposed in 2010 by the World Economic Forum (WEF) and was modified into the *Energy Sustainability Index* [70], a composite index based on a set of indicators grouped into three categories of the energy trilemma approach [66].

- Finally, the Renewable Energy Security Index (RESI) [25] is an index that assesses the impact of renewable energy technologies for electricity production. The index takes into consideration several factors based on the share of renewable energy into the electricity production mix at a national level. Decision makers are encouraged to use RESI as part of the transition from fossil-based to renewable-based power-generation technologies, as it promotes a sustainable model of electricity supply using domestic resources.

4. Perspectives on renewable energy and energy security

This section presents perspectives of specific countries toward renewable energy and energy security.

The global shift to renewable energy reflects a strengthening of the world's response to the threat of climate change. Most European countries have adopted policies toward a new energy transition with significant social, political, and economic implications. This transition to low-carbon energy is expected to alter the geopolitical landscape, shifting the dynamics between producer and consumer countries [1] and setting new energy standards for exporting countries [31]. In the words of Matsumoto and Andriosopoulos [22], "the lower the target of allowable emissions, the larger the required shifts to the energy structures will have to be."

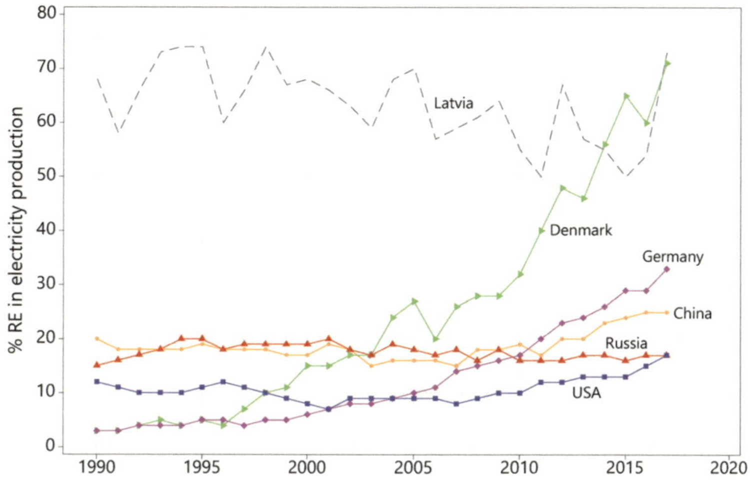

Figure 5.
Percent renewable energy in electricity production per country [71].

The energy transition is described as "a pathway toward transformation of the global energy sector from fossil-based to zero-carbon by the second half of this century" [1]. The renewable energy revolution will be one of the primary aspects that characterize and underpin the low-carbon transition. Shifts from coal to natural gas and from fossil fuels to renewable (and nuclear energy) will be critical [22].

The link between energy security and renewable energy is strengthened by the diffusion of renewable energy [10]. Data from the 2018 BP Statistical Review [14] show that renewable power grew by 14.5%, somewhat below the historical average, but near the all-time high increase of 2017 in absolute energy terms. The share of renewables in power generation increased from 8.4 to 9.3%, accounting for a third of the net increase in global power generation.

In the following paragraphs, the renewable energy situation of Denmark, Germany, China, Russia, and the United States is discussed and linked to energy security. These countries include producer, consumer, and transit, with different approaches to energy security and different usages of renewable energy sources. **Figure 5** presents the percent renewable energy in electricity production of these five countries plus Latvia, a world leader in the use of renewable energy sources.

Starting from the lowest part of the graph:

- The United States has been lingering around the 10% share of renewable energy in electricity production since 1990 and appears to have achieved a small increase since 2015. Of this share of renewable energy, 55.5% comes from hydro; 26.2% from wind; 11.8% from biomass; 3.7% from geothermal; and 2.8% from solar [72].

- Russia, with its large hydropower potential, has not been very keen to develop renewable energy, focusing its attention on exploiting its natural gas resources to maintain and strengthen its position as an energy hegemon. Almost all its shares of renewable energy in electricity production come from hydro (99.6%), with 0.3% coming from geothermal, 0.03% from biomass, and 0.02 from wind [72].

- China has been producing more electricity from renewable energy than both Russia and the United States, and it has increased its corresponding share of renewables since 2011. Of this share, 86.7% comes from hydro, 12.4% from wind, 0.6% from solar, 0.3% from biomass, and 0.02% from geothermal [72].

- While Germany started from a mere 3% of renewable energy in electricity production (like the United States), it has been increasing this 1990 share constantly, reaching a 10-fold level of 33% in 2017. Of this share, 32.1% comes from wind, 28.8% from biomass, 19.6% from hydro, 19.5% from solar, and 0.02% from geothermal [72].

- Denmark is arguably one of the most environmentally friendly energy producers in Europe, reaching a 71% share of renewable energy in electricity production in 2017. Of this share, 70.1% came from wind, 28.7% from biomass, 1% from solar, and 0.1% from hydro [72].

These five countries may be compared to Latvia, which is the undisputed historical world champion of green energy. Latvia has been producing 60–70% of its electricity from renewable energy sources (90% hydropower, 7% biomass, 3% wind, and 0.03% solar) since the 1990s [72]. Latvia is not examined in more detail because it is of less geopolitical interest than the other five countries, but it provides an interesting reference point for comparison.

Denmark, the greenest of the five countries, is indeed considered one of the most energy secure and sustainable countries among the OECD [51] and the EU27 countries [73]. Over the past 30 years, Denmark has achieved a swift decrease in its dependence on foreign energy sources from above 90% in the 1970s to practically zero and has become a net exporter of fuels and electricity [74]. At the core of Denmark's successful approach is a commitment to energy efficiency; prolonged taxes on energy fuels, electricity, and carbon dioxide; and incentives and subsidies for Combined Heat and Power (CHP) and wind turbines [75]. Denmark aims to go 100% renewable by 2050 [75, 76].

Germany, one of the largest energy importers in the EU, is in the middle of an ambitious energy transition [77]. Germany is considered the most successful country in the promotion of renewable energy [78]. In 2014, Germans had the best energy security performance among the EU countries due to the reduction of shares of oil and coal and the increase of diversification of energy imports [9]. The German energy transition (*Energiewende*) is considered the best-known renewable-based national energy policy [17]. The *Energiewende* aims to reduce the greenhouse gas emissions by 80–95% in 2050 (compared to 1990), increase the renewable share of final energy to at least 60%, and increase the renewable share of electricity demands to 80% [79]. Germany also intends to complete a nuclear phase-out by 2022 [80], a debatable move in the opinion of the authors of this chapter. Hansen et al. [79] presented a strategy for achieving 100% renewable energy for the entire German energy system and maintained that this scenario is possible with the introduction of key policies.

China is the world's largest energy consumer [14], the biggest emitter of greenhouse gases [81], the fifth largest producer of oil, the seventh largest producer of natural gas, and the largest producer of coal [82, 83]. China aims to cut the share of coal in its power mix, but coal consumption is growing, and more coal-fired power projects are under development. In fact, the percentage of fossil fuels in China's total primary energy demand is expected to exceed 90%, with coal being the main fuel source [22]. At the same time, China possesses the biggest amount of hydro resources globally, with a total theoretical hydropower potential of 694 GW [84]. By the end of 2015, China's hydro power exceeded 25% of the world's nonhydro renewable capacity, being 63.1 and 117.0% higher than the United States and

Germany, respectively [81]. Until the end of 2017, China's installed generation capacity of renewable energy was 635 million kW, which constituted 35.7% of the total installed capacity of electric power [78]. According to the BP Statistical Review [14], in 2018, China continued to lead the way in renewable growth, accounting for 45% of the global growth in renewable power generation, more than the entire OECD combined. Wang et al. [83] discussed scenarios for a Chinese sustainable energy development, with renewable energy regarded as a key emerging industry. China is reported to plan to increase the share of nonfossil fuels in primary energy consumption to around 20% by 2030 [83]. Liu [78] calls for improvements in China's renewable legal and policy framework, if the country is to achieve such an aim. All in all, China appears to be a reluctant and ambivalent actor in the renewable game.

Russia owns one of the largest fossil fuel resource stocks in the world and is the world's fourth largest emitter (after China, the United States, and India). Russia has the second largest natural gas reserves and production in the world (after the United States) [14], an endowment that allows it to play a major geopolitical role, for example, as the major supplier of natural gas to most European countries. At the same time, Russia is a country of vast geographic size and variability in terrain and climate, giving it the potential to develop virtually any kind of renewable energy [85]. Russia is accelerating the deployment of solar and wind through auctions to create benefits for employment, science, technology, and energy security for isolated populations [86]. Although the country was an early leader in the technology of renewables, Russia's strategy revolves around the development of fossil fuels and nuclear energy [87]. In 2013, the Russian government launched a Capacity-Based Renewable Energy Support Scheme (CRESS) [88]. Per CRESS implementation, a maximum installed capacity of solar, wind, and small hydro projects each year could obtain financial guarantees on Russia's wholesale electricity market [89]. The energy security of Russia will remain linked to its natural gas deposits, while its ability to continue to supply it to the European and other markets will determine its position as an energy hegemon.

In 2018, almost half of the global raise in natural gas (5.2%) came from the United States, which (as with oil production) recorded the largest annual growth

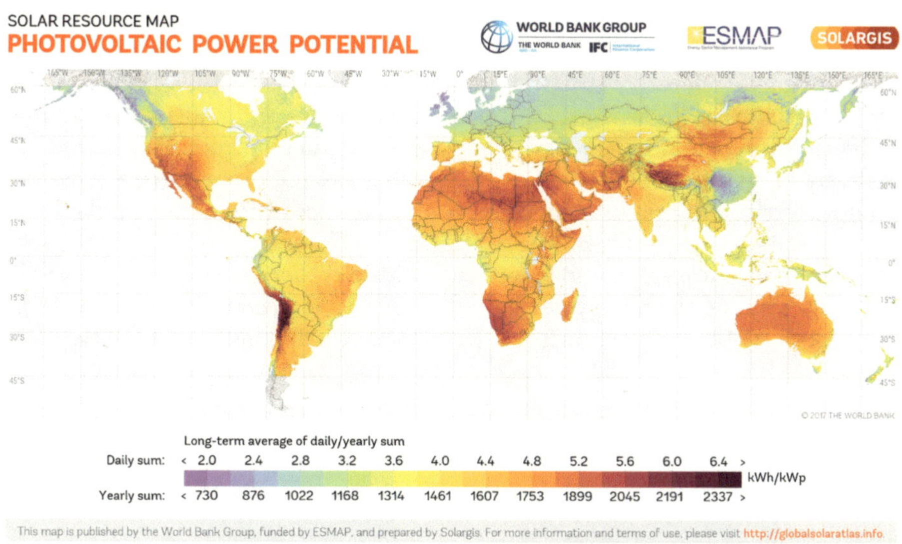

Figure 6.
Geographical distribution of photovoltaic power potential (https://solargis.com/maps-and-gis-data/download/world).

seen by any country in history [14]. The United States plans to achieve 80% renewable electricity by 2050 [81]. If this target is honored, it will be by actions of the states rather than at a federal level; in fact, most of the US states have started individually to enact policies that expand renewable energy capacity [90]. It seems unlikely that a federal-level renewable energy mandate will be passed soon, given the current (under President Trump) administration's quest for fossil fuel energy dominance.

Contrary to China and Russia, the EU and the United States are expected to develop a diversified energy mix with the aid of scientific and technological developments [66]. The EU will continue to improve renewables to limit its dependence on fossil fuel imports especially from Russia.

All in all, the introduction of more renewable energy sources into the energy mix of countries will influence their energy security favorably. As indicated in **Figure 6** for the case of solar energy, the global distribution of renewable energy sources is more evenly spread than fossil fuels. Furthermore, renewable energy is more amenable to distributed production, which is inherently more secure than the fossil fuel paradigm. Renewable energy will help usher an era of energy democracy, where a network of decentralized prosumer systems will play the role once dominated by large-scale power generation.

5. Conclusions

Geopolitical events in the twenty-first century have brought energy security and renewable energy in the forefront of political discourse and provided the motivation for this chapter. The energy security literature was reviewed, and a novel index of energy security was proposed. The index comprised the following dimensions: physical availability, technology development, economic affordability, social accessibility, governance, unconventional threats, and natural environment.

A small panel of academic and business experts in Greece rated physical availability as the most important and natural environment as the least important dimension of energy security. These experts thought that the importance of energy security has been increasing steadily since the beginning of the twentieth century.

Key geopolitical actors (the United States, Russia, China, Germany, and Denmark) have increased electricity production from renewable energy by a combination of different renewable sources. Despite any social acceptability issues and negative environmental impacts, renewable energy will help countries become more energy secure. At the same time, they will make themselves more resistant to geopolitical strife and more independent of the vagaries of fossil fuel markets.

As renewable energy sources are incorporated in the energy mix during the rest of the twenty-first century, the impact of geopolitical conflict on energy security will lessen. Furthermore, the energy generation industry will become more compatible with the original 4A energy security targets of accessibility and acceptability.

Author details

John A. Paravantis* and Nikoletta Kontoulis
Department of International and European Studies, University of Piraeus, Piraeus, Greece

*Address all correspondence to: jparav@unipi.gr

References

[1] International Renewable Energy Agency (IRENA). A New World. The Geopolitics of the Energy Transformation. Global Commission on the Geopolitics of Energy Transformation. 2019. Available from: https://geopoliticsofrenewables.org/assets/geopolitics/Reports/wp-content/uploads/2019/01/Global_commission_renewable_energy_2019.pdf [Accessed: 30 September 2019]

[2] Sovacool BK, Mukherjee I. Conceptualizing and measuring energy security: A synthesized approach. Energy. 2011;**36**(8): 5343-5355. DOI: 10.1016/j.energy.2011.06.043

[3] Kruyt B, van Vuuren DP, de Vries HJM, Groenenberg H. Indicators for energy security. Energy Policy. 2009;**37**:2166-2181. DOI: 10.1016/j.enpol.2009.02.006

[4] Asia Pacific Energy Research Center (APERC). A Quest for Energy Security in the 21st Century: Resources and Constraints. Japan: Asia Pacific Energy Research Center. 2007 Available from: http://aperc.ieej.or.jp/file/2010/9/26/APERC_2007_A_Quest_for_Energy_Security.pdf [Accessed: 30 September 2019]

[5] Chester L. Conceptualising energy security and making explicit its polysemic nature. Energy Policy. 2010; **38**:887-895. DOI: 10.1016/j.enpol.2009.10.039

[6] Winzer C. Conceptualizing energy security. Energy Policy. 2012; **46**:36-48. DOI: 10.1016/j.enpol.2012.02.067

[7] Sovacool BK. Evaluating energy security in the Asia Pacific: Towards a more comprehensive approach. Energy Policy. 2011;**39**:7472-7479. DOI: 10.1016/j.enpol.2010.10.008

[8] Ang BW, Choong WL, Ng TS. Energy security: Definitions, dimensions and indexes. Renewable and Sustainable Energy Reviews. 2015;**42**: 1077-1093. DOI: 10.1016/j.rser.2014.10.064

[9] Matsumoto K, Doumpos M, Andriosopoulos K. Historical energy security performance in EU countries. Renewable and Sustainable Energy Reviews. 2018;**82**: 1737-1748. DOI: 10.1016/j.rser.2017.06.058

[10] Hache E. Do renewable energies improve energy security in the long run? International Economics. 2018;**156**: 127-135. DOI: 10.1016/j.inteco.2018.01.005

[11] Hamed TA, Bressler L. Energy security in Israel and Jordan: The role of renewable energy sources. Renewable Energy. 2019;**135**:378-389. DOI: 10.1016/j.renene.2018.12.036

[12] United Nations (UN). Paris Agreement. 2015. Available from: https://unfccc.int/sites/default/files/english_paris_agreement.pdf [Accessed: 30 September 2019]

[13] Rogelj J, Den Elzen M, Höhne N, Fransen T, Fekete H, Winkler H, et al. Paris agreement climate proposals need a boost to keep warming well below 2°C. Nature. 2016;**534**:631-639. DOI: 10.1038/nature18307

[14] British Petroleum (BP). BP Statistical Review of World Energy. 68th ed. 2019. Available from: https://www.bp.com/content/dam/bp/business-sites/en/global/corporate/pdfs/energy-economics/statistical-review/bp-stats-review-2019-full-report.pdf [Accessed: 30 September 2019]

[15] International Renewable Energy Agency (IRENA). Renewable Energy

Statistics. 2018. Available from: https://www.irena.org/-/media/Files/IRENA/Agency/Publication/2018/Jul/IRENA_Renewable_Energy_Statistics_2018.pdf [Accessed: 30 September 2019]

[16] Moe E. Renewable Energy Transformation of Fossil Fuel Backlash—Vested Interest in the Political Economy. London: Palgrave Macmillan; 2015. DOI: 10.1057/9781137298799

[17] Harjanne A, Korhonen JM. Abandoning the concept of renewable energy. Energy Policy. 2019;**127**: 330-340. DOI: 10.1016/j.enpol.2018.12.029

[18] Gökgöz F, Güvercin MT. Energy security and renewable energy efficiency in EU. Renewable and Sustainable Energy Reviews. 2018;**96**: 226-239. DOI: 10.1016/j.rser.2018.07.046

[19] Ölz S, Sims R, Kirchner N. Contribution of Renewables to Energy Security. Paris: OECD/IEA; 2007. Available from: https://www.iea.org/publications/freepublications/publication/so_contribution.pdf [Accessed: 21 October 2019]

[20] Scholten D, Bosman R. The geopolitics of renewables: Exploring the political implications of renewable energy systems. Technological Forecasting and Social Change. 2016; **103**:273-283. DOI: 10.1016/j.techfore.2015.10.014

[21] REN21. Renewables 2019 Global Status Report. Paris: REN21 Secretariat; 2019. Available from: https://www.ren21.net/wp-content/uploads/2019/05/gsr_2019_full_report_en.pdf [Accessed: 21 October 2019]

[22] Matsumoto K, Andriosopoulos K. Energy security in East Asia under climate mitigation scenarios in the 21st century. Omega. 2016;**59**:60-71. DOI: 10.1016/j.omega.2014.11.010

[23] Bang G. Energy security and climate change concerns: Triggers for energy policy change in the United States? Energy Policy. 2010;**38**: 1645-1653. DOI: 10.1016/j.enpol.2009.01.045

[24] Valentine SV. The fuzzy nature of energy security. Sovacool BK. The Routledge Handbook of Energy Security. New York: Routledge; 2011. p. 56–73. ISBN 13: 9780415591171

[25] García-Gusano D, Iribarren D, Garraín D. Prospective analysis of energy security: A practical life-cycle approach focused on renewable power generation and oriented towards policy-makers. Applied Energy. 2017;**190**: 891-901. DOI: 10.1016/j.apenergy.2017.01.011

[26] Luft G, Korin A. Energy security: In the eyes of the beholder. In: Luft G, Korin A, editors. Energy Security Challenges for the 21st Century: A Reference Handbook. Santa Barbara, California, USA: Praeger Security International; 2009. pp. 1-17. ISBN: 978-0-275-99998-8

[27] Johansson B. A broadened typology on energy and security. Energy. 2013;**53**: 199-205. DOI: 10.1016/j.energy.2013.03.012

[28] Marquina A. The southeast–southwest European energy corridor. In: Marguina A, editor. Energy Security: Visions from Asia and Europe. UK: Palgrave Macmillan; 2008. pp. 54-68. ISBN: 978-0230595002

[29] Leonard M, Popescu N. A Power Audit of the EU-Russia Relations. Policy Paper. London: European Council on Foreign Relations; 2007. Available from: https://www.ecfr.eu/page/-/ECFR-02_A_POWER_AUDIT_OF_EU-RUSSIA_RELATIONS.pdf [Accessed: 12 March 2020]

[30] Månsson A, Johansson B, Nilsson LJ. Assessing energy security: An overview of commonly used methodologies. Energy. 2014;**73**:1-14. DOI: 10.1016/j.energy.2014.06.073

[31] Overland I. The geopolitics of renewable energy: Debunking four emerging myths. Energy Research and Social Science. 2019;**49**:36-40. DOI: 10.1016/j.erss.2018.10.018

[32] Cherp A, Jewell J. The concept of energy security: Beyond the four As. Energy Policy. 2014;**75**:415-421. DOI: 10.1016/j.enpol.2014.09.005

[33] Yergin D. Energy security in the 1990s. Foreign Affairs. 1988;**67**:110-132. DOI: 10.2307/20043677

[34] Kisel E, Hamburg A, Harm M, Leppiman A, Ots M. Concept for energy security matrix. Energy Policy. 2016;**95**: 1-9. DOI: 10.1016/j.enpol.2016.04.034

[35] Jewell J. The IEA Model of Short-Term Energy Security (MOSES): Primary Energy Sources and Secondary Fuels. Working Paper. Paris, France: OECD/IEA; 2011. Available from: https://www.oecd-ilibrary.org/energy/the-iea-model-of-short-term-energy-security-moses_5k9h0wd2ghlv-en [Accessed: 21 October 2019]

[36] International Energy Agency (IEA). Measuring Short-Term Energy Security. Paris: OECD/IEA; 2011. Available from: https://www.iea.org/publications/freepublications/publication/Moses.pdf [Accessed: 21 October 2019]

[37] Sovacool BK. Differing cultures of energy security: An international comparison of public perceptions. Renewable and Sustainable Energy Reviews. 2016;**55**:811-822. DOI: 10.1016/j.rser.2015.10.144

[38] International Energy Agency (IEA). Energy Security and Climate Policy: Assessing Interactions. Paris: IEA/OECD; 2007. Available from: https://www.iea.org/publications/freepublications/publication/energy_security_climate_policy.pdf [Accessed: 21 October 2019]

[39] European Commission (EC). Towards a European Strategy for the Security of Energy Supply. Green Paper. Luxembourg: Office for Official Publications of the European Communities; 2000. Available from: http://aei.pitt.edu/1184/1/enegy_supply_security_gp_COM_2000_769.pdf [Accessed: 21 October 2019]

[40] Yergin D. Ensuring energy security. Foreign Affairs. 2006;**85**:69-82. DOI: 10.2307/20031912

[41] Jun E, Kim W, Chang SH. The analysis of security cost for different energy sources. Applied Energy. 2009; **86**:1894-1901. DOI: 10.1016/j.apenergy.2008.11.028

[42] World Energy Council (WEC). World Energy Trilemma I 2016. Defining Measures to Accelerate the Energy Transition. London, UK: WEC; 2016. Available from: https://www.worldenergy.org/assets/downloads/World-Energy-Trilemma_full-report_2016_web.pdf [Accessed: 21 October 2019]

[43] Augutis J, Martišauskas L, Krikštolaitis R, Augutiene E. Impact of the renewable energy sources on the energy security. Energy Procedia. 2014; **61**:945-948. DOI: 10.1016/j.egypro.2014.11.1001

[44] Kopp SD. Politics, Markets and EU Gas Supply Security. Case Studies of the UK and Germany. Berlin, Germany: Springer; 2014

[45] Goldthau A, Sovacool BK. The uniqueness of the energy security, justice, and governance problem. Energy Policy. 2012;**21**:232-240. DOI: 10.1016/j.enpol.2011.10.042

[46] Sovacool BK, Rafey W. Snakes in the grass: The energy security implications of Medupi. The Electricity Journal. 2011;**24**: 92-100. DOI: 10.1016/j.tej.2010.12.002

[47] Alhajji AF. What is energy security? Definitions and concepts. Middle East Economic Survey. Oil, Gas & Energy Law (OGEL). 2008;**3**:45. Available from: https://www.ogel.org/article.asp?key=2786 [Accessed: 21 October 2019]

[48] Vivoda V. Evaluating energy security in the Asia-Pacific region: A novel methodological approach. Energy Policy. 2010;**38**:5258-5263. DOI: 10.1016/j.enpol.2010.05.028

[49] Knox-Hayes J, Brown MA, Sovacool BK, Wang Y. Understanding attitudes toward energy security: Results of a cross-national survey. Global Environmental Change. 2013;**23**: 609-622. DOI: 10.1016/j.gloenvcha.2013.02.003

[50] Zhao H. Energy security: From energy independence to energy interdependence. In: Zhao H, editor. The Economics and Politics of China's Energy Security Transition. 1st ed. London, UK: Academic Press; 2019. pp. 99-120. ISBN: 9780128151532

[51] Sovacool BK, Brown MA. Competing Dimensions of Energy Security: An International Perspective. Working Paper #45, Working Paper Series. Atlanta, GA: Ivan Allen College, School of Public Policy, Georgia Tech; 2009. Available from: https://people.iac.gatech.edu/files/publication/394_wp45.pdf [Accessed: 21 October 2019]

[52] Sovacool BK, Mukherjee I, Drupady IM, D'Agostino AL. Evaluating energy security performance from 1990 to 2010 for eighteen countries. Energy. 2011;**36**:5846-5853. DOI: 10.1016/j.energy.2011.08.040

[53] Ren J, Sovacool BK. Quantifying, measuring, and strategizing energy security: Determining the most meaningful dimensions and metrics. Energy. 2014;**76**:838-849. DOI: 10.1016/j.energy.2014.08.083

[54] Narula K, Reddy S. Three blind men and an elephant: The case of energy indices to measure energy security and energy sustainability. Energy. 2015;**80**: 148-158. DOI: 10.1016/j.energy.2014.11.055

[55] Stigka E, Paravantis JA, Mihalakakou GK. Social acceptance of renewable energy sources: A review of contingent valuation applications. Renewable and Sustainable Energy Reviews. 2014;**32**:100-106. DOI: 10.1016/j.rser.2013.12.026

[56] Paravantis JA, Stigka E, Mihalakakou G, Michalena E, Hills JM, Dourmas V. Social acceptance of renewable energy projects: A contingent valuation investigation in Western Greece. Renewable Energy. 2018;**123**: 639-651. DOI: 10.1016/j.renene.2018.02.068

[57] Caputo R. Hitting the wall: A vision of a secure energy future. Synthesis Lectures on Energy and the Environment: Technology, Science and Society. 2008;**2**:1-204. DOI: 10.2200/S00124ED1V01Y200805EGY003

[58] Thomson M. Among the energy tribes: A cultural framework for the analysis and design of energy policy. Policy Sciences. 1984;**17**:321-339. DOI: 10.1007/BF00138710

[59] Sovacool BK. Seven suppositions about energy security in the United States. Journal of Cleaner Production. 2011;**19**(11):1147-1157. DOI: 10.1016/j.jclepro.2011.03.014

[60] Brown MA, Wang Y, Sovacool BK, D'Agostino AL. Forty years of energy security trends: A comparative assessment of 22 industrialized countries. Energy Research and Social

Science. 2014;**4**:64-77. DOI: 10.1016/j.erss.2014.08.008

[61] Sovacool BK. Energy policy and cooperation in Southeast Asia: The history, challenges, and implications of the Trans-ASEAN Gas Pipeline (TAGP) network. Energy Policy. 2009;**37**(6): 2356-2367. DOI: 10.1016/j.enpol.2009.02.014

[62] Pezalla AE, Pettigrew J, Miller-Day M. Researching the researcher-As-instrument: An exercise in interviewer self-reflexivity. Qualitative Research. 2012;**12**(2):165-185. Available at https://www.ncbi.nlm.nih.gov/pmc/articles/PMC4539962 [Accessed: 21 October 2019]

[63] Greene D. Measuring energy security: Can the United States achieve oil independence? Energy Policy. 2010; **38**:1614-1621. DOI: 10.1016/j.enpol.2009.01.041

[64] Le Coq C, Paltseva E. Measuring the security of external energy supply in the European Union. Energy Policy. 2009; **37**:4474-4481. DOI: 10.1016/j.enpol.2009.05.069

[65] Löschel A, Moslener U, Rübbelke DTG. Indicators of energy security in industrialised countries. Energy Policy. 2010;**38**:1665-1671. DOI: 10.1016/j.enpol.2009.03.061

[66] Radovanović M, Filipović S, Pavlović D. Energy security measurement—A sustainable approach. Renewable and Sustainable Energy Reviews. 2017;**68**:1020-1032. DOI: 10.1016/j.rser.2016.02.010

[67] Gnansounou E. Assessing the energy vulnerability: Case of industrialized countries. Energy Policy. 2008;**36**: 3734-3744. DOI: 10.1016/j.enpol.2008.07.004

[68] Martchamadol J, Kumar S. An aggregated energy security performance indicator. Applied Energy. 2013;**103**: 653-670. DOI: 10.1016/j.apenergy.2012.10.027

[69] U.S. Chamber of Commerce. Index of U.S. Energy Security Risk: Assessing America's Vulnerabilities in a Global Energy Market. Washington D.C., USA: U.S. Chamber of Commerce; 2013. Available from: https://www.globalenergyinstitute.org/sites/default/files/Idex%20of%20US%20Energy%20Security%20Risk.pdf [Accessed: 21 October 2019]

[70] World Economic Forum (WEF). Global Energy Architecture Performance Index Report 2015. Switzerland: World Economic Forum; 2014. Available from: http://www3.weforum.org/docs/WEF_GlobalEnergyArchitecture_2015.pdf [Accessed: 21 October 2019]

[71] International Energy Agency (IEA) Energy Atlas. Available from http://energyatlas.iea.org/#!/tellmap/-1118783123/3 [Accessed: 14 February 2020]

[72] Worldwide Electricity Production from Renewable Energy Sources. Stats and Figures Series. Fifteenth Inventory - Edition 2013. 2013. Available from: http://www.energies-renouvelables.org/observ-er/html/inventaire/Eng/sommaire.asp [Accessed: 14 February 2020]

[73] Eurostat. EU27 Energy Dependence Rate at 54% in 2011. 2013. Available from: https://ec.europa.eu/eurostat/web/products-press-releases/-/8-13022013-BP [Accessed: 21 October 2019]

[74] Sovacool BK, Tambo T. Comparing consumer perceptions of energy security, policy, and low-carbon technology: Insights from Denmark. Energy Research and Social Science. 2016;**11**:79-91. DOI: 10.1016/j.erss.2015.08.010

[75] Sovacool BK. An international assessment of energy security performance. Ecological Economics.

2013;**88**:148-158. DOI: 10.1016/j.ecolecon.2013.01.019

[76] Danish Government. Energy Strategy 2050—From Coal, Oil and Gas to Green Energy. 2011. Available from: http://www.danishwaterforum.dk/activities/Climate%20change/Dansk_Energistrategi_2050_febr.2011.pdf [Accessed: 21 October 2019]

[77] Proskuryakova L. Updating energy security and environmental policy: Energy security theories revisited. Journal of Environmental Management. 2018;**223**:203-214. DOI: 10.1016/j.jenvman.2018.06.016

[78] Liu J. China's renewable energy law and policy: A critical review. Renewable and Sustainable Energy Reviews. 2019; **99**:212-219. DOI: 10.1016/j.rser.2018.10.007

[79] Hansen K, Van Mathiesen B, Skov IR. Full energy system transition towards 100% renewable energy in Germany in 2050. Renewable and Sustainable Energy Reviews. 2019; **102**:1-13. DOI: 10.1016/j.rser.2018.11.038

[80] Agora Energiewende. European Energy Transition 2030: The Big Picture. Ten Priorities for the next European Commission to Meet the EU's 2030 Targets and Accelerate Towards 2050. 2019. Available from: https://www.agora-energiewende.de/fileadmin2/Projekte/2019/EU_Big_Picture/153_EU-Big-Pic_WEB.pdf [Accessed: 21 October 2019]

[81] Yang XJ, Hu H, Tan T, Li J. China's renewable energy goals by 2050. Environmental Development. 2016;**20**:83-90. DOI: 10.1016/j.envdev.2016.10.001

[82] Zhang L, Sovacool BK, Ren J, Ely A. The dragon awakens: Innovation, competition, and transition in the energy strategy of the People's Republic of China, 1949–2017. Energy Policy. 2017;**108**:634-644. DOI: 10.1016/j.enpol.2017.06.027

[83] Wang B, Wang Q, Wei Y-M, Li Z-P. Role of renewable energy in China's energy security and climate change mitigation: An index decomposition analysis. Renewable and Sustainable Energy Reviews. 2018;**90**:187-194. DOI: 10.1016/j.rser.2018.03.012

[84] Zhang D, Wang J, Lin Y, Si Y, Huang C, Yang J, et al. Present situation and future prospect of renewable energy in China. Renewable and Sustainable Energy Reviews. 2017; **76**:865-871. DOI: 10.1016/j.rser.2017.03.023

[85] International Energy Agency (IEA). Renewables in Russia: From Opportunity to Reality. Paris: IEA/OECD; 2003. Available from: https://www.iea.org/publications/freepublications/publication/RenewRus_2003.pdf [Accessed: 21 October 2019]

[86] Gielen D, Boshell F, Saygin D, Basilian MD, Wagner N. The role of renewable energy in the global energy transformation. Energy Strategy Reviews. 2019;**24**:38-50. DOI: 10.1016/j.esr.2019.01.006

[87] Lanshina TA, "Skip" Laitner JA, Potashnikov VY, Barinova VA. The slow expansion of renewable energy in Russia: Competitiveness and regulation issues. Energy Policy. 2018;**120**:600-609. DOI: 10.1016/j.enpol.2018.05.052

[88] Smeets N. Similar goals, divergent motives. The enabling and constraining factors of Russia's capacity-based renewable energy support scheme. Energy Policy. 2017;**101**:138-149. DOI: 10.1016/j.enpol.2016.11.037

[89] Boute A. Promoting renewable energy through capacity markets: An analysis of the Russian support scheme. Energy Policy. 2012;**46**:68-77. DOI: 10.1016/j.enpol.2012.03.026

[90] Hazboun SO, Briscoe M, Givens J, Krannich R. Keep quiet on climate: Assessing public response to seven renewable energy frames in the Western United States. Energy Research and Social Science. 2019;57:101243. DOI: 10.1016/j.erss.2019.101243

Chapter 3

Evaluating Biogas Technology in South Africa: Awareness and Perceptions towards Adoption at Household Level in Limpopo Province

Solomon Eghosa Uhunamure,
Nthaduleni Samuel Nethengwe and David Tinarwo

Abstract

Despite the enormous advantages associated with biogas technology, the level of awareness and perceptions still remain very dismay. The level of adoption is relatively low compared to the potential of the technology. This chapter therefore aimed at unravelling the awareness and perceptions of the technology at household level in Limpopo Province of South Africa. In this study, 200 households were sampled; 72 households with biogas digesters and 128 without digesters. Primary data collection was elicited with the use of open- and closed-ended questionnaires. A non-parametric test of Spearman rank correlation coefficient and chi-square were employed to show the association between the variables. Empirically, the results revealed that income earned, cost of digester, lack of awareness programmes, water and feedstock availability, technical availability and assistance, and private sector participation are among the factors limiting the dissemination, awareness and perception of the technology in the province. The study recommends interventions through more elaborate awareness and promotion programmes in disseminating the technology as well as provision of technical assistance, loans, credits and subsidies to households willing to adopt the technology.

Keywords: energy, environment, digester, households, technology

1. Introduction

One critical issue confronting developing nations such as South Africa is the provision of sustainable energy, to a proportion of its population that do not have access to modern and reliable energy supply. Access to energy is viewed as a vital condition that enhances the development of a country's economic activities, in order for the people to have an improved quality of life [1]. This explains the notion why providing adequate, affordable, sustainable, clean and efficient energy remains the core interest of many countries. Despite the efforts in place to provide adequate, sustainable and modern energy, about 1.4 billion people worldwide do not have access to modern energy carriers [2]. Sadly, the majority of the people without access to modern energy subsist in Africa, with a representation of 57% of the world population [3]. In South Africa, fossil fuel dominates the energy sector, with coal accounting for 89% and crude oil accounting for 22%, thus providing much of the energy consumed in the country [4]. In Limpopo Province, the energy carriers do not differ as the energy satisfaction in the province comes from coal and oil. Although the use of fossil fuel in generating energy brings an overwhelming burden to the environment in the form of greenhouse gas emissions, water contamination, air pollution and ecosystem degradation [5].

The Limpopo Department of Economic Development, Environment and Tourism (LEDET) identified biomass and solar as the main renewable resources of energy in the province [6]. The Department of Energy has developed a programme for attracting private investment into the energy sector. The Renewable Energy Independent Power Procurement programme (REIPP) has been designed to contribute towards the national target of 3725 MW of renewable energy and towards socio-economic and environmentally sustainable growth [7]. To meet energy demands in low-to-middle income households in many developing countries, the use of biogas technology is currently being deployed. The technology does not only provide energy but also serves as a good waste management measure [8]. Limpopo Province, like many other provinces in South Africa, has seen limited growth in the dissemination of biogas technology due to awareness and perceptions of the technology. Despite the long history of biogas technology in the country, the technology has witnessed poor growth of installed domestic biogas digesters, hence the initiation of this study.

2. Household energy sources and utilisation in Limpopo Province

The energy sector is central to South Africa's economy due to its reliance on energy-intensive, large-scale coal mining activities. Limited oil and natural gas reserves are present in the country; thus, the country relies and uses large deposits of its coal to meet most of the energy required, which is principally in the power sector. In 2013, less than 1% of the energy consumed was from renewable sources; 3% from natural gas; 22% from oil while 74% of the total consumed energy was primarily from coal and more than half was consumed in the electricity sector [9]. In 2017, South Africa was rated among the 10 top producers of coal in the world [10]. Due to its dependence on coal, the country is considered one of the continent's principal emitter of carbon dioxide, accounting for about 40% and thus placing the country as the thirteenth major emitter of carbon dioxide in the world [11]. Notwithstanding the renewable energy resources endowed in the country, there has been an energy shortage, which led to the energy crisis of 2008, which still persists till date [7].

A survey by the Department of Energy [4], with the aim of gathering information related to energy behaviour in South Africa households, indicated that there are significant differences between non-electrified and electrified households in Limpopo Province. To meet the basic energy needs, households employed an array of energy sources. Electrified households reported that they use electricity for heating, lighting or cooking. Even so, it is clear that other sources of energy, such as paraffin, fuelwood, gas and candle, are relied upon by at least a fifth of all the surveyed households with electricity. On the other hand, non-electrified households, in the absence of domestic connection primarily rely on fuelwood, candles, with additional households reporting using gas and coal. The use of renewable energy, such as solar was reported by a tenth of the electrified and non-electrified households surveyed. A major factor that continues to play a significant role in domestic energy use is socio-economic differences. The use of paraffin, candles, and fuelwood was present in more than 70% of the households in the low-income bracket of less than R3000 monthly, while near-absence was almost recorded in the medium to high-income households [4]. Cooking is one of the utmost energy intensive applications in the households of Limpopo Province. Unsurprisingly, geographic variation indicated that Limpopo Province households have a lower share of electricity used for cooking purposes, which is less below the national average [4]. Although most households in the country rely on fuelwood as the second main source of energy for cooking, somewhat atypical is the case of Limpopo Province, where 44% (representing two-fifths of the households) use fuelwood as their main source of energy for cooking compared to 49% of the households using electricity for cooking [4]. Marginal share were reported for households using coal, solar electricity, gas and paraffin. In non-electrified households, paraffin and fuelwood dominate as the source of energy for cooking purposes, at 38 and 54%, respectively [4]. However, a small fraction of coal, gas, solar electricity and electricity from generators were recorded in small percentages of households as their primary sources of energy for their cooking needs. With the increases in paraffin prices, the findings are not too surprising, as fuelwood is an all-possibility compensation for the higher paraffin prices. However, the decrease in paraffin use is positive, but the increase in the use of fuelwood remains a great concern.

Domestic space heating is another intensive energy application in the households. Examination by electrification as the main source for space heating in electrified households indicated that 45% rely primarily on electricity, with a minority reporting paraffin, fuelwood and other sources of energy, at 4, 7 and 5%, respectively [4]. In non-electrified households in the province's households, fuelwood is primarily relied upon for space heating, accounting for 59%, while paraffin has a share of 11%, with other sources that consist mainly of coal stands at 5% [4]. In respect to water heating for bathing purposes, the most common electrical appliance used by electrified households in the province for water heating purposes is an electric geyser at 31%. Other appliances are the electric kettle at 23% or a combination of electric stove and kettle at 7% [4]. Conversely, in non-electrified households that rely on a single energy source for water heating, fuelwood exclusively accounts for 46%; about a quarter of the households also exclusively uses paraffin, which stands at 27 and 16% of the non-electrified households use a combination of paraffin and fuelwood [4]. The findings from the survey contend that there is a barrier in the province, which is hindering the switch to electricity as a preferred method for water heating for bathing purpose [6]. In terms of energy preferences and choice for heating water, other than for bathing purposes, the survey indicated that 93% of the households in the province, on average, depend on a single source of energy, while a small share of 5% is characterised by multiple sources. In electrified households, the use of electrical appliances for water heating, other than for bathing purposes, stands at 83%, while in non-electrified households, fuelwood exclusively accounts for 52% for the households, followed by paraffin, which is used by a further 38% of the households [4].

3. Research methodology

3.1 Description of the study area

Limpopo Province is the northern-most province of South Africa, lying within the curves of the great Limpopo River. It shares international borders with Botswana to the west, Zimbabwe to the north and to the east, Mozambique.

The province falls under the greater-savannah biome which is characterised by grassland and forest and it is sometimes referred to as the bushveld biomes. The bushveld, which comprises most part of the province, is renowned for cattle rearing. The vegetation types are of grave significance and need conservative representation in order to preserve the flora diversity, as over one-third of the forest has been reduced due to over-exploitation and utilisation of the forest resources [12]. Limpopo Province is viewed as one of the poorest provinces in the country, due to high unemployment rate that persist mostly in the rural parts of the province [13]. Most of the households in the rural parts which encompass much of the population depend on pension grants, government grants, and remittances from family members who migrate to other provinces to work. The household wealth is relatively lower, compared to other municipalities in South Africa [14].

3.2 Data collection and sampling methods

This study was centred on household survey conducted purposefully in Limpopo Province from 2018 to 2019. The province was specifically chosen because of the government promotion of pro-poor energy alternatives, transformation of organic waste-to-energy and other low carbon technologies in order to ensure energy provision and security. The primary data were elicited from respondents in the households using interviews and self-administered semi-structured open and closed-ended questionnaires. Secondary data for this study were gathered from unpublished and published research articles. For ease of understanding, the questionnaires and interviews were conducted in English language and where necessary, translated to XiTsonga or TshiVenda languages which are the local dialects of the respondents. Ethical considerations were strictly adhered to. The survey sample was drawn from households with and without biogas digesters. After an in-depth assessment of households with biogas digesters in the province, 72 households were purposively sampled, while 128 households without digesters were randomly sampled. From a household installed with biogas digester, at least one household without a digester was sampled randomly in order to elicit their opinion regarding whether a household with a digester influences their perception about the technology. The sampling technique could not be based completely on one sampling technique because in the study area, the number of households with biogas digesters were smaller, compared to households without digesters and thus the inference from the sample could not be drawn from one sampling type.

3.3 Data analysis

The generated data was analysed and simplified using Microsoft Excel spreadsheet and statistical procedures of the Statistical Package for the Social Sciences (SPSS 22.0). The data were coded, defined and labelled and fed in Microsoft Excel then, exported to the SPSS program, to generate descriptive statistics principally to identify patterns and trends. The results of the data were clearly displayed in simple pie chart, bar graph, and contingency tables. A non-parametric test of Pearson chi-square and Spearman rank correlation coefficient was also used to present a detailed analysis of the results.

4. Adoption challenges faced by biogas technology

The challenges faced by biogas technology in several developing countries including South Africa are numerous and has becloud the awareness and perception as well as the potential of the technology. These factors have hindered the general dissemination of the technology. The rate of biogas technology dissemination is low, despite its potential, thus making the share of biogas technology in the energy mix very insignificant in many households, where it is supposed to play an alternative option in fuel substituting. For example, in South Africa, the numbers of installed digesters are around 700 with less than 100 in Limpopo Province [15]. The awareness and perceptions challenges faced by the technology include the following.

4.1 Dissemination of biogas technology

Another important factor which acts as a constraint to the adoption and dissemination of biogas technology is the awareness of the technology [16]. In Ghana, for example, lack of awareness about biogas technology was mentioned as one of the barriers in adopting the technology. Some cultural viewpoints such as stigmatising the utilisation of human excreta or even cow dung as substrate to biogas digesters, has the potential of discouraging its dissemination [17]. Thus, stories of successes and failures of previous biogas installations can also aid in promoting or constraining the dissemination of the technology. According to Gitonga [16], where an installed biogas digester performed well, word of mouth from the satisfied user will encourage other potential users to own the technology. In instances where the digester fails, it will create a negative dissemination impact on the technology; thus, discouraging potential adopters in the process. In Africa, success stories of biogas demonstration plants are relatively low. Many reasons are outlined for their failure. These include absence of energy focused policy, poor design, poor construction and material used, lack of maintenance from the owner, lack of project monitoring and follow-ups and poor ownership attitude and responsibility [17].

In addition, households evaluate the awareness attributes of modern energy carrier in their adoption decisions. Identified by Rogers [18] are five attributes that can accelerate or impede the adoption rate of the technology. These attributes are relative advantages, trialability, observability, complexity and compatibility. In the relative advantage of a modern energy carrier, the technology is evaluated in economic terms; according to its social status, satisfaction and convenience. A technology that is easily tried and experimented for its appropriateness with observable results to others is expected to be rapidly adopted than others. Furthermore, a compatible technology to existing cultural norms, values and experiences of a community has a better chance of adoption compared to any technology against such values and norms. In addition, a technology that is easy in understating and utilising is likely to be adopted quicker than those that require new skills, knowledge and understanding. According to Taherdoost [19], in the traditional adoption technology model, primarily, a consumer's adoption is determined by the 'perceived ease of use' and the 'perceived usefulness/benefits' of the technology. Therefore, in the process of making and informed decision to either reject or accept the new technology, the consumers weigh the option of the technology if it is easy to utilise (perceived ease of use) and if one's productivity will improve (perceived usefulness/benefits).

4.2 Biogas technology awareness

The study findings as presented in **Figure 1** indicated that 22% of the respondents acknowledged that they have at least heard about the technology with regards to financial implications. This implies that 22% of the households in the Province are aware about the existence of the technology. This can be attributed to the few biogas projects within their locality. The presence of the technology's existence in the study area however does not imply awareness of the technology. Awareness of

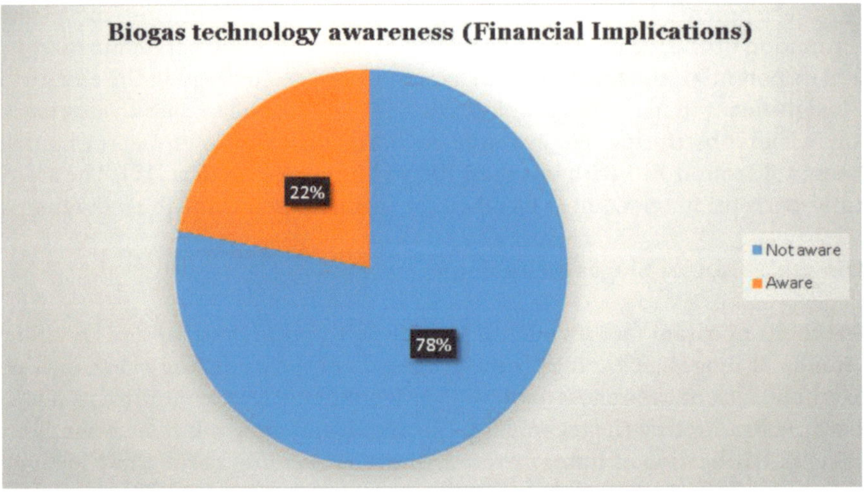

Figure 1.
Biogas technology awareness (financial implications) survey in the study area (source: field survey).

biogas technology involves households getting detailed information about the technology; from the functionality, financial implications and the numerous advantages the technology offers. The result further shows that 78% of the households have no relative idea about biogas technology in terms of the financial implications involved in respect to the technology. Further, the results (**Figure 2**) in terms of functionality, indicated that 66% of the respondents have no clue how the technology operates, while 34% revealed that they can operate the technology.

Technology awareness and perceptions are also disseminated via information channels. From the study as indicated in **Figure 3**, the identified channels of information that have helped in sensitising the households about the significances, advantages and efficiencies of biogas technology in the province include that from neighbours' with installed digesters, at 52%, and NGOs at 38%, which served as the main sources of information pertaining the technology. Others include 7% from government departments/agencies and 3% from media publications. This indicates that the role of government agencies and the media in disseminating the technology is very low. This can be improved through adequate education and dissemination, particularly in the rural areas, so that the social, economic and environmental benefits of the technology can be appreciated as against the continues use of fuelwood, which has detremental effects on their health and wellbeing [20].

4.3 Biogas technology perceptions

From the field survey, the data obtained as shown in **Table 1** clearly indicated a prevalent perception of biogas technology at household level. The responses raised on the perceptions of the technology indicated that in households with biogas digesters, 91% agreed that biogas can help solve the problem of fuelwood for cooking, as agreed by 87% from the non-users. Regarding using the slurry from biogas to improve soil fertility, 88% of the users agreed, while 86% from the non-users also concurred. Using biogas technology as a method to manage waste in order to improve environmental hygiene was at 89% from the users. The respondents believed that it is a good management method compared to 88% from the non-users category. In the province, as part of their energy mix, most households still rely on fuelwood, which is harvested from the forest, thereby creating room for degradation, which can eventually lead to deforestation. In respect to biogas technology, 90% of the households

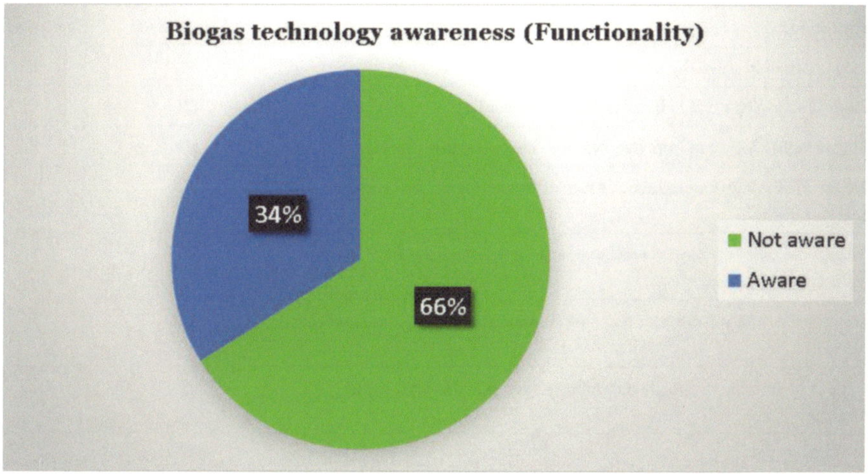

Figure 2.
Biogas technology awareness (functionality) survey in the study area (source: field survey).

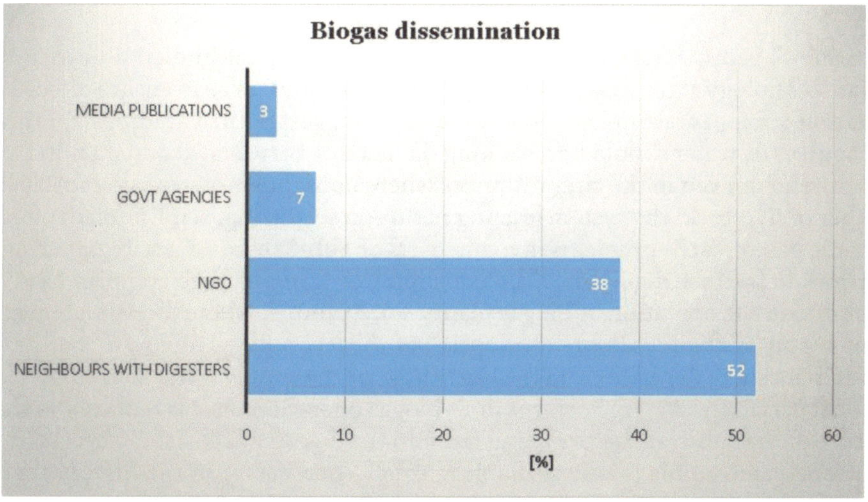

Figure 3.
Biogas dissemination in the study area (source: field survey).

using the technology have confidence that it can help reduce the rate of degradation and deforestation, while 75% from the non-users concurred to the statement. From the users and non-users, 96% from both clusters indicated that the use of biogas technology can help reduce the drudgery faced by women. On fuel consumption, compared to other cooking devices, 95% from the households using the technology agreed while from the non-users, 91% have the confidence that the technology will consume less fuel. On the general benefits of the technology, 89% from the users agreed that the benefits are worthy, while 82% from the non-users have confidence in the benefits of the technology. Further, the outcomes of the respondents were ranked and tested using Spearman rank correlation coefficient, to determine the significant correlation between the users and non-users of the technology. The Spearman rank results at $p < 0.05$, with a calculated value of 0.68, indicated that there is a positive and strong correlation in the perception of biogas technology among the users and non-users in the province. In essence, the more and better perception households have over biogas technology, the higher the chance of adopting the technology.

Statement	User (%)	Non-user (%)
Biogas can help solve the problem of fuelwood for cooking.	91	87
Biogas technology can help to improve soil fertility.	88	86
Biogas technology can improve hygiene due to the use of waste.	89	88
Biogas technology can reduce the rate of forest degradation and deforestation.	90	75
Biogas can relieve women's workload and save time used for fuelwood collection.	96	96
Biogas technology consumes less fuel than other conventional cooking devices.	95	91
Generally benefits of biogas technology over-weighs limitation/ weakness.	89	82

Source: field survey.

Table 1.
Biogas technology perceptions between user and non-user in the province.

4.4 Availability of water and feedstock

In Sub-Saharan Africa, one site-specific resource that has limited the scope of biogas technology is the availability of water that should serve to ensure effective operation of biogas technology. Studies by [21], suggest in their findings in Ethiopia that sources of water should be a walking distance of between 20 and 30 min from the household. Even in the circumstances where households own a satisfactory number of livestock, the system grazing nature, free grazing, semi-nomadic to nomadic have created problems in many parts of Sub-Sahara Africa in gathering feedstock to feed the digesters [22]. Poor supply of water has been reported as hindrance in the operation of biogas plants. For example, where there is adequate water supply, there is widespread adoption of the technology; mostly if the source of water is a short distance from the household or the supply is not altered by seasonal variation. Water shortages limit biogas operations as it is required in the mixture of the substrate before being fed into the digester [21]. Steady access to sufficient water supply is only available to small a percentage of the African region [23]. Sub-Sahara countries such as South Africa is considered as water-scare, water-stressed countries due to its climate aridity. Coupled with uneven distribution of rainfall throughout the country, most parts of the country are characterised by prolonged periods of drought between the rainy seasons with rainfall less than the world average [24]. The South African Government in 2001 approved a free basic water policy to deliver at least 6000 L of safe water to each household per month for a household of about eight persons [24]. Since the commencement of the free basic water policy, the household percentage with access to tap or piped water in their dwellings, on-site and off-site (communal taps), has improved from around 55% in 2002 to 70% in 2012. Nonetheless, general access to water by households is only improving by 4.2%, as most households still have to fetch water from dams, rivers, water pools, streams, springs and stagnant water [25].

Water is one of the critical requirements for the proper functioning of biogas technology. An equal amount of water is mixed with the required substrate before being fed into the digester. Findings from the survey indicated that households have access to water within a walking distance of 20–30 mins from the household but are still faced with acute, irregular supply and shortages that have marred most parts of the province. Water supply can be further improved by the government by ensuring adequate and regular supply of water to the households. Also, where possible, government can consider boreholes, rain water harvesting and water storage tanks to augment water scarcity. The provision of water is considered a critical factor in the perception of biogas technology which can enhance its adoption. Availability of feedstock is another requirement that is necessary in the operation of biogas technology because many digesters are failing due to unavailability of dung. Cow dung is considered the major feedstock in the study area. The findings, as portrayed in **Table 2**, revealed that 93% of the households using the technology in the province own livestock, as against 7% that do not own livestock but source for it either by buying or obtaining from neighbours who own livestock. Furthermore, 79.7% of households without the technology own livestock, while 20.3% do not own livestock. This result indicates that with proper awareness and campaign programme, biogas technology can have a foothold in the province as dung are abound for successful adoption of the technology.

4.5 Dearth of private sector participation

The private sector has key roles to play in the promotion of renewable energy, such as in biogas technology in order to make it market-oriented and commercially sustainable. Many countries have limited policies to attract renewable energy participation by private organisations [26]. For instance, in 2009, Nepal had more than 30 private organisations, which were actively involved in the biogas sector. However, only eight organisations were able to install a little over 500 biogas digesters, due to the unfavourable renewable energy policies [8]. In Limpopo Province, private sector participation in the dissemination of biogas technology is near absence. There is only one established biogas actor (Mpfuneko Biogas Project), a non-governmental organisation (NGO) that supports the development, and dissemination of biogas projects in the province. According to 82% of the sampled households with biogas digesters, the organisation (Mpfuneko Biogas Project) was responsible for the installation of their digester. Although private investment in renewable energy technology is being promoted by organisations such as the Renewable Energy Independent Producers Procurement (REIPPP) and the Department of Energy (DoE), the South African government should strengthen existing policies to support private sector energy investments and institutional mechanisms. The energy crisis being witnesses in the country provide a conducive entry point for private sector participation for an integrated biogas household level programme among other alternative renewable energy. More so, there are favourable conditions for the advancement of biogas technology in the province and the country at large; this includes availability of abundant biodegradable animals and crops waste materials.

	Users	Non-users	Total
Livestock ownership	**67** (93)	**102** (79.7)	**169** (84.5)
Do not own livestock	5 (7.0)	**26** (20.3)	**31** (15.5)
Total	72 (100)	**128** (100)	**200** (100)

Bolded faces represent frequency and brackets represent percentage frequency. Source: field survey.

Table 2.
Livestock ownership by households in the study area.

4.6 Lack of technical assistance and availability

In most Africa countries, lack of technical assistance in the form of skilled and unskilled personnel is required in the successful uptake of biogas technology. Technical assistance and availability is often cited as a reason for the impeding adoption of biogas technology. Technical knowledge ranges from the construction, maintenance and operation of the technology [27, 28]. Usually, where biogas digesters have been installed, the problem arises of reactors being of poor quality in the installed units. Poor operations and maintenance ability of users have also led to poor performance of the digester, sometimes leading to the abandonment of the technology. In some cases, due to technical availability, many demonstration plants have failed, which served to deter instead of enhancing the adoption of the technology [27, 28]. Technical availability is an integral determinant in the adoption of biogas technology at household level in the province. Available technical availability and assistance are deemed as a good support for the dissemination, adoption and utilisation of the technology. Due to inability of proper management, resulting from absence of technical expertise, several biogas projects have failed. The study reported that unreliable and unavailable technical services were common problem reported by households with installed digesters. In addition, households with interest about the technology shared the same sentiment about their perception to the technology. The question of technical support was directed to households with installed digesters and the findings show that 96% of the households complained about technical assistance of any sort. Technical issues faced by some households included blocked and leaking pipes, cracked and leaking digesters chambers, which has limit the use of the technology and sometimes leading to total abandonment. To promote the implementation and proper use of biogas technology, it is imperative to initiate long-term, biogas technology capacity-building programmes as well as training and execution of scientific work in the field through applicable research. There is the need for adequate technical expertise in the construction and maintenance of biogas digesters. Biogas technology and its implementation techniques can be introduced in the curriculum of most engineering and technical courses offered in universities, vocational and technical colleges that can train people on how to build and maintain biogas digesters.

4.7 Cost associated with installing biogas digester

One frequently cited factor limiting the development of biogas technology is financial constraints. In Ghana, for example, according to Arthur et al. [17], the findings indicated that, although the technology can solve some of the environmental and energy challenges faced in the urban and rural parts of the country, the technology requires a high initial cost of investment. In Ethiopia, one of the obstacles hindering the use of the technology by the rural cattle farmers is their inability to cover the full cost associated with installing the technology [21]. According to Bensah and Brew-Hammond [29], the principal hindrance to biogas technology expansion in Ghana is the cost of building the digesters, which most farmers have complained about. In South Africa, the average cost of mounting a smallholding biogas digester of 6 m^3 ranges from R15,000 to R40,000 [30], whereas a 10 m^3 digester costs not less than R80,000 [20]. Therefore, subsidies can enhance the relative advantages and speed up the adoption of biogas technology by those entities who would not have ordinarily adopted the technology [31]. Furthermore, some technologies have socially desired features; thus, adopting such technology is not only beneficial to the owner but to the society. In many of the Organisation for Economic Cooperation and Development (OECD) countries, companies and individual households can seek government subsidies if they adopt technology that is socially desirable. Even if the investment cost surpasses private benefits but is lower than social benefits, government provides subsidies to enhance the adoption of technologies that provides social benefits [32]. Furthermore, the size of the subsidies significantly influences the rates of adoption. In China, for instance, there was a time when interest in adopting biogas technology was fading away just after the government reduced subsidies to one-third of the investment cost from two-thirds [33]. In Nepal, it was revealed that without subsidies, most of the Nepalese farmers would not have been able to adopt the technology, due to their financial constraints [34]. Although providing subsides may also not positively increase the intended adoption rate of the technology. Individuals who adopt the technology for the sake of obtaining subsidies may be less enthusiastic to keep using the technology [35].

Additionally, households consider a variety of issues in their decisions to either to adopt or reject using modern energy technologies. Among other considerations, cost is of critical importance affecting the final decision by the consumer. Most consumers would prefer a modern technology with low initial costs compared to one that minimised cost of operations but ran over an extended period. Thus, creating a balance between initial costs alongside operation cost is important. In countries with low income, where individuals lack access to credit/ and or cash, widespread preference is often associated with low initial cost [36]. In supporting the argument, Bajgain [34] stated that in Ethiopia, high initial cost of investment remains a major obstacle in the prevalent dissemination of biogas technology. In the absence of subsidies, loans and credits, the uptake of the technology at household level can only be driven by income earned by the household. Consequently, the higher the income earned, the more likely it is for the household to adopt the technology compared to households earning lesser income. Thus, income is expected to influence the perception and thus adoption of the technology. This is because households consider a range of issues in their choice to either adopt or reject modern energy carriers. In the study area, the monthly income earned is low compared to other provinces in the country, due to the high unemployment rate that has characterised much part of the province. From the field survey results as shown in **Table 3**, only 15 households from the technology users' category, representing 20.8%, earn above R3501, with 18 households, representing 14% earning above the same amount from the non-users. Most of the users and non-users of the technology are in the monthly income bracket of

Income (ZAR[a])	Users	Non-users	Total
R0–500	**08** (11.1)	**16** (12.5)	**24** (12.0)
R501–1000	**12** (16.7)	**31** (24.2)	**43** (21.5)
R1001–1500	**17** (23.6)	**33** (25.5)	**50** (25.0)
R1501–3500	**20** (27.8)	**30** (23.5)	**50** (25.0)
R3501+	**15** (20.8)	**18** (14.0)	**33** (16.5)
Total	**72** (100)	**128** (100)	**200** (100)

[a]1 USD = ZAR 14.90.
Bolded figures represent frequency and brackets represent percentage frequency. Source: field survey.

Table 3.
Monthly income bracket of surveyed households of biogas users and non-users in the study area.

	Value	df	Asymp. sig. (two-sided)
Pearson's chi-square	43.251[a]	3	0.000
Likelihood ratio	41.598	3	0.000
Linear-by-linear association	19.917	1	0.000
No. of valid cases	200		

[a]0 cells (0.0%) have expected count less than 5. The minimum expected count is 8.91.

Table 4.
Pearson chi-square test results for income and costs of installing biogas digester.

R501 to R3500. As noted by [20, 30], the households cannot afford the average cost of installing a smallholding biogas digester.

As shown in **Table 4** using the Pearson's chi-square test, income earned by households was cross tabulated against the cost of building a digester, to determine the significant relationship between both variables. The result at $p < 0.05$ indicated that there is a statistical significant relationship between the income earned and the cost of installing a biogas digester. This implies that income earned by households in the province affects the adoption of the technology. As noted, the low income earned by the households sampled is a factor of socio-economic challenge being faced in the province, hence households finding it difficult to save and invest in a technology such as biogas. This can however be overcome by provision of loans, credits or subsidies to interested households willing to adopt the technology in order to relieve them of other households' burden as practised in other countries [36].

5. Conclusion

Drawing from the field survey, this chapter provides first-hand empirical evidence on the awareness and perceptions of biogas technology in the province by understanding the challenges in disseminating the technology. Despite the potential of biogas technology in forming part of the energy mix in households and providing environmental benefits, the level of awareness and perception of the technology remain low in the province. In any given technology, the awareness and perceptions of the users have been found to play an important role in the adoption and utilisation of the technology. Households' awareness and perceptions of biogas technology were investigated in order to get a deeper insight into the barriers to its adoption and utilisation in the province despite the prevailing conditions such as the abundance of dung to support the uptake of the technology. From the sampled households, the awareness was measured based on the financial implication, functionality and dissemination of the technology. Using the Pearson chi-square, the cost of biogas digester and income earned established a statistical significance relationship at $p < 0.05$. The perceptions of the technology was measured based on households insights regarding the role of biogas in fuel crisis, soil fertility, livestock management, burden of fuelwood collection, livestock ownership, water and feedstock availability as well as technical availability and assistance. In order to understand the in-depth perceptions of the households, the variables were further tested using a Spearman rank correlation coefficient at $p < 0.05$, with a calculated value of 0.68, indicating that there is a positive and strong correlation in the perception of biogas technology among the users and non-users households in the province. The study thus argued that the aforementioned variables are key in the dissemination and adoption of the biogas technology in Limpopo Province.

Acknowledgements

The authors acknowledge with thanks the financial supports from the Research and Publication Committee (RPC) of the University of Venda and the National Research Foundation (NRF) of South Africa.

Author details

Solomon Eghosa Uhunamure[1*], Nthaduleni Samuel Nethengwe[1] and David Tinarwo[2]

1 Department of Geography and Geo-Information Sciences, School of Environmental Sciences, University of Venda, Thohoyandou, South Africa

2 Department of Physics, School of Mathematical and Natural Sciences, University of Venda, Thohoyandou, South Africa

*Address all correspondence to: uhunamuresolomon@hotmail.com

References

[1] Scarlat N. Evaluation of energy potential of municipal solid waste from African cities. Renewable and Sustainable Energy Reviews. 2015;**50**:1269-1296. DOI: 10.1016/j.rser.2015.05.067

[2] Adeola IE, Hermish L, Emphraim O. Rethinking Biomass Energy in Africa. Berlin: Association of German Development NGOs (VENRO); 2009

[3] UNDP. United Nations Environment Programme. Emerging Issues in our Global Environment. Nairobi, Kenya: UNEP; 2013

[4] Department of Energy. A Survey of Energy-Related Behaviour and Perceptions in the Residential Sector. Pretoria, South Africa: DoE; 2013

[5] Department of Energy. IEP Planning Report Overview of Universal Energy Access Strategy. 2015. Available from: http://www.energy.gov.za/files/IEP/Mpumalanga/Overview-of-Universal-Energy-Access-Strategy.pdf [Accessed: 20 June 2018]

[6] LEDET. Limpopo Green Economy Plan—Including Provincial Climate Change Response. Polokwane: Limpopo Department of Economic Development, Environment & Tourism; 2013

[7] Department of Energy. State of Renewable Energy in South Africa. Pretoria: Department of Energy; 2015

[8] Gautam R, Baral S, Herat S. Biogas as a sustainable energy source in Nepal: Present status and future challenges. Renewable and Sustainable Energy Reviews. 2009;**13**:248-252. DOI: 10.1016/j.rser.2007.07.006

[9] International Energy Agency. World Energy Outlook. Paris: OECD/IEA; 2016. p. 2017

[10] World Atlas. The Top 10 Coal Producers Worldwide [Internet]. 2017. Available from: https://www.worldatlas.com/articles/the-top-10-coal-producers-worldwide.html [Accessed: 26 September 2017]

[11] World Bank. The Little Green Data Book, International Bank for Reconstruction and Development. Washington, DC: World Bank; 2016

[12] Limpopo Department of Finance and Economic Development. Limpopo State of the Environment Report (Phase 1). Polokwane: Limpopo Department of Finance and Economic Development; 2004

[13] Statistics South Africa. Census: Concepts and Definitions. Report No. 03-02-26, Pretoria, South Africa: StatsSA; 2011

[14] Aaron J, Muellbauer J, Prinsloo J. Estimates of household sector wealth for South Africa, 1970-2003. Review of Income and Wealth. 2006;**52**(2):285-308

[15] Southern Africa Biogas Industry Association. Biogas in South Africa German Conference. Sandton, South Africa: SABIA; 2015

[16] Gitonga S. Biogas Promotion in Kenya: A Review of Experience. Nairobi: Intermediate Technology; 1997

[17] Arthur R, Baidoo M, Antwi E. Biogas as a potential renewable energy source: A Ghanaian case study. Renewable Energy. 2011;**36**:1510-1516. DOI: 10.1016/j.renene.2010.11.012

[18] Rogers E. Diffusion of Innovations. New York: The Free Press; 2003

[19] Taherdoost HA. Review of technology acceptance and adoption

models and theories. Procedia Manufacturing. 2018;**22**:960-967. DOI: 10.1016/j.promfg.2018.03.137

[20] Mukumba P, Makaka G, Mamphweli S. Biogas technology in South Africa, problems, challenges and solutions. International Journal of Sustainable Energy and Environmental Research. 2016;**5**(4):58-69. DOI: 10.18488/journal.13/2016.5.4/13.4.58.69

[21] Eshete G, Sonder K, ter Heegde F. Report on the Feasibility Study of a National Programme for Domestic Biogas in Ethiopia. Addis Ababa, Ethiopia: SNV; 2006

[22] Winrock International. Africa Biogas Initiative: Potential for Growth and Models for Commercialization. Arkansas, USA: Winrock International; 2007

[23] Surendra KC, Takara D, Hashimoto AG, Khanal SK. Biogas as a sustainable energy source for developing countries: Opportunities and challenges. Renewable and Sustainable Energy Reviews. 2014;**31**:846-859. DOI: 10.1016/j.rser.2013.12.015

[24] Department of Water Affairs and Forestry. Free Basic Water Implementation Strategy: Consolidating and Maintaining, Version 4. Pretoria, South Africa: DWAF; 2007

[25] Statistics South Africa. In-depth analysis of the general household survey data. In: GHS Series, Volume V, Energy, 2002-2012. Pretoria, South Africa: StatsSA; 2013

[26] Ghimire PC. SNV supported domestic biogas programmes in Asia and Africa. Renewable Energy. 2013;**49**:90-94. DOI: 10.1016/j.renene.2012.01.058

[27] Amigun B, Pawawira W, Musango JK, Aboyade AO, Badmos AS. Anaerobic biogas generation for rural area energy provision in Africa. In: Kumar DS, editor. Biogas. Croatia: InTech; 2012. pp. 35-62

[28] Amigun B, Musango KJ, Stafford W. Biofuels and sustainability in Africa. Renewable and Sustainable Energy Reviews. 2011;**15**(2):1360-1372. DOI: 10.1016/j.rser.2010.10.015

[29] Bensah EC, Brew-Hammond A. Biogas technology dissemination in Ghana: History, current status, future prospects, and policy significance. International Journal of Energy and Environment. 2010;**1**(2):277-294

[30] Tiepelt M. Status quo of the biogas sector development in South Africa as well as the way forward. In: GIZ SAGEN Short-Term Biogas Training Seminar. Pretoria, South Africa: Tiepelt Mark; 2015

[31] Aalbers R, der Heijden E, Potters J, van Soest DH. Technology Adoption Subsidies: An Experiment with Managers. Discussion Paper 082/3. Amsterdam: Tinbergen Institute; 2007

[32] Rajendran K, Aslanzadeh S, Taherzadeh MJ. Household biogas digesters: A review. Energies. 2012;**5**:2911-2942. DOI: 10.3390/en5082911

[33] Shen G, Lin W, Chen Y, Yue D, Liu Z, Yang C. Factors influencing the adoption and sustainable use of clean fuels and cookstoves in China: A Chinese literature review. Renewable and Sustainable Energy Reviews. 2015;**51**:741-750. DOI: 10.1016/j.rser.2015.06.049

[34] Bajgain SS. The Nepal Biogas Support Programme: A successful Model of Public Private Partnership for Rural Household Energy Supply. Kathmandu, Nepal: Ministry of Foreign Affairs (The Netherland), SNV and Biogas Sector Partnership—Nepal; 2005

[35] Reddy S, Painuly JP. Diffusion of renewable energy technologies: Barriers and stakeholders' perspectives. Renewable Energy. 2004;**29**:1431-1447. DOI: 10.1016/j.renene.2003.12.003

[36] Gebreegziabher Z. Household fuel consumption and resource use in rural-urban Ethiopia [PhD thesis]. Wageningen, the Netherlands: Wageningen University; 2007

Chapter 4

Evaluating the Success of Renewable Energy and Energy Efficiency Policies in Ghana: Matching the Policy Objectives against Policy Instruments and Outcomes

Margaret Adobea Oduro, Samuel Gyamfi, Samuel Asumadu Sarkodie and Francis Kemausuor

Abstract

Advancement in energy policies has stimulated the adoption of instruments used in the renewable energy sector and climate change mitigation. Renewable energy policies play a crucial role in the abatement of greenhouse gas emissions, by providing access to modern energy and energy security by diversifying energy supply. There have been numerous policies developed in Ghana to improve the uptake of renewable energy for electricity production and to ensure efficient use of electrical energy. Some of the specific government policy objectives include reducing technical and commercial losses in power supply, support the modernization and expansion of the energy infrastructure to meet the growing demands, ensure reliability and accelerate the development and utilisation of renewable energy and energy-efficient technologies. These policies have defined targets and period to be implemented. This chapter outlined the renewable energy and energy efficiency policies in Ghana by matching the policy objectives against policy instruments and outcomes to measure what has been achieved. A comparative analysis was made with South Africa and Morocco's renewable energy sectorial policies on the basis of various strategies adopted to their achievements and what Ghana can learn from.

Keywords: renewable energy policies, energy efficiency, independent power producers

1. Introduction

Environmental policies have aided the adoption of superlative policy instruments profoundly used in climate change mitigation, adaptation option and renewable energy-related planning and implementation. Renewable energy policies play a crucial to abate greenhouse gas emissions, provide access to modern energy, and bring about energy security by diversifying energy supply as energy demand and its associated services to meet both social and economic development keep increasing [1]. International Renewable Energy Agency (IRENA) describes four categories of policy instruments used to promote renewable electricity generation, namely fiscal incentives; public finance, regulations; and access policies [2].

The use of renewable energy technologies has provided electricity to regions that lack electricity access, which will help create jobs as well as an increase in energy efficiency up to 30% by 2050 [3]. Also, Klein et al. [4] categorised instruments supporting renewable electricity (RES-E) generation based on whether they affect demand or supply of renewable electricity or whether they support power generation. First, policies may regulate renewable electricity price or the quantity produced and second, policies may support investment in renewable energy generation or direct subsidised generation [4]. On the other hand, categorised environmental policies instruments as regulative instruments, market-based instruments, procedural instruments, co-operative instruments and persuasive instruments [5].

Regulative instruments such as command-and-control aim at controlling the actions of firms and include mandatory regulations where the government directly intervenes in the activities of individual firms by prescribing or forbidding certain activities [6]. Market-based instruments (MBIs) are regulations that aim at providing actors or polluters with incentives to adopt low-emission technologies and encourage behaviour change through market signals and economic incentives [6, 7]. Procedural instruments aim at assessing the environmental impact of certain production processes and determine alternative arrangements that are environmentally friendly, examples are environmental impact assessment [8]. Persuasive instruments such as information provision tools work best where there is information gap-such that the missing information becomes a hindrance to behavioural change in reducing environmental impacts [9]. Co-operative instruments such as voluntary agreements and energy efficiency measure work best where there are existing incentives to behavioural change integrated with prevailing laws [8, 10].

Consistent with this, Ghana has a number of renewable energy and energy efficiency policies and initiatives to ensure the security and diversification of energy supply. **Figure 1** shows the Renewable Energy Policy Framework in Ghana that outlines the relationship between policy objectives, policy instruments, and policy institutions, while **Figure 2** shows the general energy policies. Although there are domestic and international financial incentives, policy and regulatory barriers limit the possibility of utilising renewable resources in both off-grid and grid-connected applications [13]. For instance, the percentage share of solar photovoltaic is 0.5%

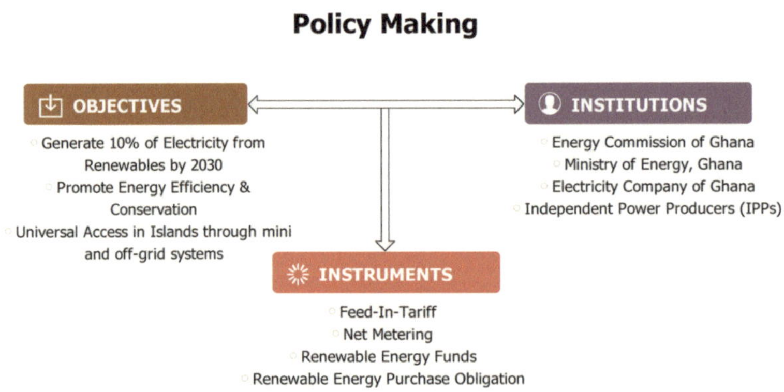

Figure 1.
Conceptual framework showing the policy making process in renewable energy sector of Ghana (source: [11, 12]).

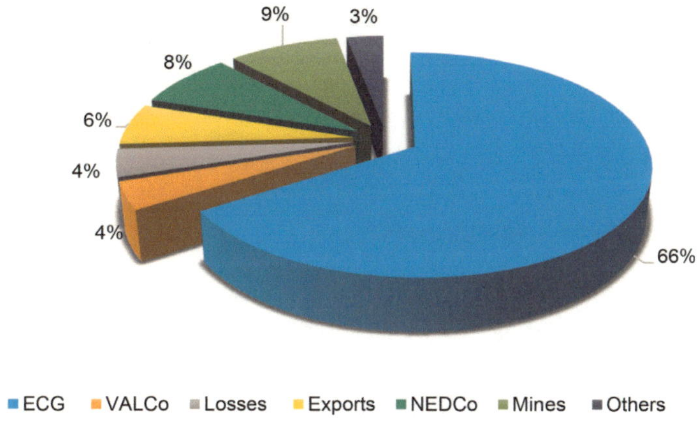

Figure 2.
2017 projected energy consumption by customers (source: [11]).

(22.6 MW) out of the 4674.85 MW of total installed energy [14]. Barriers to renewable energy application in Ghana have been classified into technical, social, environmental, economic, and policy-related [15]. The Ghana government has set up a goal to generate 10% of its electricity from renewable energy by the year 2020. Later in 2018, the goal to generate 10% of renewable energy in the national energy mix was extended to 2030. Likewise, there are numerous interventions instituted (i.e. the policy instruments) to achieve the objectives, however, the growth of renewable energy in Ghana is low. The percentage share of installed grid-connected renewable energy is a mere 22.6 MW (0.5%)—notwithstanding its potential [11]. Although, several studies have shown that network and technical barriers such as system failure and financial barriers (i.e. like the high upfront cost of renewables compared to that generated from fossil fuels) [15] have hindered the achievement of the 10% renewable energy penetration by 2020.

Another area that requires critical focus to ensure a successful transition into renewable and sustainable energy is an investment in energy efficiency—this has been given attention in Ghana in recent times. In 2007, the government of Ghana introduced the National Implementation of Incandescent Lamps Exchange programme where compact fluorescent lamps (CFLs) were freely given out nationwide. The effect of the programme resulted in cutting down the peak load demand of about 300 MW [12]. In 2012, the government of Ghana with support from the United Nations Development Programme (UNDP) and Global Environmental Facility (GEF) launched the 'rebate and turn-in' programme which encouraged consumers to trade in their old and inefficient refrigerators with new and more efficient ones for a top-up fee. The government later banned the importation of old and inefficient used refrigerators, which has led to a drop of about 63% in their import [12].

Against the background, it is essential to evaluate the success of renewable energy and energy efficiency policies in Ghana by matching the policy objectives against policy instruments and outcomes. Almost all the studies in the scope analysed the system failure and financial barriers using Ghana as a case study. However, this chapter includes a comparative analysis of Morocco and South Africa's renewable energy policies. The significance of this chapter revealed and identified the gaps in renewable energy policies as well as strategies or measures required to achieve the 10% renewable energy penetration in the national energy mix by 2030.

The remainder of this chapter includes background, methodology, results and discussion, and conclusion.

2. Background

2.1 Overview of Ghana's energy sector

As shown in **Table 1**, Ghana's installed electricity capacity currently stands at 4673.8 MW [14, 17] including large hydro, thermal and renewables of which 2334 MW is from independent power producers (IPPs).

Electricity Company of Ghana is the largest bulk purchaser and distributor, 66% of electricity consumed in Ghana while the Northern Electricity Distribution Company (NEDCo) only purchase and distribute 8% of the total energy. The rest of the energy is consumed by the Volta Aluminium Company (VALCo)—an aluminium and steel company, mining companies, exports, and sectors such as agriculture, health, and transport contributes to other energy users. **Figure 3** shows the structure of Ghana energy sector.

Type of energy	Installed capacity (MW)
Hydro	1580
Thermal	3071
Renewables	22.6
Total	4673.7

Source: [16].

Table 1.
Ghana's total energy installed capacity.

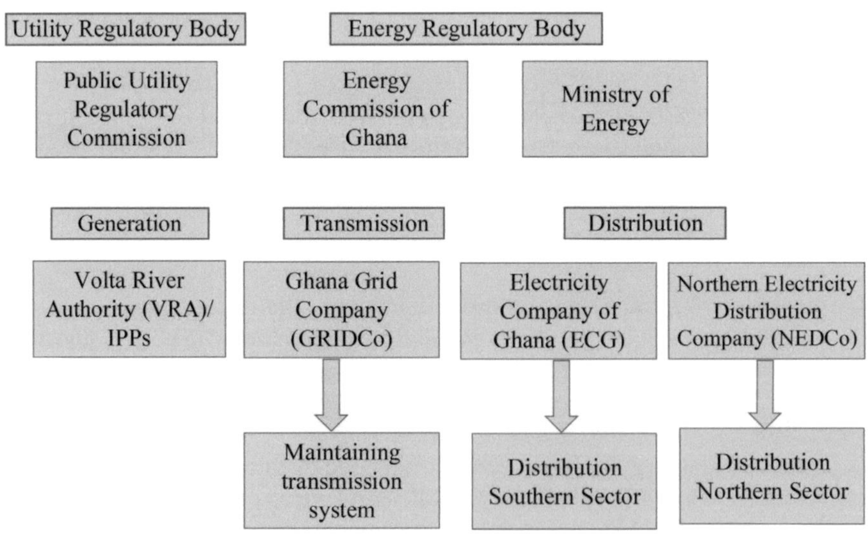

Figure 3.
Structure of Ghana's energy sector (source: [14]).

2.2 Review of policies and strategies on renewable energy deployment in Ghana

Policies facilitate the extent to which renewable energy technologies are adopted in a country. Ghana has instituted several policies and measures to help promote the development of renewable energy technologies, particularly, incentives that will attract renewable energy sector investors. **Figure 4** shows the renewable energy policies in Ghana from 2006 to 2018.

2.2.1 Strategic national energy plan, 2006–2020

The Energy Commission of Ghana has the mandate to review all investment plans to ensure energy needs are met in a sustainable manner. The commission developed the Strategic National Energy Plan for the period 2006–2020. The goal was to provide a sound energy market and to ensure the provision of sufficient energy services for Ghana. In contrast, the target of the Strategic National Energy Plan was to boost the renewable energy market. The renewable energy objective of this policy aimed at increasing the share of renewables up to 10% by 2020 while ensuring energy efficiency and conservation and achieving universal access to electricity by the year 2020 [18].

The policy sought to develop renewable energy technology regulations through the development of standards and codes. Under energy efficiency and conservation, the government continues to encourage the use of efficient appliances such as compact fluorescent bulbs (CFCs) and LED bulbs. Various measures have been instituted to ensure efficient energy consumption, adopt energy demand-side management, and set-up energy efficiency revolving fund to offer a low-interest facility for energy efficiency improvements in the country. However, the government is yet to release funds for its implementation [19].

2.2.2 National energy policy, 2010

The vision of the National Energy Policy was to develop an energy economy to ensure secure and reliable energy supply to all Ghanaians. The energy sub-sector was introduced under the National Energy Policy 2010 to increase the proportion of renewable energy in the total national energy mix and to focus on the fiscal incentives, awareness creation and regulations to promote energy efficiency and conservation practices. The policy also set the target to achieve 10% of renewables by the year 2020, reduce the consumption of woodfuels from 66 to 30% by 2020 and encourage the use of clean cooking alternatives such as LPG, and efficient cookstoves [19].

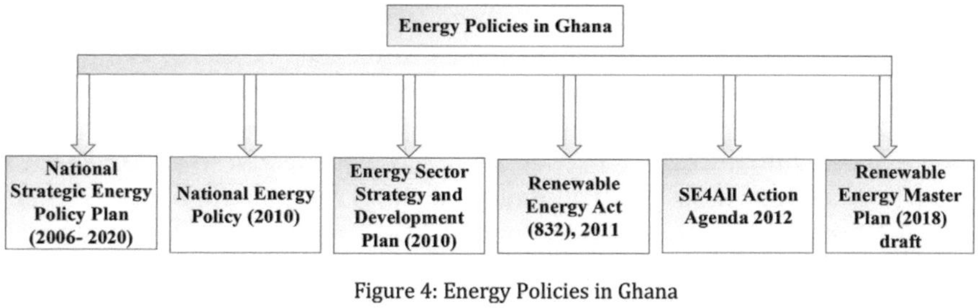

Figure 4.
Energy policies in Ghana (source: [11, 14]).

2.2.3 Energy sector strategy and development, 2010

The Energy Sector Strategy and Development Plan was introduced in 2010 alongside the National Energy Policy 2010. It covered strategies, programmes, and projects intended to support the national development agenda of Ghana in the following areas: energy sector institutions, power sub-sector, petroleum sub-sector, and renewable energy sub-sector. This policy sets the goal and strategies to increase (i) the percentage of renewable in the total national energy mix and efficient use of stoves and (ii) establish legislation to encourage the development of renewable energy technologies [19].

2.2.4 The sustainable energy for all action plan (SE4ALL), 2012

The sustainable energy for all (SE4All) action plan targets universal access to electricity to island and riverside communities in Ghana through both on- and off-grid systems and providing universal access to clean cooking solutions. The UNDP has collaborated with some partner agencies to achieve universal access to energy by 2030. The current electricity access in Ghana is approximately 84% [17].

2.2.5 The renewable energy act, 2011

The renewable energy act (832) was passed and enacted to provide for the development, management, utilisation, sustainability and adequate supply of renewable energy for the generation of heat and power, and provide an enabling environment to attract renewable energy sector investors [20]. Under this act, there are Licencing procedures, feed-in-tariff scheme, purchase obligation and rights to transmission and distribution systems for renewable energy (RE) plants, net metering and renewable energy fund [21]. The feed-in-tariff scheme was established under the renewable energy act 2011 of Ghana to guarantee the sales of electricity generated from RE sources into the national grid. The Public Utility Regulatory Commission (PURC) is responsible for setting feed-in tariff rates under the 2011 act. The feed-in tariff rates are guaranteed at a fixed rate to a registered producer for a maximum of 10 years and subject to renewal for every 2 years thereafter. PURC publish feed-in-tariff rates for energy generated from solar, wind, small hydro, waste-to-energy, biomass technology and takes into account the type of technology used, and the location of the generating facility.

2.2.5.1 Feed-in-tariff (FiT)

The feed-in-tariff rates which were gazetted by the PURC in 2016 serve as a cap for the amount at which electricity from RE sources can be purchased. An independent power producer (IPP) looking to sell power must sign a PPA with the off-taker (distribution utility or bulk customer) and the rate must not exceed that of the gazetted FiT rate. According to the private institutes, the FiT serves as a form of motivation but no one has benefitted from it, both those connected to the grid and individual photovoltaic (PV) users in the country. The policy does not work to the advantage of individual solar PV users but for those connected to the grid (but not being paid). In addition, the government of Ghana is ready to sign an agreement to pay FiT of 10 cents/kWh or below, but not above 10 cents/kWh—this is a new recommendation to generators to sign a PPA and invest in various RE (2016 FiT rate are more than 10 cents/kWh for RE technologies in Ghana). Investors are however not convinced with the change of the FiT rates—which has being one of the motivations of RE generators.

2.2.5.2 Net metering scheme

The net metering scheme was piloted by the energy commission in 2015 with 33 being installed. The scheme was gazetted by the PURC with the Electricity Company of Ghana (ECG) and NEDCo being implementing agencies. However, in 2017, the piloted project failed due to technical and financial challenges faced by the utility but the private institutes identified the problem as financial but not technical. The stakeholder agencies are working to find solutions to fully implement the scheme. However, RE school of thought argued that some energy institutions see no need to reward renewable energy generators, hence, failure to implement the scheme.

2.2.5.3 Renewable energy funds

In the past and present, various funding options have been employed to finance RE projects in Ghana. It includes loan financing, on-lending financing which will be used to create market including RE and energy efficiency interventions. The government has established the Ghana Renewable Energy Risk Capita (GRERC) as a financial instrument which seeks to assure project financiers and investor community to invest in RE with focus on small and medium-sized enterprises (SMEs) and domestic green projects. However, no RE community has benefited under the GRERC. Similarly, the government of Ghana for the past years has encountered failed projects such as the Ape Bank Project with support from the World Bank which was to install solar mini-grids in rural communities but the communities failed to pay back.

2.2.5.4 Renewable energy purchase obligations

Guidelines for the renewable energy purchase obligation (REPO) which mandates bulk customers to purchase its electricity from RE sources, has not yet been finalised. Per the RE act, electricity can be sold ONLY to either distribution utilities or bulk customers. In Ghana, there are 2 distribution utilities namely ECG (Southern sector) and NEDCo (Northern sector). Currently, the percentage of electricity being generated by RE sources and fed into the grid is from the 20 MW solar PV plant developed by BXC Company Limited, 100 kW waste-to-energy plant by Safi Sana Ghana Ltd. and 2.5 MW solar PV plant developed by VRA. BXC and Safi Sana have Power Purchase Agreements (PPAs) with ECG and the power generated by the 2.5 MW solar plant are supplied to NEDCo. According to the private institutions, the percentage of renewables sold are however not known to them. They argue that since the production is in small quantity, the percentage might be small.

2.3 Barriers to renewable energy development

Barriers to renewable energy developments have been identified as, inter alia, economic and financial, market, technical and network, policy regulatory framework, over-dependency on a centralised grid, and dependency on fossil fuel resources.

2.3.1 Cross-cutting barriers

Existing literature also identified more than one barrier to renewable energy development [1]. These barriers have been grouped under cross-cutting barriers. These include the interconnection of renewable energy barriers of financial, network, market, awareness, stakeholders' disengagement, socio-cultural beliefs, policy regulatory framework, over-dependency on fossil fuel resources.

Verbruggen et al. [22] in their study investigated the interconnection of factors affecting renewable energy supplies which pointed out that cost and prices of energy goods and services provided have effects on its development. They proposed a potentially unified taxonomy regarding policy driven as a whole. Mezher et al. [23] pointed out that not only climate change and fossil fuel consumption are the main drivers to renewable energy resource exploitation but the high cost of RE technology. Economic and political interference were also identified as barriers to renewable power generation and suggested that there should be a mixed implementation of feed-in-tariff and quota systems for the United Arab Emirates to meet its set target [23].

Bensah et al. [24] demonstrated that financial and market barriers could be removed through the provision of grants, soft loans, the flexible financial scheme as well as organising awareness and training programs. Notwithstanding the barriers such as market, technical, regulatory, social and environmental contributions to its development [25], consultation of various institutions on policy implementation, tax exemption, credit facilities, and incentive-based mechanism could minimise financial and market challenges. Although Ghana is endowed with renewable energy sources, exploitation of these sources for electricity generation is relatively low. Gyamfi et al. [15] further highlighted five main barriers to renewable energy utilisation, such as (1) technical—which includes trained personnel, voltage fluctuations (2) social (3) environmental (4) economics, and (5) policy. Kemausuor et al. [26] identified extra three main barriers such as (6) inadequate access to modern energy services, (7) inadequate information and awareness—fear on the part that the project might fail and (8) Stakeholders' involvement.

Bensah et al. [24] also conducted a survey that presented the various findings of Ghana's renewable energy policies in conjunction with China's renewable energy policy and the factors that constrain the development and the deployment of the various stakeholders. Their study concluded that poor financing of renewable energy investments, lack of affordability of renewable energy systems, cumbersome licencing processes, challenges with enabling instruments for RE investment, and unbalanced emphasis on on-grid RE systems. Bensah et al. [24] stated that one of the driving forces of access to renewable energy technology is inadequate modern energy services. Although with the high rate of electricity access, there are still a huge number of off-grid communities in the country as well as the high dependence on charcoal and wood fuels, renewable energy market size, high-interest rate to finance RE projects, among others, are the main obstacles of renewable technology transfer in Ghana. Even so, renewable energy sources do not only provide energy to the society but help in the reduction of massive indoor pollution from biomass resources [27], hence, a contribution to health improvement. Therefore, a solution that minimises the barriers to renewable energy production and consumption should be a country's priority. In addition, barriers to industrial energy efficiency improvements in some developing countries are more noticeable due to factors such as, inter alia, weak energy policy framework, financial constraints, and weak information systems.

3. Methodology

This chapter utilised both primary and secondary data. Purposive sampling technique was used to select key scientific publications and experts that have key knowledge in the field. Data were obtained from published articles and non-published papers, reports that include relevant renewable energy policy documents, questionnaires and interviewing of stakeholders namely public institutions, renewable energy private companies, universities and owners of solar rooftop installations.

To analyse the information on the implementation of RE policies, the following equation was applied and the summary of the methodology is presented in **Figure 2**.

$$\frac{n}{N} * 100\% \quad (1)$$

Country		Economic aspect
Morocco	Has the largest solar plants in the world and made improvement in their renewable energy sector	Inflation rates: 1.6%
		Steady interest rate: 2.25%
		Government taxes: 10–15%
South Africa	Largest wind installed capacity (1053 MW) in Africa (2015)	Inflation rate: 4.4%
		Steady interest rate: 6.5%
		Government taxes: 21–28%
Ghana		Inflation rate: 15%
		Interest rate: 30%
		Government taxes: 25% and above

Source: [26–29].

Table 2.
Reasons for the choice of countries.

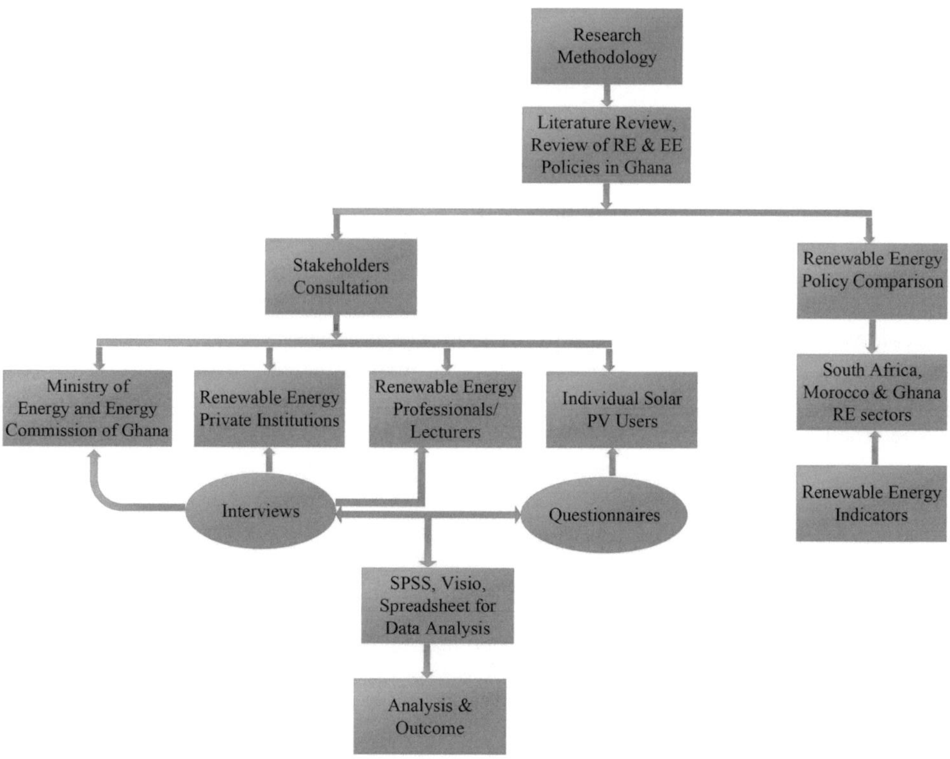

Figure 5.
Summary of research methodology.

where *n* represents the number of respondents who agreed policies are not implemented and *N* is the sample size, expressed as a percentage.

3.1 Renewable energy indicators

For business purposes, one needs to take into account the interest rate, inflation rate, exchange rate, and government taxes. These indicators are useful to examine a country's business environment. In the case of RE generator, other indicators are incorporated to assess the market attractiveness and these include; *policy incentives*: net metering scheme, feed-in-tariff, RE funds, renewable energy purchase obligation, and *economics aspects*: interest rate, government taxes, and inflation rate.

Table 2 shows the reasons for the choices of countries made for the comparative analysis between South Africa, Ghana and Morocco while **Figure 5** presents a summary of the methodology employed.

4. Results and discussion

The results of this chapter identified two major issues encountered in Ghana's renewable energy sector. These problems have been identified as:

1. Policy implementation affecting grid and off-grid solar PV generators in Ghana and

2. Barriers to renewable energy development in Ghana

4.1 Summary response on policy implementation issues identified by the stakeholders

Table 3 shows the views from stakeholders regarding RE development and policy implementation issues in Ghana.

4.2 Policies on renewable energy incentives in Ghana

The ECG is in charge of net metering implementation in Ghana and the payment of FiT for the Southern sector while the Volta River Authority (VRA) is in charge of paying feed-in-tariff for the Northern Sector. The Government of Ghana (GoG) is the main body responsible for the implementation of RE funds. There is no meter for calculating excess power fed into the grid, since feed-in-tariffs are currently not in operation because of the poor management of the scheme, although rates are published by the PURC every 2 years. **Figure 6** shows the policy incentives and the bodies responsible for implementation.

	Policy implementation instruments			
Stakeholders	**Net metering**	**Feed-in-tariff**	**RE funding**	**RE purchase obligation**
Lecturers	×	×	×	Quantity unknown
RE private companies	×	×	×	Quantity unknown
State	×	✓	×	Small amount
Individual PV users	×	×	×	Not aware

×: not implemented, ✓: implemented.

Table 3.
Stakeholders response on policy instruments.

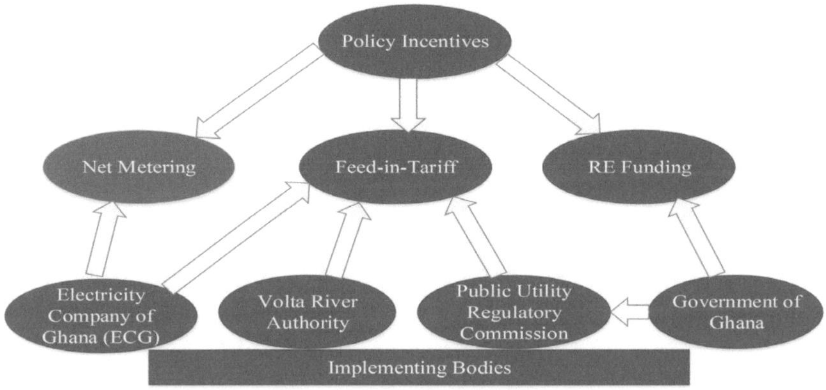

Figure 6.
Ghana's renewable energy policy incentives and implementing bodies.

4.3 Barriers identified by the stakeholders

The main policy implementation issue affecting renewable energy development in Ghana is the non-implementation of the Net Metering Scheme whereas the main barrier identified is attributed to the RE policies in Ghana. **Figure 7** shows the barriers identified by the Stakeholders whiles **Table 4** shows the percentage level of implementation issues affecting the grid and off-grid RE generators.

Figures 8 and **9** further explain **Table 4**.

The RE policies are not attractive to generators who generate excess power and feed into the national grid (i.e. there is no cash reward for injecting excess power into the grid, however, the government gets the benefit from the excess power fed charge into the grid). The implementation of the net metering scheme has been ineffective since it was passed in 2012. Thus, connecting to the national grid has become problematic—since there are no incentives for onsite power generators and the slow rate of passing legislation to back the initiative. This in effect hampers the financial status of grid-connected system developers.

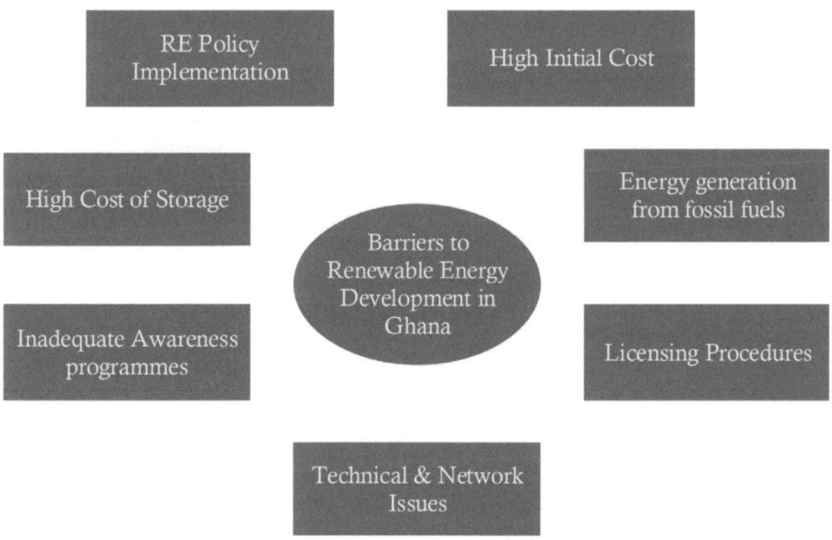

Figure 7.
Renewable energy barriers identified by the stakeholders.

RE policy implementation issues in Ghana	Net metering	Feed-in-tariffs	RE funding	Barriers to RE development in Ghana				High initial cost		
				RE policies	High initial cost	Existence of fossil fuel energy resources	Licensing procedure	High cost of storage	Technical and network	Inadequate awareness
Stakeholders' response (%)	100	85	42	80	53	21	25	42	15	25

Table 4.
Implementation issues affecting grid and off-grid renewable energy generators.

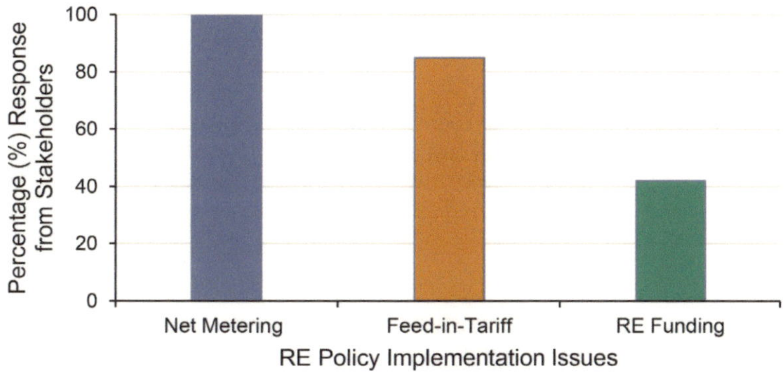

Figure 8.
Graph representing non-implementation level of renewable energy policy incentives in Ghana.

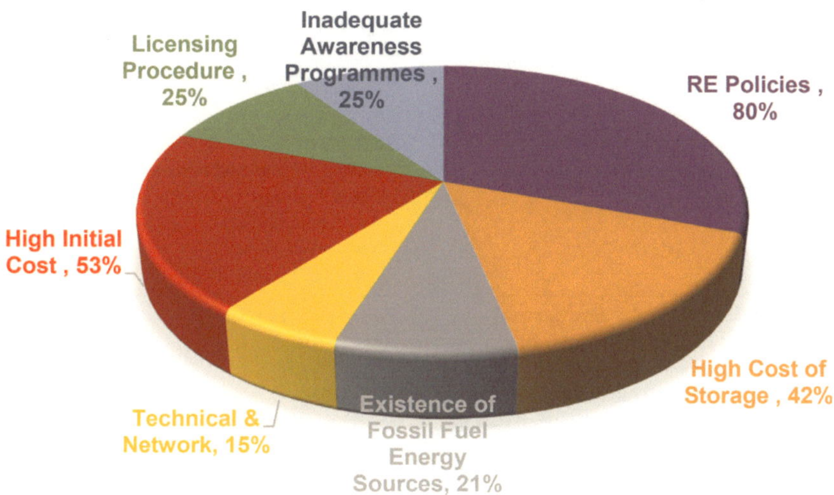

Figure 9.
Graph representing the percentage level of renewable energy barriers in Ghana.

4.4 Evaluation of renewable energy and energy efficiency policies in Ghana

In recent years, the country has been battling with energy crisis (i.e. frequent power outages) mainly attributed to the over-reliance on hydropower generation due to variability in rainfall patterns affecting water levels [30]. Measures were put in place to reduce energy consumption through energy management practices that were implemented across the country. Equally, renewable energy and energy efficiency policies have been in existence for a decade, purposely to help increase the percentage share in the national energy mix and diversify power generation from hydro sources coupled with energy efficiency measures. **Table 5** shows the areas covered by Ghana's renewable energy policies. The country has set policy instruments to match the policy objectives. In addition to the policy instrument, measures and activities were outlined to achieve the goals of the policy to increase the development of renewable energy technologies in the country. The various policies include laws and regulations that govern the sector.

4.4.1 Evaluation of renewable energy policies in Ghana

Table 6 shows the achievements of the various RE targets set in Ghana.

Energy policies	RE policy objectives	Regulations	Activities
Strategic national energy plan, (2006–2020)	Increase the share of RE up to 10% and universal access by 2020 (later changed to 10% by 2030) Promotes energy efficiency and conservation	Energy efficiency standards and labelling Energy management scheme	Standards and labelling programmes, e.g. enforcing standards for room air conditioners and CFLs, ensuring energy management practices, building codes, energy audits, load management programmes
National energy policy, 2010	Increase the proportion of renewable energy in the total national energy mix by 10% by 2020 (later changed to 10% by 2030) Promote energy efficiency and conservation in the country	**Activities** Encourage the use of clean cooking alternatives such as efficient cookstoves Use of efficient appliances such as refrigerators, CFLs, LEDs	
Energy sector strategy and development plan, 2010 Sustainable energy for all (SE4All), 2012	Increase the percentage of renewable in the total national energy mix and its efficient use Universal access to electricity to islands and riversides communities in Ghana	**Activities** To encourage the development of renewable energy technologies Provision of tax incentives on the importation of RE devices Promote off-grid and mini-grids electrification projects Solar street lightings Solar lantern project Clean cooking solutions	
Renewable energy act, 2011	Ensure the development, management, utilization, sustainability and adequate supply of renewable energy for the generation of heat and power	**Policy instruments** Net metering Feed-in-tariff RE funding Purchase obligations	

Source: [14, 19].

Table 5.
An overview of renewable energy and energy efficiency policy in Ghana.

4.4.2 Evaluation of energy efficiency policies in Ghana

Table 7 shows the achievements of the energy efficiency targets set in Ghana. With RE targets, and the different activities and measures carried out, only 22.6 MW of grid-connected RE plants have been constructed in Ghana. Although there are good wind, solar, and mini-hydro resource potentials, these resources have not been utilised to expectation [32–34]. In addition, all efforts made to boost the RE market in Ghana have not yielded much. The Government of Ghana, on the other hand, has not fully funded RE projects although the government has facilitated other energy efficiency projects. The Government of Ghana has in 2018, laid aside $230 million to cushion private companies for the development of off-grid and mini-grids across the country but the funds are yet to be released [35].

4.5 Comparative analysis: renewable energy policies in South Africa, Morocco, and Ghana

A comparative analysis was carried out to compare Ghana's RE policies and their targets to Morocco and South Africa's RE policies and targets. Morocco and South Africa have increased their renewable energy sector capacities, hence, we

Evaluation of renewable energy policies			
Energy policies	**Targets**	**Achievements**	**Analysis**
Strategic national energy plan (2006–2020)	10% of power generated from renewables by 2030 in the national energy mix	0% renewable energy as of 2010	No incentives for RE promotion, no projects to accelerate the RE development Heavily dependence on hydro dam 54.4% (2013)
National energy policy 2010	10% of power generated from renewables by 2030 in the national energy mix	0% renewable energy as of 2010	Heavily dependence on the source of hydro dam and thermal
Energy sector strategy and development plan 2010	Provision of tax incentives on the importation of RE devices 10% of power generated from renewables by 2030 in the national energy mix	The tax incentive was removed 0% renewable	Due to wrong importation of devices
Renewable energy act 2011, act (832)	10% of power generated from renewables by 2030 in the national energy mix Implementation of the RE policy instrument	0% of renewables as of 2012	Non-implementation of RE incentives Unfavourable RE environment to generators
Sustainable energy for all action plan 2012	Off grid and mini-grids electrification projects 10% of power generated from renewables by 2030 in the national energy mix	6 mini grid monitored, 3 mini grid was constructed as of 2015 16 wind and 23 mini hydro assessed as at 2015 2.5 MW solar accounting to a 0.11% of renewables as of 2013 and 22.6 MW (0.5%) as of 2015	No renewables till 2012 No wind power till date No mini hydropower constructed Unavailability of funding options Only <1% renewables achieved after a 9-year period of RE policies (2006–2015) Policies unattractive to some energy Institutions Unfavourable policies to generators

Source: [13, 15, 18, 30, 31].

Table 6.
Evaluation of renewable energy achievements in Ghana.

investigated what these two countries have done to increase their RE market and what Ghana can learn from their achievement. The Government of Morocco has instituted regulations and laws to promote the utilisation of renewable energy—institutions have been commissioned to manage and promote renewable energy investments [36]. In 2008, the National Renewable Energy and Energy Efficiency Plan were launched to achieve 42% of its energy from renewable sources by 2020 and 52% by 2030 [37]. Feed-in-tariff rates in Morocco are not fixed but an agreement of the rates is between the power producer and the government (**Figure 10**).

To promote renewable energy technologies in Morocco, the country set competitive bidding in 2006 to issue certificates to renewable energy generators which allowed them to commission a minimum of 10 MW plant from wind and other

Evaluation of energy efficiency policies in Ghana			
Energy policies	**Target**	**Achievements**	**Analysis**
Strategic national energy plan (2006–2020)	Reduce wood intensity of charcoal production from 6–5:1 to 4:1 in forest zone and 4:1 to 3:1 in savannah zone by 2015 Reduction in traditional biomass from 60 (2006) to 50% by 2015	Charcoal consumption 917 (2007)–1210 ktoe (2015) 24.2% increment within 8 years 69% of biomass consumed as of 2016	Target not met Increase in population growth Increase LPG especially 2011 (30,000,000 MMBtu) (free distribution of gas cylinders, subsidizing of LPG to wood fuel) Impacts: demand for LPG increased; increase import Effects: a switch to biomass
National energy plan 2010	Use of efficient appliances such as refrigerators, CFLs, LEDs	CFLs and LEDs 20% in 2007, 79% in 2009 Incandescent: 68% in 2007, 3% in 2009 Some population still uses inefficient refrigerator although (32,257 recycled) [12]	Target met Almost all households in Ghana use these efficient lamps A reduction in peak power electricity demand of 200–220 MW (2009) Target still in progress
Energy sector strategy and development plan 2010	Ensure the efficient use of wood fuels to reduce deforestation	There has been an increase in biomass consumption	Target not met
Sustainable energy for all action plan 2012	Install solar street lightings, solar lantern project	A number of the street solar lantern has been installed in the cities 200 each solar lantern distributed as of 2013 and 2017	Target in terms of (% cannot be determined). More of the street lighting project needs to be expanded to towns and villages Will help reduce subsidise on kerosene Projects ongoing
	Other efficiency project Solar rooftop programme 20,000 solar panels to households	1,006 units of solar panels distributed since its implementation	Target has not yet been met The problem from the service provider (energy commission)/from the government of Ghana

Source: [12, 27, 29, 31].

Table 7.
Evaluation of energy efficiency achievements in Ghana.

renewable energy sources [38]. The 10 MW bid was then increased to 50 MW in 2009, thus, investors were allowed to construct RE plants of 50 MW. This led to financial constraints on the part of the local renewable energy companies to fund the 50 MW projects. As of 2018, the country generated ~34% of its electricity from renewables [39]. In addition to the efforts made, the government of Morocco has funded its RE sector with $40 billion to help achieve its goal by 2030 [40].

In South Africa, the integrated resource plan (2010) set a target to generate 17.8 GW (9%) of electricity from renewables by 2030 [41], which has 5.2 GW of renewables as of 2016—contributing 3% of installed electricity capacity [42]. Competitive bidding was issued to investors and a private body was commissioned to oversee the bidding process for RE sector procurement, an effort attributed to the

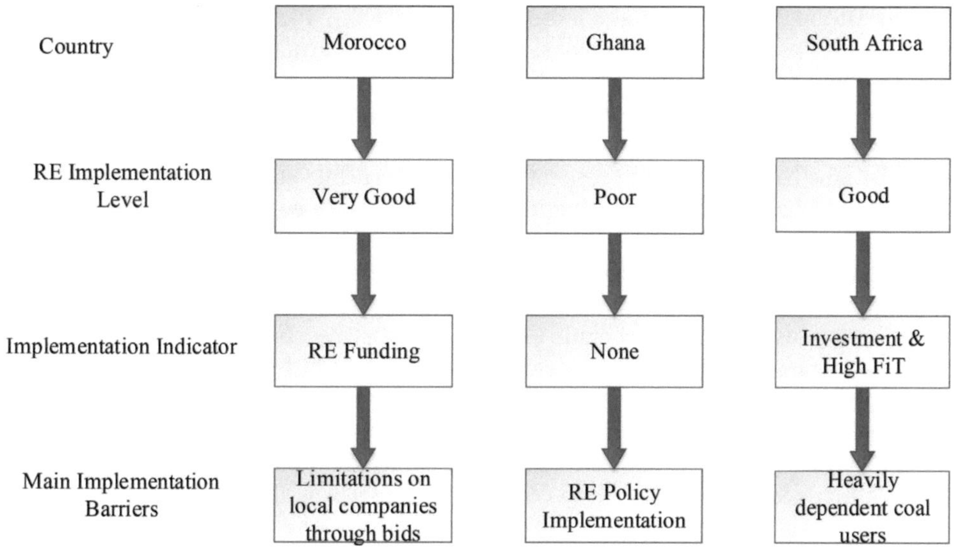

Figure 10.
Comparative analysis of RE in Ghana with Morocco and South Africa.

Renewable Energy Independent Power Producer Procurement Programme (REIPPPP) launched in 2011. Financially, an investment of \$100 million in 2011, \$5.7 billion in 2012, \$4.5 billion in 2013 [43] with a capacity of 3922 MW were injected to the various renewable energy technologies—making South Africa one of the countries with clean energy investments [44, 45]. To add up to the investment plan, the country set high FiT rates of 26, 15.6, and 46 Eurocent/kWh for solar, wind and concentrating solar power technologies. Renewable energy users are reported to have received a drastic subsidy on RE utilisation. Net metering in South Africa works for customers connected to Eskom medium-voltage and large power [46].

Comparing Ghana with Morocco and South Africa, the rate at which Morocco has increased its electricity generation from renewables is higher than that of South

Instruments	Country		
	South Africa	**Morocco**	**Ghana**
Feed-in-tariffs	Present	Agreement between the generator and the government	Present (poorly implemented)
Net metering scheme	Present (grid) For customers connected on Eskom medium voltage and power	Absent Generators allowed up to 20% annually	Absent
Renewable energy funds	In a form of Investment through bidding	Present Government funds and competitive bidding	Competitive bidding
Renewable energy purchase obligation (REPO)	Present	Present Excess purchased by ONE	Present Purchase by the distributor and/ transmitter

Source: [32, 37, 40, 47].

Table 8.
Summary on comparison analysis in South Africa, Morocco and Ghana.

Africa. Although the percentage increased from 4% in 2009 to 34% in 2017, a massive 30% increment within a period of 8 years was attained. This indicates that Morocco will achieve its remaining 20% target by 2030—if the same pace of RE development is followed. For the case of South Africa, though the generation of RE has been slower than Morocco over the last 10 years, however, their performance could be labelled as good with respect to the three countries. Referring to Ghana's situation, the various incentives instituted for renewable energy generators have not been implemented, except that the feed-in-tariff is under implementation for RE connected to the grid—creating challenges in terms of fee payment. Thus, the policies are present, but the implementation is poor. In addition to the efforts made, the government of Ghana has made plans for competitive bidding for renewable energy generators. Currently, the country has held two bids and the power plants are yet to be constructed. The lesson from Morocco and South Africa mean that Ghana would have to fund its renewable energy sector and set a private body to supervise the bidding process as well as increase the FiT rate to attract investors. The summarised renewable energy comparison analysis is shown in **Table 8**.

5. Conclusion

Renewable energy resources such as wind, hydro, solar, and biomass abound in Ghana—with policies developed to harness these resources. Policy interventions and energy efficiency measures have been instituted to achieve the objectives of increasing power generated from renewable sources by 10% in 2030. Despite the numerous attempts, these interventions have been slow in execution, notably, the percentage share of installed grid-connected renewable energy is ∼0.5%. The principal cause is attributed to energy institutional bodies responsible for implementing the policies, for they envisage it as unattractive and will lose revenue to the renewable energy generators. Likewise, the various incentives instituted for renewable energy generator have not been executed, and the proposed feed-in-tariffs have encountered challenges in fee payment and low tariff rates criticised by investors. Drastic action is required to encourage more renewable energy deployment by responsible institutions. Pushing for solar photovoltaic alone would not necessarily increase the percentage share and diversification of supply, but the incorporation of other energy technologies like wind and bioenergy in the energy portfolio.

Conflict of interest

Authors declare no conflict of interest.

Author details

Margaret Adobea Oduro[1], Samuel Gyamfi[2], Samuel Asumadu Sarkodie[3]* and Francis Kemausuor[4]

1 Pan African University Institute of Water and Energy Sciences (Including Climate Change), Algeria

2 University of Energy and Natural Resource, Sunyani, Ghana

3 Nord University Business School (HHN), Bodø, Norway

4 Kwame Nkrumah University of Science and Technology, Ghana

*Address all correspondence to: asumadusarkodiesamuel@yahoo.com

References

[1] Owusu PA, Sarkodie SA. A review of renewable energy sources, sustainability issues and climate change mitigation. Cogent Engineering. 2016;**3**(1):1167990

[2] IRENA. Evaluating Policies in Support of the Development of Renewable Power. 2012. Available from: https://www.irena.org/-/media/Files/IRENA/Agency/Publication/2012/Evaluating_policies_in_support_of_the_deployment_of_renewable_power.pdf [Accessed: 20 May 2019]

[3] IRENA. Global Energy Transformation: A Roadmap to 2050. Abu Dhabi: International Renewable Energy Agency. 2018. Available from: https://www.irena.org/-/media/Files/IRENA/Agency/Publication/2018/Apr/IRENA_Report_GET_2018.pdf [Accessed: 20 May 2019]

[4] Klein A et al. Evaluation of Different Feed-in Tariff Design Options: Best Practice Paper for the International Feed-in Cooperation. Germany: Energy Economics Group & Fraunhofer Institute Systems and Innovation Research; 2008

[5] Böcher M. Forest policy and economics a theoretical framework for explaining the choice of instruments in environmental policy. Forest Policy and Economics. 2012;**16**:14-22

[6] Bergek A, Berggren C. The impact of environmental policy instruments on innovation: A review of energy and automotive industry studies. Ecological Economics. 2014;**106**:112-123

[7] Eurosai. Market Based Instruments in Environmental Protection ISSAIs of Environmental Auditing. 2016. Available from: https://www.eurosaiwgea.org/meetings/Documents/14%20AM/14AM_report_0602.pdf [Accessed: 20 May 2019]

[8] Schmitt S, Schulze K. Choosing environmental policy instruments: An assessment of the 'environmental dimension' of EU energy policy. European Integration Online Papers. 2011;**15**(1):1-27

[9] Niles MT, Lubell M. Integrative frontiers in environmental policy theory and research. Policy Studies Journal. 2012;**40**:41-64

[10] Stavins RN. Experience with market-based environmental policy instruments. In: Handbook of Environmental Economics. Vol. 1. Amsterdam: Elsevier; 2003. pp. 355-435

[11] Energy Commission of Ghana. Energy Supply and Demand Outlook for Ghana. 2017. Available from: http://www.energycom.gov.gh/planning/data-center/energy-outlook-for-ghana?download=76:energy-outlook-for-ghana-2018 [Accessed: 20 May 2019]

[12] UNDP. In Ghana, a Victory in Energy Efficiency. 2012. Available from: https://www.undp.org/content/undp/en/home/ourwork/ourstories/in-ghana–a-victory-for-energy-efficiency.html [Accessed: 20 May 2019]

[13] Gboney W. Policy and regulatory framework for renewable energy and energy efficiency development in Ghana. Climate Policy. 2009;**9**(5):508-516

[14] Energy Commission of Ghana. Electricity Supply Plan for Ghana. 2017. Available from: http://energycom.gov.gh/files/2017%20Electricity%20Supply%20Plan%20-%20Final%20Report.pdf [Accessed: 20 May 2019]

[15] Gyamfi S, Modjinou M, Djordjevic S. Improving electricity supply security in Ghana: The potential of renewable energy. Renewable and Sustainable Energy Reviews. 2015;**43**:1035-1045

[16] VRA. Power Generation: Fact & Figures. 2017. Available from: https://www.vra.com/resources/facts.php [Accessed: 20 May 2019]

[17] Energy Commission of Ghana. Energy Supply and Demand Outlook for Ghana. 2018. Available from: http://www.energycom.gov.gh/planning/data-center/energy-outlook-for-ghana?download=76:energy-outlook-for-ghana-2018 [Accessed: 20 May 2019]

[18] Ministry of Energy. Strategic National Energy Plan, Main Report (2006–2020). 2006. Available from: http://www.energycom.gov.gh/files/snep/MAIN%20REPORT%20final%20PD.pdf [Accessed: 20 May 2019]

[19] Ministry of Energy. National Energy Plan, 2010-Final Report. 2010. Available from: http://www.petrocom.gov.gh/assets/national_energy_policy.pdf [Accessed: 20 May 2019]

[20] Appiah FK. Background to the Renewable Energy Act. 2015. Available from: http://wacee.net/Home/Documents/Implementing-the-Renewable-Energy-Act.aspx [Accessed: 20 May 2019]

[21] Energy Commission of Ghana. National Energy Statistics 2005–2014. 2015

[22] Verbruggen A et al. Renewable energy costs, potentials, barriers: Conceptual issues. Energy Policy. 2010; **38**:850-861

[23] Mezher T, Dawelbait G, Abbas Z. Renewable energy policy options for Abu Dhabi: Drivers and barriers. Energy Policy. 2012;**42**:315-328

[24] Bensah E, Kemausuor F, Antwi E, Ahiekpor J. China-Ghana South-South Cooperation on Renewable Energy Technology Transfer: Identification of Barriers to Renewable Energy Technology Transfer to Ghana. 2015. Available from: https://info.undp.org/docs/pdc/Documents/CHN/ProDoc%20-%2091276.pdf [Accessed: 20 May 2019]

[25] Painuly JP. Barriers to renewable energy penetration: A framework for analysis. Renewable Energy. 2001;**24**: 73-89

[26] Kemausuor F, Obeng GY, Brew-Hammond A, Duker A. A review of trends, policies and plans for increasing energy access in Ghana. Renewable and Sustainable Energy Reviews. 2011;**15**(9): 5143-5154

[27] Pegels A. Renewable energy in South Africa: Potentials, barriers and options for support. Energy Policy. 2010;**38**(9):4945-4954

[28] Santander. Morocco: Tax System. 2018. Available from: https://en.portal.santandertrade.com/establish-overseas/morocco/tax-system [Accessed: 20 May 2019]

[29] Van Buskirk R, Ben Hagan E, Ofosu Ahenkorah A, McNeil MA. Refrigerator efficiency in Ghana: Tailoring an appliance market transformation program design for Africa. Energy Policy. 2007;**35**(4):2401-2411

[30] Sarkodie SA, Owusu PA. A review of Ghana's energy sector national energy statistics and policy framework. Cogent Engineering. 2016;**3**(1):1155274

[31] Alberini A, Segerson K. Assessing voluntary programs to improve environmental quality. Environmental and Resource Economics. 2002;**22**:157-184

[32] Asumadu-Sarkodie S, Owusu PA. The potential and economic viability of solar photovoltaic power in Ghana. Energy Sources, Part A: Recovery, Utilization, and Environmental Effects. 2016;**38**(5):709-716

[33] Sarkodie SA, Owusu PA. A review of Ghana's solar energy potential. AIMS Energy. 2016;**4**(5):675-696

[34] Asumadu-Sarkodie S, Owusu PA. The potential and economic viability of wind farms in Ghana. Energy Sources, Part A: Recovery, Utilization, and Environmental Effects. 2016;**38**(5):695-701

[35] Graphic.com.gh. Govt Earmarks $230m for Renewable Energy Project. BusinessGhana. 2018. Available from: https://www.businessghana.com/site/news/general/162735/Govt-earmarks-230m-for-renewable-energy-project [Accessed: 20 May 2019]

[36] Leidreiter A, Boselli F. 100% Renewable Energy: Boosting Development in Morocco. 2015. Available from: https://www.worldfuturecouncil.org/wp-content/uploads/2016/01/WFC_2015_100_Renewable_Energy_boosting_Development_in_Morocco.pdf [Accessed: 20 May 2019]

[37] Moulin S. Renewable energy in Morocco. In: Road to Rio+20 a Dev. Green Econ. 2011. pp. 83-87

[38] Haas R, Resch G, Panzer C, Busch S, Ragwitz M, Held A. Efficiency and effectiveness of promotion systems for electricity generation from renewable energy sources—Lessons from EU countries. Energy. 2011;**36**(4):2186-2193

[39] Abdelbari R. Business Opportunities Report for Morocco's Renewable Energy Sector. 2018. Available from: https://www.rijksoverheid.nl/binaries/rijksoverheid/documenten/rapporten/2018/06/01/business-opportunities-report-for-morocco%E2%80%99s-renewable-energy-sector/business-opportunities-report-for-morocco%E2%80%99s-renewable-energy-sector.pdf [Accessed: 20 May 2019]

[40] Morocco World News. Morocco to Invest USD 40 Billion in Energy Sector by 2030. 2018. Available from: https://www.moroccoworldnews.com/2018/02/240667/morocco-to-invest-usd-40-billion-in-energy-sector-by-2030/ [Accessed: 20 May 2019]

[41] Oxford Business Group. South Africa's Push for Renewables. 2016. Available from: https://oxfordbusinessgroup.com/news/south-africa%E2%80%99s-push-renewables [Accessed: 20 May 2019]

[42] Power Africa. South Africa Power Africa Fact Sheet. 2007. Available from: https://www.usaid.gov/sites/default/files/documents/1860/South_Africa_-_November_2018_Country_Fact_Sheet.pdf

[43] Baker L. The evolving role of finance in South Africa's renewable energy sector. Geoforum. 2015;**64**:146-156

[44] Eberhard A, Kolker J, Leigland J. South Africa's Renewable Energy IPP Procurement Program: Success Factors and Lessons. Washington DC, USA: PPIAF; 2014. pp. 1-56. Available from: https://openknowledge.worldbank.org/bitstream/handle/10986/20039/ACS88260WP0P1482120Box385262B00PUBLIC0.pdf [Accessed: 20 May 2019]

[45] Suryapratim R, Tsidiso D, Sheila K. Clean Energy Investment in Developing Countries: Domestic Barriers and Opportunities in South Africa. International Institute for Sustainable Development. 2010. Available from: https://www.iisd.org/pdf/2009/bali_2_copenhagen_rsacase.pdf [Accessed: 20 May 2019]

[46] Eskom. Connection of Small-Scale Renewable Generation to Eskom's Network Background (Customer Small-Scale Renewable Generation). 2016. Available from: http://www.eskom.co.za/Whatweredoing/Documents/CustBulletinSPU.pdf [Accessed: 20 May 2019]

[47] Energy Commission. Final Report-CFL Exchange Programme Impact Assessment. 2009. Available from: http://www.energycom.gov.gh/files/CFL%20Report%20final.pdf [Accessed: 20 May 2019]

Chapter 5

Energy Policy Decision in the Light of Energy Consumption Forecast by 2030 in Zimbabwe

Remember Samu, Samuel Asumadu Sarkodie, Murat Fahrioglu and Festus Victor Bekun

Abstract

Sustainable energy, environmental protection, and global warming are the most discussed topics in today's world. Demand forecasting is paramount for the design of energy generation systems to meet the increasing energy demand. In this chapter, an examination of the causal nexus between energy consumption, total population, greenhouse gas emissions, and per capita GDP was carried out to forecast Zimbabwe's energy consumption by 2030. A time series data from 1980 to 2012 were employed alongside econometric techniques to explore the causal relationship among the variables under review. The stationary test revealed the integration of all the data series of interest of order one $\sim$ I(1). The autoregressive integrated moving average (ARIMA) model forecasted Zimbabwe's 2030 energy demand around 0.183 quadrillion Btu as against the current 0.174 quadrillion Btu. The empirical finding is indicative for policy- and decision makers who design the energy policy framework geared towards achieving the universal access to modern energy technologies in Zimbabwe.

Keywords: energy demand, energy policy, forecasting, greenhouse gas emissions, ARIMA, Zimbabwe

1. Introduction

The mitigation of global warming, climate change, and environmental pollution (especially greenhouse gas emissions) has been in the mainstream discussions among environmental specialist and practitioners globally. Toxic greenhouse gas emissions, especially carbon dioxide that constitutes a larger percentage of atmospheric emissions, have a long-term effect on climate change. Agricultural activities, both on large and small scales; the generation, transmission, distribution, and consumption of energy; and many other human-influenced activities have been reported to be the major causes of high carbon dioxide emissions globally. Zimbabwe has suffered a rapid increase in energy demand mainly due to economic growth and population growth. There has been an insufficient supply of electrical energy—as of 2014, ~7.25 million out of 14.6 million [1], representing 50% of Zimbabwe's population which lacked access to basic electrical energy and its related services. The deficit in electrical energy demand saw Zimbabwe importing almost 35% of its demand [2, 3]. The consumption rate has been growing rapidly, and the current generation technologies are unable to meet this increasing demand. Based on the available fact, there is an urgent need to exhaust all the possible electricity generation technologies to achieve 100% connectivity.

Due to the relationship between human development and access to energy, Zimbabwe is currently categorized among the countries with low human development (i.e. an index of 0.49) [4]. With a very low life expectancy at birth of 33.5 years as of 2002, Zimbabwe has a low GDP per capita of US$ 2400 [4] and 0.92 metric tonnes as a value for the carbon dioxide emissions per capita [5].

It is of paramount importance that an investigation is done to ascertain the causal nexus between population, greenhouse gas emissions, energy consumption, and GDP per capita and forecast Zimbabwe's energy use by 2030. The energy demand is highly driven by energy intensity (I), gross domestic product per capita (GDPC), and total population (P). The total population is highly related to the development of social and cultural changes. The degree of development in the economy is reflected by the GDPC, while the energy intensity is related to the efficiency in the usage of energy by society [6].

In literature, a couple of noteworthy efforts on energy demand forecasting have been made. Neural networks, regression models, Box-Jenkins models, and econometric models are the most frequently applied techniques for energy forecasting [7]. The constraints and applications of economic models were outlined by Finniza and Baker in which they reviewed the alternative models and their applications for strategic decisions, investment alternatives, and environment analysis [8].

An autoregressive integrated moving average (ARIMA) and spatial ARIMA (ARIMASp) models are essential for forecasting environmental and non-environmental-related variables. These projections include forecasts of electrical energy demand and consumption, greenhouse gas emissions, economic growth, and day-ahead forecast of electricity prices [9–13].

Demand forecasts can be categorized into short-, medium-, or long-term depending mostly on the time frame of the forecast. The short-term demand forecasts vary depending on what variable is under investigation—from an hour-ahead, a day-ahead, to week-ahead projections [13]. Short-term demand forecasting is important for the economic cooperation and reliability of the power systems using linear models [14]. A month-ahead forecasting can be categorized into medium-term demand forecasting. Ref. [15] did a month-ahead demand forecasting for Spain using two neural networks and concluded that the results were better than those obtained using ARIMA models.

Demand forecasting can be in small, medium, or large location size. Ref. [16] did a campus and a building electricity demand forecasting using different regression models and compared the results from these models. In their conclusions, they deduced that almost all the models performed well in the overall campus than load forecasting of a single building.

There has been an extensive analysis of the causal nexus between energy consumption and economic growth. In the seminal study of Kraft and Kraft [17], on the relationship between energy use and gross national product using cointegration test and Granger causality techniques, the empirical finding was inconclusive for that prevailing study. The same study of Kraft and Kraft [17] was an invitation to numerous studies in the energy consumption analyses as outlined in a variety of studies [18–20]. However, the energy literature can be broadly classified into three groups, namely, (a) the energy-led growth hypothesis [21, 22], which implies that energy drives economic growth, (b) feedback hypothesis that infers that economic growth stimulates energy consumption and vice versa, [23–25] and finally (c) the neutrality hypothesis [26–28], in which there is a strong assumption that energy has little or no impact on economic growth. However, there are very limited and sporadic literature documented regarding forecasting energy demand for sub-Saharan Africa (SSA) especially Zimbabwe—which is one of the fastest-growing economies in southern Africa.

As of recent, there are studies which include the empirical study by Sarkodie and Owusu [29] on carbon dioxide emissions, economic growth, energy use, and population interaction in a multivariate and causality framework, for the case of Ghana from 1971 to 2013. In this study, their empirical result revealed a cointegration relationship among all the series based on the vector error correction model (VECM) and autoregressive distributive lag model (ARDL). Their study validated the energy-induced growth hypothesis and the feedback hypothesis.

Furthermore, studies on energy forecasting in Spain precisely Asturias conducted by [30] utilized a univariate ARIMA Box-Jenkins approach from 1980 to 1996. Their study unraveled an optimum forecast with minimal forecast error. Similarly, for the case of Ghana, Sarkodie [31] estimated the electricity consumption by 2030 via an ARIMA technique. Sarkodie's [31] empirical study submitted that Ghana energy consumption will increase from 8.52 billion kWh to 9.52 billion kWh in 2030. Sarkodie's [31] findings were indicative of policymakers, which inform investments in energy infrastructure. The study also recommended the increase in energy generation to match the projected demand. In addition, Sarkodie and Owusu [32] investigated Nigeria energy use via forecast by 2030 using an ARIMA and ETS approach from 1971 to 2030. The empirical evidence showed that a 1% increase in energy use had a direct impact on carbon dioxide emissions by 3%. The ARIMA forecast prediction showed that energy use will increase from 975 kg in 2012 to 915 kg per oil equivalent by 2030.

In this chapter, a linear regression analysis is employed for the examination of the causal relationship between the variables under study. A time series data from 1980 to 2012 acquired from *World Data Atlas* [33] were used. Statistical forecasting models are then employed to project Zimbabwe's energy use by 2030. Most importantly this chapter will give information on energy policies, planning, and management of environmental pollution in order to minimize the effects of climate change and forecast Zimbabwe's energy demand and reduce the energy deficit the country is currently facing.

This chapter is of paramount importance to Zimbabwe, as it will increase the awareness of sustainable development and serve as a reference tool for integrating climate change measures into energy policies, practices, and planning by the government. Based on the findings of this chapter, Zimbabwe may be able to model an energy mix that will ensure 100% energy availability to all stakeholders. Zimbabwe is not utilizing its renewable energy resources on a large scale [2, 3]. This chapter will provide insights into how policymakers can incorporate the vast resources into the energy portfolio to ensure increased connectivity by the year 2030. Section 2 of the chapter briefly describes the methodology employed and the materials used. Results and discussion are outlined in Section 3 of the chapter. Section 4 presents the conclusions from the study, the energy policy implications, and possible recommendations for future studies.

2. Methodology

2.1 Data

The dataset employed in this present study consists of macroeconomic variables. Seven macroeconomic variables recorded yearly from 1980 to 2012 were analyzed. The data were retrieved from *World Data Atlas* [34]. These variables were then used to econometrically forecast the energy demand of Zimbabwe up to the year 2030. Time series data on total greenhouse gas emissions (kt of CO_2 equivalent), total carbon dioxide emissions (kt), total population (million), GDP per capita (2010US$), total primary energy production (quadrillion Btu), total primary energy consumption (Quadrillion Btu), and total electricity net generation (Billion kW hours) were employed. The data utilized spans from 1980 to 2012. As a preprocessing technique, missing values were imputed using MICE package in R software. Linear regression analysis was then employed to examine the causal relationship between these variables under investigation.

2.2 Model specification

The functional relationship among total greenhouse emission, total carbon dioxide emission, total population, per capita GDP, total energy production, total primary energy consumption, and total electricity net generation is based on the works of Reference [31, 32, 35]. The functional forms can be represented as follows:

Model A: ln TPEC = f (lnTGHC, lnTENG, lnTCO_2, lnTPOP, lnPGDP, lnTPEP).

Model A will help us ascertain the impact of total energy consumption on other explanatory variables:

$$\ln TPEC_t = \alpha + \beta_1 \ln TGHC + \beta_2 \ln TENG + \beta_3 \ln TCO_2 + \beta_4 \ln TPOP + \beta_5 \ln PGDP + \beta_6 \ln TPEP + \varepsilon_t \tag{1}$$

while model B seeks to verify the extent of CO_2 emission on economic growth and the impact of population growth.

Model B: lnTCO_2 = f (lnPGDP, lnTPOP, lnTENG, lnTGHC, lnTPEC, lnTPEP).

$$\ln TCO_2 = \alpha + \beta_1 \ln PGDP + \beta_2 \ln TPOP + \beta_3 \ln TENG + \beta_4 \ln TGHC + \beta_5 \ln TPEC + \beta_6 \ln TPEP + \varepsilon_t \tag{2}$$

where t is time trend, also $\alpha, \beta_1, \beta_2 \ldots \beta_6$ are unknown coefficients of repressors, and ε_t is the stochastic error term for the formulated models.

The empirical route of this study proceeds as follows: first, determination of the order of integration of series; second, estimation of the ordinary least squares (OLS) regression; and lastly, the forecast estimation.

2.3 Model estimation

Based on relevant studies [31, 32, 36] and our long-term forecasting using macro variables, an autoregressive integrated moving average (ARIMA) and spatial ARIMA (ARIMASp) models were utilized. These models are useful in forecasting greenhouse gas emissions, economic growth and electrical energy demand, consumption, and electricity prices [9, 10, 12, 32]. Some studies have utilized neural networks for a medium-term demand forecasting and concluded that the results were better than those obtained using ARIMA models [15]. Based on further analysis of the data variables and available literature and resources, a suitable model will be chosen for the continuation of this study. The ARIMA model [ARIMA (p. d, q)] was conducted in this chapter given as

$$\varphi(B)\nabla^d z_t = \phi(B)\alpha_t or Z_t = \sum_{i=0}^{p} \gamma Z_{t-1} + \alpha_t - \sum_{k=1}^{q} \gamma_i \alpha_{t-k} \tag{3}$$

where

$$\varphi(B) = 1 - \varphi_1 B - \varphi_2 B^2 \ldots - \varphi_k B^k \tag{4}$$

3. Results and discussions

3.1 Descriptive statistical analysis

This section outlines the descriptive statistical analysis of the study variables. **Figure 1** displays the trend of the variables after data imputation. It is visible from the trend that population increases rapidly, while the trend of GDP, total greenhouse gas, and carbon dioxide emissions exhibits similar feature, but fluctuations are observed in the trend of energy consumption.

Table 1 presents a summary of the descriptive statistical analysis of the study variables. Further analysis of the parameters indicates that total population and energy generation has long left tails (negative skewness), while CO_2 emissions, GDP, and energy consumption have long right tails (positive skewness). Total primary energy production and total greenhouse gas emission exhibit a positive skewness. Furthermore, energy production shows a leptokurtic distribution since its excess kurtosis is greater than zero, while the rest of the variables have an excess kurtosis less than zero, thus presenting a platykurtic distribution.

Grubbs' test was then used to estimate outliers in the study variables. Evidence from **Table 2** reveals the highest values of all the variables, except total population,

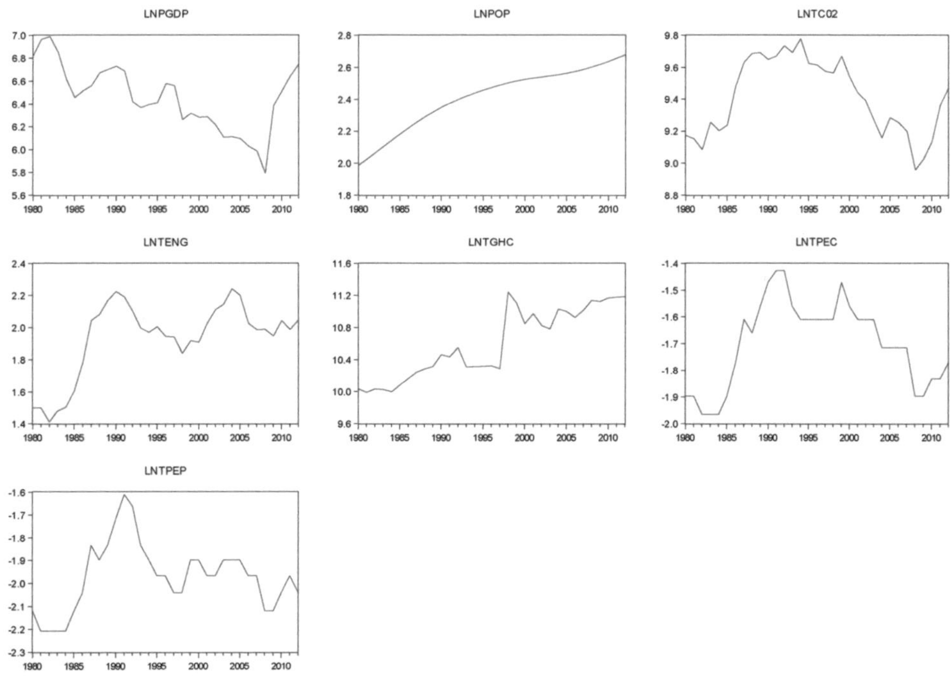

Figure 1.
Plot of series in natural logarithm.

	LNPGDP	LNPOP	$LNTCO_2$	LNTENG	LNTGHC	LNTPEC	LNTPEP
Mean	6.457	2.413	9.414	1.936	10.599	−1.699	−1.971
Median	6.455	2.475	9.441	1.989	10.466	−1.661	−1.966
Maximum	6.989	2.679	9.778	2.242	11.244	−1.427	−1.609
Minimum	5.791	1.987	8.958	1.411	9.991	−1.966	−2.207
Std. dev.	0.289	0.195	0.238	0.234	0.436	0.162	0.151
Skewness	−0.208	−0.722	−0.160	−0.980	0.075	−0.166	0.341
Kurtosis	2.524	2.407	1.696	2.913	1.450	1.946	2.921
Jarque-Bera	0.549	3.352	2.481	5.288	3.333	1.678	0.646
Probability	0.760	0.187	0.289	0.071	0.189	0.432	0.724
Sum	213.073	79.641	310.662	63.873	349.764	−56.073	−65.037
Sum sq. dev.	2.671	1.219	1.806	1.754	6.078	0.842	0.725
Observations	33	33	33	33	33	33	33

Table 1.
Summary statistics.

Variable	G	U	P-value	Alternative hypothesis
GDP	2.0	0.9	0.4	Highest value 1084.21 is an outlier
Population	2.0	0.9	1	Lowest value 5.39 is an outlier
CO_2 emissions	2.0	0.9	1	Highest value 17645.6 is an outlier
GHG emissions	2.0	0.9	1	Highest value 76391.8 is an outlier
Energy production	3.0	0.8	0.1	Highest value 0.2 is an outlier
Energy consumption	2.0	0.9	0.9	Highest value 0.24 is an outlier
Energy generation	2.0	0.9	1	Highest value 9.41 is an outlier

Table 2.
The Grubbs test for outliers.

are outliers. The Anderson-Darling test was done to test for the normality of the data variables. Testing at a 5% significance level, the null hypothesis is rejected if the p-value is less than or equal to 5%; hence, it can be concluded that the data do not follow a normal distribution. However, if the p-value is greater than 5%, then the test fails to reject the null hypothesis of normal distribution.

Table 3 presents the correlation matrix that exists between the variables.

The results of the correlation coefficient estimation show a positive significant relationship between per capita GDP and the total population. Thus, this implies that a higher population increases national income for the study country. Similarly, negative association but significant relationship exists among PGDP and TENG as well as TPEC but insignificant for TPEC and PGDP. This revelation implies that energy intensity impedes economic growth at certain threshold validating the environmental Kuznets curve hypothesis (EKC).

3.2 Anderson-Darling normality test

Table 4 shows that except GDP with a p-value greater than 5%, the entire variables do not follow a normal distribution. It is therefore evident that we fail

	LNPGDP	LNPOP	$LNTCO_2$	LNTENG	LNTGHC	LNTPEC	LNTPEP
LNPGDP	1						
t-stat	—						
P-value	—						
No. obs.	33						
LNPOP	−0.667	1					
t-stat	−4.987	—					
P-value	0.00	—					
No. obs.	33	33					
$LNTCO_2$	0.158	0.045	1				
t-stat	0.893	0.250	—				
P-value	0.379	0.8039	—				
No. obs.	33	33	33				
LNTENG	−0.486	0.721	0.404	1			
t-stat	−3.095	5.788	2.459	—			
P-value	0.004	0.000	0.020	—			
No. obs.	33	33	33	33			
LNTGHC	−0.635	0.886	−0.177	0.550	1		
t-stat	−4.578	10.642	−0.999	3.665	—		
P-value	0.0001	0	0.3255	0.0009	—		
No. obs.	33	33	33	33	33		
LNTPEC	−0.217	0.336	0.823	0.697	0.163	1	
t-stat	−1.238	1.987	8.077	5.415	0.917	—	
P-value	0.225	0.056	0.000	0.000	0.366		
No. obs.	33	33	33	33	33	33	
LNTPEP	−0.186	0.314	0.700	0.787	0.164	0.890	1
t-stat	−1.055	1.843	5.451	7.103	0.925	10.878	—
P-value	0.299	0.075	0.000	0.000	0.362	0.000	
No. obs.	33	33	33	33	33	33	33

Note: Table reports the estimates of the Pearson correlation coefficient between the pairs of variables. t-stat is the t-statistics for the significance of the correlation coefficient, and p-value is its marginal probability.

Table 3.
Correlation coefficient estimates.

to reject the null hypothesis for GDP. Further analysis of the GDP distribution from the fitting is shown in **Figure 2**, while the Cullen and Frey graph in **Figure 3** concludes that the data for GDP follows a normal distribution. The remaining distributions of the variables were decided using Cullen and Frey graph.

Further evidence from the Cullen and Frey graph support the previous evidence that these variables do not follow a normal distribution. The PDF plots presented in **Figures 4** and **5** additionally support that GDP follows a normal distribution. Energy consumption was used in this chapter as a dependent variable for the forecasting. The relationship between energy consumption and population shown

Variable	A	P-value
GDP	0.2	0.9000
Population	0.9	0.0300
CO_2 emissions	2	0.0004
GHG emissions	3	0.0000
Energy production	1	0.0020
Energy consumption	2	0.0003
Energy generation	2	0.0001

Table 4.
Anderson-Darling normality test.

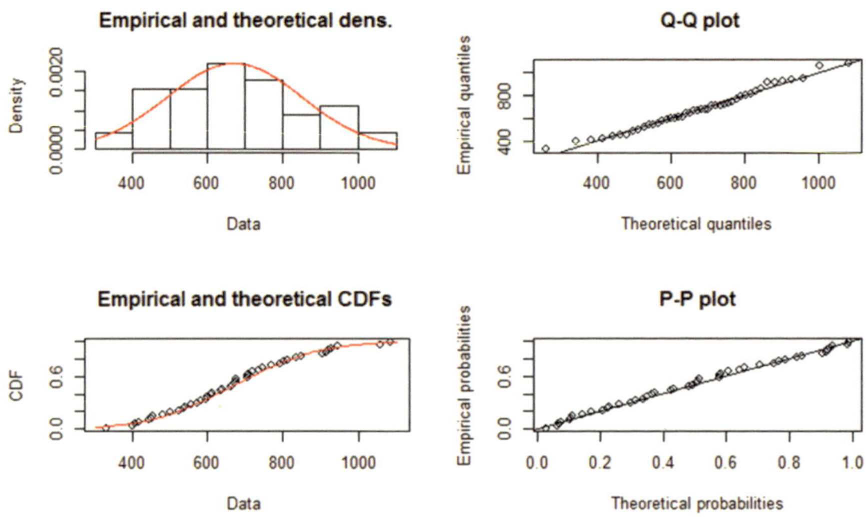

Figure 2.
Normal distribution fitting for GDP.

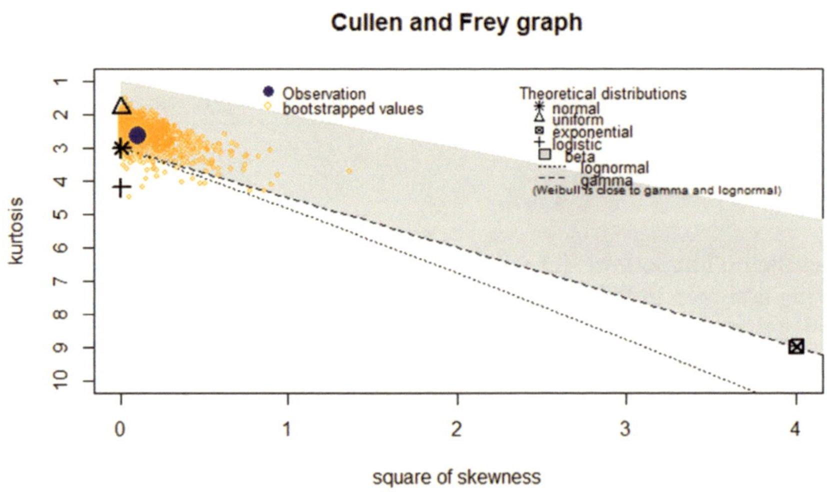

Figure 3.
GDP fits normal, lognormal, gamma, and beta distributions.

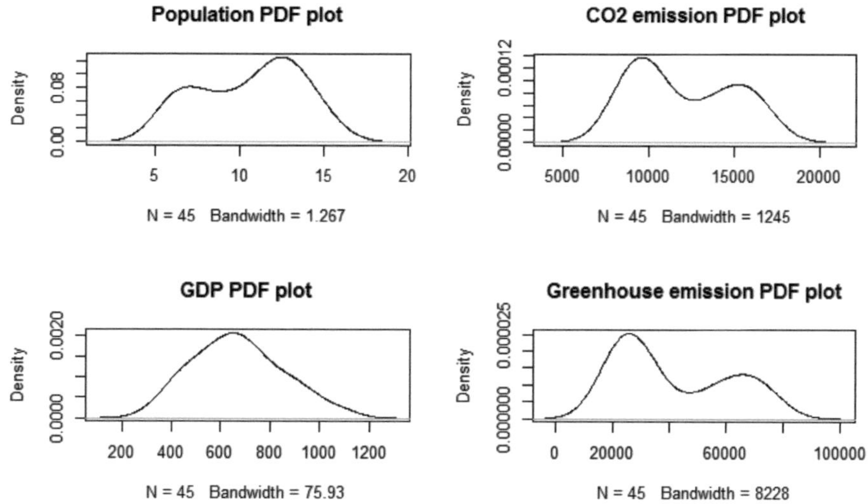

Figure 4.
PDF plots for the study variables (a).

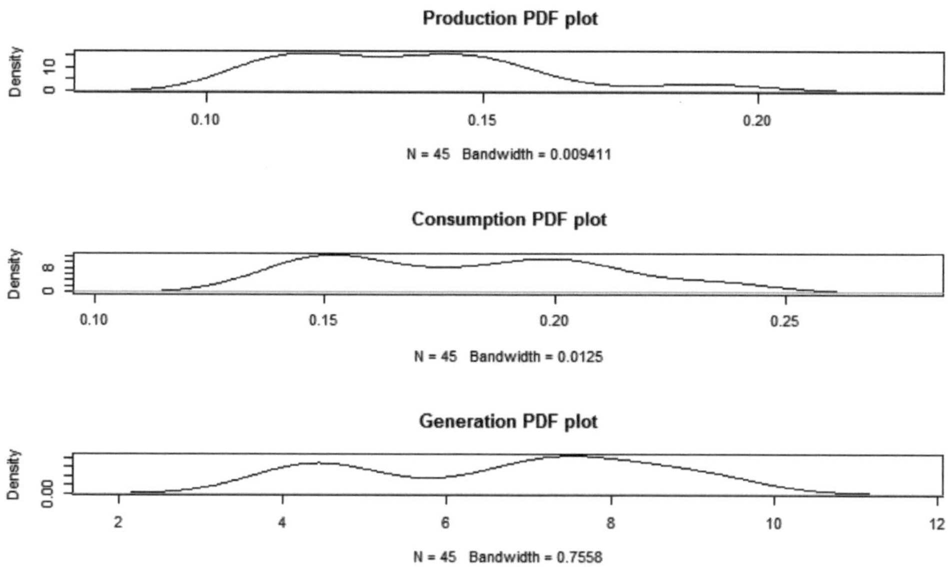

Figure 5.
PDF plots for the study variables (b).

in **Figure 6** reveals that energy consumption increases with an increase in population.

3.3 Stationarity test

It is well established that most macroeconomic variables possess trends/seasonality; thus, the need to know the order of integration of such series is pertinent to avoid spurious regression and misleading policy implication. This current chapter employed augmented Dickey-Fuller (ADF) and Phillips-Perron (PP) unit root test to ascertain the stability traits and asymptotic properties of the variables under consideration. These tests are conducted with the null hypothesis of a unit root

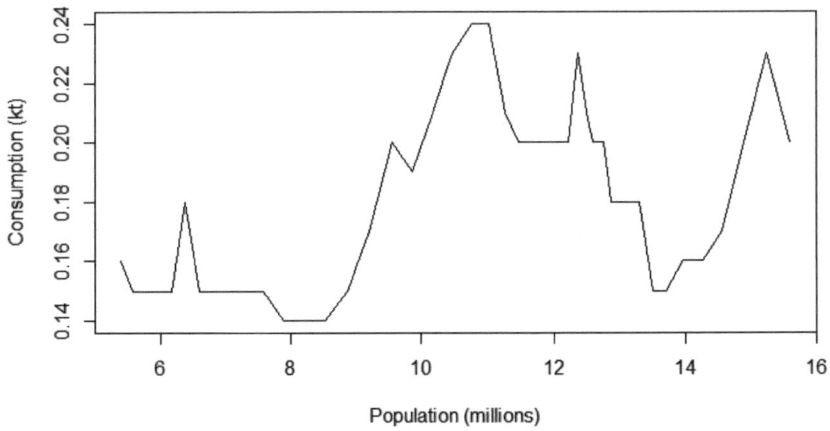

Figure 6.
A relation between energy consumption and total population.

	Level				First difference			
Variables	ADF		PP		ADF		PP	
	τ_μ	τ_T	τ_μ	τ_T	τ_μ	τ_T	τ_μ	τ_T
LNGDPC	−2.01	−2.08	−2.19	−3.18	-5.21^{***}	-5.14^{***}	-5.09^{***}	-5.17^{***}
$LNTCO_2$	−1.69	−0.67	−6.23	−6.17	-6.23^{***}	-4.19^{***}	-6.18^{***}	-6.18^{***}
LNTGHC	0.99	−3.63	−7.55	−7.47	-0.99^{**}	-8.20^{**}	3.64^{**}	-8.14^{***}
LNTPOP	−2.50	−3.10	−2.18	−3.08	-3.68^{**}	-1.29^{**}	1.17^{***}	1.11^{***}
LNTPEP	−2.45	−3.72	−3.64	−3.82	1.72^{***}	1.73^{***}	-4.63^{***}	-4.65^{***}
LNTPEC	−1.57	−1.42	−4.57	−4.67	-1.69^{***}	-1.41^{**}	-4.57^{***}	-4.62^{***}
LNTENG	−3.79	−3.61	−3.74	−3.81	-1.99^{***}	1.76^{***}	-6.18^{***}	-6.23^{***}

Note: $\tau\mu$ represents a model with intercept, while τ_T denotes model with intercept and trend.
***Significant at 5% level.*
****Significant at 1% level.*

Table 5.
Unit root results.

against the alternative of stationarity [37, 38]. **Table 5** presents the unit root test. The general form of the unit root test is given as

$$\Delta Y_t = \beta_1 + \beta_2 t + \gamma Y_{t-1} + \sum_{i=1}^{k} \alpha_i \Delta Y_{t-i} + \epsilon_t \quad (5)$$

where ϵ_t denotes the Gaussian white noise term which is asymptotically characterized by zero mean and constant variance. The null hypothesis of the unit root test is nonstationary against the alternative of stationarity.

The unit root test reported in **Table 5** reveals that all series are integrated of order one ~ I (1), that is, it has a unit root. However, all variables turn stationary at first difference, thus integrated of order one ~ (1). Subsequently, this study proceeded with the ordinary least squares (OLS) estimation.

Tables 6 and 7 present the OLS regression estimates for models A and B, respectively. **Table 6** shows a tradeoff between total population and total primary consumption. That is, a 1% increase in the total population decreases the total

Variable	Coefficient	Std. error	t-statistic	Prob.
C	−3.7982	1.1311	−3.3580	0.0024
LNTGHC	0.0489	0.0555	0.8803	0.3868
LNTENG	−0.0245	0.1185	−0.2071	0.8376
$LNTCO_2$	0.3689	0.0705	5.2314	0.0000
LNPOP	−0.0486	0.1586	−0.3065	0.7617
LNPGDP	−0.1020	0.0473	−2.1592	0.0402
LNTPEP	0.5421	0.1672	3.2420	0.0032
R-squared	0.9091			
F-statistic	43.3549			
Prob (F-statistic)	0.0000			

Model A: lnTPEC = f(lnTGHC, lnTENG, $lnTCO_2$, lnPOP, lnPGDP, lnTPEP).

Table 6.
Regression estimation for Model A.

Variable	Coefficient	Std. error	t-statistic	Prob.
C	11.7740	1.2570	9.3668	0.0000
LNPGDP	0.2396	0.0878	2.7282	0.0113
LNPOP	0.5429	0.2895	1.8754	0.0720
LNTENG	−0.1941	0.2270	−0.8551	0.4003
LNTGHC	−0.2367	0.0991	−2.3897	0.0244
LNTPEC	1.3901	0.2657	5.2314	0.0000
LNTPEP	−0.0148	0.3846	−0.0386	0.9695
R-squared	0.8404			
F-statistic	22.8174			
Prob (F-statistic)	0.0000			

Model B: $lnTCO_2$ = f(lnPGDP, lnTPOP, lnTENG, lnTGHC, lnTPEC, lnTPEP).

Table 7.
Regression estimation for Model B.

energy consumption by 0.05%. Similarly, a negative trend was seen among per capita GDP total energy consumption with a magnitude of 0.10%. Thus, we can infer that population does not increase CO_2 emission in Zimbabwe. However, a positive and significant relationship is observed among TPEP and TGHC with the dependent variable at a magnitude of 0.54 and 0.05%. The fitted model has a robust coefficient of determination (R^2) of 90%, implying that 90% of the variation in total primary energy consumption was explained by the explanatory variables, while the rest 10% are left uncaptured in this model. The joint significance of the model by the F-statistic was also significant at all levels (1, 5, and 10%). In the same way, **Table 7** targeted for model B. The model has a coefficient of 84%. That is, 84% of the variation in CO_2 was explained by another explanatory variable with F-statistic significance indicating joint significance among all variables. Interestingly, the fitted model shows that a 1% increase in PGDP increases CO_2 by 0.24%. Similarly, there is also a positive trend between CO_2 and TPOP with over 0.54%.

Year	TPECF (predicted)	TPEC
1980		0.15
1981		0.15
1982	0.151	0.14
1983	0.151	0.14
1984	0.152	0.14
1985	0.153	0.15
1986	0.153	0.17
1987	0.154	0.2
1988	0.155	0.19
1989	0.155	0.21
1990	0.156	0.23
1991	0.157	0.24
1992	0.157	0.24
1993	0.158	0.21
1994	0.159	0.2
1995	0.159	0.2
1996	0.160	0.2
1997	0.161	0.2
1998	0.161	0.2
1999	0.162	0.23
2000	0.163	0.21
2001	0.163	0.2
2002	0.164	0.2
2003	0.165	0.2
2004	0.165	0.18
2005	0.166	0.18
2006	0.167	0.18
2007	0.167	0.18
2008	0.168	0.15
2009	0.169	0.15
2010	0.169	0.16
2011	0.170	0.16
2012	0.171	0.17
2013	0.171	
2014	0.172	
2015	0.173	
2016	0.173	
2017	0.174	
2018	0.175	
2019	0.175	

Year	TPECF (predicted)	TPEC
2020	0.176	
2021	0.177	
2022	0.177	
2023	0.178	
2024	0.179	
2025	0.179	
2026	0.180	
2027	0.181	
2028	0.181	
2029	0.182	
2030	0.183	

Table 8.
Forecast (ARIMA) for total energy consumption.

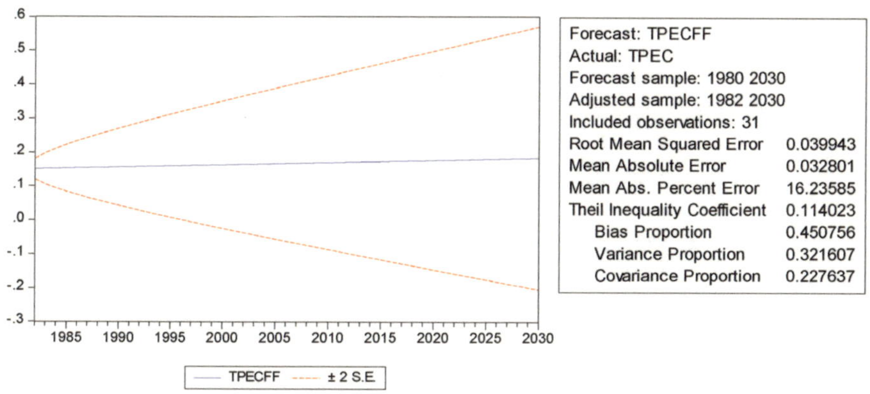

Figure 7.
ARIMA forecast for Zimbabwe electricity consumption.

Table 8 reports the ARIMA (1,1,1) which is the best fit and parsimonious model for the choice regression fit. For brevity, other simulations and OLS regression can be made available on request as well as a forecast for other energy-related variables. The study mainly focuses on energy demand forecast. The estimation for the forecast reveals that electricity consumption for Zimbabwe as reported in **Table 5** was conducted utilizing the dataset from 1980 to 2012 after the imputation of missing data in order to avoid spurious estimation. Empirical evidence shows that in 2030 energy consumption will reach ~0.18 quadrillion Btu against the currently available ~0.17 quadrillion Btu.

The estimation affirms the goodness of fit with a coefficient of determination R^2 of over 80%, with a corresponding F-statistic rejected at $p < 0.01$—indicating joint significant of the selected model. Finally, the study forecast also displays high parsimony with harmony among the root mean square error (RSME) of ~0.04, while the mean absolute error was ~0.03. Similarly, the Theil inequality coefficient was ~0.11.

Figure 7 reports the diagrammatic view with relatively fair deviation from the forecast variable. All forecast indicators resonate with **Figure 7**.

4. Conclusion and policy implications

This study employed econometric techniques to forecast Zimbabwe's energy consumption by 2030. Using the rule of thumb (i.e. less than 20% of the dataset), it was possible to impute the NA values in the dataset using MICE package in R. The unit root tests revealed that all the variables are integrated of order one—which informed our choice of ARIMA model. Using an ARIMA (1,1,1) model with data spanning from 1980 to 2012, the empirical analysis showed Zimbabwe's energy consumption by 2030 will increase to ~0.18 quadrillion Btu from ~0.17 quadrillion Btu in 2017. Thus, the need to diversify and intensify into clean energy sources is crucial among policymakers. This is in order to meet the energy demands given the dynamic fast-growing nature of the study area. The current energy policy in Zimbabwe is found to lack a large-scale utilization of solar and wind resources. Such policy suggests the following measures: encourage the generation of electricity from biomass cogeneration and mini-hydro projects and bagasse from sugar cane—Hippo Valley and Triangle sugar estates generate for their own consumption. However, the existing energy policy suggested the following strategies which have not been implemented: extension of Kariba south by the end of 2016 and 800 MW Batoka hydro by 2020 and mandate the installation of solar geysers by 2013 and fix (REFIT) renewable feed-in tariffs.

Zimbabwe's energy policy currently lacks research on energy consumption forecast; hence, this chapter is indicative for policymakers who design the energy policy framework. The OLS regression revealed a positive relationship between carbon dioxide emissions (CO_2), population (POP), and gross domestic product (GDP). Thus, it implies that population triggers economic growth; however, there is a negative deteriorating effect on environmental quality. It means that policymakers are enjoined to bring forth environmentally friendly regulations to combat the excesses of pollution. Such regulations include renewable energy policy that promotes large-scale utilization of renewable energy resources.

Conflict of interest

Authors declare no conflict of interest.

Appendix A

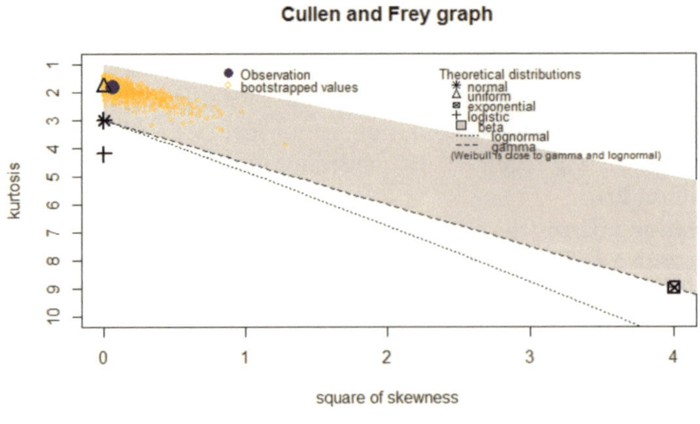

Figure 8.
Population follows a uniform distribution.

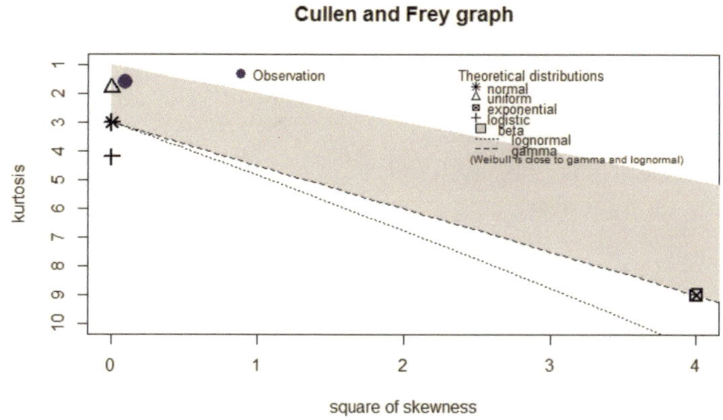

Figure 9.
CO_2 emissions follow uniform and beta distributions.

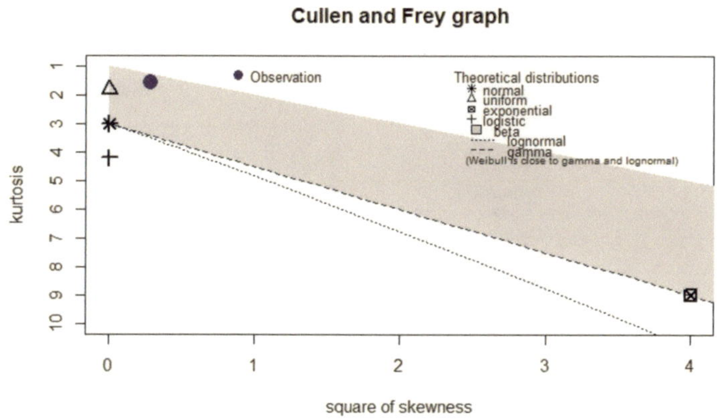

Figure 10.
GHG emissions follow a beta distribution.

Appendix B

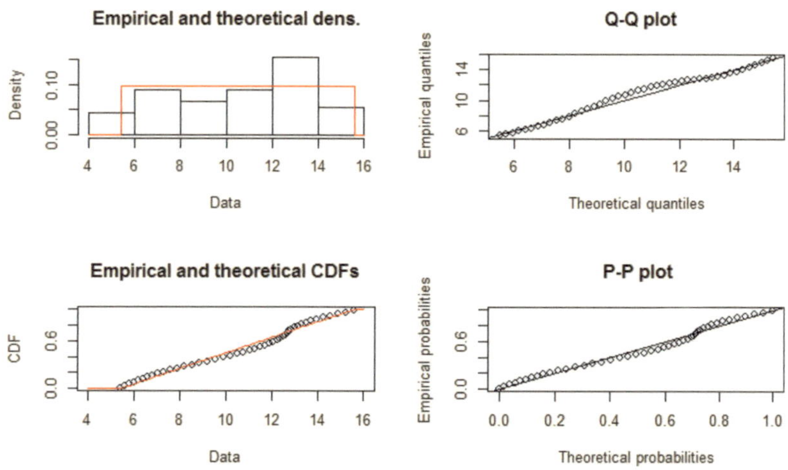

Figure 11.
Population distribution.

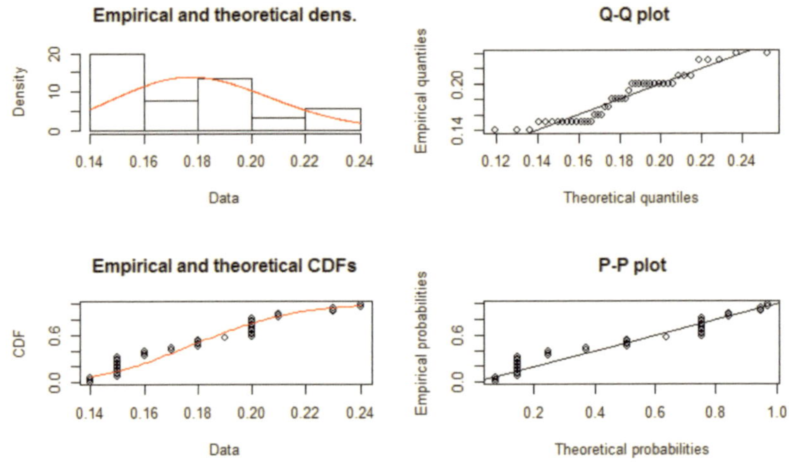

Figure 12.
Energy consumption distribution.

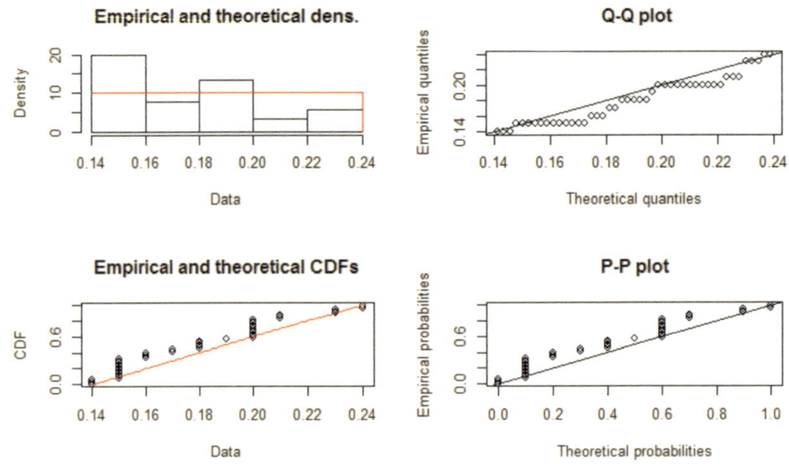

Figure 13.
Energy generation distribution.

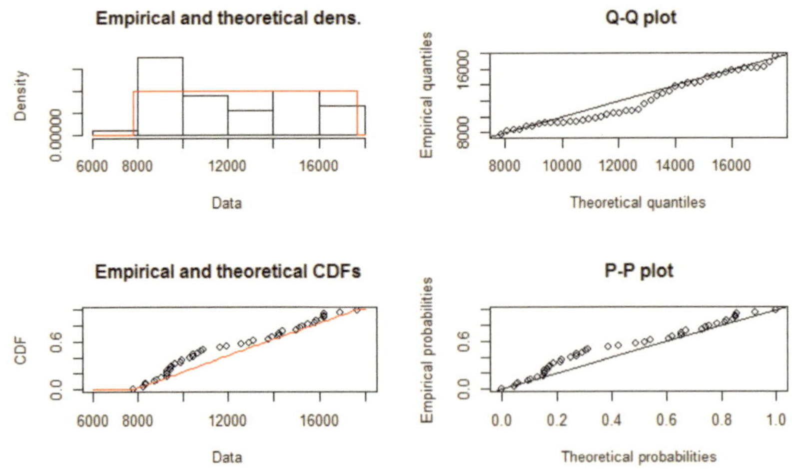

Figure 14.
CO_2 emission distribution.

DOI: http://dx.doi.org/10.5772/intechopen.87249

Author details

Remember Samu[1], Samuel Asumadu Sarkodie[2*], Murat Fahrioglu[3] and Festus Victor Bekun[4]

1 Discipline of Engineering and Energy, College of Science, Health, Engineering and Education, Murdoch University, Murdoch, Australia

2 Nord University Business School (HHN), Bodø, Norway

3 Electrical and Electronics Engineering Department, Middle East Technical University Northern Cyprus Campus, Mersin, Turkey

4 Department of Economics, Eastern Mediterranean University, Famagusta, Northern Cyprus, Turkey

*Address all correspondence to: asumadusarkodiesamuel@yahoo.com

References

[1] Knoema. World Energy Electricity Data Base, 2014. 2017. Retrieved from: https://knoema.com/WEEDB2015/world-energy-electricity-data-base-2014?tsId=1002040 [30 March 2017]

[2] Samu R, Fahrioglu M, Taylan O. Feasibility study of a grid connected hybrid PV-wind power plant in Gwanda, Zimbabwe. In: 13th HONET-ICT International Symposium on Smart MicroGrids for Sustainable Energy Sources Enabled by Photonics and IoT Sensors, HONET-ICT 2016. 2016. DOI: 10.1109/HONET.2016.7753434

[3] Samu R, Fahrioghlu M. An analysis on the potential of solar photovoltaic power in Zimbabwe. Energy Sources, Part B: Economics, Planning, and Policy. 2017

[4] World ONO. Human Development Index—Countries with Low Human Development. 2017. Retrieved from: http://www.nationsonline.org/oneworld/human_development_low.htm

[5] World Bank. Zimbabwe CO_2 Emissions Per Capita. 2013. Retrieved from: http://www.multpl.com/zimbabwe-co2-emissions-per-capita [30 March 2017]

[6] Morales-Acevedo A. Forecasting future energy demand: Electrical energy in Mexico as an example case. Energy Procedia. 2014;**57**:782-790. DOI: 10.1016/j.egypro.2014.10.286

[7] Jebaraj S, Iniyan I. A review of energy models. Renewable and Sustainable Energy Reviews. 2006

[8] Baker AB, Finizza JB. Corporate point of view modelling. Energy. 1990

[9] Sen P, Roy M, Pal P. Application of ARIMA for forecasting energy consumption and GHG emission: A case study of an Indian pig iron manufacturing organisation. Energy. 2016;**116**:1031-1038. DOI: 10.1016/j.energy.2016.10.068

[10] Cabral JA, Legey LFL, De Freitas Cabral MV. Electricity consumption forecasting in Brazil: A spatial econometrics approach. Energy. 2017; **126**:124-131. DOI: 10.1016/j.energy.2017.03.005

[11] Pao H-T, Tsai C-M. Modelling and forecasting the CO_2 emissions, energy consumption, and economic growth in Brazil. Energy. 2011;**36**(5):2450-2458. DOI: 10.1016/j.energy.2011.01.032

[12] Barak S, Sadegh SS. Forecasting energy consumption using ensemble ARIMA–ANFIS hybrid algorithm. International Journal of Electrical Power & Energy Systems. 2016;**82**:92-104. DOI: 10.1016/j.ijepes.2016.03.012

[13] Campillo J, Wallin F, Daniel Torstensson IV. Energy demand model design for forecasting electricity consumption and simulating demand response scenarios in Sweden. In: International Conference on Applied Energy. 2012

[14] Yukseltan E, Yucekaya A, Bilge AH. Forecasting electricity demand for Turkey: Modelling periodic variations and demand segregation. Applied Energy. 2017;**193**:287-296. DOI: 10.1016/j.apenergy.2017.02.054

[15] González-Romera E, Jaramillo-Morán MÁ, Carmona-Fernandez D. Monthly electric energy demand forecasting based on trend extraction. IEEE Transactions on Power Systems. 2006:21

[16] Yildiz B, Bilbao JI, Sproul AB. A review and analysis of regression and machine learning models on commercial building electricity load forecasting. Renewable and Sustainable Energy

Reviews. 2017;**73**:1104-1122. DOI: 10.1016/j.rser.2017.02.023

[17] Kraft J, Kraft A. On the relationship between energy and GNP. Journal of Energy Development. 1978;**3**(2): 401-403

[18] Narayan PK, Smyth R. Energy consumption and real GDP in G7 countries: New evidence from panel cointegration with structural breaks. Energy Economics. 2008;**30**(5): 2331-2341

[19] Yoo SH, Kim Y. Electricity generation and economic growth in Indonesia. Energy. 2006;**31**(14): 2890-2899

[20] Stern PC. New environmental theories: Toward a coherent theory of environmentally significant behaviour. Journal of Social Issues. 2000;**56**(3): 407-424

[21] Azlina AA. Energy consumption and economic development in Malaysia: A multivariate cointegration analysis. Procedia-Social and Behavioral Sciences. 2012;**65**:674-681

[22] Jumbe CB. Cointegration and causality between electricity consumption and GDP: Empirical evidence from Malawi. Energy Economics. 2004;**26**(1):61-68

[23] Hu X, Lin X. A study of the relationship between electricity consumption and GDP growth in Hainan international tourism island of China. Research in World Economy. 2013;**4**(1):109

[24] Lee CC, Chang CP, Chen PF. Energy-income causality in OECD countries revisited: The key role of capital stock. Energy Economics. 2008; **30**(5):2359-2373

[25] Zou G, Chau KW. Short-and long-run effects between oil consumption and economic growth in China. Energy Policy. 2006;**34**(18):3644-3655

[26] Apergis N, Payne JE. Energy consumption and economic growth in Central America: Evidence from a panel cointegration and error correction model. Energy Economics. 2009;**31**(2):211-216

[27] Soytas U, Sari R. Energy consumption and income in G-7 countries. Journal of Policy Modeling. 2006;**28**(7):739-750

[28] Soytas U, Sari R. Energy consumption and GDP: Causality relationship in G-7 countries and emerging markets. Energy Economics. 2003;**25**(1):33-37

[29] Asumadu-Sarkodie S, Owusu PA. Carbon dioxide emissions, GDP, energy use, and population growth: A multivariate and causality analysis for Ghana, 1971–2013. Environmental Science and Pollution Research. 2016; **23**(13):13508-13520

[30] Chavez SG, Bernat JX, Coalla HL. Forecasting of energy production and consumption in Asturias (northern Spain). Energy. 1999;**24**(3):183-198

[31] Sarkodie SA. Estimating Ghana's electricity consumption by 2030: An ARIMA forecast. Energy Sources, Part B: Economics, Planning, and Policy. 2017;**12**(10):936-944

[32] Asumadu-Sarkodie S, Owusu PA. Forecasting Nigeria's energy use by 2030, an econometric approach. Energy Sources, Part B: Economics, Planning, and Policy. 2016;**11**(10):990-997

[33] Knoema. World Data Atlas. 2017. Retrieved from: https://knoema.com/atlas/zimbabwe [Available from: 14 March 2017]

[34] World Data Atlas. World Data Atlas. 2017. Retrieved from: https://knoema.com/atlas/Zimbabwe

[35] Alam MM, Murad MW, Norman AHM, Ozturk I. Relationships among carbon emissions, economic growth, energy consumption and population growth: Testing environmental Kuznets curve hypothesis for Brazil, China, India and Indonesia. Ecological Indicators. 2016;**70**:466-479

[36] García-Martos C, Rodríguez J, Sánchez MJ. Modelling and forecasting fossil fuels, CO_2 and electricity prices and their volatilities. Applied Energy. 2013;**101**:363-375

[37] Dickey DA, Fuller WA. Likelihood ratio statistics for autoregressive time series with a unit root. Econometrica: Journal of the Econometric Society. 1981:1057-1072

[38] Phillips PC, Perron P. Testing for a unit root in time series regression. Biometrika. 1988;**75**(2):335-346

Chapter 6

Renewable Energy in Ukraine-Poland Region: Comparison, Critical Analysis, and Opportunities

Lyubomyr Nykyruy, Valentyna Yakubiv, Grzegorz Wisz, Iryna Hryhoruk, Zhanna Zapukhlyak and Rostyslaw Yavorskyi

Abstract

Fundamental and applied research on renewable energy is actively supported for the development of world science and maintaining the energy independence and security of different countries. This section analyzes the publications of scientists from two countries—Ukraine and Poland—in the field of "thermoelectricity," "photoelectricity," and "bioenergy" to find regularities in each state and to determine the prospects for joint research. Ukraine and Poland share a common border and have similar climatic conditions and historical heritage, but Poland is a member of the EU, and its legislation in the field of renewable energy complies with the regulations of the European Community. Ukraine is making every effort to develop renewable energy. Comparison of the state of research in these countries is also an example of the analysis of the situation at the borders of EU countries and may answer questions related to sustainable development, the mass transition to renewable energy, and the refusal to use fossil fuels and nuclear power plants. The analysis is based on the results of data published in the international scientific databases Web of Science and Scopus. The most advanced areas of research in each country are identified, analyzed, and aimed at practical application.

Keywords: renewable energy, h-index, thermoelectricity, photovoltaic, bioenergy, Poland, Ukraine

1. Introduction

Increasing global warming, with frequent storms, melting ice, droughts, etc., indicates the fatal impact of fossil fuels on the planet's ecosystem.

Unfortunately, the lack of a proper energy policy among the states, in particular those whose economy is based on carbon fuels, often has detrimental effect on the environment [1]. Reducing the dependence on carbon-containing fuels is an important step in transforming a sustainable energy system, being required by Statements on Paris climate agreement [2], which foresees the complete abandonment of fossil fuels by 2050. For this reason, most states have started to support the development of renewable energy sources at the legislative level and to encourage the transition to their widespread use. For example, Directive 2009/28/EC of the European Parliament of 23 April 2009 obliges Member States to provide a specific share of energy from renewable sources for final gross energy consumption in 2020 [3]. On average, 20% of energy should come from renewable sources. For Poland, this figure is about 15%.

However, the challenge of such EU policy is possible noncompliance of these initiatives by states bordering the EU.

Thus, this section analyzes the renewable energy market in two countries—Poland and Ukraine. These countries have a close mentality, driven by a long shared history. The length of the common border is 535 km. The population of Poland and Ukraine is 38 million and 42 million, respectively, with an area of 312.3 km^2 and 603.6 km^2. The most important factor for qualitative and quantitative analysis of the development of renewable energy sources in these countries is their similar geographical location (climate, common mountain system, geographical zone, etc.).

However, the situation with the development of renewable energy in these countries is different. According to 2017 data, the total amount of renewable energy in Poland was 383,168 TJ [4]. It comes primarily from solid biofuels (67.9%), wind (14.0%), and liquid biofuels (10.0%). In Ukraine, according to the State Agency on Energy Efficiency and Energy Saving of Ukraine, renewable energy sources have produced 7566 TJ of energy in the same period [5]. Such significant differences in the amount of energy produced are related to public policy in this area. For Poland, this policy is unified with relevant EU acts, where support is extended from the largest megaprojects to the smallest solar project in the community, leading to the effective development of future renewable energy [6]. Ukraine has only recently started to actively support green energy. The Polish experience can be very helpful for the development of renewable energy in Ukraine because of the centuries-old mental similarity between the populations of both countries. Therefore, the experience of the transition to renewable energy in Poland and the challenges that they face and overcome are beneficial for Ukraine.

This paper attempts to analyze three promising areas of renewable energy development in Ukraine and Poland, such as *solar photovoltaics*, *thermoelectricity*, and *bioenergy*. These directions are chosen for the following reasons. Both countries are in similar latitudes and have relatively close sunny days. Both countries have similar infrastructures, including scientific ones. Academic institutes and universities in these countries have devoted considerable attention to the development of materials science for energy (possibly energy materials science). Solar photovoltaics is one of the areas where it is possible to test the properties of new materials quickly and bring them into production. And, given the powerful global industry in this direction, it is very easy to compare the results obtained with the industry and to select new objects for research flexibly.

Thermoelectricity in these countries does not have a strong practical application but has powerful world-class scientific results obtained by scientists from the universities of Ukraine and Poland [7, 8]. Therefore, in the future, the production and use of thermoelectric devices in these countries may become the world's foremost.

Bioenergy is a specific feature of these countries. Both countries have well-developed agriculture, the waste of which is a direct source for biofuel or biogas [9].

2. Methodology

Scientific results can be analyzed if they are in the form of a publication. To exclude few influential publications, only peer-reviewed ones were selected for analysis. The scientific literature was searched through academic libraries, Web of Science and Scopus, reviewed in English, and published books related to the topic of this section. Given that Ukraine became an independent state in the early 1990s, it is possible to make a clear identification of publications linked specifically to Ukrainian scientific groups since then.

Thus, the search has been performed since 1991. Thanks to advanced tools of scientometric systems, scientific papers were selected, indicating the affiliation of Poland or Ukraine. The procedure is described in detail in [10].

In applying to this section, there were certain features in the proposed methodology. In particular, the review of sources was performed on the basis of analysis of both the most important or the most cited (sorting by number of citations) and the most recent (sorting by date).

In the second stage, analysis of the type of publications, their financial support, as well as the areas specific to each industry, such as "materials science," "physics," "technology," etc., has been done. At the same time, patterns and relationships between the areas of renewable energy studied in this work and within each direction have been found.

At the last stage, a critical analysis of the most prominent publications and regularities was performed to predict the options for the development of a particular direction and to generate the key results for supporting theoretical development within the direction or certain practical technologies.

3. Analysis of results and discussion

For the convenience of performing analysis of research impact conducted by scientists from Poland and Ukraine, the collected data are summarized in **Table 1**.

Even before the 1990s, the number of publications in particular renewable energy areas was one or tens of publications per year. However, since the early 2000s, there has been a rapid nonlinear increase in the number of publications. Are there certain reasons that could explain the sharp changes in the number of publications? Apparently so. One of them is the Treaty of Amsterdam signed in 1997, which laid down the principle of sustainable development for the EU, the essence of which was to improve the production of renewable energy. Another reason may be the awareness of the scientific community of the instability of the existing energy state, the exhaustion of natural resources, and the need to find alternative sources of fuel to reduce emissions into the environment. Another condition that led to a new step in the development of renewable energy was the "forced policy" of thermal collectors in some states, which obliged people to put heat collectors on their homes. Accordingly, the search for cheaper collectors and photoelectric and hybrid systems has become a new cause for the increasing number of publications in this field.

As for Ukrainian publications, the number of papers is much smaller than the number of publications of the world scientific community. The economic situation in the country has a significant impact on the number of publications.

Below is a summary of information available in the international science databases on the state of research in the three renewable energy fields. These three areas are chosen for a number of reasons. This chapter examines studies conducted in two bordering countries. However, Poland is a member of the EU, complies with the relevant directives adopted by the European Community, and anticipates developing its economy and science in terms of forming a single European research area. Ukraine, however, is a neighbor of the EU and the closest neighbor of Poland with related culture and climatic conditions, but is not a member of the EU. For a long time, scientific research in Ukraine was closed to the world scientific community (until the first decade of the post-USSR),

Ukraine	Comparison	Poland
Solar photovoltaics		
1370	Number of publication	2284
38	h-index	66
• Taras Shevchenko National University of Kyiv • NU "Lviv Polytechnics" • Yuriy Fedkovych Chernivtsi National University • Institute of Physics, National Academy of Sciences in Ukraine • National technical University "Kharkiv Polytechnic Institute"	Leading institutions (five items)	• Politechnika Warszawska • Wroclaw University of Science and Technology • AGH University of Science and Technology • Silesian University of Technology • Lodz University
• Engineering (21.63%) • Physics and astronomy (20.71%) • Materials sciences (19.11%) • Energy (7.98%) • Chemistry (7.64%)	Most presented scientific areas	• Physics and astronomy (29.5%) • Materials sciences (25.57%) • Engineering (16.73%) • Chemistry (6.68%) • Energy (5.6%)
• Ministry of Education and Science of Ukraine • National Academy of Sciences of Ukraine • Science and Technology Center in Ukraine • State Fund for Fundamental Research of Ukraine • European Regional Development Fund	Most finance support	• Narodowe Centrum Nauki • Narodowe Centrum Badan I Rozwoju • European Regional Development Fund • European Commission • Fundacja na rzecz Nauki Polskiej
• RF • Germany • Poland • United States • France	Partner country	• Germany • France • United States • United Kingdom • Ukraine
Thermoelectricity		
901	Number of publication	1145
28	h-index	40
• Institute of Thermoelectricity NAS • Ivan Franko LNU • NU "Lviv Polytechnics" • Yu. Fedjkovych ChNU • NTU "Kharkiv Polytechnic Institute"	Leading institutions (five items)	• AGH University • Włodzimierz Trzebiatowski Institute of Low Temperature and Structure Research PAS • Wrocław Branch of PAS • Politechnika Warszawska • Institute of Molecular Physics PAS
• Engineering (26.6%) • Physics (26.6%) • Materials science (24.5 %) • Energy (4.9%) • Computer sciences (4.5%)	Most presented scientific areas	• Physics (29.1%) • Materials science (27.7 %) • Engineering (21.7%) • Chemistry (6.1%) • Energy (3.9%)
• NAS of Ukraine • MES of Ukraine • SFFR of Ukraine	Most finance support	• NCN • SC of Antarctic Research • KBN
• Poland • United States	Partner country	• Germany • Ukraine

Ukraine	Comparison	Poland
• RF • Germany • Austria		• United States • France • Moldova
Bioenergy (energy from biomass and biofuel)		
71	Number of publication	324
13	h-index	37
• National University of Life and Environmental Sciences of Ukraine • NU "Lviv Polytechnics" • Taras Shevchenko National University of Kyiv • Vasyl Stefanyk Precarpathian National University	Leading institutions (five items)	• Polish Academy of Sciences • Uniwersytet Jagielloński w Krakowie • Uniwersytet Warminsko-Mazurski w Olsztynie • Wrocław University of Science and Technology • Uniwersytet im. Adama Mickiewicza w Poznaniu
• Energy • Agricultural and biological sciences • Environmental science • Biochemistry, genetics, and molecular biology • Engineering	Most presented scientific areas	• Biochemistry, genetics, and molecular biology • Agricultural and biological sciences • Environmental science • Energy
• European Commission • Ministry of Education and Science of Ukraine	Most finance support	• Narodowe Centrum Nauki • Ministry of Higher Education
• Germany • Italy	Partner country	• United States • Germany

Table 1.
Comparative statistic characteristics of the researches of Ukraine and Poland, carried out in the directions "solar photovoltaics," "thermoelectricity," and "bioenergy."

and later for a long time did not integrate into it due to the significant lack of funds caused by the poor economy. However, these two countries are now implementing the desire to develop together and are an example of how a competitive scientific environment can be started at the EU's border.

3.1 Solar photovoltaics

Photoelectricity is one of the most popular types of renewable energy today. The reason for this is a free source—solar radiation, which, by getting on properly prepared semiconductor materials, is converted into direct electric current due to the known phenomenon, photo effect. Interest in photovoltaics arose a long time ago; however, only with the development of silicon technology, it has become widespread in almost all countries of the world. On the other hand, industrial silicon production is a technologically complex, expensive, and environmentally hazardous process. Therefore, the scientific community of the world today is trying to find new, cheaper, and environmentally friendly materials that would significantly reduce the cost of photovoltaic energy produced. These are, first and foremost, thin-film materials of different types of heterostructure, materials with nanoinclusions, etc. Thus, these studies have evolved from the usual design decision toward materials science.

Analyzing publications by the tag "solar photovoltaics" requires careful study. For the entire period (1991–2019), there are 2292 publications of Polish researchers in the Scopus database. However "open access" for research results has only 403 of 2292 publications, being available publicly. By citation number h = 66 is the total number of solar photovoltaic publications in the period of 1991–2019.

However, it should be borne in mind that not all publications are concerned with the production, storage, or conversion of solar energy and only a fraction of the publications are relevant to the search query. The observations were made on the first hundred of the most cited publications and the most recent publications. The most interesting of them were selected for analysis.

The most cited is the paper [11] of h = 2122. It presents a brief overview of well-established multilevel converters strongly oriented to their current state in industrial applications and the review of new converters that have made their way into the industry. In addition, new promising topologies and nontraditional applications powered by multilevel converters were discussed.

One of the most cited papers is [12], discussing specific chemical and physicochemical requirements for organic compounds to be applied in organic or hybrid electronic devices such as photodiodes, light-emitting diodes, photovoltaic cells, etc.

One of the most cited (184 times) and new publications (2019) is [13] which reports a new non-fullerene n-type organic semiconductors that have attracted significant attention as acceptors in organic photovoltaics (OPVs) due to its great potential to realize high-power conversion efficiencies. OPVs made exhibited a high efficiency of 15.7%.

Among other publications, papers related to "organic solar cells," "polymer solar cells," "semiconductor heterojunctions solar cells," "silicon-based solar cells," "dye-sensitized solar cells" (DSSCs), "perovskite photovoltaics," and other materials, due to their high power conversion efficiencies, can be found. These papers describe possible physicochemical processes and phenomena that occur during the preparation of materials, testing of properties and their approbation as metastabilities of electrical properties, photoelectrical parameters, light sensitivity and absorbance, chemical treatment and deposition methods, bands, defects, grain boundaries, exiton binding energies, factors affecting conversion efficiency, etc.

In particular, since 1991 dye-sensitized solar cells (DSSCs) have attracted considerable interest from the scientific and commercial communities due to their promising characteristics as solar light converters. About 8% of the first 200 publications are related to the topic of materials for solar cells.

By the tag "solar photovoltaics," the most cited are also the studies on the properties of TiO2 [14, 15] and ZnO [16, 17] and heterojunctions based on them.

Other highly cited publications are research papers about polymeric materials in photovoltaic device application for polymer solar cells [18] or investigation and modeling of metastabilities in chalcopyrite-based thin-film solar cells, for example, Cu(In,Ga)Se2- [19] or ZnO/CdS/Cu(In,Ga)Se2-based [20] thin-film solar cell.

The newest publications are related to the issues of ecology, economics, etc., in particular the implementation and application of hybrid energy conversion systems in Poland [21, 22]. There have also been publications on non-silicon PV modules [23], studies of weather and climate conditions on the efficiency of solar energy conversion [24], etc.

The results presented in these papers became the basis for the development of the production of renewable energy sources and the effective commercialization of certain scientific results.

The solar photovoltaics sector is one of the fastest growing renewable energy sectors in Poland and in the world. The photovoltaic market in Poland has enormous, but so far highly unused, development potential.

The total installed capacity in photovoltaic sources at the end of 2018 was about 500 MW, and already in May 2019, it exceeded 700 MW. The growth of new PV installations is dynamic. In 2018, Poland finally began to stand out from the other EU countries and with an annual increase of 235 MW was already in the ninth place. Considering current and real investments in progress, in 2019 Poland may be in fourth place in the EU in terms of annual increases in new solar power sources. The authors of the report estimate that in 2019 there will be even 1 GW of new PV installations and the cumulative power of solar installations in Poland at the end of the year will be 1.5 GW.

Solar energy in Ukraine has been actively developing since the end of 2008 with the adoption of a “green” tariff at the legislative level, which made financially attractive investments in industrial grid solar power plants.

According to the State Agency on Energy Efficiency data (http://saee.gov.ua), the theoretically possible potential of solar energy in Ukraine is more than 730 billion kWh per year, and technically possible is only 34.2 billion kWh per year.

The use of solar panels on rooftops by private households by 2050 can reach 40–50%. In addition, the use of solar collectors for water heating will be more cost-effective. These technologies will provide hot water demand in private households for 70–100% during summer and 15% in winter.

For the investigation period, there are 1370 publications in Scopus, which are 922 fewer than in the same period in Poland. Only 114 of 1370 publications are in “open access.” By the number of citations, h = 38 is the total number of solar photovoltaic-related publications during this period.

The most cited publication with the participation of Ukrainian researchers [25] with h = 207 is concerned with efficient photocatalytic water splitting.

The paper [26] was one of the fundamental at the time (2000s) in which design and photovoltaic performance of solar cells based on various semiconductor nanorod materials, such as TiO_2, ZnO, CdS, CdSe, CdTe, CuO, and Si, were presented and compared with respective solar cells based on semiconductor nanoparticles; specific of synthesis and application of carbon nanotubes in photovoltaic devices were also reviewed in these papers.

The largest number of publications in Ukraine by the tag “solar photovoltaics” relates to photoelectric materials science, mainly, these are high-tech results aimed at optimizing the parameters of existing materials for photovoltaics, improving methods for their production and research, as well as creating new efficient and low-cost materials for competing with existing ones.

The highly cited are the studies on the properties of CdTe-based heterojunction solar cells. The first works devoted to the study of CdTe properties in Ukraine since its independence appeared in 1997 [27]. The first mention of the concept of quantum efficiency of CdS/CdTe SCs appeared in 2012 [28].

Solar modules based on the developed flexible solar cells ITO/CdS/CdTe/Cu/Au on polyimide films were mentioned in 2009 for the first time [29].

The researches in [30, 31] were mainly focused on CdTe-based compounds. The publications with the tag “solar photovoltaics” on CdTe-containing compounds make up ~12.5% of the first 200 publications by researchers from Ukraine, while the percentage of publications related to silicon and its possible modifications for use in solar power is ~10%. However in recent years, there has been an increase in the number of publications of other scientific groups in this field, which demonstrate the modern competitive achievements [32, 33].

Publications looking at the prospects of using quantum dots (QD) in solar cells are very popular. Their photophysical and electrophysical properties can be varied by different particle size and shape, and QD can provide absorption of solar energy in a much wider spectral range compared to conjugated organic compounds [34].

At the forefront of science are promising techniques for the development of technologies of second-generation PV systems, the efficiency of which is now comparable to that of silicon and the cost and consumption of material are significantly lower [35].

Analyzing recent publications as a whole, we can observe a tendency to the interest in modeling of weather condition influences on properties of photovoltaic installations, methods of its identification [36], modeling of building energy supply based on PV modules [33, 37], microgrid with hybrid renewable energy system, and the concept of energy accumulation from photovoltaic and wind power plants [38].

However, the structure of most publications has the character of fundamental research in semiconductor physics. Semiempirical approaches that combine materials science research, the development of low-dimensional structure fabrication technology, the testing of the obtained parameters to work as a photocell, and point to the prospect of further practical application are the main features of Ukrainian scientists.

3.2 Thermoelectricity

Thermoelectric energy conversion, like photovoltaics, is based on known effects discovered more than a hundred years ago. The principle of thermoelectric power generation devices is very simple: the thermoelectric device (usually a thermoelectric generator or module) is placed so that one side is at higher and the other at lower temperatures. Accordingly, due to the temperature difference, thermoelectric driving force arises, or, in other words, it is possible to record the potential difference or the presence of direct current. Despite such ease of use, thermoelectricity has long been considered inefficient due to its low efficiency. However, when semiconductors were proposed to be used as thermoelectric materials in the mid-twentieth century, it was an important step to start the rapid development of this field. Industrial thermoelectric devices have relatively low efficiency. This is about 4–8%, which is significantly inferior to the efficiency of photovoltaics (up to 28%). However, such small values make it possible to create a whole series of different autonomous energy sources, which are used in medicine, space applications, automotive technology, etc. The undeniable advantage of thermoelectricity is the reliability of the devices, their quietness, and the extremely long service life.

The thermoelectric module is based on a sequence of n- and p-type conductivity materials. It is the quality of these materials that determines the effectiveness of the device. That's why researchers are trying to create new, low-cost, high-efficiency materials. These are nanomaterials, multicomponent compounds, and materials with the inclusion of different phases, etc.

Studies in thermoelectricity are closely related to semiconductor materials science. The most cited papers, with the participation of Polish authors, relate to fundamental studies of the electronic properties of a wide range of semiconductors. These are, in particular, the study of the first principles of resonance states or transport phenomena of carriers [39, 40]; the study of fundamental processes in modern materials, in particular, nanostructures and materials with nanoinclusions [41]; and the engineering of new ones, including multicomponent structures that are promising for practical use in thermoelectric because of their unique properties, environmental friendliness and relatively low cost [42, 43]. However, attention is paid to the study of classical thermoelectric materials, and the obtained results correspond to a high world level. This is, in particular, a study taking into account current approaches, both experimental and theoretical, of compounds of types IV–VI [44] and II–VI [45].

However, the most recent publications refer to the same directions [46, 47]. It is important to note here that the most cited papers by Polish scholars refer to those published mainly in the last 5–10 years. That is, at this time there is a peak of qualitative research, recognized by the scientific community of the world. For the same reason, the latest publications have the authorship of the same scientific teams. These are representatives of well-known scientific centers, such as AGH University (Krakow), two institutions in Wrocław, and Institute of Molecular Physics PAS in Poznan (see **Table 1**). The geography of the research institutions is quite broad and not concentrated in a particular region, which indicates the systematic support of such high-tech research in Poland by the state. In terms of quality, it is worth noting the list of the most popular publications in which scientists from Poland are published. Among the most quoted, the unambiguous leader is *Physical Review*, which testifies to the development of fundamental research. The *Journal of Alloys and Compounds*, the *Journal of Electronic Materials*, and the *Solid State Ionics* are also very popular. Regarding publications that are newest at the time of analysis, they have been published in the *Journal of Alloys and Compounds*, the *Journal of Electronic Materials*, and the *Journal of Applied Physics* (**Table 1**). The impact factors (IF) of each of these journals range from three to seven. That is, the average journals in which Polish scholars are published also belong to the flagship publications.

The most cited papers by researchers from Ukraine concerned with new classes of materials that are in line with global trends—the search for new, environmentally friendly and inexpensive materials. It includes theoretical work concerning the modeling of new classes of multicomponent compounds [48] and modeling of performance with respect to new classes of thermoelectric materials—skutterudite [49], half-Heusler alloys [50], or graphene [51]—as elements of end devices.

However, the study of classical materials is also carried out at a high level. Moreover, in some cases the effects investigated in certain materials for the first time are demonstrated. The publications are devoted to the type II–VI materials (Bi-Sb crystals, Bi2Te3) [52] and IV–VI compounds and studies of thermoelectric parameters oscillations for PbSe [53], new compounds of LATT-PbAgSbTe type [54], or multilayer heterostructures based on them [55].

Moreover, a number of papers published by Ukrainian scientists contribute significantly to the development of the theory of thermoelectric phenomena [56].

A feature of research in thermoelectricity is the inexpensive opportunity to test ready devices. The largest thermoelectricity center in Eastern Europe located in Chernivtsi (Ukraine), namely, the Institute of Thermoelectricity of NAS and MES of Ukraine with ALTEK production company, based of which thermoelectric generators are developed and tested [57], or in Odesa (Thermion Co.) [58]. There are also a large number of citations concerning the cooling systems or the use of thermoelectric measuring devices [59].

Ukrainian researchers usually publish their results in such leading editions as *Applied Physics Letters*, *Acta Materialia*, and *Journal of Alloys and Compounds*. These are journals with an impact factor of 3–7. However, unlike the journals in which Polish scholars usually publish their most cited papers, there is a certain balance between applied results and purely material studies.

An analysis of scientific papers published recently shows that the trend remains to investigate multicomponent bulk and low-dimensional compounds and nanomaterials [54, 60, 61].

But, at the same time, there is a large number and very clearly directed applied research concerning the development of specific thermoelectric devices, in particular for medicine [62] or hybrid power systems that combine different types of renewable energy generation in a single device [63].

Also, the International Thermoelectric Academy has been established on the base of the Institute of Thermoelectricity in Chernivtsi, which unites the efforts of not only Ukrainian scientists but also representatives of practically all known scientific centers of the world engaged in thermoelectricity. The *Journal of Thermoelectricity*, published by them, is one of the most popular journals for Ukrainian scientists in the field of thermoelectricity.

3.3 Biofuel energy

Unlike the two previous areas covered in this section, the development of bioenergy is more specific and needs the presence of certain sectors of the economy in the country. Ukraine and Poland, among other countries, are favorably distinguished by the presence of developed agriculture. Accordingly, the rational disposal of biomass waste from crop and livestock products is very important. At the same time, along with the environmental issue, the issue of renewable electricity generation is partially addressed. In the agricultural regions of both countries, various biogas stations have been actively introduced. Thus, there is a different approach to this area of renewable energy: the realities of the economy require its development to a new, more effective scientific level. And the presence of a common border and similar geographical conditions is a factor that creates the same technological conditions for the development of bioenergy for both countries.

Among all renewable energy resources, one of the most promising and strategically important is the bioenergy resource. The role and importance of bioenergy for the development of the economy have been repeatedly emphasized in the reports of scientists, experts, practitioners, and all those who are in one way or another involved in energy problems.

Bioenergy is closely connected with notions “biomass” and “biofuel.” Biomass means organic matter of plant origin and waste materials obtained through natural or artificial transformation that can be used for energy purpose. Biofuel is a renewable energy source derived from plant or animal biomass. Although in many studies the terms biomass” and “biofuel” are used interchangeably, we consider it appropriate to differentiate them. Biomass is a raw material, whereas biofuel is a product of biomass processing.

The analysis of publications shows that bioenergy research is highly relevant in terms of environmental improvement, namely, the study of biocomposites (or biopolymers) as a replacement for plastics [64] and biodiesel production [65]. The most interesting renewable energy sources for the Polish region are agricultural biomass wastes, which are also useful for improving the environment and for medical purpose [66]. However, the most interesting is the production of biogas [67–71], which requires cheap agricultural raw materials. The review [67] is the most important in terms of process technology development for biogas (i) production, (ii) conditioning, (iii) utilization, and (iv) industrial symbiosis. And given the high capacity of biogas plants in Poland, attention has been paid to researching the quality of compost for the production of quality biogas [71–73].

As far as biomass is available as by-product of many industrial and agricultural processes almost everywhere, easy to be obtained, and is a carbon-neutral energy source, biomass represents a growing renewable energy source with high growth potential in the economic analysis of bioenergy in Europe [74–77].

Nowadays, bioenergy is a field of great interest to the scientific community. We can observe growth in the amount of publications, starting from the mid-2000s. The most important reason for that was oil peaking at over $136 a barrel in 2008.

The most cited papers related to biomass and biofuel have been published in the following journals: *Renewable and Sustainable Energy Reviews* (h-index is 193), *Biomass & Bioenergy* (h-Index is 143), *Bioresource Technology* (h-Index is 229), *Renewable Energy* (h-index is 143), and *Energy* (h-index is 146).

Also according to the Scopus database, the most prominent and influential scientists in the sphere of bioenergy have been determined. They are Omer, A.M. from the United Kingdom (122 publications on biomass as renewable energy source); Pari, L. from Italy (103 publications); and Kaltschmitt, M. from Germany (102 publications).

Using the Scopus database, it was determined that h-index of publications related to biomass and biofuel for Poland is 37, and for Ukraine it is 13 (**Table 1**).

A couple of years ago, Poland and Ukraine had the same problems with energy efficiency, being dependent on the old gas transmission infrastructure. The energy systems of two countries are based on large-scale installations for electricity production and distribution. Research in this field, developing renewables, in particular bioenergy resources, are crucial for both countries.

Most of the papers from Polish authors are focused on general prospects of biomass development. Thus, in [78] the largest establishments producing bioethanol and biodiesel have been analyzed. The comparative analysis of different renewable energy resources for rural areas has been done [79].

The influence of energy efficiency in biofuel production on the potential fulfillment of agricultural energy demand has been investigated in [80]. The mathematical model shows the results of exclusion of crops from food production aimed at satisfaction of the energy purposes. Ref. [81] presents a range of products which can be obtained from agricultural production and used for energy purposes. The paper [82] analyzes the potential volume of raw materials that can be obtained from agriculture in Poland for biogas production.

A lot of publications are aimed at technical characteristics of biomass production. Ref. [83] presents results of the study on biomass processing technologies for willow and black locust biomass.

Moreover, according to the forecast in 2020, approximately 80% of the final energy from renewable sources will come from biomass, and almost all of it will be generated from agriculture [84]. The changes in the structure of biomass use are shown. Thus the use of biomass for heating is decreasing, while the share of biomass for electricity and biofuel is getting larger. At the same time, there is a tendency in the publications about the environmental friendliness of bioenergy, in particular, its contribution to reducing the greenhouse effect [85, 86].

A number of papers of Ukrainian scientists are also devoted to general trends in bioenergy production. The analysis of biological resources for biofuel production in Ukraine has been done in [87]. Aspects of transition to agrobioenergy in Ukraine, as well as strategies and recommendations to a variety of stakeholders to facilitate this transition, are suggested in [88, 89].

The positive trends in bioenergy are emphasized in a number of studies. The estimated energy potential of existing biomass waste is about 25 million tons, and the energy potential of biomass which can be grown on unused agricultural lands is about 13 million tons [90]. The paper [91] analyzes the Ukrainian and European Union rapeseed markets being horizontally integrated.

The authors [92] calculated the profit from the biogas installation for poultry farm, and in [93] the main practical steps of establishing a Ukrainian biogas market have been given.

A joint publication [94] summarizes the studies on bioaerosols which were carried out in the years 1972–2009 in the following branches of agricultural industry in the Ukraine and Poland: animal farms, feed facilities, production of biofuel from rape, etc. Another joint publication [95] assesses the potential of biomass, obtained from by-products of crop production and animal breeding, which can be used for energy purposes.

4. Results and discussion

Considering the three selected areas of renewable energy analysis in Ukraine and Poland, it is possible to distinguish certain common features. First of all, the cooperation of Polish and Ukrainian scientists, expressed through a large number of joint publications, is a priority of both countries. Also, scientists from both countries successfully cooperate with their colleagues from Germany and the United States in all the areas of research (**Table 1**). That is, it is not only for photovoltaics and thermoelectricity where it is necessary to synthesize new materials and study their fundamental properties but also for bioenergy, the practical development of which is determined solely by the economic situation in the agricultural sector. It is logical to explain that there are joint studies of both neighboring states and high tech, which possess the most advanced equipment.

Regarding the publications of Polish scientists in recent years, their applied orientation should be noted. In particular, for thermoelectricity these are the problems of creating cooling systems [96, 97], creating different types of thermoelectric elements for sensors [98], and developing classical thermoelectric generators [99]. However, there are several publications concerning hybrid systems, such as the combination of thermoelectricity and photoelectricity [100].

The number of publications on thermoelectricity is comparable in both countries: 1145 submitted by Polish scientists and 901 by Ukrainian. But the h-indexes are different here. It is almost 50% higher for Polish publications (**Figure 1**, blue line). It means that publications of Ukrainian scientists, which are often of a very high scientific level, are still published in less well-known world scientific community journals and, thus, less cited. There is an even greater difference in this regard for the photoelectricity direction. The number of publications of Polish scientists is 2284, while for Ukrainian scientists is 1370. h-indexes for Polish and Ukrainian scientists are 66 and 38, respectively (**Figure 1**, blue line).

There is a difference in the number (324 and 71, for Poland and Ukraine, respectively) and the "impact index" (37 and 13, respectively) of publications in the bioenergy direction.

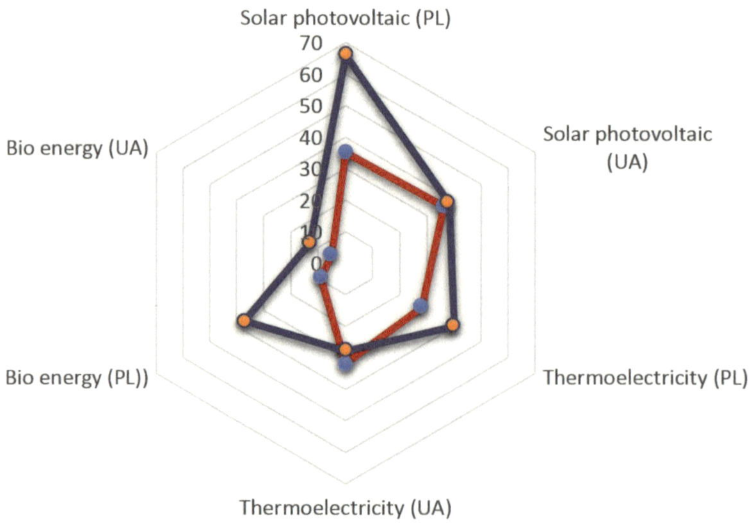

Figure 1.
h-index (blue line) and normalized h-index (red line) for publications of investigation directions for Ukrainian and Poland cases.

Apparently, in all these areas, the number of publications is very different, as well as h-indexes. For more efficient analysis, their normalized dependencies are constructed (**Figure 2**):

$$h_{norm} = h_i/N_i, \qquad (1)$$

where h_i is the h-index of publications in the relevant direction and N is the total number of publications in this direction.

As can be clearly seen from **Figure 1** (red line), the normalized values of the h-index of publications of Ukrainian and Polish scientists are practically the same in magnitude for the directions of "photoelectricity" and "thermoelectricity." That is, we can conclude that the quality of scientific publications in high-tech fundamental directions is almost the same for scientists of both countries.

For "bioenergy," the h-index of Polish scientists is significantly higher. This indicates a stimulation of such research at the state level in Poland and too little support from Ukraine at this stage. However, given that publications by Ukrainian scientists have begun to be published in this field in the last few years and through partnerships between scientific institutions of both countries, it is possible to predict an increase in both the quantity and quality of such materials within the next 5 years.

Comparison of highly cited publications of Polish and Ukrainian scientists in the field of thermoelectricity indicates that the emphasis of Polish scientists is on the study of the properties of materials; however, the development of devices has actively begun to develop only in recent years. For Ukraine, a feature is the parallel existence of whole series of works, devoted to thermoelectric materials science and

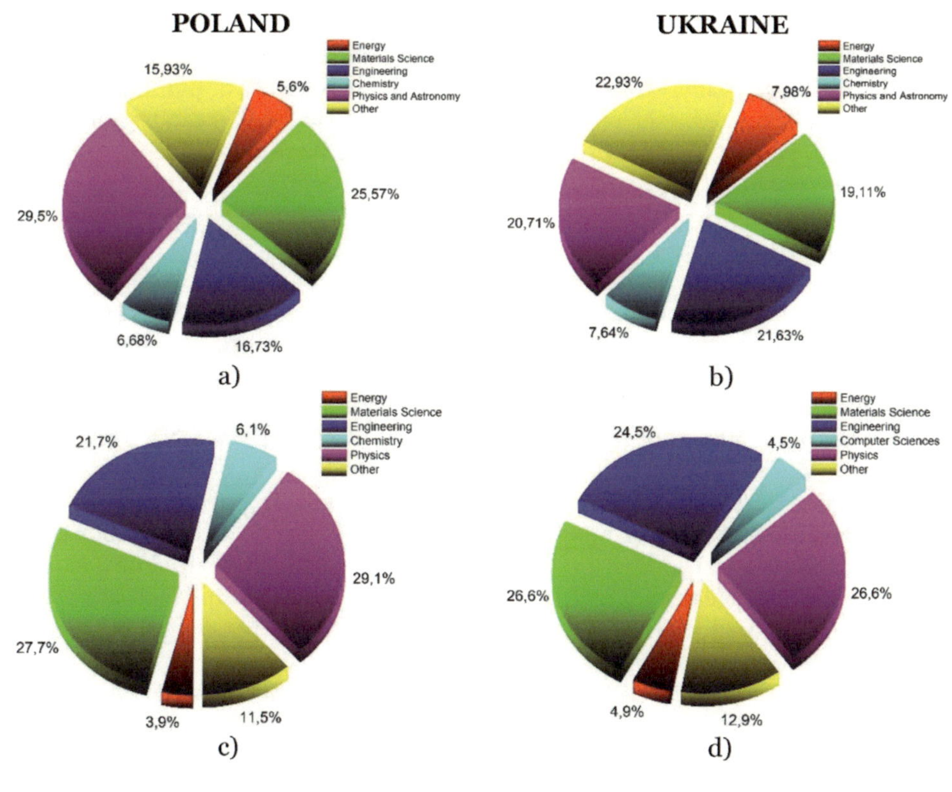

Figure 2.
h-indexes of publications by research areas in solar photovoltaics (a and b) and thermoelectricity (c and d) used (1). Left column for Poland research papers and right column for Ukrainian research papers.

thermoelectric application systems. One of the conclusions following from this analysis is that there has always been a good basis in Ukraine for the industrial production of thermoelectric systems or elements. In Poland, it is only in the initial stages of creation. However, taking into account the quality of research carried out by Polish scientists, the emerging industrial base will be focused on the new, environmentally friendly thermoelectric materials. Today, they are cheaper but have significantly lower values of thermoelectric efficiency. The progress that has been made in recent years gives hope for a breakthrough in this promising direction.

The main scientific directions on the keywords "photoelectricity" and "thermoelectricity," which are technologically similar, are shown in **Figure 2**. For ease of analysis, each direction is indicated by the same color in each figure. It is important to note that under the direction of photoelectricity, Polish scientists' research is mainly focused on fundamental research (physics and materials science), while for Ukraine there is a certain shift of priority to applied research (engineering and physics). With regard to "thermoelectricity," here such a shift is less noticeable, but still it is. In other words, significant and recognized technical solutions are more specific for Ukraine, while more fundamental results are observed in Polish scientists.

For photovoltaics, the situation is similar. In Ukraine it used to be related to photoelectric research for military or space purposes. They often had an applied aspect and were not aimed at publishing scientific achievements. Following the reorientation of the state to peaceful tasks, often representatives of the most famous institutions (Kharkiv, Kyiv) demonstrated a considerable number of publications on technical aspects of the operation of photoelectric systems or design of the lines for their production. But so far collaborative work has emerged, and as a result high-level collaborative publications appear.

Therefore, since 2016 the number of joint Ukrainian-Polish publications has been increasing annually. If we consider separately the number of published papers in Poland and Ukraine by the tag "solar photovoltaics" for 2019, it can be noted that in Poland 158 papers were published, while in Ukraine there are only 64.

5. Conclusions

For the effective development and implementation of renewable energy, different factors must be considered simultaneously. In the short term, the introduction of renewable energy sources is determined by the economic factor and the availability of certain investments in green energy. However, the long-term perspective requires the development of new, environmentally friendly and effective research and the creation of opportunities for their implementation.

Globally, no state can address the issue of renewable green energy on its own. In order to develop international cooperation, a critical analysis of two neighboring countries, Poland and Ukraine, has been made, but, apart from their close geographical location, there are significant differences in economic and legislative systems. Such an analysis is a typical example of the development of cooperation between states at the EU border.

The quality of research can be proved on the basis of quality scientific publications in peer-reviewed journals. The number and impact of such scientific publications indicate the potential of scientific teams, their relevance, and the possibility of implementation.

The normalized h-index indicates that, in the spheres of basic materials science research for energy, this value is practically at the same level, whether or not it is an

EU Member State. The corresponding value of hnorm is 35 and 36, respectively, for studies in Poland and Ukraine in the direction of "photoelectricity" and 28 and 32 in the direction of "thermoelectricity." However, the difference of almost two times (9 and 5, respectively) for the direction of "bioenergy" indicates a much greater progress of Polish scientists in this direction. Such values can be transferred with some accuracy to the other countries at EU borders. That is to say, fundamental research has a high priority almost everywhere, whereas applied research that require sophisticated installations for testing them or expensive simulation software are more effectively implemented in EU countries.

Conflict of interest

The authors declare no conflict of interest.

Author details

Lyubomyr Nykyruy[1*], Valentyna Yakubiv[1], Grzegorz Wisz[2,3], Iryna Hryhoruk[1], Zhanna Zapukhlyak[1] and Rostyslaw Yavorskyi[1]

1 Vasyl Stefanyk Precarpathian National University, Ivano-Frankivsk, Ukraine

2 University of Rzeszow, Rzeszow, Poland

3 Subcarpathian Renewable Energy Cluster, Rzeszow, Poland

*Address all correspondence to: lyubomyr.nykyruy@gmail.com

References

[1] Burke MJ, Stephens JC. Political power and renewable energy futures: A critical review. Energy Research and Social Science. 2018;**35**:78-93. DOI: 10.1016/j.erss.2017.10.018

[2] Available from: https://corporate.exxonmobil.com/energy-and-environment/environmental-protection/climate-change/statements-on-paris-climate-agreement#exxonMobilStatementOnCOP21

[3] McCollum DL, Zhou W, Bertram C, De Boer HS, Bosetti V, Busch S, et al. Energy investment needs for fulfilling the Paris Agreement and achieving the Sustainable Development Goals. Nature Energy. 2018;**3**(7):589. DOI: 10.1038/s41560-018-0179-z

[4] Available from: https://stiloenergy.pl/2019/05/07/energia-odnawialna-w-polsce-dane-dotyczace-wykorzystywania-oze/#2

[5] Available from: http://saee.gov.ua/sites/default/files/IV%202017.pdf

[6] Adil AM, Ko Y. Socio-technical evolution of decentralized energy systems: A critical review and implications for urban planning and policy. Renewable and Sustainable Energy Reviews. 2016;**57**:1025-1037. DOI: 10.1016/j.rser.2015.12.079

[7] Filin SO. State of thermoelectricity development in Poland. Journal of thermoelectricity. 2009;**2**:7-11

[8] Anatychuk LI. On the development of thermoelectricity in Ukraine. Visnik Nacional'noi' Academii' Nauk Ukrai'ni. 2016;**11**:23-30. DOI: 10.15407/visn2016.11.023

[9] Yakubiv V, Zhuk O, Prodanova I. Model of region's balanced agricultural development using the biomass energy potential. Economic Annals-XXI. 2014;**3**(4):1

[10] Wisz G, Nykyruy L, Yakubiv V, Hryhoruk I, Yavorskyi R. Impact of advanced research on development of renewable energy policy: Case of Ukraine. International Journal of Renewable Energy Research (IJRER). 2018;**8**(4):2367-2384

[11] Kouro S, Malinowski M, Gopakumar K, Pou J, Franquelo LG, Wu B, et al. Recent advances and industrial applications of multilevel converters. IEEE Transactions on Industrial Electronics. 2010;**57**(8):2553-2580. DOI: 10.1109/TIE.2010.2049719

[12] Pron A, Gawrys P, Zagorska M, Djurado D, Demadrille R. Electroactive materials for organic electronics: Preparation strategies, structural aspects and characterization techniques. Chemical Society Reviews. 2010;**39**(7):2577-2632

[13] Yuan J, Zhang Y, Zhou L, Zhang G, Yip HL, Lau TK, et al. Single-junction organic solar cell with over 15% efficiency using fused-ring acceptor with electron-deficient core. Joule. 2019;**3**(4):1140-1151. DOI: 10.1016/j.joule.2019.01.004

[14] Hoseinzadeh T, Solaymani S, Kulesza S, Achour A, Ghorannevis Z, Țălu Ş, et al. Microstructure, fractal geometry and dye-sensitized solar cells performance of CdS/TiO_2 nanostructures. Journal of Electroanalytical Chemistry. 2018;**830**:80-87. DOI: 10.1016/j.jelechem.2018.10.037

[15] Sawicka-Chudy P, Starowicz Z, Wisz G, Yavorskyi R, Zapukhlyak Z, Bester M, et al. Simulation of TiO_2/CuO solar cells with SCAPS-1D software. Materials Research Express. 2019;**6**(8):085918. DOI: 10.1088/2053-1591/ab22aa

[16] Skompska M, Zarębska K. Electrodeposition of ZnO nanorod

arrays on transparent conducting substrates—A review. Electrochimica Acta. 2014;**127**:467-488. DOI: 10.1016/j.electacta.2014.02.049

[17] Witkowski BS. Applications of ZnO nanorods and nanowires—A review. Acta Physica Polonica A. 2018;**134**(6): 1226-1246. DOI: 10.12693/APhysPolA.134.1226

[18] Iwan A, Boharewicz B, Parafiniuk K, Tazbir I, Gorecki L, Sikora A, et al. New air-stable aromatic polyazomethines with triphenylamine or phenylenevinylene moieties towards photovoltaic application. Synthetic Metals. 2014;**195**:341-349. DOI: 10.1016/j.synthmet.2014.07.004

[19] Koen D, Zabierowski P, Burgelman M. Modeling metastabilities in chalcopyrite-based thin film solar cells. Journal of Applied Physics. 2012;**111**(4):043703. DOI: 10.1063/1.3686651

[20] Igalson M, Zabierowski P. Transient capacitance spectroscopy of defect levels in CIGS devices. Thin Solid Films. 2000;**361**:371-377. DOI: 10.1016/S0040-6090(99)00822-6

[21] Ceran B, Szczerbowski R. Energy cost analysis by hybrid power generation system. IOP Conference Series: Earth and Environmental Science. 2019;**214**(1):012001

[22] Paska J, Biczel P, Kłos M. Hybrid power systems–An effective way of utilising primary energy sources. Renewable Energy. 2009;**34**(11): 2414-2421. DOI: 10.1016/j.renene.2009.02.018

[23] Latała H, Kurpaska S, Kwaśniewski D. Theoretical and real efficiency of non-silicon PV modules in a 3-year cycle. In: 2019 Applications of Electromagnetics in Modern Engineering and Medicine (PTZE); June 2019. IEEE; 2019. pp. 93-96

[24] Kurz D, Nawrowski R. Thermal time constant of PV roof tiles working under different conditions. Applied Sciences. 2019;**9**(8):1626. DOI: 10.3390/app9081626

[25] Pihosh Y, Turkevych I, Mawatari K, Uemura J, Kazoe Y, Kosar S, et al. Photocatalytic generation of hydrogen by core-shell $WO_3/BiVO_4$ nanorods with ultimate water splitting efficiency. Scientific Reports. 2015;**5**:11141. DOI: 10.1038/srep11141 (2015)

[26] Kislyuk VV, Dimitriev OP. Nanorods and nanotubes for solar cells. Journal of Nanoscience and Nanotechnology. 2008;**8**(1):131-148. DOI: 10.1166/jnn.2008.N16

[27] Boiko BT, Khripunov GS, Yurchenko VB, Ruda HE. Photovoltaic properties in CdS/CdTe thin-film heterosystems with graded-gap interfaces. Solar Energy Materials and Solar Cells. 1997;**45**(4):303-308. DOI: 10.1016/S0927-0248(96)00068-2

[28] Brus VV. On quantum efficiency of nonideal solar cells. Solar Energy. 2012; **86**(2):786-791. DOI: 10.1016/j.solener.2011.12.009

[29] Khrypunov GS, Chernykh EP, Kovtun NA, Belonogov EK. Flexible solar cells based on cadmium sulfide and telluride. Semiconductors. 2009;**43**(8): 1046-1051. DOI: 10.1134/S1063782609080156

[30] Kosyachenko LA, Savchuk AI, Grushko EV. Dependence of efficiency of thin-film CdS/CdTe solar cell on parameters of absorber layer and barrier structure. Thin Solid Films. 2009;**517**(7): 2386-2391. DOI: 10.1016/j.tsf.2008.11.012

[31] Kosyachenko LA, Grushko EV, Motushchuk VV. Recombination losses in thin-film CdS/CdTe photovoltaic devices. Solar Energy Materials and Solar Cells. 2006;**90**(15):2201-2212. DOI: 10.1016/j.solmat.2006.02.027

[32] Saliy YP, Nykyruy LI, Yavorskyi RS, Adamiak S. The surface morphology of CdTe thin films obtained by open evaporation in vacuum. Journal of Nano- and Electronic Physics. 2017;**9**: 05016. DOI: 10.21272/jnep.9(5).05016

[33] Wisz G, Virt I, Sagan P, Potera P, Yavorskyi R. Structural, Optical and electrical properties of zinc oxide layers produced by pulsed laser deposition method. Nanoscale Research Letters. 2017;**12**:253-259. DOI: 10.1186/s11671-017-2033-9

[34] Shmid V, Kuryliuk V, Nadtochiy A, Korotchenkov O, Li PW. Improving photoelectric energy conversion by structuring Si surfaces with Ge quantum dots. In: 2019 IEEE 39th International Conference on Electronics and Nanotechnology (ELNANO); April 2019. IEEE; 2019. pp. 92-96

[35] Nykyrui L, Saliy Y, Yavorskyi R, Yavorskyi Y, Schenderovsky V, Wisz G, et al. CdTe vapor phase condensates on (100) Si and glass for solar cells. In: 2017 IEEE 7th International Conference Nanomaterials: Application & Properties (NAP2017); September 2017; Odessa. Ukraine: IEEE; 2017. p. 01PCSI26-1

[36] Dyvak M, Górecki K, Zarebski J, Porplytsya N, Dqbrowski J, Krac E. Mathematical model of weather conditions influence on properties of photovoltaic installation and method of its identification. In: 2019 9th International Conference on Advanced Computer Information Technologies (ACIT); June 2019. IEEE. pp. 35-39

[37] Nykyruy LI, Yavorskyi RS, Zapukhlyak ZR, Wisz G, Potera P. Evaluation of CdS/CdTe thin film solar cells: SCAPS thickness simulation and analysis of optical properties. Optical Materials. 2019;**92**:319-329. DOI: 10.1016/j.optmat.2019.04.029

[38] Vasko P, Verbovij A, Moroz A, Pazych S, Ibragimova M, Sahno L. Concept of accumulation of energy from photovoltaic and wind power plants by means of seawater pumped hydroelectric energy storage. In: 2019 IEEE 6th International Conference on Energy Smart Systems (ESS); April 2019. Vol. 2019. IEEE. pp. 188-191

[39] Heremans JP, Wiendlocha B, Chamoire AM. Resonant levels in bulk thermoelectric semiconductors. Energy & Environmental Science. 2012; **5**(2):5510-5530. DOI: 10.1039/c1ee02612g

[40] Kutorasinski K, Wiendlocha B, Kaprzyk S, Tobola J. Electronic structure and thermoelectric properties of n-and p-type SnSe from first-principles calculations. Physical Review B. 2015;**91**(20):205201. DOI: 10.1103/PhysRevB.91.205201

[41] Świrkowicz R, Wierzbicki M, Barnaś J. Thermoelectric effects in transport through quantum dots attached to ferromagnetic leads with noncollinear magnetic moments. Physical Review B. 2009;**80**(19):195409. DOI: 10.1103/PhysRevB.80.195409

[42] Paschen S, Carrillo-Cabrera W, Bentien A, Tran VH, Baenitz M, Grin Y, et al. Structural, transport, magnetic, and thermal properties of Eu 8 Ga 16 Ge 30. Physical Review B. 2001;**64**(21): 214404. DOI: 10.1103/PhysRevB.64.214404

[43] Wojciechowski KT. Effect of tellurium doping on the thermoelectric properties of $CoSb_3$. Materials Research Bulletin. 2002;**37**(12):2023-2033. DOI: 10.1016/S0025-5408(01)00758-9

[44] Jaworski CM, Wiendlocha B, Jovovic V, Heremans JP. Combining alloy scattering of phonons and resonant electronic levels to reach a high thermoelectric figure of merit in PbTeSe and PbTeS alloys. Energy & Environmental Science. 2011;**4**(10): 4155-4162. DOI: 10.1039/c1ee01895g

[45] Wojciechowski K, Godlewska E, Mars K, Mania R, Karpinski G, Ziolkowski P, et al. Characterization of thermoelectric properties of layers obtained by pulsed magnetron sputtering. Vacuum. 2008;**82**(10): 1003-1006. DOI: 10.1016/j.vacuum.2008.01.039

[46] Parashchuk T, Dashevsky Z, Wojciechowski K. Feasibility of a high stable PbTe: In semiconductor for thermoelectric energy applications. Journal of Applied Physics. 2019; **125**(24):245103. DOI: 10.1063/1.5106422

[47] Radchenko M, Lashkarev G, Baibara O, Bugaiova M, Stelmakh Y, Krushynska L, et al. Electronic transport and magnetic properties of Co/SiO_2 magnetic nanocomposites. Physica Status Solidi (b). 2019;**256**(11):1900145. DOI: 10.1002/pssb.201900145

[48] Parasyuk OV, Voronyuk SV, Gulay LD, Davidyuk GY, Halka VO. Phase diagram of the CuInS2–ZnS system and some physical properties of solid solutions phases. Journal of Alloys and Compounds. 2003;**348**(1-2):57-64. DOI: 10.1016/S0925-8388(02)00860-5

[49] Scherrer H, Vikhor L, Lenoir B, Dauscher A, Poinas P. Solar thermolectric generator based on skutterudites. Journal of Power Sources. 2003;**115**(1):141-148. DOI: 10.1016/S0378-7753(02)00597-9

[50] Gürth M, Rogl G, Romaka VV, Grytsiv A, Bauer E, Rogl P. Thermoelectric high ZT half-Heusler alloys Ti1−x−yZrxHfyNiSn ($0\leq x\leq 1$; $0\leq y\leq 1$). Acta Materialia. 2016;**104**: 210-222. DOI: 10.1016/j.actamat.2015.11.022

[51] Sharapov SG, Varlamov AA. Anomalous growth of thermoelectric power in gapped graphene. Physical Review B. 2012;**86**(3):035430. DOI: 10.1103/PhysRevB.86.035430

[52] Zemskov VS, Belaya AD, Beluy US, Kozhemyakin GN. Growth and investigation of thermoelectric properties of Bi–Sb alloy single crystals. Journal of Crystal Growth. 2000;**212** (1-2):161-166. DOI: 10.1016/S0022-0248(99)00587-4

[53] Rogacheva EI, Tavrina TV, Nashchekina ON, Grigorov SN, Nasedkin KA, Dresselhaus MS, et al. Quantum size effects in PbSe quantum wells. Applied Physics Letters. 2002; **80**(15):2690-2692. DOI: 10.1063/1.1469677

[54] Nykyruy L, Ruvinskiy M, Ivakin E, Kostyuk O, Horichok I, Kisialiou I, et al. Low-dimensional systems on the base of PbSnAgTe (LATT) compounds for thermoelectric application. Physica E: Low-dimensional Systems and Nanostructures. 2019;**106**:10-18. DOI: 10.1016/j.physe.2018.10.020

[55] Rogacheva EI, Nashchekina ON, Meriuts AV, Lyubchenko SG, Dresselhaus MS, Dresselhaus G. Quantum size effects in n-PbTe/p-SnTe/n-PbTe heterostructures. Applied Physics Letters. 2005;**86**(6):063103. DOI: 10.1063/1.1862338

[56] Snarskii AA, Bezsudnov IV, Sevryukov VA, Morozovskiy A, Malinsky J. Transport processes in macroscopically disordered media. In: From Mean Field Theory to Percolation. LLC New York: Springer Science +Business Media; 2007

[57] Anatychuk LI, Luste OJ, Kuz RV. Theoretical and experimental study of thermoelectric generators for vehicles. Journal of Electronic Materials. 2011; **40**(5):1326-1331. DOI: 10.1007/s11664-011-1547-7

[58] Semenyuk V. Miniature thermoelectric modules with increased cooling power. In: 2006 25th International Conference on Thermoelectrics. IEEE; 2006. pp. 322-326

[59] Tkatc VI, Limanovskii AI, Denisenko SN, Rassolov SG. The effect of the melt-spinning processing parameters on the rate of cooling. Materials Science and Engineering: A. 2002;**323**(1-2):91-96. DOI: 10.1016/S0921-5093(01)01346-6

[60] Martynova KV, Rogacheva EI. Thermoelectric properties of cold pressed samples of semiconductor (Bi). Functional Materials. 2018;**1**:55. DOI: 10.1407/fm25.01.54

[61] Horichok IV, Prokopiv VV, Zapukhlyak RI, Matkivskyj OM, Semko TO, Savelikhina IO, et al. Effects of oxygen interaction with PbTe surface and their influence on thermoelectric material properties. Journal of Nano- and Electronic Physics. 2018;**10**(5): 05006(5pp). DOI: 10.21272/jnep.10(5).05006

[62] Anatychuk L, Vikhor L, Kotsur M, Kobylianskyi R, Kadeniuk T. Optimal control of time dependence of temperature in thermoelectric devices for medical purposes. International Journal of Thermophysics. 2018;**39**(9): 108. DOI: 10.1007/s10765-018-2430-z

[63] Prokopiv VV, Nykyruy LI, Voznyak OM, Dzundza BS, Horichok IV, Yavorskyi YS, et al. The thermoelectric solar generator. Physics and Chemistry of Solid State. 2017; **18**(3):372-375. DOI: 10.15330/pcss.18.3.372-375

[64] Faruk O, Bledzki AK, Fink HP, Sain M. Biocomposites reinforced with natural fibers: 2000–2010. Progress in Polymer Science. 2012;**37**(11):1552-1596. DOI: 10.1016/j.progpolymsci.2012.04.003

[65] Antczak MS, Kubiak A, Antczak T, Bielecki S. Enzymatic biodiesel synthesis—Key factors affecting efficiency of the process. Renewable Energy. 2009;**34**(5): 1185-1194. DOI: 10.1016/j.renene.2008.11.013

[66] Bhatnagar A, Sillanpää M, Witek-Krowiak A. Agricultural waste peels as versatile biomass for water purification—A review. Chemical Engineering Journal. 2015;**270**:244-271. DOI: 10.1016/j.cej.2015.01.135

[67] Budzianowski WM. A review of potential innovations for production, conditioning and utilization of biogas with multiple-criteria assessment. Renewable and Sustainable Energy Reviews. 2016;**54**:1148-1171. DOI: 10.1016/j.rser.2015.10.054

[68] Dauber J, Brown C, Fernando AL, Finnan J, Krasuska E, Ponitka J, et al. Bioenergy from "surplus" land: Environmental and socio-economic implications. BioRisk. 2012;**7**:5. DOI: 10.3897/biorisk.7.3036

[69] Adamczak M, Bornscheuer UT, Bednarski W. The application of biotechnological methods for the synthesis of biodiesel. European Journal of Lipid Science and Technology. 2009;**111**(8): 800-813. DOI: 10.1002/ejlt.200900078

[70] Budzianowski WM. Negative carbon intensity of renewable energy technologies involving biomass or carbon dioxide as inputs. Renewable and Sustainable Energy Reviews. 2012; **16**(9):6507-6521. DOI: 10.1016/j.rser.2012.08.016

[71] Godlewska P, Schmidt HP, Ok YS, Oleszczuk P. Biochar for composting improvement and contaminants reduction. A review. Bioresource Technology. 2017;**246**:193-202. DOI: 10.1016/j.biortech.2017.07.095

[72] Urbaniec K, Bakker RR. Biomass residues as raw material for dark hydrogen fermentation—A review. International Journal of Hydrogen Energy. 2015;**40**(9):3648-3658. DOI: 10.1016/j.ijhydene.2015.01.073

[73] Grembecka M. Sugar alcohols—Their role in the modern world of

sweeteners: A review. European Food Research and Technology. 2015;**241**(1): 1-14. DOI: 10.1007/s00217-015-2437-7

[74] Budzianowski WM, Budzianowska DA. Economic analysis of biomethane and bioelectricity generation from biogas using different support schemes and plant configurations. Energy. 2015;**88**: 658-666. DOI: 10.1016/j.energy. 2015.05.104

[75] Li Y, Rezgui Y, Zhu H. District heating and cooling optimization and enhancement—Towards integration of renewables, storage and smart grid. Renewable and Sustainable Energy Reviews. 2017;**72**:281-294. DOI: 10.1016/j.rser.2017.01.061

[76] Hurmak N, Yakubiv V. Efficiency of intermediary activity of agricultural enterprises: Methods and assessment indicators. Bulgarian Journal of Agricultural Science. 2017;**23**(5):712-716

[77] Al-Mansour F, Zuwala J. An evaluation of biomass co-firing in Europe. Biomass and Bioenergy. 2010; **34**(5):620-629. DOI: 10.1016/j. biombioe.2010.01.004

[78] Igliński B, Iglińska A, Kujawski W, Buczkowski R, Cichosz M. Bioenergy in Poland. Renewable and Sustainable Energy Reviews. 2011;**15**(6):2999-3007. DOI: 10.1016/j.rser.2011.02.037

[79] Szafranko E. Comparative analysis of renewable energy resources potentially accessible in rural areas. In: 18th International Scientific Conference Engineering for Rural Development. 2019

[80] Wasiak AL. Effect of biofuel production on sustainability of agriculture. Procedia Engineering. 2017; **182**:739-7467. DOI: 10.1016/j. proeng.2017.03.192

[81] Marks-Bielska R, Kurowska K, Kryszk H. The role of agriculture in ensuring the energy security in Poland Research for Rural Development. In: Annual 20th International Scientific Conference Proceedings. Vol. 2. 2014. pp. 191-198

[82] Bielski S, Marks-Bielska R. The potential for agricultural biogas production in Poland. In: International Multidisciplinary Scientific GeoConference Surveying Geology and Mining Ecology Management (SGEM). Vol. 1(4). 2015. pp. 575-580

[83] Wrobel M, Mudryk K, Jewiarz M, Knapczyk A. Impact of raw material properties and agglomeration pressure on selected parameters of granulates obtained from willow and black locust biomass. In: 17th International Scientific Conference Engineering for Rural Development. Jelgava, Latvia; 23-25 May 2018. DOI: 10.22616/ERDev2018. 17.N542

[84] Baum R, Wajszczuk K, Pepliński B, Wawrzynowicz J. Potential for agricultural biomass production for energy purposes in Poland: A review. Contemporary Economics. 2013;**7**(1): 63-74

[85] Frank S, Havlík P, Soussana JF, Levesque A, Valin H, Wollenberg E, et al. Reducing greenhouse gas emissions in agriculture without compromising food security? Environmental Research Letters. 2017;**12**(10):105004. DOI: 10.1088/1748-9326/aa8c83

[86] Ivanov Y, Pyatnichko O, Zhuk H, Onopa L, Soltanibereshne M. Extraction of carbon dioxide from gas mixtures with amines absorbing process. Energy Procedia. 2017;**128**:240-247. DOI: 10.1016/j.egypro.2017.09.062

[87] Panchuk M, Kryshtopa S, Shlapak L, Kryshtopa L, Panchuk A, Yarovyi V, et al. Main trends of biofuels production in Ukraine. Transport Problems. 2017;**12**(4): 15-26. DOI: 10.20858/tp.2017.12.4.2

[88] Voytenko Y. Pathways for agro-bioenergy transition in Ukraine. Biofuels, Bioproducts and Biorefining. 2012;**6**(2):124-134. DOI: 10.1002/bbb.347

[89] Raslavičius L, Grzybek A, Dubrovin V. Bioenergy in Ukraine—Possibilities of rural development and opportunities for local communities. Energy Policy. 2011;**39**(6):3370-3379. DOI: 10.1016/j.enpol.2011.03.032

[90] Karpenko V, Burliai O, Mostoviak I. Economy's agricultural sector potential in Ukrainian energy self-sufficiency forming. Economic Annals-XXI. 2015; **155**(11-12):55-58

[91] Hamulczuk M, Makarchuk O, Sica E. Searching for market integration: Evidence from Ukrainian and European Union rapeseed markets. Land Use Policy. 2019;**87**:104078. DOI: 10.1016/j.landusepol.2019.104078

[92] Yevdokimov Y, Chygryn O, Pimonenko T, Lyulyov O. Biogas as an alternative energy resource for Ukrainian companies: EU experience. Innovative Marketing. 2018;**14**:7-15. DOI: 10.21511/im.14(2).2018.01

[93] Cebula J, Chygryn O, Chayen SV, Pimonenko T. Biogas as an alternative energy source in Ukraine and Israel: Current issues and benefits. International Journal of Environmental Technology and Management. 2018;**21**: 421-438. DOI: 10.1504/IJETM.2018.100592

[94] Tsapko VG, Chudnovets AJ, Sterenbogen MJ, Papach VV, Dutkiewicz J, Skórska C. Exposure to bioaerosols in the selected agricultural facilities of the Ukraine and Poland—A review. Annals of Agricultural and Environmental Medicine. 2011;**18**:19-27

[95] Lypchuk V, Syrotiuk K. Estimation of energy potential of agricultural enterprise biomass. In: E3S Web of Conferences: Energy and Fuels 2017; 15 March 2017. Vol. 14. EDP Sciences. p. 02017. DOI: 10.1051/e3sconf/20171402017

[96] Gołębiowska J, Żelazna A. Experimental investigation of thermoelectric cooling system with heat recovery. In: E3S Web of Conferences. Vol. 100. EDP Sciences; 2019. p. 00020. DOI: 10.1051/e3sconf/201910000020

[97] Worsztynowicz B. The influence of the drive type in city bus on the cooling system parameters. In: IOP Conference Series: Materials Science and Engineering. Vol. 421(4). IOP Publishing; 2018. p. 042084. DOI: 10.1088/1757-899X/421/4/042084

[98] Kocemba I, Rynkowski J, Arabczyk W. The thermoelectric sensor for controlling the gas nitriding process. Sensors and Actuators A: Physical. 2019; **288**:144-148. DOI: 10.1016/j.sna.2019.02.005

[99] Ziółkowski A, Fuć P, Dobrzyński M. Analysis of the construction of TEG thermoelectric generator using CFD tools. In: AIP Conference Proceedings. Vol. 2078(1). AIP Publishing; 2019. p. 020052. DOI: 10.1063/1.5092055

[100] Aizpurua ML, Leonowicz Z. Advanced solar energy systems with thermoelectric generators. In: Proceeding of 2018 IEEE International Conference on Environment and Electrical Engineering and 2018 IEEE Industrial and Commercial Power Systems Europe (EEEIC/I&CPS Europe). IEEE; 2018. pp. 1-4. DOI: 10.1109/EEEIC.2018.8493685

Section 2

Sustainability and Reducing Emissions

Chapter 7

Operational Challenges towards Deployment of Renewable Energy

Pankaj Kumar, Kumar Avinash Chandra, Sanjay Patel, Nitai Pal, Mohit Kumar and Himanshu Sharma

Abstract

This chapter focuses on the feasibility analysis and different challenges toward deployment of renewable energy to achieve global sustainability. The analysis emphasizes that the technological advancement, cost, and efficiency are the basic elements for mass adaptation of renewable energy. At the same time, huge available resources, favorable economies, and large social-economic benefits attract major parts of the globe toward the transition from conventional to renewable energy. The proposed chapter also indicates the major options and barriers toward the deployment of different renewable energies in India, which will act as a catalyst to achieve the India's dream renewable energy target of 175 GW by 2022. In the current era of modern technologies, highly CO_2 releasing countries like India and China demand a wide range of renewable energy integration into their power generation portfolios to meet the requirements of global sustainability. Therefore, the proposed chapter will also provide a strong base of energy security for upcoming generations.

Keywords: solar energy, wind energy, hydro power, challenges, deployment

1. Introduction

Energy is the one of the most important building blocks of human development and as such acts as the key factor in determining the economic development of all the countries. It also acts as one of the most critical components of infrastructure development, which becomes crucial for the economic growth and welfare of any nation. The existence and development of adequate infrastructure are indispensable for sustained development and growth of the economy of a country. To meet the energy demands of developing nations, the energy sector has observed rapid extension. It is important to note that the nonrenewable sources are notably consumed by human use, whereas renewable resources are produced by continuous process that can sustain indefinite human exploitation.

In the COP21 agreement, India agreed to transform its entire energy system to a zero emission-based energy system. To work on the same, the government has done a radical transformation in its green energy sector to achieve the target. In this regard, government has more focused on the exploitation of solar energy. The government of India has also launched a solar international alliance, which will help to transfer the various solar technologies with each member country [1].

Most of the renewable energy (RE) is derived directly or indirectly from the sun. Sunlight can directly be captured using solar photovoltaic cells to produce electrical energy. The Sun's heat is also driving the wind, which in turn runs the turbine to produce the energy [2]. Necessarily not all REs depend on the sun. The geo-thermal energy relies on the earth's internal heat to produce heat, while tidal energy occurs due to the moon's gravitational pull.

The nonconventional energy systems are the hot and emerging wing for world's future energy requirement and global climate change for the following two reasons:

i. RE system produces less greenhouse effect than produced by the fossil fuels.

ii. RE system provides the energy source that will not ever deplete.

The nonconventional energy is in booming phase due to the recent innovations, which has brought down the cost to a significantly low level and started to deliver on the promise of clean energy future. The environmental advantage of RE including lower carbon emission and reduced air pollution has been widely known for decades. Its numerous socioeconomic benefits, however, have only been apparent in recent decades as the development of RE technology has been more widespread.

However, RE systems are better for the environment and produce less emission than conventional energy sources [3]. But, many of these resources still face difficulties in being deployed over at large scale due to technological barriers, high initial investment, and intermittency challenges [4]. Moreover, other different obstacles have also been broadly discussed in upcoming sections.

It is important to clarify that terms "RE," "green energy," and "clean energy" are not interchangeable in all cases, for example, a "clean coal plant" is simply a coal plant with emission reduction technology. The coal plant itself is still not a "RE source." "Green energy" is a subset of RE, which boasts low or zero emissions and low environmental impacts to system as land and water [5].

RE systems are increasingly displacing dirty fossil fuels in the power sector, offering the benefits of low carbon emission and diminishing the other adverse environmental impacts. But not all sources of energy marketed as RE are beneficial to the environment. Biomass and large hydroelectric dams create difficult tradeoffs when considering the impact on wildlife, climate change, and other issues [6].

2. Types of RE sources

RE referred to as clean energy, which comes from natural sources or processes that are constantly replenished. RE is often thought of as new technology, harnessing nature's power that has long been used for heating, transportation, lighting, and so on.

2.1 Solar energy

Humans have been harnessing solar energy for thousands of years to grow crops, to stay warm, or to dry foods. According to national RE laboratory report, "More energy from the sun falls on the earth in an hour than is used by everyone in a year." Modern advance technologies offer utilization of solar energy in many ways: (a) to heat buildings, (b) to warm water, and (c) to generate power using solar photovoltaic cells made of Silicon.

Solar energy contributes roughly 3% of total electrical generation of India. But, nearly fourth of all new generating capacities came from solar in the fiscal year

Resources	Total installed capacity (MW)	2022 target (MW)
Wind	36,368	60,000
Solar	29,549	1,00,000
Biomass	9806	10,000 (biomass and waste to power)
Waste to power	138	
Small hydro	4604	5000
Total	80,467	1,75,000

Table 1.
Current installed grid interactive renewable capacity of India [13].

2018–2019. A total figure of 39,2668 GWh of solar energy has been produced in the last year using solar energy technology [7].

Solar energy systems do not produce air pollutants or greenhouse gases, and as long as they are responsibly sited, most solar panels have few environmental impacts beyond the manufacturing process.

2.2 Wind energy

Wind, which accounts for a little more than 4% of country's total electrical generation, has become the cheapest energy source in parts of the country. Top wind power states in India including Gujarat, Andhra, and Tamilnadu exploit the wind energy through various technologies like off- and on-shore technologies. The wind turbines can be placed anywhere with high wind speeds such as hilltops, open plains, or offshore in open water [7]. India has significantly increased his installed wind energy capacity and has claimed to be the fourth largest wind power producer of the world [8]. **Table 1** shows current installed capacity of wind energy in India.

2.3 Hydropower

Hydropower is the largest renewable energy source for electricity in India, which accounts for the around 11% of country's total electricity generation. Typically, hydropower relies on fast moving water in a large river or rapidly descending water from a high point and converts the force of that water into electricity by spinning a generator's turbine blades.

Large hydroelectric plants or mega dams are considered often to be a part of the RE source. Mega dams divert and reduce natural flows, restricting access for animals and human populations that rely on rivers. Small hydroelectric plants of rating less than 40 MW are less harmful for the environmental damage because it diverts only fraction of flow [9].

2.4 Biomass energy

Biomass is organic material that comes from plants and animals, which includes crops, waste wood, trees, manure, and so on for its energy resources. When biomass is burned, the chemical energy is released as heat. The produced heat can be used to generate electricity with a steam turbine. Biomass is often mistakenly described as a clean renewable fuel and a green alternative to coal and other fossil fuels because recent science shows that many forms of biomass especially from forests produce higher carbon emissions than fossil fuels [10].

2.5 Geothermal energy

The earth's core is considered to be as hot as the sun's surface, due to the slow decay of radioactive particles in rocks at the center of the planet. Drilling deep wells brings very hot underground water to the surface as a hydrothermal resource, which is then pumped through a turbine to create electricity. Geothermal plants have low emissions. There are ways to create geothermal plants where there are not underground reservoirs. But, the major barrier is the risk of earthquake in concerned areas. Somewhere, geological hot spots have been already observed [11].

2.6 Ocean energy

The ocean energy is usually deployed by four technologies: current energy technology, wave energy technology, tidal energy technology, and ocean thermal energy technology (OTET). Marine current technology is used to move the blades of the rotor of the machine to produce electricity. Wave energy technologies are generally used to convert the wave energy into electricity [12].

Tidal cycle can be observed due to the moon's gravitational force. These technologies are in their early stage of development. The drawback of the tidal energy plants appears in its initial investment because it needs to construct high civil infrastructure. Western part of India has huge potential of tidal energy in which Gulf of Kutch and Gulf of Cambay are the prime locations. Some tidal energy approaches may harm wildlife, such as tidal barrages, which work like dams and are located in an ocean bay or lagoon. Like tidal power, wave power relies on dam-like structures or ocean floor–anchored devices on or just below the water's surface. OTET utilizes ocean temperature difference to mine energy. The minimum temperature difference required for this purpose is 20°C [12].

3. RE resources with the Indian prospective

The association of mankind with energy is as historical as discovery of fire or the steam. Different battles have been fought for the energy security ever since the industrial revolution or the discovery of the scientific evolution. The nineteenth and twentieth centuries' entanglement with the oil and coal is slowly tapping feet with solar and wind or any other renewable resources for the energy. Global climate change is also one cause for the switchover from conventional energy resources to nonconventional energy resources such as solar and wind. However, the transformation or switchover would not happen disruptively as expected due to technological lack in conventional generation methods and fuels. It will require huge investments and technology transfer, and the equation is unbalanced specially when in reference with the developing nations such as India.

The rapid increase of the Indian population demands the rapid expansion of energy sector. The advancement of human life has directly affected the rate of consumption of energy. The application of advance luxury equipment in the field of transportation, agriculture, domestic application, and industries has significantly increases the demand of electricity. In order to meet the energy demand, the developing country like India is mainly dependent on fossils fuels. In recent years, India has faced many terrible conditions in terms of weather change. In such conditions, RE has been playing an important role in India's energy planning. The importance of RE sources in the transition to a sustainable energy base was recognized in the early 1970s. In the current era, RE is being used increasingly in four distinct areas: (a) power generation, (b) heating and cooling, (c) transport, and (d) rural/off-grid energy services. The ministry of new and renewable energy (MNRE) in India has been facilitating the renewable energy implementation of many programs including harnessing renewable power, RE for village areas for lighting, cooking and motive power, use of RE in cities for lightening, industrial and commercial applications, and so on.

RE resources account for 13.5% of the world's total energy supply and 22.5% of the world's electricity [3]. However, the power generation from renewable sources is on the rise in India, with the share of RE in the country's total installed capacity rising from 7.8% in 2008 to around 13% in 2014 [13]. **Table 1** shows the total installed capacity of RE in India till March 31, 2019. It shows that India has an installed capacity of 80.474 GW energy. Among these, wind is the largest contributor and stands at around 36.625 GW of installed capacity making India the world's fourth largest wind energy producer. Moreover, small hydropower, bioenergy, and solar energy constitute the remaining capacity of 29.55 GW.

The geographical location of India is very much favorable for generation of wind energy. Wind power capacity is mainly spread across the south, west, north, and eastern region of India. The development of wind energy system in India has started a many decade ago, when Maneklal Sankalchand Thacker, a distinguished power engineer, initiated a project with the Indian council of scientific and industrial research (CSIR) to explore the possibilities of harnessing wind power in the country. The heart of wind energy generation in India exists in the state of Tamilnadu, Gujarat, Maharashtra, Karnataka, Rajasthan, Madhya Pradesh, and Andhra Pradesh [14].

Solar energy is being realized as a one of the fastest growing clean energy sectors in India. A rapid growth of 0.522 GW of installed capacity has been observed in last 4 months. At the same time, India has the lowest per MW capital cost for the installation of the solar power plants. Looking at the geographical scenario, most part of India covers 300 clear sunny days in a year. The calculated solar energy incidence on India's land area is about 5000 trillion kilowatt-hours (kWh) per year. The available solar energy in a single year exceeds the possible energy output of all of the fossil fuel reserves in India. The daily average solar power generation capacity in India is 0.20 kWh/m^2 of used land area, which is equivalent to 1400–1800 peak rated capacity operating hours in a year with available commercially proven technology. Therefore, India has a huge opportunity for the deployment of solar energy [15].

India is the seventh largest producer of hydroelectric power in the world. As of April 30, 2017, India's installed utility-scale hydroelectric capacity was 44,594 MW, which was 13.5% of its total utility power generation capacity. This shows that hydroelectric sector has a big contribution in the total installed power generation capacity. Additionally, smaller hydroelectric power units with a total installed capacity of 4380 MW, which were 1.3% of its total utility power generation capacity, have been added in its capacity. The National Hydroelectric Power Corporation (NHPC), Northeast Electric Power Company (NEEPCO), Satluj Jal Vidyut Nigam Ltd. (SJVNL), THDC, and NTPC-Hydro are some of major public sector companies producing hydroelectric power in India. Bhakra Beas Management Board (BBMB), a state-owned enterprise in north India, has a vast potential to generate hydropower, but rivers such as Godavari, Mahanadi, Nagavali, Vamsadhara, and Narmada river basins have not been developed on a major scale due to protest from the tribal population. This shows that hydroelectric sector is highly underutilized in India [16].

The RE sector has emerged as a significant role in the country affecting the power generation capacity and in good view. This RE sector supports the government's agenda of sustainable development while becoming an integral part in meeting nation's energy demand. For recent years in past, the Government of India (GoI) has taken several initiatives such as introduction of solar parks, organizing global investor's meet, launching a massive grid connected roof top solar program for green energy corridor, and program to train 50,000 people for solar installation as Solar Mitra Scheme (SMS). Also, steps have been taken toward renewable generation obligations on new thermal and lignite plants and so on.

Apart from many obstacles of the deployment of RE resources in India, the country has the following advantages:

a. *Robust demand:* With the growing India economy, the electricity consumption is projected to reach 15,280 TWh by 2040.

b. *Increasing investments:* With GoI's ambitious targets, the sector has become attractive to various investors from foreign as well as Indian.

c. *Competitive advantage:* India is blessed with plenty of sunlight throughout the year, huge hydropower potential, various wind power generation sites, and so on.

4. Dream energy project of India

One of the reasons of increase in the demand for power in India is the growing population. Consequently, it has become necessary to explore all the possible nonconventional sources of energy to meet the issue of global climate change and to overcome the problem of energy security in the country. Energy systems around the world are witnessing toward transformations at an unprecedented rate. There has been a very rapid increase in RE technologies in last few decades. The rapid increase of energy demand forces every country to make efforts to move away from a fossil fuel-based energy generation portfolio. In this regard, India has made a remarkable dream energy project of 175 GW power generation from RE by 2022.

As depicted in **Table 1**, India has targeted an installed capacity of 175 GW by 2022. In this project, large hydropower project has not been included. If we will add the large hydropower project, the figure will rise to 225 GW, which will be a praiseworthy achievement for the country. Due to the huge feasibility of solar energy, more emphasis has been given to solar energy, which includes 100 GW of total installed capacity. It will comprise 60 GW from ground-mounted, grid-connected projects, and 40 GW from solar rooftop projects. Wind power projects will contribute 60 GW of total installed capacity. The ministry is implementing a wide range of schemes with financial support and conducive policies to achieve this target. The largest ever wind power capacity addition of 3423 MW and solar power capacity addition of 3019 MW were made in the fiscal year 2015–2016. For the first time in the year 2015–2016, the largest solar power project capacity of 20,904 MW was tendered, and 31,472 solar pumps were installed, which are higher than the total number of pumps installed during the last 24 years [17]. The installed pumps have a feature of powering through solar energy. It runs on the electricity produced through solar panel [18]. The ministry has been facilitating the implementation of broad spectrum programs for the accelerated deployment of RE. It includes the use of RE to rural and urban areas for lighting, cooking, motive power, industrial and commercial applications, and so on. To maximize the use of solar power, the ministry is making all efforts in rural and urban areas to create awareness of the benefits of solar power.

In India, the increasing addition of RE sources to its energy mix and the imminent advent of electric mobility have driven the market to look at grid-scale energy storage solutions. The grid-scale energy storage would support the sustainable growth of renewable integration and aid grid balancing efforts by increasing energy security and reliability of the nation. Most global efforts toward climate change adaptation have happened in the form of RE additions to the energy sector.

The Indian government's vision of greening the power sector will require large-scale adoption of energy storage technologies. Unfortunately, it is being met by a host of hurdles relatively high cost of technology and a lack of sector experience. The development of the energy storage sector was imminent, and unsurprisingly, the global market in 2018 was estimated at 12 GWh. This sector is very important if India is to meet its 2022 target of 175 GW. The first tender was released in 2015, and since then, the frequency of tenders with battery energy storage systems (ESS) has steadily increased, highlighting India's motivations. Several tenders for solar plants with ESS were released by the Solar Energy Corporation of India (SECI). For a grid-scale storage, cost incurred per unit energy stored is highly dependent on ramping time, efficiency, and life of storage. Technologies such as pumped-hydro, compressed air, and gravity storage are sound alternatives to battery storage [19].

Thus, it is not easy to achieve the targeted dream energy project of the country. India has to face a number of technical and nontechnical barriers for the completion of its project in the estimated time limit.

5. Challenges in the deployment of RE

The barriers or challenges can be broadly classified into two categories: (a) nontechnical challenges and (b) technical challenges. Apart from this, there also exists some other challenges toward the advancement of RE in India.

5.1 Nontechnical challenges

5.1.1 Initial investment

The investment requirement for wind and solar power-based plants is significantly higher than that of the coal-based plant. The development of a coal-based power plant requires around INR 4 crore per MW. A wind power plant with a capacity utilization of 25% requires an investment of INR 6 crore per MW. The actual investment, at more efficient capacity utilization of 80%, works out to be INR 18 crore per MW. Similarly, the investment in a solar-based plant, with a capacity utilization of 15%, is INR 18 crore. The actual investment, at 80% capacity utilization, is around INR 98 crore. Apart from this, it also requires to invest in research and development field to get new and efficient technologies for better performance.

5.1.2 Land acquisition

It is a very major issue faced by the companies or government prior to the installation of new power plant and transmission line. The factor had slowed down many RE projects in India. For example, the solar power developer company like Mercom in India faced the same problem during the expansion of its large utility scale solar initiatives. Many of the south Indian states such as Karnataka and Tamilnadu take approximately 1 year for the clearance under the act of Land (Ceiling and Regulation). Moreover, the time for the completion of such projects remains only 1.5–2 years. Similar problems arise, if any developer wants to commission a RE project near proximity to airport area. The airport authorities have divided the proximity zones in three levels: red, yellow, and green. For the yellow and green zones, companies have to certify their documents from the survey of India as well as from MNRE, which take more than 1.5 years [20].

The RE sector developer Mercom has also given in his statement in April 2019 that 1.2 GW wind project of SECI was decreased by 50% due to this reason only [20].

5.1.3 Social acceptance

Social acceptance of renewable-based energy system is still not very encouraging in urban India. Despite heavy subsidy being provided by the government for installation of solar water heaters and lighting systems, its installation is still very low. At the same time, rural part of India is facing the lack of adequate knowledge, which leads to the decrease in social acceptance of clean energy sources [21, 22].

There are six major elements of social acceptance of a RE system. These are knowledge of technology, cost, risk, perk, local situation, and decision making. Based on these factors, RE technologies are accepted and rejected in societies. For example, human manure-based biogas is highly unacceptable in the current modern Indian era. People of India treat it as dirty fuel. At the same time, social acceptance of solar energy is on the rise in all parts of India. In this regard, government has projected a target of 50,000 solar photovoltaic-based lighting system and 20,000 solar irrigation pumps for rural India. Government is also supplying 2 lakh advance solar cook stoves for rural India [23].

5.1.4 Lack of skilled manpower

The trained skilled manpower is another area, where developing countries like India need to work. Currently, the Indian RE power sector is facing severe shortage of trained personnel [21]. Due to the lack of skilled personnel, time of completion of project increases, which in turn causes cost overrun.

As per the joint report of the Council of Energy, Environment and Water (CEEW), and Natural Resource Defense Council (NRDC) of India, skilled manpower is the biggest problem for the hiring in recruits. The report also said that approximately 624,000 personnel will be either semiskilled or unskilled, which will work in industry for the completion of India's Dream RE by 2022 [24].

5.2 Technical challenges

5.2.1 Intermittent nature of renewable

The energy generation of traditional fossil-fuel plants is majorly fuel dependent in nature. Hence, consistency and predictability of amount of electricity make it more reliable than nonconventional energy-based plants. Energy provided by conventional energy plants is easily controlled by the control units. For example, energy output from a solar panels can drop without warning due to clouds. Similarly, wind speeds cannot be reliably forecasted. To prepare for this fluctuation in advance, research and investment into energy storage systems are on the rise. Moreover, wind power ramp events are also a major challenge. Therefore, developing energy storage mechanisms is essential for the efficient deployment of RE sources [22]. The grid may not always be able to absorb surplus wind power generated by the uncertain wind speed hike.

The intermittency occurs not only in energy generation but also in equipment's cost. **Table 2** shows a range of solar panel prices of different leading solar companies in India. It widely effects on the initial investment of companies [25].

Companies in India	Price range in INR/W
Adani Solar	18–35
Luminous	24–58
Vikram Solar	19–30
Microtek Solar	25–60
Waaree Solar	19–28
Tata Power Solar	20–62

Table 2.
Range of solar panel prices of different leading manufacturers in India [25].

We can easily analyze that minimum and maximum prices of solar panels are highly intermittent in nature.

5.2.2 Integration of distributed energy systems

For controlling and monitoring purpose of a renewable-based energy park, system requires intelligent tools/software for its efficient operation. But, due to the occurrence of different types of distributed energy resources in the system, it becomes very complex to monitor it during its operation. It results in the difficulty during its integration. Many RE generation sites, such as solar PV and wind farms, are distributed across a wide geographical scope. Therefore, it becomes very difficult to control and monitor without a sophisticated tools in the system. For example, the new project for offshore wind farm in India is currently under construction, which is away from coastline. For a precise and accurate energy management of the aforesaid generation sites, the data from each asset need to be combined into a singular entity [21–26]. The tool should be efficient enough to combine many items of distributed equipment into one system to provide a complete visualization of the grid.

There are many technical concerns that arise during the integration of DG into the system. The first one is the stability. When integration is done, it effects on the rotor angle of the generator. At the same time, it also effects frequency and voltage stability of the system. Hence, DGs either increase or decrease the stability of the system based on the generator rating. However, local voltage perturbation of the grid voltage can be observed due to the intermittency in the injected power from the wind and solar units [27].

The second challenge is the optimal number of placement of the DG units in the system, so that the demand reactive power will be equal to the supply reactive power, which is very important to maintain a healthy voltage level of the system [27].

The another drawback in this regard is the need of protecting devices for DG units, which increases the overall capital cost of the system [27].

5.2.3 Location dependency

Most of the RE power generations are location specific. The feasibility of wind energy system can be seen to the locations having more than the cut-off wind speed. Sometimes, generation sites and load center are far away from each other [26]. Also, transmission of power from generating station to load center produces huge cost overrun. The transmission costs become very high especially in case of offshore wind resources. Hence, these types of technology are not much feasible in land-based transmission lines.

For example, the major load center of the northern part of India lies in Delhi. But the onshore wind power plants are far away from this region. Also, the transmission of hydropower to this location causes cost overrun.

5.2.4 Lack of transmission line infrastructure

Transmission line infrastructure is also a major problem behind the expansion of RE resources in India. Although government has decided to made green energy corridor for this purpose, the biggest challenge is to guarantee that transmission systems must be ready for operation before the completion of RE projects. As the complete execution of transmission projects may take up to 5 years or even more. At the same time, the solar energy-based projects require less time. Therefore, availability of transmission infrastructure is a big problem for the rapid deployment of green energy sources.

As per the CEA, GoI report 2015, Lakshadweep island in India has its 100% energy production through RE sources. This shows a huge possibility of RE from this region, which can be utilized to the other parts of the nation. But the lack of transmission line infrastructure causes its underutilization. Similar example occurs in the states of Himachal Pradesh, Jammu Kashmir, Sikkim, Arunachal Pradesh, and Meghalaya, which have the renewable energy production of 94, 73.41, 71.08, 78.49, and 75.93%, respectively. At the same time, the states such as West Bengal, Jharkhand, and Delhi have their installed capacity of 14.43, 8.42, and 10.26%, respectively [28].

5.2.5 Mismatch in load demand centers and available corridors

There is some unsuitability in the number of load demand centers and the available corridors for RE due to the lack of an effective plan to design a dedicated infrastructure for RE transmission. For example, the 1 GW substation project at Kayathar in Tamil Nadu, India, was scheduled to be start in early 2018. Unfortunately, it could not be commissioned on time because of the independent power producers who are more inclined to transmit the power to Gujarat and Maharashtra, which have adequate load demand centers as compared to the north-eastern states.

5.2.6 Necessity of energy storage system (ESS) and associated challenges

Due to the intermittency of various REs, it is very important to integrate a ESS to get an uninterrupted power supply to the consumers. The key criteria for the choice of ESS are size, application, initial cost, and durability. The major drawback of ESS is maintenance cost, which increases the net maintenance cost of the system and increases the payback period for the system. The operation of ESS is primarily based on its charging and discharging time [27].

There are many challenges associated with the integration of ESS to a renewable integrated power supply system. Among them, most important is drawback of power electronic converter over the power quality of the system. To integrate a ESS in the system, power electronic interface unit becomes necessary. It not only increases the initial investment of the system but also decreases the power quality of the system by injecting harmonics to the system [27].

5.2.7 Other challenges

The lack of proper financial system is also a major obstacle for the expansion of RE in India. Though government provides remarkable subsidies for the solar

Year	2009–2010	2010–2011	2011–2012	2016–2017
Energy requirement (in MU)[a]	820,920	891,203	968,659	1,392,066
Share of RE as mandated under NAPCC (in %)[b]	5%	6%	7%	12%
Quantum of RE required (in MU)	41,046	53,472	67,806	167,048
RE capacity addition targeted by MNRE (in MW)[c]	15542[c]	20,376	25,211	57,000
Solar capacity targeted under JNNSM (in MW)			1000	10,000
Quantum of RE available (in MU)[d]	29,952	39,269	50,514	129,122
Additional RE required to meet RE share mandated under NAPCC (in MU)	11,094	14,203	17,292	37,926

[a]*As per 17th EPS.*
[b]*5% in 2009–2010 and 1% increase each year.*
[c]*As on 31.10.2009.*
[d]*Assuming a capacity utilization factor of 22%.*

Table 3.
Mismatch between RE capacity envisaged under different policies and capacities addition targeted 2009–2010.

PV systems, consumers are still lacking for an appreciable financial platform at the domestic level. The mega RE projects need appreciable amount of time for its completion. At the same time, the investors and stakeholders face problem due to the abrupt annulment of the RE policies by the government. Moreover, participation of private organization increases the project completion time because huge liability is not a full-recourse finance in this situation [14].

In case of development of biomass energy, obstacles occur in the process of transportation of biomass, which increases its cost of production. Also, efficiency is very low compare to conventional fossil fuel-based energy system.

The hindrance in the adaptation of new technologies based on RE sources can also be due to the lack of proper policies and regulations favoring the development of these technologies. Clear policies and legal procedures are required for the RE market to increase the interests of the investors. Additionally, standards and codes are some of the regulatory measures that enhance the adaptation of renewable energy technologies by diminishing the risk factor that comes along with investments in these projects.

Countries like India have been lacking with complete policy declaration on RE as most of the technologies are on its early advancement stage. Policies have been issues as and when necessary to facilitate the growth of specific RE technology. As per **Table 3**, targeted capacity of JNNSM is inadequate to meet the target for RE generation mandated by National Action Plan on Climate Change (NAPCC) [29].

6. Recommendations to eliminate the proposed challenges

In the previous part, different challenges have been identified and discussed, which act as a barrier toward the accomplishment of the India's dream energy project of RE. The challenges must be eradicated to stimulate global sustainable development prospective in the field of RE. Following are some recommendations to eliminate the proposed challenges:

a. Zonal grid of each state should give plots prepared by the state electricity board that hold the transmission necessities and its implementation as per load centers and available corridors, for the execution of RE-based power generation system in grid connected mode.

b. RE sector should be given priority in the rolling plan of the government, so that financial problem can be eliminated.

c. Government must frame extra budget for the research and development purpose of RE technologies, so that the nation can compete with the advance technologies.

d. There must be an attractive subsidy structure for the RE technologies at the domestic use level, so that social acceptance can be increased.

e. The necessary social awareness programs should be conducted at different parts of the country to increase the market size of the consumer for RE.

f. The problem of high initial investment can only be compensated through the advanced technologies. It is recommended to upgrade the existing technologies and adopt the superior one.

g. The use of efficient and durable energy storage system can overcome the issue of intermittent nature of RE because it will reduce the cost of maintenance, which will further decrease the operational cost of the system. At the same time, hybrid utilization along with the energy storage technology can also eliminate the problem of intermittent nature of RE.

h. The government has to promote the small-scale off-grid energy system for the rapid deployment of RE sources.

i. GoI must formulate a comprehensive policy or action plan for all-round development of the sector, encompassing all the key aspects. The action plan should be prepared in consultation with the state governments. It is understood that the energy coordination committee of GoI has approved the preparation of an umbrella RE law to provide a comprehensive legislative framework for all types of RETs, their usage, and promotion. However, GoI has fixed no timeframe for the formulation and enactment of such a law. The GoI must speed-up this task and ensure that the desired law be enacted expeditiously.

j. There is a need for stronger initiatives at local body levels for the promotion of RE. For example, local bodies must be discouraged from granting municipal approvals for commercial building in urban areas unless it houses a solar application. Solar installations should be a precondition for a power connection from the utility.

k. The commercial success of RETs depends significantly on adoption and enforcement of appropriate standards and codes. GoI must prescribe minimum performance standards in terms of durability, reliability, and performance for different RETs to ensure greater market penetration.

l. There is an urgent need for clarity on the Renewable Purchase Obligation (RPO) framework. It may be better to specify the overall RPO percentage rather than technology-specific percentages. This in turn would encourage investments in RE on the basis of technoeconomic analysis. Further, there should be no cap on RPO.

m. To meet the objective of RPO, it is imperative that an enforcement mechanism be introduced in all states.

n. RE should be declared as a priority sector. At present, the priority sector broadly comprises agriculture, small-scale industries, and other activities/borrowers (such as small business, retail trade, small transport operators, professional and self-employed persons, housing, education loans, and microcredit). The inclusion of RE in priority sectors will increase the availability of credit to this sector and lead to larger participation by commercial banks in this sector.

o. GoI should ask banks to allow an interest rebate on home loans if the owner of the house is installing a RE application such as solar water heater, solar lights, or PV panel. This would incentivize people to integrate RE technologies into their home, thereby encouraging the use of RE. The rebate could vary depending on the number of applications installed or the type of technology installed.

p. To achieve low capital costs and to capitalize on its inherent advantages in the solar sector, India needs to consider revamping and upgrading its solar R&D and manufacturing capabilities. In this regard, GoI may consider promoting a core company to produce wafer and silicon. This will enable substantial reduction in the costs of solar technologies.

q. There is an urgent need for technical assistance programs designed to increase the planning skills and understanding of RETs by utilities, regulators, local and municipal administrations, and other institutions involved.

r. Information specific to viable RETs needs to be made easily accessible. It will increase general awareness and acceptability as well as aid potential investors and sponsors of such projects.

s. Capacity building initiatives should be undertaken to train people/workers to operate and maintain RE facilities.

t. There is a need to improve the maintenance support mechanism for RE products/plants for redressing the post-installation problem faced by the users.

It can be easily suggested that the government has to mainly work in the field of financial mechanism, policy and regulation system, transmission system, social awareness programs, and technology sector to accomplish the India's dream energy project by 2022.

7. Conclusion

The chapter has been focused on to study the feasibility of different RE sectors in the Indian prospective. A analysis has been done about the government dream energy project of 175 GW by 2022. The potential of RE in India shows enormous opportunity toward its deployment, but the aforesaid technical, nontechnical, and other barriers need to be abated for the accomplishment of the project by 2022. Based on the identified obstacles, a brief recommendation has been proposed to overcome the problems. The completion of project will not only give a boost to its economy but also set a remarkable benchmark step for other nations around the globe to conquer the issue of global climate change.

Author details

Pankaj Kumar[1*], Kumar Avinash Chandra[2], Sanjay Patel[1], Nitai Pal[1], Mohit Kumar[3] and Himanshu Sharma[4]

1 Department of Electrical Engineering, IIT(ISM), Dhanbad, India

2 Department of Electrical Engineering, Dr. APJAKWIT, Darbhanga, India

3 Department of Electrical Engineering, BTKIT, Dwarahat, India

4 Department of Electrical and Electronics Engineering, SRM Institute of Science and Technology, Ghaziabad, India

*Address all correspondence to: pankajkumar903106@gmail.com

References

[1] Gulagi A, Bogdanov D, Breyer C. Demand of storage technologies in energy transition pathways towards 100% renewable energy for India. Energy Procedia. 2017;**135**:37-50. DOI: 10.1016/j.egypro.2017.09.485

[2] Renewable Energy World [Internet]. 2019. Available from: http://www.renewableenergyworld.com/rea/tech/home [Accessed: 11 July 2019]

[3] IEA-Renewable Energy-International Energy Agency [Internet]. 2019. Available from: http://www.iea.org/aboutus/faqs/renewableenergy [Accessed: 16 July 2019]

[4] Painuly JP. Barriers to renewable energy penetration; a framework for analysis. Renewable Energy. 2001;**24**:73-89. DOI: 10.1016/S0960-1481(00)00186-5

[5] Green Power Partnership [Internet]. 2019. Available from: http://www.epa.gov/greenpower/gpmarket/ [Accessed: 23 July 2019]

[6] Hydropower and the environment [Internet]. 2019. Available from: https://www.eia.gov/energyexplained/hydropower/hydropower-and-the-environment.php [Accessed: 23 July 2019]

[7] Electricity Sector in India [Internet]. 2019. Available from: https://en.wikipedia.org/wiki/Electricity_sector_in_India#Renewable_energy [Accessed: 03 August 2019]

[8] Global Wind Statistics 2017 [Internet]. 2018. Available from: https://gwec.net/wp-content/uploads/vip/GWEC_PRstats2017_EN-003_FINAL.pdf [Accessed: 07 January 2020]

[9] Mishra MK, Khare N, Agarwal AB. Small hydro power in India: Current status and future prospectives. Renewable and Sustainable Energy Reviews. 2015;**51**:101-115. DOI: 10.1016/j.rser.2015.05.075

[10] Singh R, Setiawan AD. Biomass energy policies and strategies: Harvesting potential in India and Indonesia. Renewable and Sustainable Energy Reviews. 2013;**22**:332-345. DOI: 10.1016/j.rser.2013.01.043

[11] Geothermal Energy [Internet]. 2020. Available from: https://www.irena.org/geothermal [Accessed: 08 February 2020]

[12] Ocean Energy [Internet]. 2020. Available from: https://mnre.gov.in/ocean-energy [Accessed: 08 February 2020]

[13] Renewable 2019 [Internet]. 2019. Available from: https://www.iea.org/reports/renewables-2019 [Accessed: 07 February 2020]

[14] Thacker MS. Natural Resources and their Planned Utilization. 1st ed. Maharaja Sayajirao University of Baroda; 1963

[15] Tiwari GN. Solar Energy—Fundamentals, Design, Modelling and Application. 2nd ed. Delhi: Narosa Publishing House; 2013 . pp. 27

[16] Ministry of Power Central Electricity Authority New Delhi [Internet]. 2017. Available from: http://www.cea.nic.in/reports/monthly/executivesummary/2017/exe_summary-02.pdf [Accessed: 09 August 2019]

[17] Renewable Energy Overviews and Targets [Internet]. 2019. Available from: https://en.wikipedia.org/wiki/Renewable_energy_in_India#cite_note-nmreAnnualRpt2016-13 [Accessed: 21 August 2019]

[18] Dermott M, Horne JE, Moura. The Next Green Revolution: Essential Steps to Sustainable to a Healthy Sustainable Agriculture. New York: Food Products Press; 2001. pp. 226

[19] Banguero E, Correcher A, Pérez-Navarro A, García E, Aristizabal A. Diagnosis of a battery energy storage system based on principal component analysis. Renewable Energy. 2020;**146**:2438-2449. DOI: 10.1016/j.renene.2019.08.064

[20] Land is Still the Biggest Impediment for Large Scale Solar Development [Internet]. 2020. Available from: https://mercomindia.com/land-impediment-large-scale-solar/ [Accessed: 19 February 2020]

[21] Kumar N, Pal N. The existence of barriers and proposed recommendations for the development of renewable energy in Indian perspective. Environment, Development and Sustainability. 2018;**20**:1-19. DOI: 10.1007/s10668-018-0284-y

[22] Kumar P, Sharma H, Pal N, Sadhu PK. Comparative assessment and obstacles in the advancement of renewable energy in India and China. Problemy Ekorozwoju [Problems of Sustainable Development]. 2018;**14**:191-200

[23] Renewable Energy and Green Growth in India [Internet]. 2020. Available from: https://www.teriin.org/projects/green/pdf/National-RE.pdf [Accessed: 09 February 2020]

[24] Renewable Energy Sector Faces Acute Manpower Shortage in India [Internet]. 2020. Available from: https://economictimes.indiatimes.com/industry/energy/power/renewable-energy-sector-faces-acute-manpower-shortage-in-india/articleshow/50961060.cms?from=mdr [Accessed: 10 February 2020]

[25] Solar Panels Price, Manufacturers, Technology for home in India 2020 [Internet]. 2020. Available from: https://www.bijlibachao.com/solar/solar-panel-cell-cost-manufacturer-price-list-in-india.html [Accessed: 07 February 2020]

[26] Sharma H, Kumar P,Pal N, Sadhu PK. Problems in the accomplishment of solar and wind Energy in India. Problemy Ekorozwoju [Problems of Sustainable Development]. 2018;**13**:41-48

[27] Energy Storage Systems [Internet]. 2020. Available from: https://www.oreilly.com/library/view/distributed-generation-systems/9780128042632/B9780128042083000078.xhtml [Accessed: 09 February 2020]

[28] All India Installed Capacity (in MW) of Power Stations [Internet]. 2020. Available from: https://web.archive.org/web/20160304051607/http://www.cea.nic.in/reports/monthly/installedcapacity/2015/installed_capacity-11.pdf [Accessed: 10 February 2020]

[29] Barriers to Development of Renewable Energy in India & Proposed Recommendations [Internet]. 2020. Available from: http://www.idfc.com/pdf/publications/Discussion-paper-on-Renewable-Energy.pdf [Accessed: 08 March 2020]

Chapter 8

Bioinspired Nanocomposites: Functional Materials for Sustainable Greener Technologies

Sarmad Ahmad Qamar, Muhammad Asgher and Nimrah Khalid

Abstract

This chapter presents a broad overview of the current advancements in bioplastics and bioinspired nanocomposites with nanoscale reinforcements that are being applied for a broad range of applications, that is, biomedical, electronics, durable goods and packaging materials. The production of nanocomposites by completely and/or partially renewable and biodegradable materials has helped in a range of different applications. Several drawbacks of conventional materials such as hydrophilicity, low-heat deflection, poor conductivity, and barrier properties can be efficiently overcome using biohybrid nanomaterials. *Nano*-reinforcements in composite materials deliver remarkably improved properties such as decrease in hydrophilicity and increase in mechanical properties as compared with neat biopolymer, which fails to exhibit these properties on its own. This approach can be used for other natural polymers to induce desired functionalities. This chapter covers the recent trends in nano-functional materials, renewable materials that are being applied for the production of nanobiocomposites and their applications especially in biomedical and healthcare sectors, which are discussed in detail. This emerging concept will definitely enhance the scope of nanohybrid materials for sustainable products development with improved properties than previously applied synthetic polymer-based or natural polymer-based materials.

Keywords: bioplastics, nanobiocomposites, multifunctional materials, biomedical applications

1. Introduction

Synthetic polymers are widely being used in everyday life for various applications. They can meet industrial and commercial market requirements, for example, durability, convenience, good performance, low cost, and high variability in regard to mechanical and barrier properties [1]. A significant amount of plastics is being used for packaging applications, which has grown rapidly from previous two decades [2]. These synthetic polymers/plastics of petrochemical origin are highly resistant to biodegradation, causing serious threat to the environmental sustainability because of the accumulation of nonbiodegradable wastes, which is increasing every year. Overdependence of fossil resources can be reduced by the development of bio-based materials using renewable resources. Currently, bioplastic market is progressing with an annual growth rate of 30% of synthetic plastic market [3]. Many scientists are working on the production of new compounds of biological origin either by chemical modifications or by industrial biotechnological processing. Efforts are being made for the production of biopolymers or polymer building blocks using microorganisms and/or plants such as exopolysaccharides and other polyesters [4]. For the betterment of material characteristics, different types of polymers are blended together, which is known as composite material, and the materials with nanoscale reinforcement (i.e., at least one nanoscale dimension) are called nanocomposite materials.

Polyhydroxyalkanoates (PHA) production has significantly progressed; recently it has been demonstrated that the lignocellulosic components of residual of sugarcane bagasse are effective fermentation biomaterials for PHA production. The concept of utilization of waste-based biomass is promoting sustainable, bio-based economy [5]. Bio-based and/or biodegradable plastics may include some biopolymers derived from and/or returned to the nature. The terms "biodegradable" and "bio-based" are used interchangeably, but it is not correct. Bioplastics can be manufactured from biodegradable petro-based polymers, renewable materials, or some combination of these. The various types of plastics available in the market are presented in **Figure 1**.

The development of novel nanohybrid materials for the induction of desired characteristics among polymer matrix is an emerging area among life sciences, material sciences, and nanotechnology. During the previous decade, "nanobiotechnology" became a familiar term, used to indicate nanohybrid materials involving natural-based or a biopolymer conjugated with inorganic moieties [6]. Since the development of nanocomposite materials, huge efforts were made by the scientists because of outstanding characteristics of these nanohybrid materials for both functional or structural materials, comprising amazing applications as electrochemical devices, and heterogeneous catalysts [7].

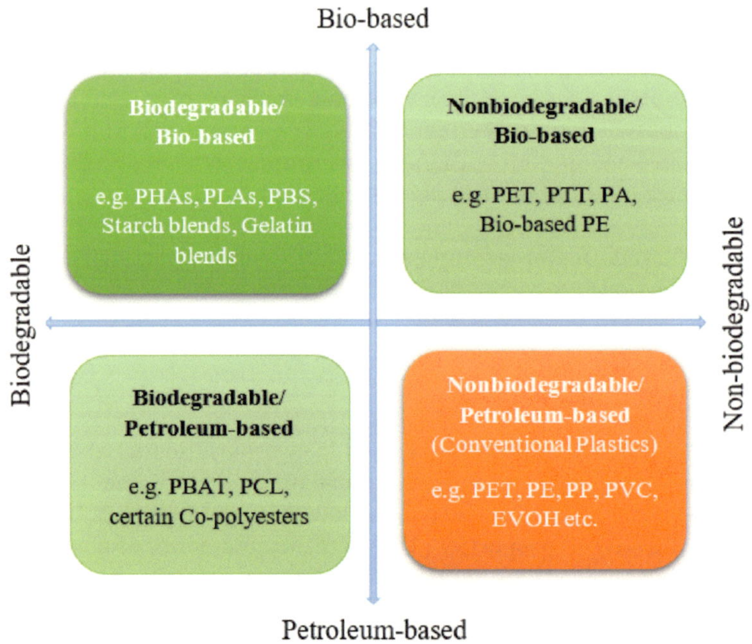

Figure 1.
Various types of plastics available in the market from origin and degradability point of view.

Researchers in nanotechnology are now focusing on the development of biopolymer-based nanocomposites that present outstanding characteristics similar to synthetic polymer-based materials (i.e., better thermal stability with improved mechanical and barrier properties) [8, 9]. In addition to these properties, nanobiocomposites also present remarkable advantage of biodegradability, biocompatibility, and, sometimes, functional characteristics provided by inorganic or biological moieties. The increasing interest in nanobiocomposites can also be imagined by the number of publications in previous two decades as per Web of Science, ISI database (**Figure 2**).

Several research groups are making efforts to replace petroleum-based polymers by natural, biodegradable, and abundant products synthesized from renewable biomass [10, 11]. Various biomacromolecules are present in nature, which could be utilized as renewable biomass for the production of nanohybrid materials such as starch, cellulose, lignin, polylactic acid (PLA), and other polyesters for the development of "greener" materials [12, 13]. Their blends with natural inorganic materials, for example, nanocellulosic-clay and carboxymethylcellulose, provide enhanced biodegradability and biocompatibility among matrix molecules.

Microbes are able to decompose biologically originated molecules, giving CO_2, which is utilized by the plants during the process of photosynthesis. The applications of these bio-based nanohybrid materials in the agricultural, biomedical, and/or in other areas will definitely help in the maintenance of environmental sustainability. Biomacromolecules or biopolymers bearing functional moieties representing highly specific catalytic properties, for example enzymes, present significant role in the production of nanobiocomposites aiming to produce nanohybrid materials with required characteristics. In nanobiocomposites that are based on enzymes, the inorganic portion is considered as the protective matrix for the immobilization of macromolecules and imparts multifunctionality to the nanohybrid matrix [14, 15]. The production of inorganic hybrid enzymes is an alternative way toward enzyme immobilization, which is a useful method for the development of enzymatic reactors and biosensors.

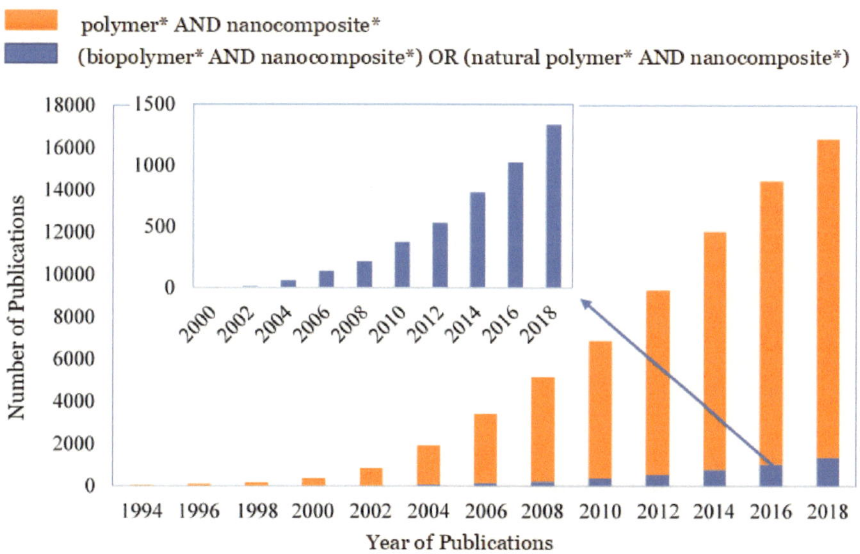

Figure 2.
Graphical representation of year-wise number of scientific publications related to synthetic polymer composites and bio-based nanocomposites (Web of Science, ISI statistics).

2. Nanocomposites from renewable resources

Currently, a growing concern among industrialists and researchers is to use environmentally friendly substances, aiming to replace nondegradable substances, thereby reducing the long-term accumulation of plastic waste in the environment. Biocompatible and biodegradable materials having applications in agricultural, food, or healthcare sectors are the major goals of several scientific studies. Petroleum-based materials are being replaced by natural/biological and/or biodegradable materials, which are also renewable in natural environment; for example, cellulose, starch, polycaprolactone, and PLA are being used to synthesize biodegradable packaging materials [16, 17]. These renewable materials consist of nontoxic compounds that are capable of biological degradation by several soil microorganisms. This emerging concept will definitely help in the reduction in environmental damage due to petrochemical dependence.

Because of huge benefits of renewable materials with environmentally sustainable nature and a broad spectrum of various industrial and healthcare applications, several scientific studies are focusing on the development of bio-based materials with improved characteristics [18, 19]. This has led toward the production of biodegradable nanocomposites that can exhibit more improved properties than nonreinforced bioplastics. Biomacromolecules, for example cellulose, starch, and their derivatives, are natural polymers used for the production of nanobiocomposites [20, 21]. These materials include synthetic or natural clay minerals or modified clay minerals such as nanofiller, providing exfoliation or intercalation compounds. Cloisite and montmorillonite are commonly applied silicates in these researches, having function of nanocharges that can act as reinforcement in the biopolymer material, resulting in improved mechanical strength of biopolymeric films.

Plasticizers are the substances added to synthetic resins to increase flexibility and plasticity to make the resulting plastic less brittle. Typically, glycerol, vegetable oil, or tryethylcitrate are added as plasticizers to bioplastic films with melting temperature near decomposition to prevent them from degradation, resulting in good-quality thermoplastic polymers. Plasticizers also contribute to better nanofiller dispersion in the matrix, giving amazing mechanical properties. Thermoplastic PLA, produced by cornstarch fermentation, is a most frequently used biopolymer for the production of bioplastic blended with organically altered silicates [22, 23]. The addition of titanate as a nanofiller to PLA bioplastics results in improvement in biodegradation, comparable to TiO_2 [24].

Although various researches comprising recent available data on nanobiocomposites have been explained above, the production of nanobased biocomposites is still in the developing phase. Further progress lies in the development of new materials by using novel biopolymers, to increase their compatibility with inorganic moieties. Polysaccharides and other natural macromolecules, and their integration with several nanofillers other than silicates and silica, for example, LDHs, would help in the improvement of mechanical and barrier properties of nanobiocomposites. Besides the improvement in mechanical properties, clay films also exhibit improved gas barrier and thermal stability that can be used for food packaging applications [25, 26].

Nanocomposites that comprise synthetic polymers and inorganic reinforcements, the distribution of silicates in biopolymer matrix initiates the "tortuous" pathway, leading to reduction in gas diffusion property of nanohybrid materials. In addition to silicates, several different inorganic solids have been added as reinforcements to biopolymer materials; for example, the distribution of sepiolite in natural rubber causes improvement in mechanical properties [27]. Tensile strength and elastic modulus of natural rubber are increased by the addition of single walled carbon nanotubes (SWCNTs) and SiC nanoparticles-based reinforcements, have become improved that those with just SWCNTs-based materials [28]. Multiwalled carbon nanotubes (MWCNTs) dispersion in natural rubber materials also represented a similar effect, for example, improved physical, mechanical, and chemical properties of biopolymer [29, 30] as presented in **Figure 3**.

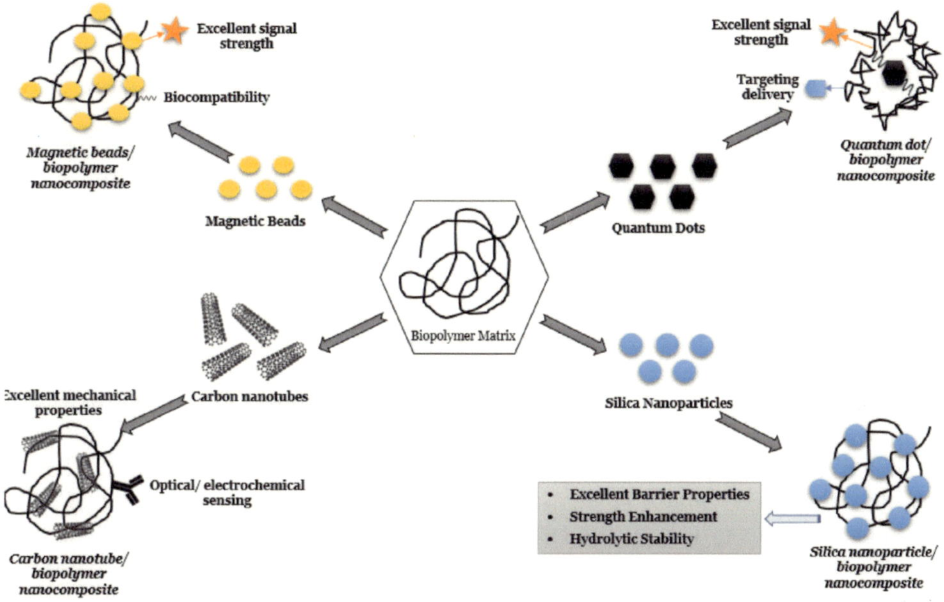

Figure 3.
Different types of nanostructured reinforcements among biopolymer matrix to induce desired functionality.

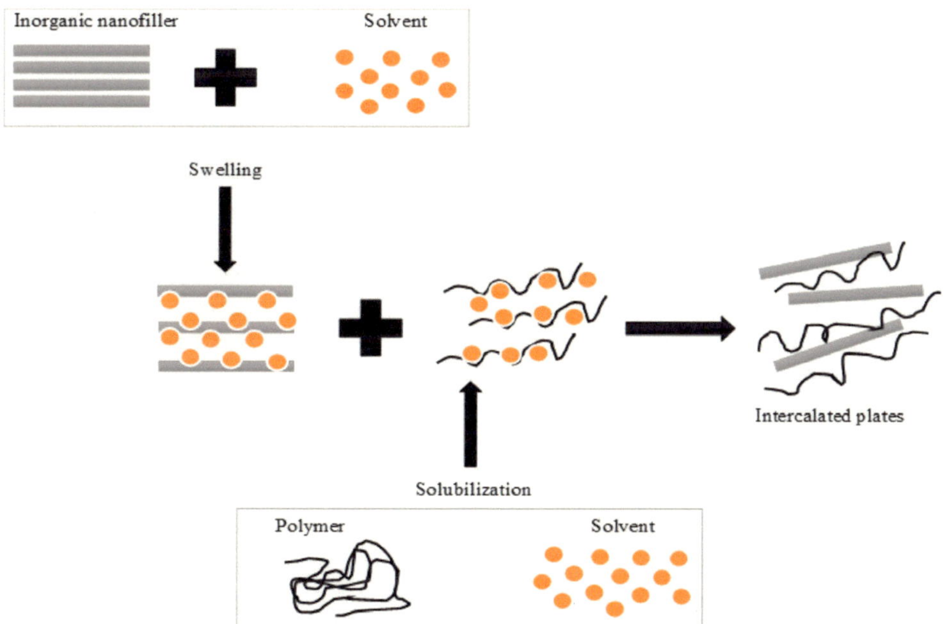

Figure 4.
Crosslinking between inorganic nanofiller and polymer matrix to form intercalated plates with improved tensile strength and modulus.

Organic reinforcement of starch and cellulose whiskers has become a sustainable replacement to other inorganic fillers; for example, nanocrystals of maize starch have been utilized as nano-reinforcement in glycerol plasticized maize starch [31, 32] leading toward improvement in mechanical properties of nanocomposites (**Figure 3**). Improvement in mechanical characteristics was also observed by the use of sodium carboxymethyl cellulose whiskers synthesized from cotton linter pulp when employed as reinforcement [33]. Enhancement in both Young's modulus and the tensile strength was observed, caused by the nanofiller and polymer matrix crosslinking resulting from intermolecular hydrogen bonding as shown in **Figure 4**.

The interest in using environmentally friendly, that is, biodegradable, products is increasing among companies; for example, NEC and Fujitsu have started to commercialize environmentally friendly mobile phones and notebook computers based on PLA-chips, reinforced with kenaf fibers or petrochemically derived polymers. The electronic applications will require more researches on enhancement of characteristics regarding distribution of biodegradable whiskers in polymeric materials.

3. Development of nanocomposite materials

Development of nanohybrid materials is a stepwise approach such as breakdown of intermolecular bonds comprising less energy, shaping new orientation and arrangement, and the production of new 3D network of polymeric substance by new interaction and bonds. Formation of new intermolecular forces relies upon polymer shape (length/diameter, ratio) and also the conditions provided. The material formed is stabilized by electrostatic, hydrophobic, covalent, and hydrogen bonds. Dry and wet processing of polymers is frequently reported useful for the synthesis of biopolymer-based nanocomposites [34]. Dry processing depends upon the thermoplastic characteristics of polymer, in which mechanical and thermal treatments cause induction of disulfide/sulfhydryl exchange reactions, while wet process depends on solubilization, type of solvent used, and pH, which can alter the polymer conformation [35].

3.1 Wet processing

Wet processing, also referred to as continuous spreading or casting method, is commonly used for the manufacture of bio and nanocomposites from natural resources, such as carbohydrates, proteins, and lipids (**Figure 5A**). Wet processing is based on polymer solubilization in a suitable solvent for the production of film forming solution. Desired additives (filler, plasticizers, antioxidant, antimicrobial compounds, nano-/microparticles, cross-linking agents) are added in the resultant solution. The method is followed by film spreading and solvent evaporation. Plasticizer addition is useful as it decreases intermolecular attractions and stiffness by giving flexibility and smoothing handling. This method is useful for packaging material development and it improves the mechanical properties of the resulting material [36].

3.2 Dry processing

As described above, this process is based upon thermoplastic properties of polymers that have an outstanding role in the synthesis of composite material. It can be correlated with glass transition theory, in which a glassy material is changed into a viscous state at a specific temperature. Transition state basically induces disorder, mobility, and free volume by changing physicochemical as well as mechanical properties of a substance [37]. In general, polymers can be shaped into desired material

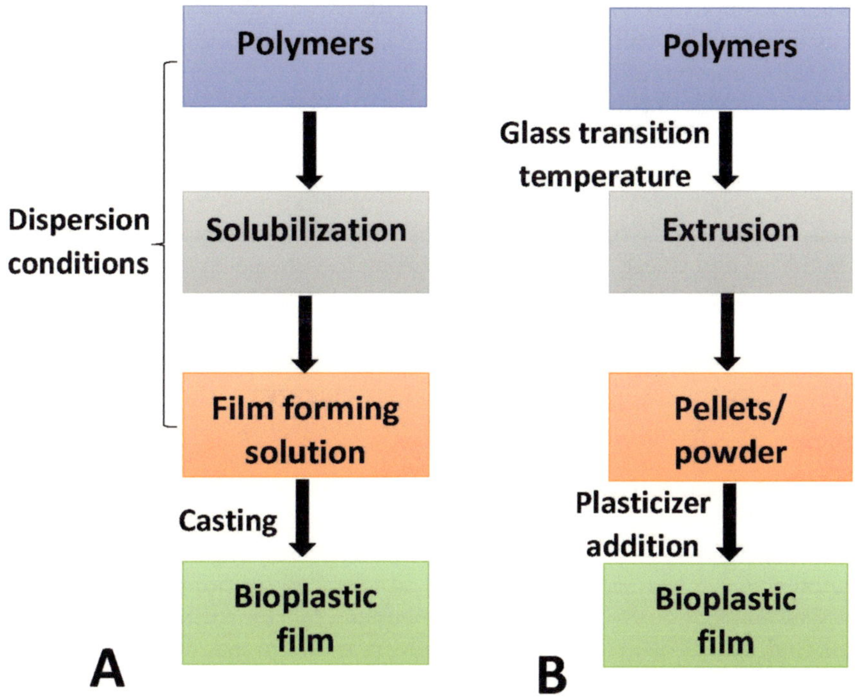

Figure 5.
Developmental strategies of bio/nanocomposites for functional applications: (A) wet processing; (B) dry processing.

by addition of plasticizer at high temperature and providing shearing force. Proteins denature at high temperature and the bonds in their molecules break, and new bonds and links establish in their molecules causing change in material properties [38]. Materials based on polymeric dry processing can be manufactured by several ways, for example, thermal processing or extrusion technique (**Figure 5B**). These processes can be used independently or both at the same time, in which extrusion is used for mixing and limited modification and thermal processing for the synthesis of final product.

4. Nanobiocomposites in healthcare sector

Nanobiocomposites present various applications especially in biomedical sciences like tissue engineering. The development of nanocomposites for regenerative medicine with bone implants and tissue engineering is still considered an emerging field [39, 40]. PLA and collagen are the most widely studied biopolymers for tissue regeneration as they provide artificial support for growth of the cells. This bioresorbable scaffold requires suitable mechanical properties and sufficient macroporosity with interconnected pores to avoid collapse of implantation and to allow the transportation of metabolic substances and the nutrients, and to control biodegradability [41]. Most of the articles published are related to bone repair. Thyroid hormones have important role in proper metabolism and functioning of the body such as cardiovascular homeostasis [42] and normal kidney function [43, 44]. Abnormalities in thyroid hormone production can cause serious health issues. Recent progress in the development of nanoscale biocomposites has led toward the development of catalase immobilized nanotubes *graft*-poly (L-lysine) for the diagnosis of iodate and H_2O_2 [45].

Nanobiocomposites tested and implanted for tissue regeneration include hydroxyapatite (HAp/collagen) to reproduce biocompatibility, composition, and mechanical properties of bones [46]. Other biopolymers, for example, chitosan [47], PLA [48], silk fibroin [49], and alginate [50] have also been studied in combination with HAp for the development of suitable bone regeneration scaffold. These implants mimic the surface roughness, porosity, and nanostructure of natural bones, as this facilitates the propagation of osteoblasts and helps in the regeneration of bones. Various synthetic techniques, for example, phase separation, gas foaming, fiber bonding, and freeze-drying/emulsification have been used to synthesize foam-like biocomposites with interconnected pores and suitable porosity [51, 52].

Future improvements in this area could be the replacement of HAp in natural polymers with some inorganic or the combination of organic/inorganic reinforcements. Sepiolite comprising microfibrous morphology has been blended with polymers, for example, collagen, giving rise to high-quality multifunctional hybrid materials [53]. High affinity between sepiolite and collagen biopolymer leads toward alignment with sepiolite fibers. Degradation rate can be reduced by the treatment of this biomaterial with a crosslinker, for example, glutaraldehyde, that increases mechanical properties, enhancing persistence after tissue implantation [54].

Nanobiocomposites also have a range of different applications, for example, drug delivery system [55] due to reduced dimensions and biocompatibility (**Figure 6**). Various studies have been reported in past few years about nanobiocomposites in targeted drug delivery system [55, 56]. The use of layered double hydroxide nanostructure (LDH) transporter as a non-viral vector for gene therapy has also been studied [57]. DNA intercalation in environment of Mg-Al/LDH by ion-exchange chromatography has also been confirmed. Analysis by XRD showed the increase of interlayer distance, revealing LDH parallel conformation to DNA double helical structure. The DNA transfer mechanism relies upon the shielding effect induced by the negative charge of DNA structure. This conformation facilitates the transportation of hybrid

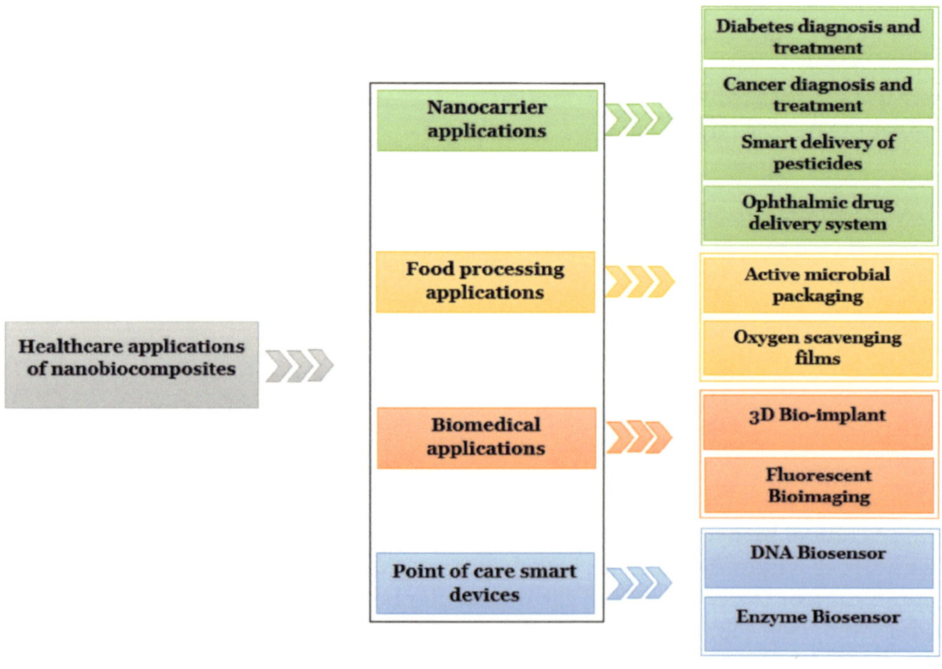

Figure 6.
Applications of nanobiocomposites in healthcare sector.

structure through the cell membrane, leading to LDH dissolution at acidic pH in lysosomes, the movement of DNA to the nucleus [58]. Nanosized hybrid materials, suitable for drug delivery purposes, have also been extensively studied for the treatment of leukemia and diabetes using gene therapy [59, 60].

5. Summary and future perspectives

There has been an explosion of scientific interest among nanotechnologists and material scientists to use biomass as a source of renewable materials and energy. For this purpose, the utilization of neat biopolymers comprises several limitations, that is, poor mechanical and barrier characteristics, which can be efficiently overcome using nanomaterials as reinforcing agents. The term "nanomaterials" covers a range of different materials with at least one dimension in nanoscale, that is, nanocrystals, nanoparticles, nanotubes, dendrimers, and several other inorganic nanoparticles. The use of "green chemistry" approach for the development of nano/biocomposite materials comprises several advantages over conventional materials processing strategies, that is, their environmentally friendly, biocompatible, and biodegradable nature. Biocompatibility is an important property for the application of these nanohybrid materials in healthcare sector including regenerative medicine, tissue engineering, or food industry.

Efforts are being made for the development of HAP-based nanocomposites for bone-engineering purposes. Another most important use of nanohybrid materials is targeted drug delivery, and the development of non-viral DNA vectors for gene therapy. Several functional nanohybrid materials working as optical and electronic gadgets are also being developed. Another promising application is the production of bio-based nanohybrid products, integrating natural-based polymers like chitosan, that have strong ion-exchange ability and effective electrochemical sensors. Enzyme entrapment by using several inorganic materials has led toward the production of active nanobiocomposites that can be efficiently used in bioreactor and biosensor devices. The development of novel nanobiocomposites with multifunctionality and improved characteristics can be considered as a developing area for scientific research.

Author details

Sarmad Ahmad Qamar*, Muhammad Asgher and Nimrah Khalid
Department of Biochemistry, University of Agriculture, Faisalabad, Pakistan

*Address all correspondence to: sarmad_qamar@uaf.edu.pk

References

[1] Vert M, Doi Y, Hellwich KH, Hess M, Hodge P, Kubisa P, et al. Terminology for biorelated polymers and applications (IUPAC Recommendations 2012). Pure and Applied Chemistry. 2012;**84**:377-410

[2] Sommerhuber PF, Welling J, Krause A. Substitution potentials of recycled HDPE and wood particles from post-consumer packaging waste in wood–plastic composites. Waste Management. 2015;**46**:76-85

[3] Shen L, Haufe J, Patel MK. Product overview and market projection of emerging bio-based plastics PRO-BIP 2009. Report for European Polysaccharide Network of Excellence (EPNOE) and European Bioplastics; 2009. p. 243

[4] Asgher M, Urooj Y, Qamar SA, Khalid N. Improved exopolysaccharide production from *Bacillus licheniformis* MS3: Optimization and structural/functional characterization. International Journal of Biological Macromolecules. 2020;**151**:984-992

[5] Dietrich K, Dumont MJ, Del Rio LF, Orsat V. Sustainable PHA production in integrated lignocellulose biorefineries. New Biotechnology. 2018;**49**:161-168

[6] Qamar SA, Asgher M, Khalid N, Sadaf M. Nanobiotechnology in health sciences: Current applications and future perspectives. Biocatalysis and Agricultural Biotechnology. 2019;**22**:101388

[7] Zhang WD, Xu B, Jiang LC. Functional hybrid materials based on carbon nanotubes and metal oxides. Journal of Materials Chemistry. 2010;**20**:6383-6391

[8] Sanchez-Garcia MD, Lopez-Rubio A, Lagaron JM. Natural micro and nanobiocomposites with enhanced barrier properties and novel functionalities for food biopackaging applications. Trends in Food Science and Technology. 2010;**21**:528-536

[9] Bordes P, Pollet E, Averous L. Nano-biocomposites: Biodegradable polyester/nanoclay systems. Progress in Polymer Science. 2009;**34**:125-155

[10] Han D, Wen TJ, Han G, Deng YY, Deng Y, Zhang Q, et al. Synthesis of Janus POSS star polymer and exploring its compatibilization behavior for PLLA/PCL polymer blends. Polymer. 2018;**136**:84-91

[11] Xu Y, Lin L, Xiao M, Wang S, Smith AT, Sun L, et al. Synthesis and properties of CO_2-based plastics: Environmentally-friendly, energy-saving and biomedical polymeric materials. Progress in Polymer Science. 2018;**80**:163-182

[12] Laadila MA, Hegde K, Rouissi T, Brar SK, Galvez R, Sorelli L, et al. Green synthesis of novel biocomposites from treated cellulosic fibers and recycled bio-plastic polylactic acid. Journal of Cleaner Production. 2017;**164**:575-586

[13] Darni Y, Dewi FY, Lismeri L. Modification of Sorghum starch-cellulose bioplastic with Sorghum stalks filler. Jurnal Rekayasa Kimia & Lingkungan. 2017;**12**:22-30

[14] Saba N, Jawaid M, Asim M. Nanocomposites with nanofibers and fillers from renewable resources. In: Green Composites for Automotive Applications. Oxford, UK: Woodhead Publishing; 2019. pp. 145-170

[15] Cai Z, Zhang H, Wei Y, Cong F. Hyaluronan-inorganic nanohybrid materials for biomedical applications. Biomacromolecules. 2017;**18**:1677-1696

[16] Mangaraj S, Yadav A, Bal LM, Dash SK, Mahanti NK. Application of biodegradable polymers in food packaging industry: A comprehensive review. Journal of Packaging Technology and Research. 2018;**2018**:1-20

[17] Rhim JW, Park HM, Ha CS. Bio-nanocomposites for food packaging applications. Progress in Polymer Science. 2013;**38**:1629-1652

[18] Meite N, Konan LK, Bamba D, Goure-Doubi BI, Oyetola S. Structural and thermomechanical study of plastic films made from cassava-starch reinforced with kaolin and metakaolin. Materials Sciences and Applications. 2018;**9**:41

[19] Moro TM, Ascheri JL, Ortiz JA, Carvalho CW, Melendez-Arevalo A. Bioplastics of native starches reinforced with passion fruit peel. Food and Bioprocess Technology. 2017;**10**:1798-1808

[20] Syafri E, Kasim A, Abral H, Sulungbudi GT, Sanjay MR, Sari NH. Synthesis and characterization of cellulose nanofibers (CNF) ramie reinforced cassava starch hybrid composites. International Journal of Biological Macromolecules. 2018;**120**:578-586

[21] Yunus M, Fauzan R. Mechanical properties of bioplastics cassava starch film with zinc oxide nanofiller as reinforcement. IOP Conference Series: Materials Science and Engineering. 2017;**210**:012015

[22] Zamir SS, Frouzanmehr MR, Nagalakshmaiah M, Ajji A, Robert M, Elkoun S. Chemical compatibility of lactic acid-grafted starch nanocrystals (SNCs) with polylactic acid (PLA). Polymer Bulletin. 2018;**2018**:1-9

[23] Tabasum S, Younas M, Zaeem MA, Majeed I, Majeed M, Noreen A, et al. A review on blending of corn starch with natural and synthetic polymers, and inorganic nanoparticles with mathematical modeling. International Journal of Biological Macromolecules. 2018;**122**:969-996

[24] Sun J, Shen J, Chen S, Cooper M, Fu H, Wu D, et al. Nanofiller reinforced biodegradable PLA/PHA composites: Current status and future trends. Polymers. 2018;**10**:505

[25] Tang MC, Agarwal S, Alsewailem FD, Choi HJ, Gupta RK. A model for water vapor permeability reduction in poly (lactic acid) and nanoclay nanocomposites. Journal of Applied Polymer Science. 2018;**135**:46506

[26] Opelt CV, Coelho LA. Reinforcement and toughening mechanisms in polymer nanocomposites–reinforcement effectiveness and nanoclay nanocomposites. Materials Chemistry and Physics. 2016;**169**:179-185

[27] Zaini NA, Ismail H, Rusli A. Tensile, thermal, flammability and morphological properties of sepiolite filled ethylene propylene diene monomer (EDPM) rubber composites. Iranian Polymer Journal. 2018;**27**:287-296

[28] Dolati S, Azarniya A, Azarniya A, Eslami-shahed H, Hosseini HR, Simchi A. Toughening mechanisms of SiC-bonded CNT bulk nanocomposites prepared by spark plasma sintering. International Journal of Refractory Metals and Hard Materials. 2018;**71**:61-69

[29] Wang X, Yang C, Jin J, Li X, Cheng Q, Wang G. High-performance stretchable supercapacitors based on intrinsically stretchable acrylate rubber/MWCNTs@ conductive polymer composite electrodes. Journal of Materials Chemistry A. 2018;**6**:4432-4442

[30] Abraham J, Kailas L, Kalarikkal N, George SC, Thomas S. Developing highly conducting and mechanically durable styrene butadiene rubber composites with tailored microstructural properties by a green approach using ionic liquid modified MWCNTs. RSC Advances. 2016;**6**:32493-32504

[31] Garcia NL, Ribba L, Dufresne A, Aranguren M, Goyanes S. Effect of glycerol on the morphology of nanocomposites made from thermoplastic starch and starch nanocrystals. Carbohydrate Polymers. 2011;**84**:203-210

[32] Angellier H, Molina-Boisseau S, Dole P, Dufresne A. Thermoplastic starch–waxy maize starch nanocrystals nanocomposites. Biomacromolecules. 2006;**7**:531-539

[33] Oun AA, Rhim JW. Preparation and characterization of sodium carboxymethyl cellulose/cotton linter cellulose nanofibril composite films. Carbohydrate Polymers. 2015;**127**:101-109

[34] Oksman K, Aitomäki Y, Mathew AP, Siqueira G, Zhou Q, Butylina S, et al. Review of the recent developments in cellulose nanocomposite processing. Composites Part A: Applied Science and Manufacturing. 2016;**83**:2-18

[35] Blanco-Pascual N, Fernández-Martín F, Montero MP. Effect of different protein extracts from *Dosidicus gigas* muscle co-products on edible films development. Food Hydrocolloids. 2013;**33**:118-131

[36] Farris S, Introzzi L, Piergiovanni L. Evaluation of a bio-coating as a solution to improve barrier, friction and optical properties of plastic films. Packaging Technology and Science: An International Journal. 2009;**22**:69-83

[37] Wang YH, Wang WH, Zhang Z, Xu L, Li P. Study of the glass transition temperature and the mechanical properties of PET/modified silica nanocomposite by molecular dynamics simulation. European Polymer Journal. 2016;**75**:36-45

[38] Miaudet P, Derre A, Maugey M, Zakri C, Piccione PM, Inoubli R, et al. Shape and temperature memory of nanocomposites with broadened glass transition. Science. 2007;**318**(5854):1294-1296

[39] Huang T, Fan C, Zhu M, Zhu Y, Zhang W, Li L. 3D-printed scaffolds of biomineralized hydroxyapatite nanocomposite on silk fibroin for improving bone regeneration. Applied Surface Science. 2019;**467**:345-353

[40] Zhang W, Chang Q, Xu L, Li G, Yang G, Ding X, et al. Graphene oxide-copper nanocomposite-coated porous CaP scaffold for vascularized bone regeneration via activation of Hif-1α. Advanced Healthcare Materials. 2016;**5**:1299-1309

[41] Hasnain MS, Ahmad SA, Chaudhary N, Hoda MN, Nayak AK. Biodegradable polymer matrix nanocomposites for bone tissue engineering. In: Applications of Nanocomposite Materials in Orthopedics. Woodhead Publishing; 2019. pp. 1-37

[42] Qamar SA, Mahmood Z, Munir N, Jahangeer M, Basharat A. Thyroid hormones and cardiovascular homeostasis: A review. Pakistan Heart Journal. 2019;**51**:264-272

[43] Katz AI, Emmanouel DS, Lindheimer MD. Thyroid hormone and the kidney. Nephron. 1975;**15**:223-249

[44] Basharat A, Munir B, Jahangeer M, Qamar SA, Mahmood Z, Ghaffar A. Biochemical profile of patients with

chronic kidney disease (CKD) undergoing regular hemodialysis. Life Science Journal of Pakistan. 2019;**1**:3-9

[45] Vilian AE, Chen SM, Lou BS. A simple strategy for the immobilization of catalase on multi-walled carbon nanotube/poly (L-lysine) biocomposite for the detection of H_2O_2 and iodate. Biosensors and Bioelectronics. 2014;**61**:639-647

[46] Zhou Y, Yao H, Wang J, Wang D, Liu Q, Li Z. Greener synthesis of electrospun collagen/hydroxyapatite composite fibers with an excellent microstructure for bone tissue engineering. International Journal of Nanomedicine. 2015;**10**:3203

[47] Jahangeer M, Qamar SA, Mahmood Z, Asgher M, Basharat A. Applications and perspectives of chitosan as functional biopolymer: An extended review. Life Sciences Journal of Pakistan. 2019;**2**:33-40

[48] Thanh DT, Trang PT, Huong HT, Nam PT, Phuong NT, Trang NT, et al. Fabrication of poly (lactic acid)/ hydroxyapatite (PLA/HAp) porous nanocomposite for bone regeneration. International Journal of Nanotechnology. 2015;**12**:391-404

[49] Behera S, Naskar D, Sapru S, Bhattacharjee P, Dey T, Ghosh AK, et al. Hydroxyapatite reinforced inherent RGD containing silk fibroin composite scaffolds: Promising platform for bone tissue engineering. Nanomedicine: Nanotechnology, Biology and Medicine. 2017;**13**:1745-1759

[50] Naik K, Chandran VG, Rajashekaran R, Waigaonkar S, Kowshik M. Mechanical properties, biological behaviour and drug release capability of nano TiO_2-HAp-Alginate composite scaffolds for potential application as bone implant material. Journal of Biomaterials Applications. 2016;**31**:387-399

[51] Yazdimamaghani M, Razavi M, Vashaee D, Moharamzadeh K, Boccaccini AR, Tayebi L. Porous magnesium-based scaffolds for tissue engineering. Materials Science and Engineering: C. 2017;**71**:1253-1266

[52] Dziadek M, Stodolak-Zych E, Cholewa-Kowalska K. Biodegradable ceramic-polymer composites for biomedical applications: A review. Materials Science and Engineering: C. 2017;**71**:1175-1191

[53] Cavallaro G, Lazzara G, Fakhrullin R. Mesoporous inorganic nanoscale particles for drug adsorption and controlled release. Therapeutic Delivery. 2018;**9**:287-301

[54] Grant SA, Zhu J, Gootee J, Snider CL, Bellrichard M, Grant DA. Gold nanoparticle-collagen gels for soft tissue augmentation. Tissue Engineering Parts A. 2018;**24**:1091-1098

[55] Rani A, Asgher M, Qamar SA, Khalid N. Nanostructure-mediated delivery of therapeutic drugs—A comprehensive review. International Journal of Chemical and Biochemical Sciences. 2019;**15**:5-14

[56] Patwekar SL. Nanobiocomposite: A new approach to drug delivery system. Asian Journal of Pharmaceutics (AJP). 2016;**2016**:10

[57] Andrea KA, Wang L, Carrier AJ, Campbell M, Buhariwalla M, Mutch M, et al. Adsorption of Oligo-DNA on magnesium aluminum-layered double-hydroxide nanoparticle surfaces: Mechanistic implication in gene delivery. Langmuir. 2017;**33**:3926-3933

[58] Wang J, Zhu R, Gao B, Wu B, Li K, Sun X, et al. The enhanced immune response of hepatitis B virus DNA vaccine using SiO_2@ LDH nanoparticles as an adjuvant. Biomaterials. 2014;**35**:466-478

[59] Huang S, Kamihira M. Development of hybrid viral vectors for gene therapy. Biotechnology Advances. 2013;**31**:208-223

[60] Calin GA, Dumitru CD, Shimizu M, Bichi R, Zupo S, Noch E, et al. Frequent deletions and down-regulation of micro-RNA genes miR15 and miR16 at 13q14 in chronic lymphocytic leukemia. Proceedings of the National Academy of Sciences. 2002;**99**:15524-15529

Chapter 9

Road Transportation Industry Facing the Energy and Climate Challenges

Brahim Mebarki, Belkacem Draoui, Boumediene Allaoua and Abdelghani Draoui

Abstract

On a worldwide scale, vehicles, which constitute the transportation sector, play a major role in supporting the other productive sectors. In addition, the automobiles help promoting the living standards of human beings by satisfying the needs of most people in their daily activities. Nowadays, for the traction or repulsion of vehicles, the overwhelming majority of these automobiles are equipped with internal combustion engines (ICE); however, the automotive industry is moving steadily toward the adoption of new technologies because of the rapid depletion of fossil fuels and climate challenges caused by the transport sector, which accounts for the 27% of global energy consumption and for 33.7% of pollutant emissions and green house gas (GHG). For road transportation sector, the fuel-cell electric vehicle is one of the promising solutions advocated by car manufacturers and research entities to replace gradually conventional vehicles.

Keywords: energy and transportation, electric vehicles, fuel cell vehicles, battery

1. Introduction

The internal combustion engine is considered to be one of the greatest inventions in the human era. It is exploited in most road and rail transport solutions. This engine operates by burning products originating from fossil fuels. But due to the limited reserve of fossil fuels and because of the harming effects of burning these resources on the environment, the stakeholders of the automotive industries have opted for the development of efficient and high performance substitutions to this type of engines. Furthermore, the economic development and population growth experienced by the world during the last 15 years have led to a sharp increase in demand for energy in this sector, which will increase the rate of depletion of fossil fuels and lead to major air pollution and global warming. To address these concerns, transportation-related research and development focus on developing viable renewable and clean solutions.

2. Road transportation: energy and environmental challenges

The global transport sector will face a number of unprecedented challenges over the next four decades (2010–2050). It is expected that the world population will increase from 2.2 to 9.2 billion, with over two thirds of the population living in cities, compared to about half of today's population. In addition, it is expected that the number of megacities increased from 22 to between 60 and 100 megacities today in 2050. Many of these megacities, emerging mainly in Asia, Africa, and Latin America, will face high levels of traffic congestion, pollution, and noise. Furthermore, this effect will be amplified by the 2 or 3 billion cars and trucks that could be outstanding. During the same period, travel and road freight will at least double because of the increased demand for transport, as well as economic development and improvement of living standards [1].

2.1 World population, vehicle fleet, and mobility

To understand the evolution of the vehicle market, it is necessary to compare the increase in the world population. We start with the investigation on the growth of the world population. During most of human existence, population growth was so slow that it was imperceptible within a single generation. To achieve a world population of 1 billion people, it took until 1804 for those modern humans to appear on the world stage. To add the second billion, it was not until 1927, a little over a century. Thirty-three years later, in 1960, the world population reached 3 billion. Then the pace accelerated, since we added a billion every 13 years or more. In 2000, the worldwide population has reached 6 billion with 700 million vehicles; by the year of 2050, it has been estimated that the population will reach 10 billion with 2.5 billion vehicles (**Figure 1**) [2, 3].

Between 1950 and 1990, the number of road vehicles has increased ninefold in the world from 75 to 675,000,000. The vehicles predominantly for personal transport (cars and motorcycles) accounted for nearly 80% of all of them during these four decades. During the same period, the world population doubled from approximately 2.5 to nearly 5 billion. During the period 1990–2030, the number of registered vehicles increased from 675,000,000 to 1,624,000,000 and mileage driven 10.7 billion kilometers to 26.6 billion kilometers, most of this increase occurring in countries that do not currently belong to the Organization for Economic Co-operation and Development (OECD) (**Table 1**). This table shows that despite the already high levels of use of motor vehicle in the OECD countries, the number of vehicles and the amount of displacement is called to increase significantly over the coming decades [5, 6].

The study [5] summarized in **Table 1**, predicts that all indicators related to transportation via lightweight vehicles in OECD countries—at the exception of fuel consumption—will increase significantly between 1990 and 2030 [5].

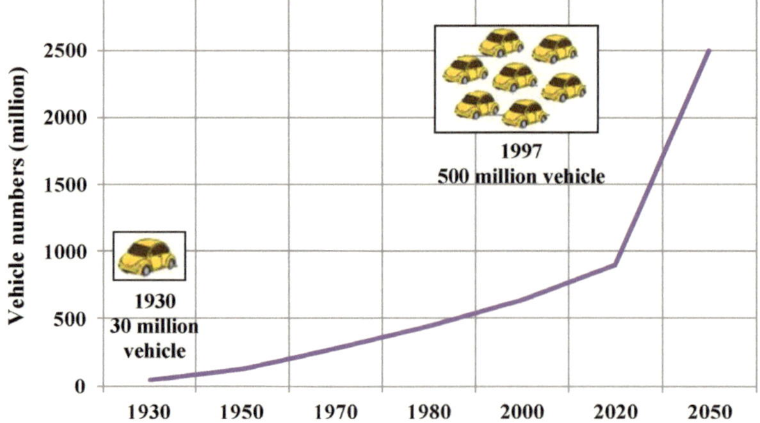

Figure 1.
Vehicle world fleet growth [4].

	Light vehicles			Heavy vehicles		
	Totals		Δ%	Totals		Δ%
	1990	2030		1990	2030	
OECD countries						
Number of vehicles (millions)	468	811	73	16	31	94
Mileage traveled (billion)	7.057	12.448	76	687	1.377	100
Weight of fuel consumed (megatons)	563	520	8	182	359	97
Other countries						
Number of vehicles (millions)	179	725	305	14	56	300
Mileage traveled (billion)	2.380	9.953	318	647	2.512	288
Weight of fuel consumed (megatons)	167	394	136	142	552	289
All countries						
Number of vehicles (millions)	648	1.537	137	30	87	190
Mileage traveled (billion)	9.437	22.400	137	1.334	3.889	192
Weight of fuel consumed (megatons)	730	914	25	324	911	181

Table 1.
Evolution of global fleet and the distance covered, 1990–2030 [5].

In order to avoid dramatic climate change, climatologists advised for aiming to reduce the emissions of greenhouse gases by 60% from current levels by 2050 [7]. These two predicted scenarios are completely contradictory: a significant increase in the number of vehicles creates a huge demand for fossil energy day after day, fossil energy, mostly originating from oil, which goes against environmental the objectives of mitigating the greenhouse gases emissions. So the dilemma to solve is how to reduce pollution through toxic emissions combined with a significant increase in the number of vehicles? [3].

2.2 Energy resources

2.2.1 Nonrenewable energy reserves

It has been stated that the rate of depletion of nonrenewable energies, namely oil, gas, and carbon, is faster than the time of regeneration [8]. Crude oil is a result of the transformation of organic (animal and vegetal) debris from marine populations, under great pressure and in the absence of oxygen. This waste, mixed with sediments and gradually buried by new layers which are deposited, undergoes molecular changes under the combined effect of an increase in temperature and pressure. It thus becomes a liquid or a paste made up essentially of hydrocarbons, molecules made of hydrogen, and carbon assembled in chains which vary in their degree of complexity; as well as hydrocarbons, variable proportions of sulfur, nitrogen, oxygen, and traces of various metals are present [9].

According to data reported in 2013 by British Petroleum Company, **Figures 2–4**, respectively, illustrate the reserves estimation of oil, gas, and coal at the end of 1992, 2002, and 2012. On aggregate, the reserves can be estimated to 1075 billion tons of oil equivalent (Gtoe).

Although new deposits of oil and natural gas are discovered regularly, with the ever increasing rate of consumption, it is predicted that the aforementioned reserves will be exhausted in the twenty-first century. Coal provides more distant prospects. In addition, it is an abundant and inexpensive resource compared to other fossil fuels. However, coal is also the most energy-emitting source of CO_2 gas, recognized for its harming effects on global warming, and this is the main issue related to its use [8–11].

Table 2 shows the ratio of energy reserves on consumption. It is calculated by dividing the remaining reserves at the end of a given year by the consumption of that year. The result represents exploitation duration at the current consumption rates.

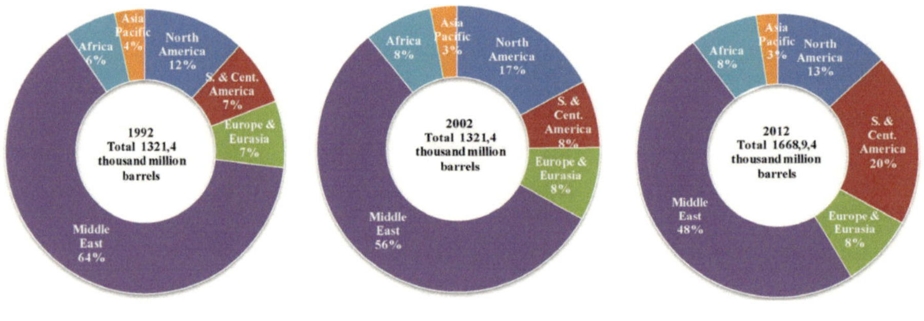

Figure 2.
Distribution of proven reserves for oil [10].

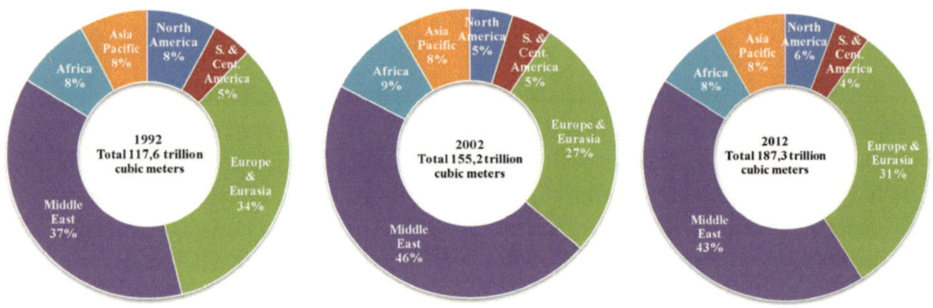

Figure 3.
Distribution of proven reserves for gas [10].

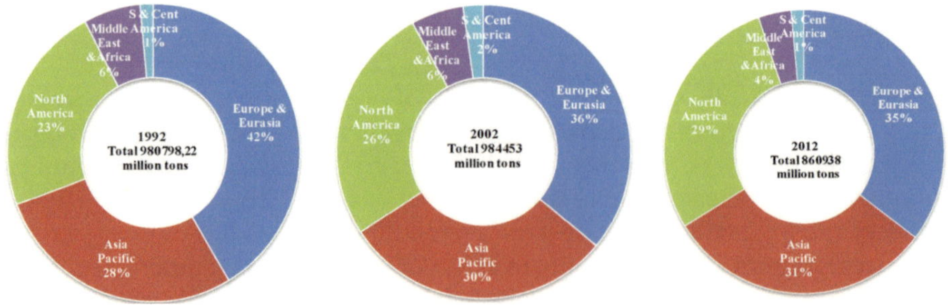

Figure 4.
Distribution of proven reserves for coal [10].

Fossil energy	Petrol	Gas	Carbon
Operating life	52.9	55.7	109

Table 2.
Operating life of fossil resources [10].

Regarding nuclear energy, according to the IAEA and the World Nuclear Association, the current uranium reserves can be exploited for 30 years provided that the price remains less than 40 $ per kg of uranium and over 60 years, if the production cost rises to $ 80 per kg. However, by adding all the proven reserves (not extracted today), the duration of exploitation is forecasted to have a slightly more than 200 years of consumption (depending on the price of uranium) [12].

Oil is the main source of energy, providing 33% of global demand, followed by coal (27%) and gas (21%). Renewables energies account for 13% of demand where 10% is supplied by hydraulic energy. The nuclear power contribution fluctuates around 6% [13].

The exploitation of the first three nonrenewable energies is due to the fact that these sources have a high specific energy density (around 40 MJ/kg for oil, 20 MJ/kg for coal, and 60 MJ/kg for methane). On the other hand, the oil has the advantage of being a liquid fuel, which makes it easy to transport, store, and use once refined; this explains its widespread use in many applications [14].

Given that oil resources are limited (**Figure 5**), leading to an increase in the cost of oil. A study showed that if the rate of discovery of fossil resources should continue at current levels and consumption were to increase, then the oil resources would be exhausted by 2038 [9].

2.2.2 Evolution of consumption

Historically, global energy demand has grown steadily over the last 40 years, starting from 5000 Mtoe in 1970 to 14414.4 Mtoe in 2015. It has tripled in size in 45 years, which corresponds to an average annual growth rate of about 2.22% [13–16]. This growth is slightly lower than the 2.22% observed between 2010 and 2015. However, given the difference in economic growth between the two periods, the energy intensity is expected to improve at a constant rate of 1.5% per year [16]. **Figure 6** shows that the primary energy consumption is expected to grow on average 1.7% per year by 2040 to 2050 to reach 20 billion tons of oil equivalent of which the share of developing countries (China, India, Brazil, etc.) will be over 70% [5]. More than 85% of the increase in global demand for energy from 2010 to 2040 occurs in developing countries outside the OECD [16].

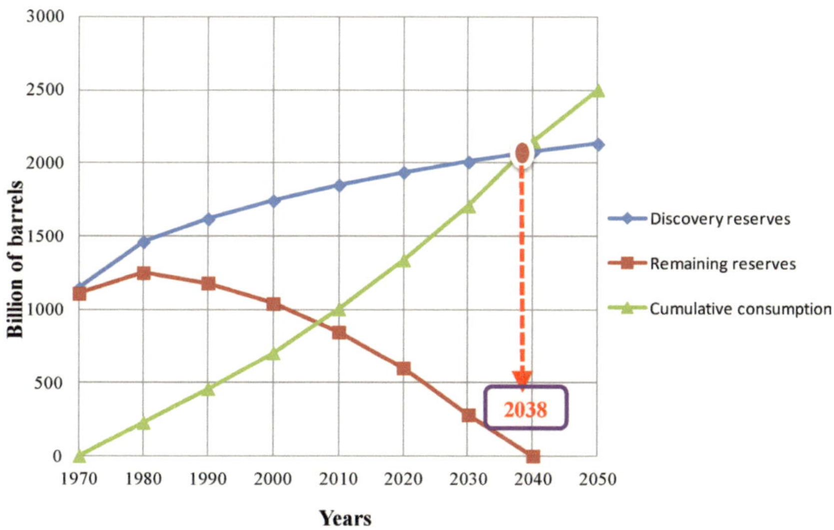

Figure 5.
World oil discovery, remaining reserves, and cumulative consumption [9–15].

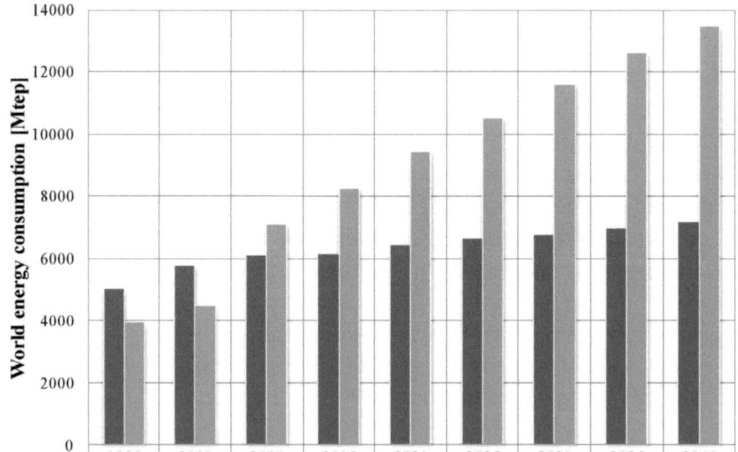

	1999	2000	2010	2015	2020	2025	2030	2035	2040
OECD	5018,656	5766,113	6098,4	6148,8	6426	6627,6	6778,8	6955,2	7182
Non-OECD	3950,831	4484,743	7106,4	8265,6	9450	10533,6	11592	12625,2	13482

Years

Figure 6.
World energy consumption [16].

Among the factors explaining the growth of developing countries, there are of course the macroeconomic fundamentals: strong economic growth, driven by sustained industrial development and population growth, coupled with a broader access to energy sources, and in addition to this, in many developing countries, a policy of low energy prices and the frequent existence of subsidies.

2.2.3 Energy consumption in the transportation sector

Energy consumption in the transportation sector includes energy used for the movement of people and goods by road, rail, air, and water. The latter has experienced a real explosion during this century and is increasing on average by 1.1% per year (**Figure 7**). Most of the growth in the use of energy transport is in non-OECD

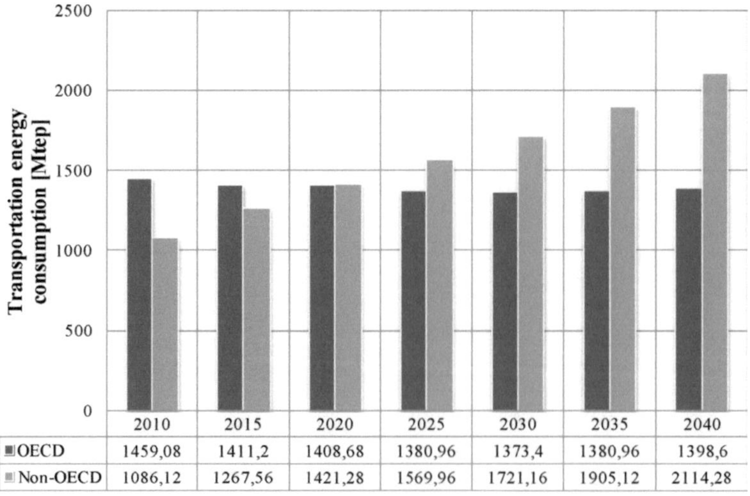

	2010	2015	2020	2025	2030	2035	2040
OECD	1459,08	1411,2	1408,68	1380,96	1373,4	1380,96	1398,6
Non-OECD	1086,12	1267,56	1421,28	1569,96	1721,16	1905,12	2114,28

Years

Figure 7.
Transportation sector energy consumption.

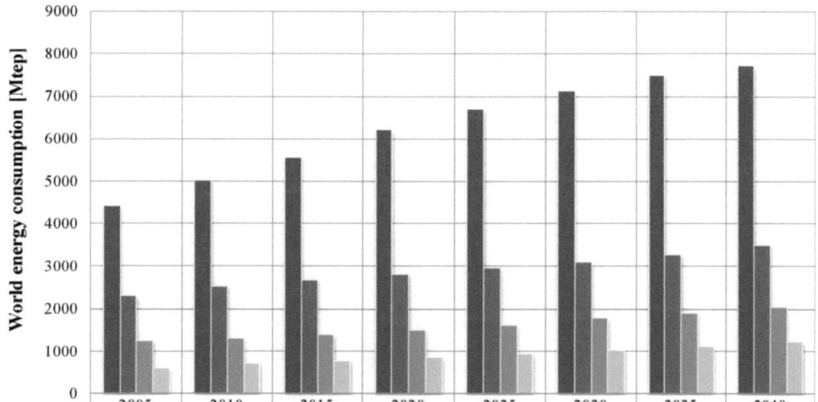

	2005	2010	2015	2020	2025	2030	2035	2040
Industrial	4450,2192	5040	5579,28	6216,84	6713,28	7144,2	7507,08	7733,88
Transport	2305,53288	2545,2	2676,24	2827,44	2948,4	3094,56	3286,08	3515,4
Residential	1260	1310,4	1388,52	1506,96	1638	1784,16	1922,76	2061,36
Commercial	616,59612	728,28	776,16	846,72	934,92	1030,68	1128,96	1234,8

Figure 8.
World consumption by sector [16, 17].

countries that experienced strong economic growth at the same time, leading to improved living standards and a corresponding increase in personal and commercial travel request. Demand in these countries is almost doubled, from 43.1 quadrillion Btu in 2010 to 83.9 quadrillion Btu in 2040. That is to say, it increased by an average of 2.2% per year. On the other hand, the energy consumption of the OECD countries decreased on average by 0.1% per year, that is to say, 58 quadrillion Btu in 2010 to 56 quadrillion Btu in 2040 due to the relatively slow economic growth, improving energy efficiency, and stable population growth levels.

Compared to other economic sectors, transportation is ranked second after the industrial sector with a rate of 25.68% of final world consumption in 2015 [16, 17]. Under this scenario, the consumption will be reduced and will represent 24.16% of world consumption in 2040 [16] (**Figure 8**).

2.2.4 Transportation dependence on oil

Since its origins, the transport sector remains totally dependent on oil production. This results in an increased demand for the latter [9–18].

In 2015 the global transport sector consumes about 2,676,240,000 tons of oil equivalents of energy annually. Of this quantity, over 96% originates from oil, representing more than 60% of the total oil production in the world (**Figure 9**). Road transport accounts for the majority of this energy consumption. Light vehicles (LV) (including light trucks, light commercial vehicles, and minibusses) represent about 52% of the total means of road transportation, while busses and trucks represent a share of 4% and 17%, respectively. While the air and marine transportation account for about 10% of world consumption of transport energy, aviation is by far the most dynamic sector, with an increase in revenues-ton-kilometers of around 5.1% by year 2030. The railway sector represents only about 3% of the total energy consumption related to transportation [19].

However, the consumption patterns of industrialized and developing countries are very different (**Figure 10**). Oil consumption is almost three times higher in industrialized countries than in developing countries. Global demand will reach 106.5 million barrels per day in 2020 compared to 96 million barrels a day today.

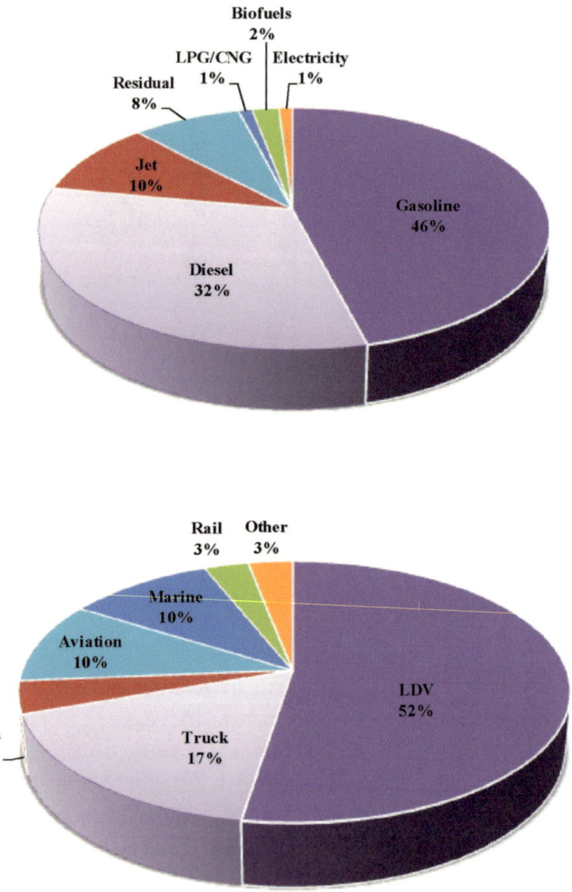

Figure 9.
Distribution of fuel consumption by source and mode [19].

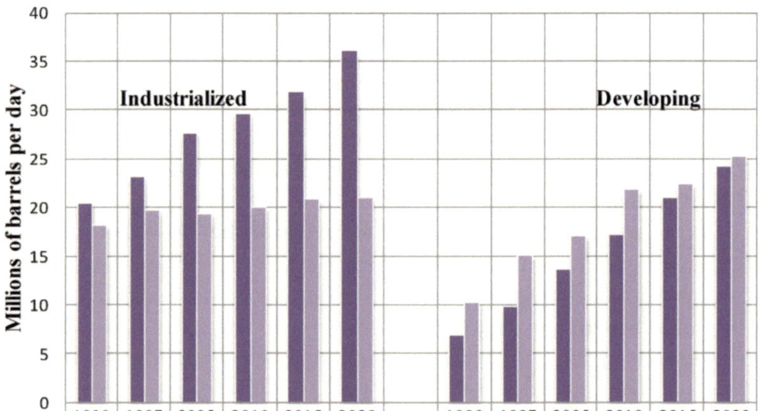

	1990	1997	2005	2010	2015	2020	1990	1997	2005	2010	2015	2020
Transportation	20,46	23,07	27,66	29,68	31,89	36,09	6,859	9,85	13,63	17,22	21	24,19
Other	18,09	19,7	19,34	19,96	20,79	21,02	10,24	15,01	17,01	21,79	22,4	25,19

Years

Figure 10.
Increase in world oil demand, 1997–2020 [9].

In other sectors, oil consumption should stabilize or even decline in industrialized countries. Oil consumption will increase relatively slowly in all sectors for developing countries as compared to transportation sector. In developing countries, the transport sector also shows the highest expected growth in oil consumption. Currently, over 55% of the oil in the world (about 53 million barrels per day) will be consumed in the transport sector [9].

2.2.5 Consumption standards

Currently, only the United States and Japan have laws establishing fuel consumption standards. Across the Atlantic, the index of average consumption models is produced by the entity [Corporate Average Fuel Economy (CAFE)] and California standards for greenhouse gases. In Japan, the Top Runner program is adopted [20].

2.2.5.1 US standards

CAFE standards date back to the first oil crisis. They regulate the average consumption of constructors fleets on a combined city/highway use. Their initial goal was to double the energy efficiency of vehicles between 1975 and 1985. The first objective set for 1978 was to achieve a consumption of 18 miles per gallon (mpg) for cars or 13.1 l/100 km. In 1985, these latter were to reach a level of consumption of 27.5 mpg in 1985 (8.6 l/100 km). Subsequently, these standards have been facilitated. From 1990 and till the present time, the standard of 27.5 mpg has been re-applied [20].

In 2007, the US Congress set a new objective in the matter: 35 mpg (6.7 l/100 km) by 2020. In addition, the Department of Transportation (DOT) and the Environmental Protection Agency (EPA) have jointly proposed to apply this new standard on the period 2012—2016, which would allow, at national level, to meet the standards of all states. This program would save 5% of fuel per year, 1.8 million barrels of oil, and reduce CO_2 emissions to 950 million m^3 (1.8 million tons of the nearly 6000 produced) or 21% in 2030 compared to the situation that would prevail without the new standards [20].

2.2.5.2 Japanese standards

Japan is the country where the GHG emission standards are the most strict ones. According to the Top Runner program, introduced in 1999, emissions from gasoline light vehicles should be less than 22.8% of their 1995 level. For diesel vehicles, the goal was set at 2005 and the reduction achieved was 8.8% (the required improvement was, originally, 14.9%) [20].

2.2.5.3 European Union

In Europe, the European Automobile Manufacturers Association (ACEA) committed in 1998 to reduce CO_2 emissions from new cars to 140 g/km in 2008. Since 1998, the average reduction was 2.5% per year. Subsequently, in 2007, the European Commission proposed to limit to 130 g/km by 2012, and non-compliance of any car manufacturer would expose it to financial penaltiess. The goal is to eventually reach 120 g/km through complementary measures. However, reaching the threshold of 130 g/km has been delayed for 3 years, in 2014 [20].

3. Impact of transport on the environment

During the last decades, and with the industrialization of many countries, the world has experienced an increase in the number of individual vehicles in which their use is considered a good indicator of economic growth. But oil consumption used currently in a vehicle results in various impacts on the environment: emissions pollute the air and cause to climate change, noise causes harm and leads to health issues, and infrastructure affect landscapes and ecosystems. The further social impacts are as follows: hundreds of thousands of people are killed or injured each year in accidents, and congestion levels achieved in many densely populated areas are sources of wasted time. These problems today are important issues for governments and international organizations.

Pollutants are formed during the combustion of fuel in vehicles equipped with internal combustion vehicles. The products emitted by combustion vehicles are still the same; only the relative amounts vary with the type of combustion, the geometry of the engine, and the operating conditions of the vehicle [6–21].

The combustion of hydrocarbon liquids rejects pollutants called "primary" directly from the exhaust pipe and pollutants "secondary" formed by the chemical conversion of the first in the atmosphere. In addition to gas emissions, road transport is the source of emissions of pollutants in the form of fine particles, heavy metals, noise, etc. These pollutants have a direct impact on the public health and are responsible for climate change due to the greenhouse effect.

Pollutants from automobile exhaust are as follows [22]:

- Carbon compounds: CO, CO_2
- Nitrogen compounds: NO, NO_2 commonly referred to under the generic name of nitrogen
- Oxides NOx, N_2O, and more rarely NH_3, HCN, nitrosamines, etc.
- Organic compounds, volatile, irritating, or odor, such as hydrocarbons (HC), benzene, polycyclic aromatic hydrocarbons (PAHs), carbonyl compounds (aldehydes, etc.), carboxyl (organic acids)
- The sulfur compounds, particularly SO_2 and SO_3 and rarely H_2S
- Halogenated compounds, mostly in leaded fuels disappearing over
- Metal compounds, especially the fuel lead and zinc lubricants associated with its impurity cadmium
- Volatile organic compounds (VOC) and soot appointed as particulate matter (PM_{10}, $PM_{2.5}$), derived almost exclusively diesel engines

3.1 Climate change and air pollution

The potential harms of climate change is well established. Average temperatures on the surface of the earth and the oceans have risen, causing climatic disturbances that are already present in almost all regions of the world. At the global level, it is estimated that average temperatures have risen by 0.7°C over the twentieth century. The last decade (2001–2011) was also the warmest ever recorded by meteorological services, and the current warming is accelerating [23, 24].

Climate change is already resulting in a multitude of visible and measurable phenomena: the gradual disappearance of Arctic ice cover in summer accelerated melting of glaciers, ocean acidification, etc. Climate models predict a substantial increase in sea levels, which then lead to disastrous consequences for low-lying coastal areas and island states, and in the near horizon for the youngest among us are living the consequences. Greater frequency of extreme weather events such as floods and droughts is also anticipated. Climate change will directly affect ecosystems, infrastructure, economy, and well-being of people worldwide. Emissions of GHG in the past, present, and future will destabilize the climate system for a long time. This means that even if global emissions were reduced and stabilized now, climate change will continue to affect us for several more decades. Beyond the gradual effects approached due to the increase in average global temperatures, the Intergovernmental Panel on Climate Change (IPCC) estimates that a global warming of more than 2°C above the temperature of the preindustrial exposes us to the risk of abrupt and irreversible changes (tipping points) in the functioning of the climate system. To limit global warming to a maximum increase of 2°C, the IPCC estimates that global GHG emissions must be stabilized by 2015 and be reduced by half by 2050. In summary, there is still time to act but the time window in which we can do it quickly narrows [24, 25].

3.1.1 Greenhouse effect and global warming

The greenhouse effect is a natural phenomenon of partial retention of solar radiation and the earth's heat in the atmosphere, described in 1827 by the physicist Jean Baptiste Fourier. This phenomenon essential to our existence gives an average temperature of 15°C on earth. If it did not exist, the average surface temperature of the earth would be −18°C. The water is in ice and life would probably never appear on earth. What is dangerous is not the phenomenon itself, but its rapid increase due to human activities is worrying [18–26, 27].

The Intergovernmental Panel on Climate Change predicts an average warming of 1–3.5°C by 2100. This would raise the sea level by 15–95 cm. This threatens to flood completely some islands in the Pacific and Indian oceans and it will amplify the frequency and severity of weather events such as floods and droughts [28–31].

3.1.1.1 Greenhouse effect mechanism

The climate is governed by the heat balance of the earth. The essential energy source for the planet is the incident solar flux (short wavelength) whose absorption or not by the earth system depends primarily on the constituents of the atmosphere that absorb and reflect about 50% of the flow incident (H_2O, CO_2 and O_3, aerosols and clouds) (**Figure 11**). The other half of the incident flux is absorbed by the land surface and is re-emitted to the atmosphere as infrared wavelengths (IR, wavelengths). The greenhouse gases absorb IR and retransmit them to the surface of the earth and into space. So, greenhouse gases act like the glass in a greenhouse; increasing the concentration of these gases in the atmosphere causes an increase of the temperature of the greenhouse.

3.1.1.2 Gas contributing to the greenhouse effect

Gases contributing to the greenhouse effect are mainly water vapor, carbon dioxide, methane, and nitrous oxide. Industrial greenhouse gas includes fluorinated compounds, namely, hydrofluorocarbons or CFCs, perfluorocarbons or PFCs, and sulfur hexafluoride or SF6; these gases are also responsible for the degradation on the ozone layer. These are the six gases covered by the Kyoto Protocol (**Table 3**). The gases do not all cause to the same intensity to the green house effect. Indeed,

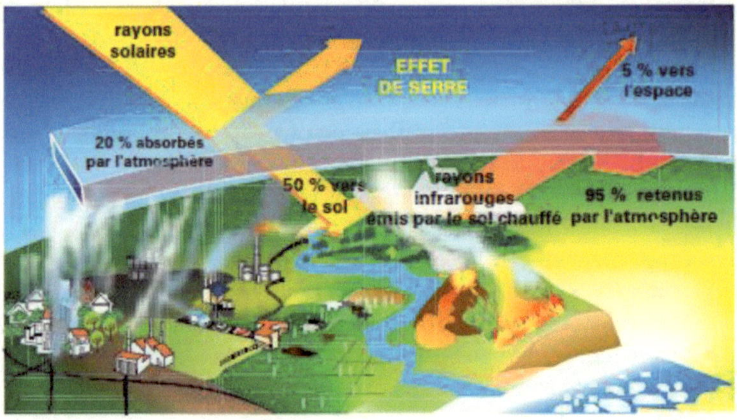

Figure 11.
Greenhouse effect mechanism.

Greenhouse gas		Chemical formula	GWP	Atmospheric lifetime (years)
Carbon dioxide		CO_2	1	200 (variable)
Methane		CH_4	25	12
Nitrous oxide		N_2O	298	114
Fluorochemicals	Dichlorodifluoromethane (CFC-12)	CCl_2F_2	10,900	102
	Chlorodifluoromethane (HCFC-22)	$CHClF_2$	1810	12,1
	Perfluoromethane	CF_4	7390	50,000
	Sulfur hexafluoride	SF_6	22,800	3200

Table 3.
Length of stay and global warming potential of greenhouse gases [32, 33].

some have a greater warming potential than others and/or length of stay (residence) longer.

It is noted that the duration of stay in the atmosphere of the different green house gases varies widely: 12 for CH_4, approximately 200 years for CO_2 and more than 50,000 for CF_4. This means that the carbon dioxide produced today will still affect in a century.

The contribution to the green house gas effect of each gas is measured by the Global Warming Potential (GWP). The global warming potential of a gas quantify the radiative forcing (i.e., the power that radiative greenhouse gas returns to the ground) accumulated over a period of 100 years (that is how we increase the greenhouse when it emits one kilogram of the gas). This value is measured in relation to CO_2. If 1 kg of methane is emitted into the atmosphere, it has the same effect on a century if we had issued 25 kg of carbon dioxide. If 1 kg of sulfur hexafluoride is emitted in the atmosphere, it has the same effect on a century if we had issued 22,800 kg of carbon dioxide. That is why greenhouse gases are measured in carbon equivalent. By definition, 1 kg CO_2 contains 0.2727 kg carbon equivalent, i.e., the weight of carbon only in the carbon dioxide compound. For other gases, the carbon equivalent is given by the equation:

$$\text{Carbon Equivalent} = 0.2727.\ \text{GWP relative} \tag{1}$$

Greenhouse gases are not very abundant naturally. But because of the human activity since the Industrial Revolution, the concentration of these gases in the atmosphere has changed significantly. According to the IPCC, if no measures are taken, the CO_2 content in the atmosphere rose from 260 to 400 ppm today to 1000 ppm at the end of the century, leading to a rise in temperature between 2 and 6°C [34].

The following figure shows the concentration of CO_2. First observed seasonal net CO_2 concentration. It varies in the range of 2% per year, with a maximum in month of May (at the beginning of the growing season) and a minimum in October (end of season) [34].

Global emissions of greenhouse gases are increasing steadily, despite various policy initiatives such as the Kyoto Protocol. The situation varies considerably from one sector to another.

In the reference scenario IEO 2013 [16], global emissions of carbon dioxide produced from burning fossil fuels increased by 31.2 billion tons in 2010 to 36.4 billion tons in 2020 and 45.5 billion tons in 2040 (**Figure 12**).

The increase in CO_2 emissions was 40% from 1990 to 2010, reaching 31.7 billion tons of carbon dioxide in 2012. The increase of 1.3% in CO_2 emissions worldwide is largely caused by 300 million tons of emissions in China and 70 million tons in Japan (since the Fukushima disaster in 2011, Japan uses more fossil energy). The United States and some European countries reduced their emissions. Much of the increase in emissions is attributable to developing countries that are non-OECD members who continue to rely heavily on fossil fuels to meet the fast pace of growth in energy demand. Regarding emissions from non-OECD countries, it is predicted to reach 31.6 billion tons in 2040, or 69% of the world total, in comparison to emissions from OECD countries which would be around 13.9 billion tons in 2040 or 31% of the world total (**Figure 13**).

CO_2 emissions from transportation modes, should, if no action is taken, continue to rise at current rates. The graph in **Figure 14** shows the evolution observed and projected

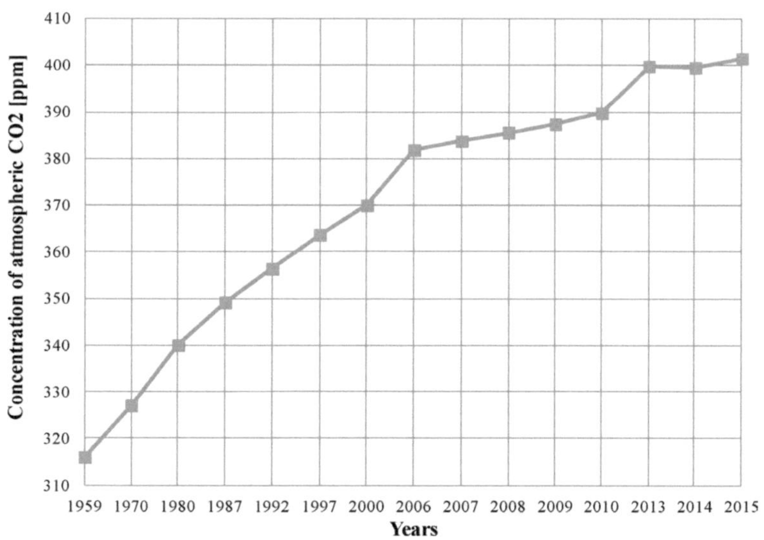

Figure 12.
Global CO_2 concentration [34, 35].

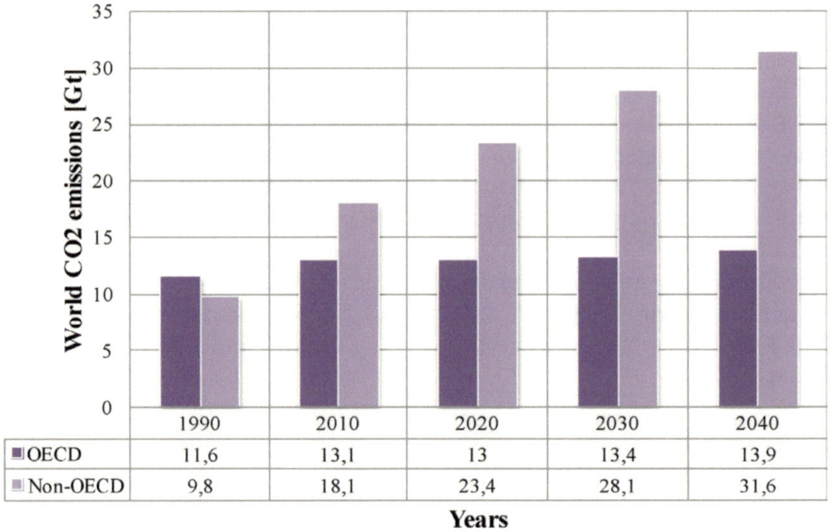

Figure 13.
Global CO_2 emissions.

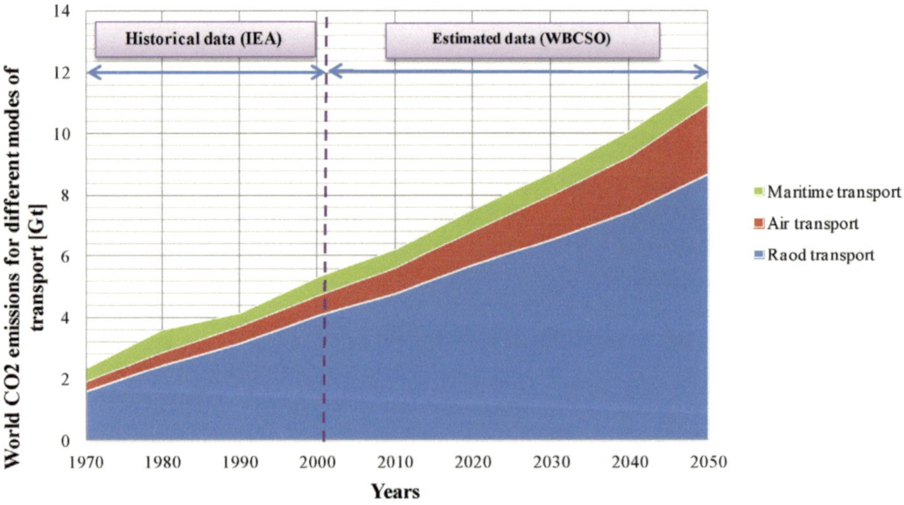

Figure 14.
Evolution observed and projected global CO_2 emissions in different modes of transport [36].

of emissions from different transport modes between 1970 and 2050. CO_2 emissions from transport are expected to double between 2000 and 2050, the largest part of the increase being related to road and air transport. Freight transport has grown faster than passenger transport, and progression seems set to continue in the future [36].

Globally, the transport sector is the second largest emitting sector with 7 Gt of CO_2 emitted in 2011, behind the energy sector, comprising the production of electricity and heat. According to projections by the International Energy Agency, these emissions will continue to rise, reaching 18 Gt in 2050 in the baseline scenario [37].

Beyond this overall finding, disparities are hiding across geographical areas. According to the reference scenario of the International Energy Agency, while the CO_2 emissions of the transport sector in developed countries will grow steadily in 2050, the same emissions in developing countries will, in turn, grow exponentially. At the head of these, China and India contributions reach, respectively, 4 and 1.5 Gt in 2050 (against 0.332 Gt and 0.1 Gt in 2005, respectively) [38].

3.2 Local pollution

Unlike greenhouse gases, local pollution has a direct impact on the health and comfort of the car users and others. The impact is most noticeable in areas of heavy traffic, especially in cities. Pollutants are due to fuel combustion or simply the running of the vehicle, such as the following:

- **Fine particles:** Rated PM for particulate matter, they are inorganic or organic particles or a mixture thereof [39]. They are classified according to their size; finer less than 2.5 μm named PM2.5 and coarser that have a diameter between 2.5 μm and 10.mu.m called PM10 [40, 41]. In 2007 in France, road transport occupied the fourth place of emission sources of these two types of particles, with 12% and 11%, respectively [40]. Tests on a vehicle "light" diesel, produced according to the European procedure (cycle Motor Vehicle Emissions Group (MVEG)), measured emissions of 1010 particles/km much higher than the limit (6106 particles/km) specified in the standard EuroV 2009 [42].

- **Heavy metals:** Other fine particles of heavy metals such as copper (Cu) and lead (Pb) are also issued by the transport sector. Copper is linked to the wear of brake pads for road vehicles, but it mainly comes from the wear of overhead lines in rail transport. Copper emissions are continuously increasing with the growth in traffic. As for lead, it is due to the use of leaded petrol and consumption of part of the "engine" oils containing traces of lead. Lead emissions are related to its content in the fuel [43]. Upon the arrival of unleaded gasoline, emissions have dropped drastically. Since 2000, road transport contributes 4% against 91% in 1990 of total Pb emissions in France [42–44].

- **Noise:** Noise pollution from transport, contrary to releases of gases and particles, has little adverse impact on public health. However, they represent a source of annoyance for 40% of French and 25% of the European population. 17% of these noises are caused by cars [45]. In fact, emissions of conventional cars are measured at an average intensity of 70dBA9. The electric vehicle emits the same intensity noise than conventional vehicles beyond 30 km/h. Nevertheless electric vehicles fell by 10dBA and 6.5dBA for speed 5 and 10 km/h, respectively [46]. Knowing that the noise is characterized by its intensity and duration [41], people in urban areas are the most affected, and the noise reduction can be one more argument for the electrification of transport [47–49].

The traffic noise has multiple negative effects on health. The World Health Organization recognized the ambient noise, especially the noise originating from car traffic, as a serious public health issue. The traffic noise has various kinds of harm. The most common effect is a mere annoyance, but it seems also proven that traffic noise causes serious health problems, including sleep cycle disturbances affecting cognitive function (especially in children), and helps in the emergence of certain cardiovascular diseases. It also seems increasingly obvious that it raises blood pressure. It has been estimated that over 245,000 people in the European Union acquired every year cardiovascular disease attributable to traffic noise. Some 20% of these patients (nearly 50,000 people) are victims of a heart attack, leading to their premature death. There are no comparable estimates for other parts of the world, but there is no reason not to think that much of the population suffers from traffic noise elsewhere [36].

Pollutant emissions are a source of health costs, damage to buildings and materials, crop losses, and other damage to ecosystems (biosphere, soil, water). Each impact is driven by one or more types of pollutants [36]:

- Impact on health: this impact comes from inhaling fine particles ($PM_{2.5}$/PM_{10} and other air pollutants). The particles contained in the exhaust gas may be considered the most important of these pollutants. Ozone (O_3) also affects health. The impact essentially translates into a worsening health problems for people with respiratory diseases and an increased risk of contracting these diseases.

- Degradation of buildings and materials such as damage mainly consists in two aspects: the first in the form of dirt and other surfaces of the facades of the buildings due to particles and dust and the second largest in the form of degradation due the corrosive action of acidic air pollutants such as NO_x and SO_2.

- Losses of crops and impacts on the biosphere: acid rain, ozone, and SO_2 damage crops, forests and other ecosystems.

4. Regulatory and technological aspects

Since the 1970s, the issue of preserving the environment has become a concern. International authorities have begun to take concrete steps to reduce the discharge of pollutants. The Kyoto Protocol adopted in 1997 is one of these measures. It had set a target average 5.2% reduction in greenhouse gas emissions of industrialized countries. And, although all sectors are concerned, the emissions from transport and automotive in particular have been the target of regulatory mechanisms put in place by governments. Thus, automakers are constantly forced to improve their technologies while maintaining low costs.

4.1 Regulatory aspects

The regulatory framework for emissions is different from one country to the other. However, all standards adopted by industrialized countries such as the European Union and its member states, the United States and Japan, are aimed at limiting emissions of CO, NO_x, HC, and PM (**Figure 15**). Each standard is based on specific test procedures. For information, the Tier-2 standard adopted in 1999 in the United States does not distinguish between fuels, while the Japanese standard takes into account the state (hot or cold) of the vehicle. **Figure 15** shows a comparison of these standards. It may be noted that Japanese standards are the strictest. They are followed by the European standard Euro V applied from 2009 and finally in US standards. European standards are, like all standards, whenever revisited and emission thresholds are revised continually declining.

CO_2 emissions have been subject to a voluntary agreement between the European Community and the European Automobile Manufacturers Association. This agreement aimed to reduce emissions from passenger cars to 120 g/km by 2012 and 95 g/km by 2020. To achieve this, France and other countries have imposed, since May 2006, car manufacturers labeling their classes 10 new vehicles to encourage the purchase of less polluting models. Financial incentives to buy cleaner vehicles were also implemented through environmental taxes such as the Bonus/Malus ecological established in France since January 2008.

4.2 Technological aspects: alternative to the internal combustion engine

The scarcity of oil coupled with increasing demand (primarily bound to the development of emerging countries) causes an inevitable increase in the price of oil. This increase in oil prices gives researchers and industry the opportunity to

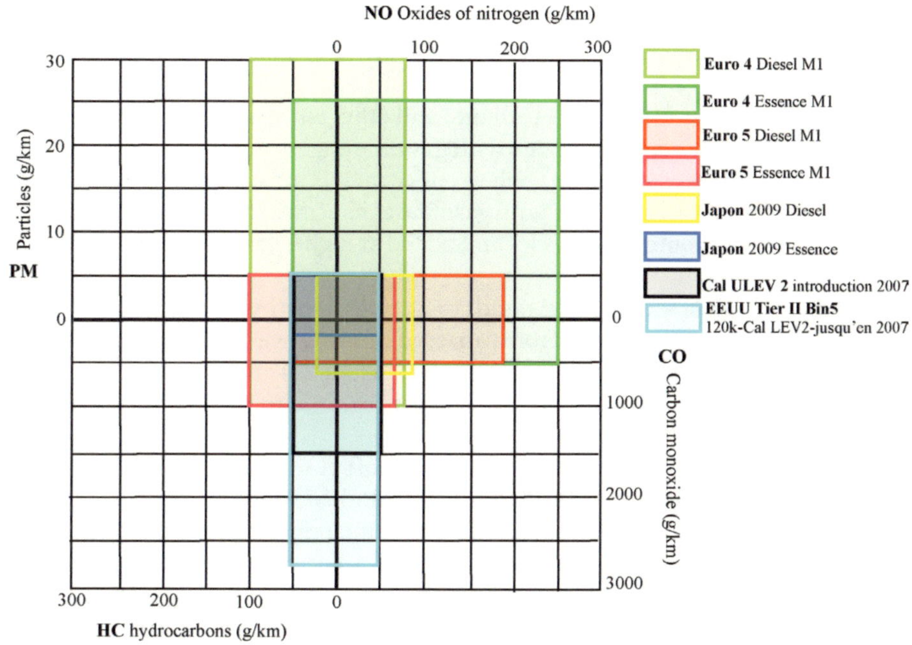

Figure 15.
Comparison of car emission limits in the European Union, Japan, and the United States.

explore new technological avenues that were not possible before. In this context, the return of investment on these technology is almost guaranteed. The era of the post-oil is actually launched. The urgent need to reduce CO_2 emissions worldwide, combined with the rise in oil prices, requires decision-making advocated by new political commitments. New commitments lay the foundations for a new global orientation in terms of energy management. The goal is to enable research and commercialization of new solutions, which involve the following:

- The optimization of combustion engines (improved technologies used currently in vehicles):
- Improving the efficiency of engines (downsizing, supercharging, injection, post processing, aerodynamics, etc.)
- Hybridization (electrical and thermal) and electrification of the drive train
- The use of new energies (diversification of energy sources used for traction vehicles)
- Biofuels
- Liquefied Petroleum Gas (LPG)
- Natural gas vehicles (NGV)
- Hydrogen (a few decades)

4.2.1 Biofuels

They were encouraged by the government following the two oil shocks in the 1970s. Today, after over 20 years of industrial development, these plant-derived fuels have their place in the energy mix. The European Union has set a particular incorporation target of 10% of renewable fuels in road transport by 2020 [50]. The two large biofuel production chains are the petrol sector and the diesel industry.

- The biofuel sector includes fuel ethanol and ethyl tertiary butyl ether (ETBE). The ethanol to be supplied to cars with gasoline engine is derived mostly from sugar plants (beet, sugar cane), wheat, or corn. It is mostly used in Brazil and the United States. It can be mixed with gasoline at concentrations ranging from 5 to 10% for the conventional vehicles and added at higher rates for adapted vehicles.

- The diesel biofuel sector corresponds to fatty acid methyl esters (FAME), which are made from vegetable oils from rapeseed or sunflower (Europe), animal fats, or recycled used oil and soy (United States). Biodiesel can be incorporated into the diesel fuel in amounts up to 7%. This rate can rise to 30% for certain fleets.

To increase the availability of biofuels, new biofuel production chains are being studied. They are used as feedstock lignocellulosic biomass, agricultural residues (corn stalks, cereal straw), and forestry, including dedicated poplar crops and organic waste such as sewage sludge. Ethanol processing by the biochemical pathway is studied. Similarly, for diesel, the transformation of this synthetic liquid fuels from biomass by the Fischer-Tropsch process is the subject of ongoing R&D around the world. These channels have many advantages: potentially lower costs, no competition with the food chain, and no limits to production volumes, no co-products to sell. But research is still needed for their development. LPG is a mixture of 80% butane (C_4H_{10}) and 20% of propane (C_3H_8) and heavier than air. It can directly come from stripping operations (oil extraction) during production on fields or crude oil refining. Its use has advantages compared to diesel fuel to reduce [51, 52]:

- 50% of nitrogen oxide emissions

- 50% those of carbon monoxide

- 90% of those hydrocarbons and particles

LPG is mainly distributed in gas cylinders for domestic heating, cooking food, and chemical industry, transport representing only a minority share of its consumption. Currently, It is consumed in significant quantities for the automobile only in few European countries and Australia, but China is a growing market, and the United States is considered to be a potential consumers in the years to come. If it is in the gaseous state at ambient conditions, LPG is usually stored in liquid format a pressure of 10 bars [52]. Currently, vehicles using this fuel are mostly equipped with a "dual-fuel" system (operation on petrol or LPG), which allows to adapt to the refueling station density available. This approach involves the LPG as an added feature to the existing petrol vehicles and makes operation in fuel optimal [50].

4.2.2 GNV/GNL

The largest deposits of natural gas (methane), whose reserves are larger than those of oil, are in the following countries: the Unites States, Russia, Canada, Iran and Qatar, Algeria, Nigeria, South Africa, Argentina, Australia, etc. It is fed by a major pipeline network or, alternatively, at high cost to the liquid, chilled to −162°C between very specific terminals special water (LNG); it gives the appellation liquefied natural gas (LNG). For automotive applications, called NGV. Methane is stored on board with a pressure of 200 bars in the tanks, the volume of which corresponds to 5 times that of a fuel tank of equivalent energy content. A distribution network already exists in some cities, but a compression facility is necessary for refueling stations because the pressure of the network generally does not exceed 30 bar. The acceptability of the presence of reservoirs and the specific filling operation are also issues to consider [52].

NGV can be easily fitted to a motor provided to run on gasoline. It reduces by nearly 25% the emissions of greenhouse gases. The engines used offer good energy efficiency (greater than 10 to 15% that of a gasoline engine) combined with a very low potential for regulated toxic emissions. Its exhaust will emit no sulfur oxides, lead, or particles [50–52]. 11.3 million CNG vehicles are circulating in the world. These are mostly light and commercial vehicles; trucks and busses do in fact represent only 5–6% of the park. The park is primarily developed in the following areas: Pakistan (20% of global NGV fleet), Argentina (16%), Iran (15%), and Brazil (14.5%). These four countries alone account for 2/3 of global NGV fleet. In Europe, CNG has developed mainly in Italy (6% of global NGV fleet) [50].

4.2.3 Synthetic fuels

Synthetic fuels are liquid fuels made from natural gas, coal, or biomass. They are called gas to liquids (GTL), coal to liquids (CTL), and biomass to liquids (BTL). Their production takes place in two steps:

- Conversion of the synthesis gas energy source formed by a mixture of carbon monoxide and hydrogen
- Chemical conversion of synthesis gas (process "Fischer-Tropsch") into liquid hydrocarbons

4.2.3.1 Fuels from natural gas (GTL)

They can be integrated into "fuel pool" current and distributed through existing channels. The fuels produced are of excellent quality. They contain no sulfur or aromatics and result in net reductions in particulate emissions, unburned hydrocarbons, and CO (carbon monoxide). The cost of production has been reduced in recent years, and a new generation of catalysts is used to maximize yields. This sector accounts for natural gas, an outlet which in the future could become major.

4.2.3.2 The path from coal (CTL)

More expensive, it is interesting for countries with large coal resources (China and India). Research efforts are still needed and the problem of CO_2 emitted must be paid by his capture and geological storage.

4.2.3.3 The solution biomass (BTL)

Liquid fuels made from it, in this case, lignocellulosic biomass: agricultural residue (stalks, straw) and forestry, including dedicated poplar crops and organic waste such as sewage sludge. This solution has two advantages: reduced energy dependence and reduced CO_2 emissions. But the costs are still high because the sector is still at the stage of research and development. New technologies are expected in 2015.

4.2.4 Hydrogen (H_2)

Hydrogen is the most abundant element in the universe. On our planet, it is mainly found in water and in hydrocarbons which are the sources of industrial

Hydrogen	Natural gas	LPG	Coal
34 kWh/kg	17 kWh/kg 11 at 12 kWh/m^3	13 kWh/kg	7.2 kWh/kg
Fuel	**Gasoline**	**Wood**	**Natural uranium**
11,6 kWh/kg	12 kWh/kg	2 at 4 kWh/kg	120.10^3 kWh/kg

Table 4.
The average energy values of PCI main fuels [54].

hydrogen. It is widely used in the chemical industry and oil refining, among others. Due to its clean combustion, it is also considered as one of the energy vectors of the future [53]. The H_2 gas is about 8 times lighter than methane and must be compressed at very high pressure or liquefied for storage in significant amounts: under 700 bars, 1 kg of H_2 still occupies a volume of 23 l. Its liquefaction for cryogenic storage at −253°C consumes at least 30% of the initial energy content of the hydrogen, but it allows to concentrate 1 kg in a volume of 14 l [52]. A kilogram of hydrogen releases about three times more energy than a kg of gasoline and fuel oil (**Table 4**).

Indeed since hydrogen does not exist on earth, it took and it will in the future develop less expensive methods and more profitable products. Here are the different ways to produce hydrogen; now some have already reached technological maturity, and others are still in development and study.

- We can create it from fossil fuels (several methods): steam reforming and partial oxidation.
- But also from the electrolysis of water (commonly called crack water), electricity would come from renewable energy.
- From biomass.

4.2.4.1 Production of hydrogen from fossil fuels

It is the most prevalent currently. But this technique is not a viable solution because hydrocarbons have a limited lifespan. There are two different methods for this style of manufacture:

4.2.4.2 Steam reforming

Steam reforming or steam reforming is to convert hydrocarbons by reaction of synthesis gas with water vapor and in the presence of a nickel catalyst, at high temperature (840–950°C) and moderate pressure (20–30 bar). The fillers conventionally used are light hydrocarbons which include natural gas, LPG, and naphtha to boiling points of 200–220°C. Alcohols such as methanol or ethanol may also be used. Natural gas, however, is the reference load [55–56]. The steam reforming of the general reaction is:

$$C_nH_m + 2nH_2O \rightarrow (m/2 + 2n)\,H_2 + nCO_2 \qquad (2)$$

4.2.4.3 Partial oxidation

The partial oxidation is to transform the hydrocarbon by oxidation synthesis gas formed in the presence of oxygen. This reaction takes place at high temperature

(1200–1500°C) and high pressure (20–90 bars) and does not require the presence of a catalyzer [53]. The chemical reactions may be summarized in the case of hydrocarbons by the following formulas:

$$C_nH_m + (n/2)\,O_2 \rightarrow nCO + (m/2)\,H_2 \tag{3}$$

$$nCO + nH_2O \rightarrow nCO_2 + nH_2 \tag{4}$$

The reaction is exothermic, for example, the enthalpy of the reaction with methane is -35.7 kJ.mol^{-1}. The advantage of the partial oxidation reaction is its exothermic (unlike the steam reforming reaction) for assisting catalysis (temperature rise). The main drawback lies in the fact that the H_2 percentages are lower than those obtained by steam reforming, due to the majority presence of the nitrogen from the air. In addition, it is possible to obtain NOx.

4.2.4.4 The water electrolysis

Electrolysis of water, if it covers only a few percent of all hydrogen produced [57–59] is nevertheless of great interest because it represents a fashion own production (especially if the electricity is nuclear, hydro, solar, and wind) which provides high purity hydrogen [53]. This technology works by passing an electric current through water to obtain the dissociation of water molecules into oxygen and hydrogen gas. The electrolysis of water involves two chemical reactions taking place separately on two electrodes. At the cathode, the electrolysis of water occurs according to Eq. (1), and at anode, the oxidation of hydroxide ions occurs according to:

$$2H_2O + 2e^- \leftrightarrow H_2 + 2OH^- \tag{5}$$

$$2OH^- \leftrightarrow 1/2O_2 + H_2O + 2e^- \tag{6}$$

The global reaction is written as:

$$H_2O \rightarrow H_2 + 1/2O_2 \tag{7}$$

There are other means of hydrogen production, some of which are still under study. There are particular biological processes set works by algae or microorganisms (digestion, photosynthesis).

The hydrogen can be used either as:

- Heat engine
- Fuel cell

4.2.5 The hydrogen combustion engine

Internal combustion means combustion of engine motors (commonly called "explosion") which, according to Beau de Rochas' Otto cycle or Diesel cycle, operates from the combustion of petrol, diesel or gas coal, natural gas, or distillates derived from fermentation of organic matter. All internal combustion engines can be converted to operate on hydrogen. Precautions are to be taken to avoid a flashback to the intake manifold. The loss on the power level is 20–25% compared to a gasoline engine of the same capacity, but the energy efficiency is equivalent. There is more need to select suitable materials for hydrogen (corrosion, lubrication). Such engines do not emit CO_2 but only few nitrogen oxides; they are well suited to hybrid gasoline-hydrogen as are biofuel prototypes developed since 1979 by the German manufacturer BMW. The BMW manufacturer who has subsequently improved its engines by the adopting the technique of direct injection of hydrogen at high pressure [60] has now abandoned this type of hydrogen thermal engine. Other manufacturers such as MAN, Ford, Mazda, and Quantum have also conducted research on these types of hydrogen thermal engines and have built prototype cars (cars, vans, and busses) [61], but eventually, these manufacturers has stopped at the prototyping phase. For its part, Mercedes has designed an engine operating with a mixture of hydrogen and natural gas [62].

4.2.6 Fuel cell

Car manufacturers seem to focus more on the torque hydrogen/fuel cell for the long term. The couple hydrogen/fuel cell appears, indeed, in principle, as a strong candidate to succeed the torque petroleum fuel/combustion engine. Hydrogen is then used to power a fuel cell which generates electricity to enable operation of an electric motor which will move the vehicle. Hydrogen is by definition the best energy carrier for the cell: no CO_2 emissions and better performance for the battery, including a performance about twice a petrol engine urban cycle. Fueled by a mixture of air and hydrogen, the cell converts the chemical energy of hydrogen into electrical energy according to the reverse principle of electrolysis. By reacting hydrogen with oxygen on the electrodes, fuel cells can produce electricity without programming other than water vapor. The idea dates back to 1839! It has long been used to generate electricity onboard rockets. The proton exchange membrane fuel cells (PEMFC) are best suited as a carrier. It is this type of battery that automakers concentrate most of their research.

4.2.7 Hybrid vehicles

It is a vehicle equipped with a system of thermal/electric motorization mixed and two energy storage systems: a fuel tank and a battery. The main advantage of the hybrid car is to get to optimize the integration of the two types of engines to make the most of each of them. Any kind of combinations is theoretically possible, the combustion engine can be used both for charging the batteries for the vehicle drive and the electric motor can be used both to move the vehicle to retrieve his braking energy. The dual petrol (or diesel)/electricity allows the optimization of energy use in the vehicle, reducing emissions of pollutants, including CO_2, and a consumer economy (10–50% depending the degree of hybridization).

4.2.8 Electric vehicles

The electric car will appear environmentally as "zero emission" solution. The electric vehicle has a reduced autonomy and a time of significant recharge, which could limit its use (on relatively short trips or for captive fleets) and its commercial development. The current batteries offer only a hundred kilometers of autonomy against nearly 1000 km for diesel vehicles. Though, a good progress is made in this research area. Advanced lithium ion or lithium polymer batteries, offered by Dassault and Bolloré groups, will enable a range of approximately 250–300 km (always with several hours cooldown). However, some manufacturers now positioning their electric vehicles as urban perished vehicles with a modest autonomy (between 80 and 160 km), which is no longer a handicap for commuting home/work. There is, in addition, a quick battery charging solutions for use in an emergency; these refills would restore in minutes between 50 and 80% of autonomy to allow the vehicle to return to its final destination.

5. Conclusion

The all-electric vehicles have again become a topical issue that seems to be the modern answer to deal with the explosion in the price of oil and environmental constraints, including increased green house gas emission effect linked to road transport especially automotive. Moreover, these means which are independent of fossil fuels and friend to the environment itself have many other benefits; they are silent, featured with relatively constant consumption and have good overall energy efficiency.

Author details

Brahim Mebarki[1*], Belkacem Draoui[1], Boumediene Allaoua[2] and Abdelghani Draoui[3]

1 ENERGARID Laboratory, Tahri Mohammed University of Bechar, Bechar, Algeria

2 Laboratory of Smart Grids and Renewable Energies, Tahri Mohammed University of Bechar, Bechar, Algeria

3 Laboratory of Environmental and Energetics Systems (LSEE), Institute of Science and Technology, University Center Ali KAFI Tindouf, Algeria

*Address all correspondence to: brahimo12002@yahoo.fr

References

[1] Energy Council. Global Transport Scenarios 2050. London: World Energy Council; 2011. Available from: https:/ worldenergy.org

[2] Chan C. The state of the art of electric and hybrid vehicles. Proceedings of the IEEE. 2002;**90**(2):247-275

[3] Guemri M. Heuristiques optimisées et robustes de résolution du problème de gestion d'énergie pour les véhicules électriques et hybrides [Thèse de doctorat]. Université de Toulouse; 2013

[4] Available from: http://www.cybersciences.com/

[5] Centre National du Transport Avancé. Commission sur les enjeux énergétiques du Québec: Electrification des Transports. 2010

[6] Marie-Lou P. Evaluation environnementale du véhicule électrique: Méthodologies et application [Thèse de doctorat]. École Nationale Supérieure des Mines de Saint-Étienne; 2010

[7] Çagatay Bayindir K, Gözüküçük M, Teke A. A comprehensive overview of hybrid electric vehicle: Powertrain configurations, powertrain control techniques and electronic control units. Energy Conversion and Management. 2011;**52**(2):1305-1313

[8] Damien P. Dimensionnement et gestion d'énergie de systèmes d'entrainement électriques hybrides: Application a un ascenseur avec récupération d'énergie [Thèse de doctorat]. Université de technologie de Belfort-Montbéliard; 2010

[9] Ehasani M, Gao Y, Emadi A. Modern Electric, Hybrid Electric and Fuel Cell Vehicles-Fundamentals, Theory, and Design. CRC Press Taylor & Francis; 2010

[10] BP Statistical Review of World Energy. 2013

[11] IFP New Energies: What are the raw material resources for a sustainable energy system. Panorama briefing notes; 2010

[12] Available from: http://www.energiepositive.info

[13] Jacques P, Claude M. Rapport énergies 2050. Ministère de l'économie des finances et de l'industrie: République Française; 2012

[14] Contribution à l'Etude d'Electro-générateurs à Pile à Combustible. Conceptions d'Architectures et de Leurs Commandes [Thèse de doctorat]. Université Paris-Sud XI; 2010

[15] Samantha L. Modélisation et commande d'une chaîne de conversion pour véhicule électrique intégrant la fonction de charge des batteries [Thèse de doctorat]. Université Paris-Sud; 2013

[16] U.S. Energy Information Administration. International Energy Outlook 2013; 2013

[17] Tie SF, Tan CW. A review of energy sources and energy management system in electric vehicles. Renewable and Sustainable Energy Reviews. 2013;**20**:82-102

[18] Victor M. Conception Optimale Systémique des Composants des Chaînes de Traction Electrique [Thèse de doctorat]. Ecole Centrale de Lille; 2007

[19] World Economic Forum. Repowering Transport. 2011

[20] Yves S. Impacts de la voiture électrique sur les émissions de CO_2 liées à la mobilité individuelle Le cas de l'agglomération franco-valdo-genevoise [Mémoire master]. Université de Lausanne; 2010

[21] Mesures des émissions polluantes automobiles—Application à la modélisation eulérienne 3D de la formation des oxydants photochimiques dans la troposphère. [Mémoire de thèse]. USTL; 2005

[22] Paul D. Pollution atmosphérique Origine et réduction des émissions au niveau moteur. Techniques de l'ingenieur. 1994;**B2710**

[23] World Meteorological Organization

[24] Gouvernement du Québec. Plan d'action 2013-2020 sur les changements climatiques, 2012

[25] International Energy Agency. Key World Energy Statistics. 2011

[26] Christian N, Conseil E. Analyse et perspectives énergétiques mondiales. Techniques de l'ingenieur. 2009;**BE8515**

[27] Le changement climatique. Agence Internationale de la Maitrise d'Energie; 2012

[28] Source Documentaire PNUE. L'avenir de l'environnement mondial, GEO5. 2005

[29] Pachauri RK, Reisinger A, et al. Bilan 2007 des changements climatiques. [Rapport de synthèse PNUE]. 2007

[30] Jaecker-Voirol A. Emissions polluantes des moteurs thermiques—Valeurs limites. Dossier Techniques de l'ingénieur, ref. BM 2 507. 2010

[31] Redha S. Intérêt d'une Source d'Energie Electrique Hybride pour véhicule électrique urbain—Dimensionnement et tests de cyclage [Thèse de doctorat]. Ecole centrale de Lille, Université Lille Nord-de-France; 2013

[32] Pachauri RK et al. Rapport du Groupe de Travail 1 du Groupe Intergouvernemental d'Experts sur l'Evolution du Climat (GIEC) – Résumé Technique. 2007

[33] Guillaume ALIX. Comparaison de Concepts de Moteur Diesel en Vue de Leur Hybridation: Bilan des émissions de CO_2 Suivant l'Usage en Véhicule. Ecole centrale de LYON; 2008

[34] Available from: http://www.planetoscope.com/co2/261-emissions-mondiales-de-co2-dans-l-atmosphere.html [Accessed: 23 June 2015]

[35] Available from: http://co2now.org/current-co2/co2-now/ [Accessed: 25 April 2015]

[36] OCDE. Mondialisation, Transport et Environnement. OCDE; 2011

[37] International Energy Agency. CO_2 Emissions from Fuel Combustion Highlights. 2013

[38] Simon V, Paula C. IFP, Innovation Energie Environnement. Panorama. 2009

[39] Caplain I. Mesure des émissions polluantes automobiles—Application à la modélisation eulérienne 3D de la formation des oxydants photochimiques dans la troposphère [thèse de doctorat]. Université de Lille; 2005

[40] Documentary source, Sétra, Cete de Lyon, CeteNormandie-Centre, Atmospheric pollutants emissions in roads, graphs and influence factors. Information notes. 2009

[41] Cottignies M. Transport et développement durable. Dossier technique de l'ingénieur, ref. AG 8 230

[42] Sassi A, Rohart E, Belot G, Post traitement des émissions polluantes des moteurs thermiques à combustion interne—Moteurs à allumage commandé, Dossier techniques de l'ingénieur, ref. BM 2 508

[43] Fulton L, Eads G. IEA/SMP Model Documentation and Reference Case Projection. 2004

[44] Source Documentaire CITEPA. Les polluants atmosphériques. Available from: http://www.citepa.org/fr/pollution-et-climat/polluants/metaux-lourds/plomb [Available from: 09 December 2012]

[45] Lambert J. La gêne due au bruit des transports terrestres, Acoustique et Techniques n°28. Paris; 2002

[46] Gruson A. Etude de la détectabilité du véhicule électrique en milieu urbain, PFE-IRCAM. 2012

[47] Lambert J, Vincent B, Vallet M. Bruit des transports, effets sur l'homme, indices et réglementations, Dossier techniques de l'ingénieur, BR2005, 2007

[48] Wolfram M. Sustainable Urban Transport Plans (SUTP) and Urban Environment: Policies, Effects, and Simulations, Review of European References Regarding Noise, Air Quality and CO_2 Emissions. 2005

[49] Mobility and transport in Ile-de-france, noise issued from transport in Ile-de-france

[50] Available from: http://www.ifpenergiesnouvelles.fr/

[51] Available from: http://www.actu-environnement.com

[52] Valeo. Le Véhicule Zéro Emission. 2010

[53] Christophe B. Hydrogène. Techniques de l'ingenieur. 2012;**J 6368**

[54] Bernard M, Gael R, Marie R, Hamid BA. Situation énergétique mondiale à l'aube du 3eme millénaire. Perspectives offertes par les ressources renouvelables. Revue 3EI. 2004. pp. 20-33

[55] Bourbonneux G. Production d'hydrogène, Le Raffinage du Pétrole, T3 Procédés de Transformation. 1998. pp. 463-515

[56] Les techniques de production de l'hydrogène et les risques associés. Rapport d'étude N° DRA-08-95313-07833B. Institut National de l'Environnement Industriel et des Risques; 2008

[57] Hydrogen generation by water electrolysis. L'actualitéchimique. Janv-févr 1995. pp. 39-42

[58] Shultz P. Production d'hydrogène par électrolyse de l'eau. Mémento de l'hydrogène, Fiche 3.2.1, AFH2

[59] Andreassen K. Hydrogen Production by Electrolysis, Hydrogen Power: Theoretical and Engineering Solutions. The Netherlands: Kluwer Academic Publishers; 1998

[60] Rottengruber H, Berkmuller M, Elsässer G, Brehm N, Schwarz C. A high-efficient combustion concept for direct injection hydrogen internal engines. In: 15th World Hydrogen Energy Conference (WHEC), Yokohama, Japan, 27 June–2 July, 2004

[61] Hichlseder H, Grabner P, Heindl R. Hydrogen internal combustion engines. In: Stolten D, editor. Hydrogen and Fuel Cells, Chapter 39. Weinheim, Germany: Wiley-VCH; 2010

[62] Darkrim-Lamari F, Malbrunot P. Combustible hydrogène: utilisation. Techniques de l'ingenieurm. 2013;**BE 8566**

Chapter 10

Solid Green Biodiesel Catalysts Derived from Coal Fly Ash

Miroslav Stanković, Stefan Pavlović, Dalibor Marinković, Marina Tišma, Margarita Gabrovska and Dimitrinka Nikolova

Abstract

Coal fly ash (CFA) is generated during the combustion of coal for energy production. Many studies are based on its utilization as the most abundant, cheap aluminosilicate industrial residue, which is recognized as a risk for the environment and human health. The present review is focused on CFA origin, chemical properties, and its catalytic application for biodiesel production. The aluminosilicate nature and the presence of rare earth elements make CFA suitable for different adsorption, catalytic, and extraction processes for obtaining valuable products including alternative fuels and pure elements. However, the presence of toxic elements is a potential environmental problem, which should be solved in order to avoid soil, water, and air pollution. The most used modification methods are alkali activation, hydrothermal, and thermal treatment that improve the structural, morphological, and textural properties. The active catalytic form could be obtained by impregnation or ion exchange method. It was found that such synthesized materials have significant catalytic potential in the biofuel chemistry. In the case of biodiesel production, the high values of conversion or yield can be achieved under mild low-energy reaction conditions in the presence of low-cost waste-based catalysts.

Keywords: coal fly ash, waste materials, zeolites, catalyst, transesterification, biodiesel

1. Introduction

In recent decades, we have witnessed more and more stories about energy, energy efficiency, fossil fuels reserves, and alternative energy sources. The dependence of the world on the fossil fuels is a topic of discussion by many scientists, researchers, and environmental activists worldwide. Biodiesel, as a biofuel, with numerous advantages (biodegradability, lower content of CO_2, SO_2, and hydrocarbons during combustion, high flash point, high lubricant properties, and high octane number) is a serious competitor to fossil diesel [1, 2]. Also, very important is the fact that biodiesel obtained in accordance to standard does not require modification of existing diesel engine [3]. Modern science and chemical technology know the following concepts of biodiesel production: base-catalyzed transesterification (homogeneous or heterogeneous), acid-catalyzed esterification and transesterification [4–7], biodiesel synthesis catalyzed by bifunctional heterogeneous solid catalysts [8–10], enzyme-catalyzed transesterification [11–13], deoxygenation [14, 15], and supercritical methanolysis [16, 17]. In order to intensify biodiesel production, existing processes are modified in terms of treatment of reaction mixture by ultrasound [18–20] and microwave [21–23]. However, the modern concept of biodiesel production is focused on synthesis of the new catalytic systems, use of different triacylglycerol (TAG) feedstock, and improved batch and continuous reactor systems. Also, it is very important that nowadays investigations are based on the concept of low-cost production, i.e. that waste materials from various production processes are basis for catalysts, and waste feedstock as the main source of TAG.

The numerous disadvantages of homogenous base and acid catalysts, such as soap formation, catalyst recovery, high corrosion, and inhibition by water [1, 24] can be replaced using heterogeneous catalysts such as alkaline [25] and alkaline earth metal oxides [2, 26, 27], mixed oxides [28–30], modified layered double hydroxides [31–34], zeolites [35–37], sulfonated solids [38], ion exchange resins [39–41], supported heteropolyacids [42, 43], etc. In order to design new catalysts based on the modern concept of environmental protection, greatest attention of many scientists is directed at investigation of different waste materials (fly ash from coal-fired power stations (CFPSs), biomass fly ash, agricultural and animal waste, industrial waste reach in calcium such as mud and slug, and natural sources) for potential catalyst synthesis, which can often be very dangerous and leave a lasting impact on the environment. Using such materials has double benefit. The environmental and financial problems of disposal of hazardous materials can be solved, while such material can be used as a catalyst for biofuel production.

A particular challenge in the production of biofuels, primarily biodiesel, is the adaptation of the aforementioned catalytic systems in biodiesel production from waste TAG feedstock (non-edible oil, waste frying oil, and oil with high free fatty acid content), and also from TAG from sources (microalgae) related with the modern generation of biodiesel.

This review will be focused on the valorization of coal fly ash as a waste material in order to synthesize catalyst support or catalyst for biodiesel production using various modification techniques such as alkali activation, hydrothermal and thermal treatment, impregnation, and ion exchange.

2. Coal and coal fly ash: types, resources and utilization

2.1 Coal

Coal is a solid fossil fuel derived from fossilized plant matter by the process of coalification. As a geological process, over millions of years under suitable conditions (high pressure and temperature), coalification starts with dead plant matter firstly decaying into peat, and then converted into lignite, sub-bituminous, bituminous, and finally anthracite coals.

The classification into four main ranks, or types, namely lignite (lowest rank of coal), sub-bituminous, bituminous, and anthracite (highest rank of coal), is based on the content of carbon as primary coal constituent. Coal is a complex organic–inorganic system composed of mostly organic matter (non-crystalline carbon compounds) associated with petrographic maceral components, and, to a lesser extent, of inorganic matter. The inorganic constituents in coal include mineral (crystalline) matter, glassy (amorphous) matter and gas–liquid inclusions (fluid) matter [44]. The elements present in the coal are classified into three groups depending on their concentration: (1) major elements (>0.1%): C, H, O, N, S; (2) minor elements

(0.01-0.1%): Si, Al, Ca, Mg, K, Na, Fe, Mn, Ti (ash-forming elements), and occasionally Ba, Sr, P, and halogens (F, Cl, Br, I); (3) trace elements (<0.01%): As, B, Cd, Hg, Mo, Pb, which are considered hazardous pollutants [45].

Coal is the second most important fossil fuel resource for energy production, covering around 30% of global primary energy consumption [46]. The world currently consumes over 7.7 billion tons of coal which are used primarily in coal combustion processes for power generation. Bituminous, sub-bituminous and lignite are the principal energy resources in power generation with 40% of globally generated power [47]. It is reported [48] that total proved reserves at the end of 2018 were 1.055 trillion tons, enough to last about 137 years at the current rates of consumption. Nowadays, coal is simultaneously the fossil fuel with the highest carbon content (anthracite: 90–95% C, bituminous: 76–90% C, sub-bituminous: 72–76% C, lignite: 65–72% C) per unit of energy and the fossil fuel with the most abundant resources in the world.

As the most abundant low-cost energy resource, coal has various applications in many commercial processes, including power generation, iron and steel production, cement manufacturing, and production of liquid fuels. The most common and important use of coal is thermal (steam) coal utilization by pulverized coal combustion for the production of electricity and heat in CFPSs. However, coal utilization technologies generate considerable amounts of greenhouse gases, primarily carbon dioxide (CO_2) due to the higher carbon content of coal, pollutants (NO_x, SO_x), and solid particulates [49]. Therefore, the utilization of thermal coal without or with reduced CO_2 emissions is a major technological challenge [50]. Accordingly, to obtain future benefits from enormous low-cost coal reserves, various efforts are necessary in order to avoid environmental risks. The promising technological solutions are clean coal technologies: cleaner and more efficient technologies for coal combustion, including supercritical coal plants, more efficient industrial boilers, fluidized bed combustion, as well as coal gasification, and various "end-of-pipe" pollution abatement technologies for CO_2 capture and storage [51].

2.2 Coal fly ash

Coal ash, an industrial solid waste, is generated from pulverized coal combustion during electricity production in CFPSs. Over 70% of coal combustion residues (fly ash, bottom ash, boiler slug, and solid flue-gas desulfurization residues) contain CFA, captured by electrostatic precipitators (particulate collection equipment of flue emissions) [52], and bottom ash from the hoppers under the economizers and air preheaters of large pulverized coal boilers [53]. CFA is the most massive lightweight ash particulates, ranging from 0.5 to 300 μm dominantly spherical in shape-solid or hollow (cenospheres) [54, 55]. The major parameters affecting the characteristics of CFAs are phase-mineral and chemical composition of parent coal and coal combustion conditions in pulverized CFPSs (boiler temperature and its configuration, particulate control equipment, and size of feed coal) [56]. CFAs are a complex inorganic–organic mixture (316 individual minerals and 188 mineral groups are found in coals and CFAs) and as such complicated for identification and characterization of their constituents [57]. CFAs are a complex system with the unique, multicomponent, heterogeneous and variable composition of their inorganic, organic, and fluid constituents.

The principal CFAs inorganic components (90–99%) are silicon dioxide (SiO_2) both amorphous and crystalline, aluminum oxide (Al_2O_3), ferric oxide (Fe_2O_3), and calcium oxide (CaO), the main mineral constituents of coal-bearing rock strata (coal seams) [58]. CFAs are also composed of variable amounts of some rare earth elements (Ce, Gd, La, Nd, and Sm) [59], and trace elements (e.g. As, Se, Cd, and Cr) originating from a parent coal that make it potentially toxic [60] (Section 2.1). The bulk chemical composition and loss on ignition of CFAs (expressed as oxides) collected from various countries is shown in **Table 1**.

CFAs have a bulk chemical composition containing various metal oxides in the order: $SiO_2 > Al_2O_3 > CaO > Fe_2O_3 > MgO > Na_2O > K_2O > TiO_2$ (**Table 1**). The bulk chemical composition suggests that the CFAs are aluminosilicate with higher concentration of calcium oxide than ferric oxide.

Table 2 shows the content of rare earth elements (REEs) in different countries. It is noticeable that content of some elements varies from region to region. The most abundant REEs are cerium, lanthanum, and yttrium. REEs play an important role in many areas from household products to materials used in high technologies due to their adequate properties (luminescent and magnetic). The major industries that use REEs are catalysis, metallurgy, ceramics and polishing industry. On the other hand, the wide application is focused on catalysts, high technology products, health care devices, and rechargeable batteries [55].

The content of toxic elements in CFAs is shown in **Table 3**. These elements present serious problem, causing air, soil and water pollution. From the data presented, it can be seen that in some ashes (**Table 3**) the content of some toxic element is very high. For example, the content of the arsenic in some ashes is even 0.2%. That is the exact reason such material should be utilized in order to avoid negative impact on environment and human health.

	Chemical composition and loss on ignition (%)									
	SiO_2	Al_2O_3	CaO	Fe_2O_3	K_2O	MgO	TiO_2	Na_2O	LOI	Ref.
Australia	31.1–68.6	17–33	0.1–5.3	1–27.1	0.1–2.9	0–2	1.2–3.7	1.2–3.7	na	[56]
Bulgaria	30.1–57.4	12.5–25.4	1.5–28.9	5.1–21.2	0.8–2.8	1.1–2.9	0.6–1	0.4–1.9	0.8–32.8	[61]
Canada	35.5–62.1	12.5–23.2	1.2–13.3	3–44.7	0.5–3.2	0.4–3.1	0.4–1	0.1–7.3	0.3–9.7	[62]
China	35.6–57.2	18.8–55	1.1–7	2.3–19.3	0.8–0.9	0.7–4.8	0.2–0.7	0.6–1.3	na	[56]
Europe	28.5–59.7	12.5–35.6	0.5–28.9	2.6–21.2	0.4–4	0.6–3.8	0.5–2.6	0.1–1.9	0.8–32.8	[56]
France	47–51	26–34	2.3–3.3	6.9–9.8	na	1.5–2.2	na	2.3–6.4	0.5–4.5	[63]
Germany	20–80	1–19	2–52	1–22	0–2	0.5–11	0.1–1	0–2	0–5	[64]
India	50.2–59.7	14–32.4	0.6–9	2.7–16.6	0.2–4.7	0.1–2.3	0.3–2.7	0.2–1.2	0.5–7.2	[56]
Italy	41.7–54	25.9–33.4	2–10	3–8.8	0–2.6	0–2.4	1–2.6	0–1	1.9–9	[61]
Japan	53.9–63	18.2–26.4	2–8.1	4.2–5.7	0.6–2.7	0.9–2.4	0.8–1.2	1.1–2.1	0.5–2.1	[65]
Korea	50–55.7	24.7–28.7	2.6–6.2	3.7–7.7	1.1	0.7–1.1	na	na	4.3–4.7	[66]
Poland	32.2–53.3	4–32.2	1.2–29.9	4.5–8.9	0.2–3.3	1.2–5.9	0.6–2.2	0.2–1.5	0.5–28	[67]
Russia	40.5–48.6	23.2–25.9	6.9–13.2	na	1.9–2.6	2.6–4	0.5–0.6	1.2–1.5	na	[68]
Serbia	53.5–59.7	17.4–21	5.8–8.7	6–10.5	0.6–1.2	2–2.7	0.5–0.6	0.4–0.5	1.8–4.9	[69]
S. Africa	46.3–67	21.3–27	6.4–9.8	2.4–4.7	0.5–1	1.9–2.7	1.2–1.6	0–1.3	na	[70]
Spain	41.5–58.6	17.6–45.4	0.3–11.8	2.6–16.2	0.2–4	0.3–3.2	0.5–1.8	0–1.1	1.1–9.7	[71]
Turkey	37.9–57	20.5–24.3	0.2–27.9	4.1–10.6	0.4–3.5	1–3.2	0.6–1.5	0.1–0.6	0.4–2.7	[72]
USA	34.9–58.5	19.1–28.6	0.7–22.4	3.2–25.5	0.9–2.9	0.5–4.8	1–1.6	0.2–1.8	0.2–20.5	[56]
Min	20.0	1.0	0.1	1.0	0.0	0.0	0.1	0.0	0	
Max	80.0	55.0	52.0	44.7	4.7	11.0	3.7	7.3	32.8	

na = not available, LOI = loss on ignition—measure for unburned carbon.

Table 1.
Bulk chemical composition and loss on ignition of CFAs worldwide.

	Rare earth elements content (ppm)									
	La	**Ce**	**Sm**	**Eu**	**Dy**	**Ho**	**Er**	**Y**	**Pr**	**Ref.**
Austria	31.0	78.0	13.0	3.1	15.0	2.9	8.3	—	10.0	[73]
Bulgaria	37.7–40.2	82.4–87.6	7.0–7.6	1.6–1.8	5.4–5.8	1.1–1.2	2.9–3.3	30.7–35.1	9.3–9.8	[74]
Canada	25.0–95.0	43.2–173.0	4.4–14.2	1.1–4.2	3.7–19.5	1.0–2.7	2.0–7.0	—	5.1–18.1	[75]
China	79.6–81.5	191.3–195.1	17.0–17.7	3.2–3.4	13.2–13.9	2.5–2.6	7.0–7.4	64.5–66.0	21.9–23.4	[76]
Croatia	13.0	27.9	2.3	0.43	2.2	0.2	1.1	13.3	3.3	[77]
Greece	22.1	55.9	6.9	1.6	6.3	1.11	3.3	33.2	11.9	[78]
India	50–88.7	100–200	3.5–9.8	1.8–3.5	5.3–7.3	2.1–2.3	4.0–4.6	30–40	14.3–48	[59]
Japan	148	310	17.4	5.1	—	—	—	—	—	[79]
Korea	9.6–86.5	16.1–115	1.6–12.6	0.5–2.5	1.5–10.6	0.3–2.2	0.9–6.2	9.2–60.5	2.3–18.9	[80]
Poland	15.5–81.7	30.7–172.5	2.8–17.0	0.6–3.8	2.6–12.2	0.6–2.6	1.8–5.0	17.9–73.8	3.3–14.7	[65]
Russia	33.6–114.3	71.0–203.8	11.7–45.3	2.7–9.3	—	—	—	70.0–330.0	—	[81]
S. Africa	85.4	141.0	10.6	1.8	8.6	1.7	4.9	42.1	17.3	[82]
Spain	21.0–42.0	64.7–113	19.9–22.9	4.9–6.3	16.4–25.1	2.8–4.5	7.6–11.4	95.0–126.0	9.9–15.3	[83]
Turkey	36.0–41.0	72.0–85.0	6.1–7.2	1.6–1.8	5.4–5.8	1.1–1.2	2.9–3.3	30.7–35.1	9.3–9.8	[74]
USA	64.6–86.9	137–190	15.0–19.0	3.3–4.2	14.3–18.3	2.7–3.3	8.1–9.9	72.5–85.7	16.9–22.3	[84]
Min	9.6	16.1	1.7	0.4	1.5	0.2	1.1	9.2	2.3	
Max	148.0	310.0	45.3	6.3	18.3	4.5	9.9	330.0	48.0	

Table 2.
Rare earth elements content of CFAs in different countries.

	Toxic elements content (ppm)									
	As	**Cd**	**Co**	**Cr**	**Cu**	**Mn**	**Ni**	**Pb**	**V**	**Zn**	**Ref.**
Brazil	127–1915	11–33	—	74–181	31–88	219–714	48–95	66–627	207–293	434–2453	[85]
Bulgaria	1–76	0–1	16–43	71–93	74–207	174–821	40–73	25–60	119–262	87–174	[86]
Canada	17.5–52.0	0.5–1.9	—	31–101	—	—	30–41	32–84	—	—	[87]
China	—	0–2.3	—	0–78	4–60	13–772	4–41	3–40	2.2–81.1	2.4–76.7	[88]
Croatia	—	0.1–0.9	—	14–38	28–120	—	—	2–144	11.2–624	5.73–229	[89]
EU	69	—	41	153	101	—	123	88	255	161	[90]
Greece	—	0.2–0.9	10–60	127–1502	19–227	400–1700	85–1075	4–79	49–121	12.4–35.9	[91]
India	—	—	9–18	54–103	40–83	47–182	26–63	10–56	—	29–124	[92]
Korea	0.6–25.3	0–0.4	5–19	31–119	22–73	—	16–49	12–51	—	14.4–95.0	[93]
Poland	10.2–50	0.1–2.7	—	—	13–73	—	20–72	3–101	—	11–210	[65]
Serbia	—	0–1.1	6–26	12–63	10–29	200–1270	22–1148	7–70	30–123	25–208	[94]
Slovakia	—	0.1–2.6	—	9.4–32	10–81	—	10–32	14–142	—	35–375	[95]
S. Africa	16.6	0.16	5.45	73.0	18.8	148.10	14	24	104	20.03	[96]
Spain	57–726	<0.5	22–60	167–279	72–103	225–315	89–141	54–115	225–352	53–189	[97]
Turkey	—	<5	5–13	22–252	18–141	—	30–326	2–82	—	22–270	[98]
USA	—	—	—	14–61	30–290	—	9–23	6–1600	—	163–1512	[99]
Min	0.6	0	5.0	0.25	3.9	12.8	9.0	2.0	2	2.4	
Max	1915	33	60.0	1501.8	290.0	1700.0	1075	627	624	375.0	

Table 3.
Toxic elements content of CFAs in different countries.

CFAs are commonly categorized into two chemical types for their industrial applications, by name Class C and Class F. The American Society for Testing and Materials (ASTMs) classified CFAs as Class C and Class F on the basis of chemical composition and coal origination. According to the ASTM standard C618, Class F CFA has a combined SiO_2, Al_2O_3, and Fe_2O_3 content of greater than 70% compared to greater than 50% for Class C CFA. Fly ash of Class F is regarded as a true pozzolanic material exhibiting cementitious properties [100]. Class C CFAs derived from lignite and sub-bituminous coals with a high CaO content of above 20% possesses self-cementitious properties. The pozzolanic (Class F CFA) and cementitious (Class C CFA) properties of CFAs may allow their use as a binding agent or as raw material to produce clinker, replacing cement in concrete manufacturing. Apart from ASTMs, the European body has devised standard EN 450-1 defining CFA as a fine powder containing mostly spherical, glassy particulates derived from burned pulverized coal, with or without co-combustion material, which has pozzolanic properties and consists essentially of SiO_2 and Al_2O_3 [56]. Vassilev and Vassileva [101] have devised a new chemical classification system in accordance to the contents of ash-forming elements in CFAs using three composition-based criteria: (1) sum of Si, Al, K, and Ti oxides; (2) sum of Ca, Mg, S, and Na oxides; (3) ferric oxide [101]. This approach resulted in four chemical CFAs types, namely sialic, calsialic, ferrisialic, and ferricalsialic. Classifying CFAs in this way should simplify the choice of utilization for each unique CFAs composition.

The abundant availability and low price of coal, rising global energy demand, and the unsteadiness of alternative energy resources launched a growth in coal-based energy use, generating large amounts of CFAs. Increase in coal production to meet the growing demand for energy has resulted in an exponential increase in the generation of CFAs from 500 in 2005 to about 750 million tons in 2015 [54, 102]. Contrarily, global use of CFA for various applications is only lower part (about 25%) of the total production while the larger part (about 75%) is disposed or stored in different ways (landfills or lagoons) depending on the processes at CFPSs, and regulations the CFPSs have to follow. CFAs are harmful if released into the environment due to the presence of metal(loid)s, toxic substances and organic pollutants [103]. The presence of toxic waste contaminants in the ash requires it to be stored appropriately. It is known that otherwise landfills can decay, causing many environmental concerns and serious troubles to local communities [104]. In the view of the imminent strict disposal restriction, the disappearing availability of landfill space and the increasing cost of disposal, demands the need for economical and green CFAs utilization technologies. Therefore, maximizing the valorization of CFA into valuable products, rather than of its storage and disposal, is the optimal solution to preserve the environment and open new economic opportunities [105]. The utilization of CFA as an industrial waste residue or by-product has received a great deal of attention over the past two decades, as more sustainable solutions to waste problems have been searched for. However, it must be emphasized that extending the CFAs utilization to various valuable products in the future, imposes the necessity of detoxifying CFA and converting extracted toxins into valuable materials to create conditions for safe conversion into new products. Contrarily, the direct utilization of CFAs leads to hazardous effects on the environment. CFAs as a pozzolanic material has been prevalently employed in manufacturing cement either as a raw material or as a supplement to save its consumption [106]. CFA has been used in different geotechnical applications such as grouting, asphalt filler, sub-grade stabilization, pavement base course, general engineering fill, structural fill, soil amendment, and infill [52]. Extensive research has been carried out for use of fly ash-based adsorbents in both gaseous and aqueous applications. CFA has been found to be effective for removing different metal ions [107, 108] and aqueous pollutants or gaseous pollutants [58] from wastewaters. Currently, application of CFAs in wastewater treatment (WWT) is broad but still inadequate. CFAs have tremendous potential for WWT. The utilization of CFAs in water treatment in the near future is quite promising [109]. In the last few years, CFA usage as a cheap source of aluminosilicate has attracted scientists who have shown the successful transformation of this waste material into zeolites [110]. These synthetic fly ash-based zeolites are synthesized by various chemical processes (hydrothermal, alkaline fusion-assisted hydrothermal process, multi-step treatment method, microwave irradiation and sonication approach) resulting in a more uniform and cleaner state than natural types in terms of their lattice structures, pore size, and cages in their aluminosilicate frameworks [111–116].

The type of zeolites formed is a function of several reaction parameters such as temperature, pressure, the concentration of the reagent solutions, pH, process of activation and aging period, SiO_2 and Al_2O_3 contents of the CFAs. Zeolite of type A, X, Y, P, and Na-P1 are well known synthetic zeolites synthesized from CFA which have a wider range of industrial applications than their natural counterparts. The utilization of fly ash-based zeolites not only brings more revenue for CFPSs but also reduces the costs associated with the disposal of coal ashes.

The use of CFA in catalytic applications was examined for its potential to reduce the consumption of materials that have limited reserves or are expensive to produce. The application of CFA as a material to be used in heterogeneous catalysis has attracted much attention. Heterogeneous catalysis is attractive because it is often easier to recover catalysts upon completion of the reaction as compared to homogeneous catalysts. For heterogeneous catalysis, catalytic materials can be supported on other materials; their activity depends on both the active component and its interaction with the support. Typically, catalyst supports include various metal oxides such as SiO_2, Al_2O_3, MgO, and TiO_2. Since the CFA consists primarily of SiO_2 and Al_2O_3, CFAs offer desirable properties such as thermal stability for use as a support. Also, CFAs are often used as the catalytically active component.

3. CFA catalyst synthesis

In order to obtain a suitable form of CFA, it can be modified by various techniques using different synthesis conditions presented in **Table 4**.

It can be seen from the results that the main techniques, which can be used for CFA modification are alkali activation, hydrothermal and thermal treatment, wet impregnation, and ion exchange. The influence of various parameters, such as temperature, the concentration of alkali agent, the synergism of alkali agents, reaction time, aging period are crucial for obtaining the material with a suitable structure. Babajide et al. [116] studied the synthesis of Na-X zeolite from CFA, which was used as a catalyst for biodiesel production in the K-ion exchanged form. Such synthesized material exhibits higher activity than non-ion exchanged. It can be noted, that the main goal of CFA modification is the total or partial destruction of CFA crystalline cenosphere structure, with very low specific surface area and inaccessible pore system. Depending on CFA alkali activation and hydrothermal conditions (temperature, sodium-aluminate addition, and time), zeolite materials of different compositions and characteristics can be obtained. Bhandari et al. [120] optimized CFA alkali fusion process in order to obtain high crystalline zeolite and reported that zeolite with suitable structural, morphological, and textural properties can be obtained at alkali activation and hydrothermal temperature 550 °C and 90 °C, respectively and NaOH/CFA ratio of 1.5. By adding sodium aluminate in the range 10–20%, zeolite X was obtained, whereas further adding leads to the formation

Catalyst type	Catalyst synthesis		CFA and catalyst characteristics	Ref.
(A-CFA or B/CFA)	Method(s)	Conditions	Catalyst efficiency	
Sodalite (A-CFA) CFA-derived sodalite zeolite	Hydrothermal	Activators: (a) NaOH (b) $NaAlO_2$; Aging time (AT): 6 days; Temperature (T): 100°C; Time (t): 24 h	• Diffraction peaks: quartz (SiO_2), mullite ($3Al_2O_3{\cdot}2SiO_2$), small amounts muscovite and sodalite ($Na_8Al_6Si_6O_{24}Cl_2$) (XRD); rounded particles of sodalite agglomerates (SEM); mesoporous sodalite: N_2 ads/des type IV isotherm; S_{BET}: 9.7 m^2/g	[117]
High potential of zeolite sodalite as a low-price product to be used as a catalyst for biodiesel production on an industrial scale.			Soybean oil transesterification: Catalyst loading/FAME • 4 wt%/95.5 wt% FAME	
Eggshell/CFA (B/CFA) CaO/CFA catalyst Impregnated CFA-based catalyst	Drying starting material	(a) Eggshell T: 105°C; t: 24 h (b) Fly ash T: 100 ± 5°C; t: 24 h	• Crystalline: α-quartz, hematite, mullite, calcium oxide (CaO), dicalcium silicate (Ca_2SiO_4) (XRD); agglomerated structures of calcined metal oxides (SEM); mesoporous solid: N_2 ads/des type III isotherm; S_{BET}: 0.7 m^2/g	[118]
	Impregnation (wet)	T: 70°C; t: 4 h; pH: 12.10; AT: 24 h		
	Calcination	T: 1000°C; t: 2 h		
Effective waste valorization is procreated through the preparation of a novel low-cost catalyst for synthesis of fuel-grade biodiesel.			Soybean oil transesterification: Catalyst loading/FAME • 1.0 wt%/96.97 wt% FAME	
Modified^{MW} CFA (A-CFA) MW modified	Alkali fusion	CFA:NaOH = 1:1.5; T: 600°C; t: 1.5 h;	**Microwave irradiation (MW)** • Amorphous glassy phase (untreated CFA), SiO_2 and mullite (calcined CFA), new crystal phases $NaAlO_2$, Na_2SiO_3 upon MW (XRD); OH- and SO_4^{2-} supported on CFA upon MW (FTIR)	[23]
	Hydrothermal	30 wt% Na_2SO_4; T: 60°C; t: 10 h		
	Microwave (MW)			
	Calcination	T: 600°C; t: 1.5 h		
Modified coal fly ash catalyst improved biodiesel yields under the microwave irradiation system.			Waste cooking oil transesterification: Catalyst loading/FAME 3.99 wt%/94.91 wt% FAME	
Modified^{US} CFA (A-CFA) US modified	Alkali fusion	CFA:KOH = 1:1; T: 550°C; t: 2 h;	**Ultrasound-assisted (US)** • Amorphous glassy phase (untreated CFA), SiO_2 and $Al_6Si_2O_{13}$ (calcined CFA), new crystal phases $KAlO_2$, K_2SiO_3 upon US (XRD); OH- and NO_3^- supported on CFA upon ultrasound-assisted (FTIR)	[119]
	Hydrothermal	30 wt% KNO_3; T: 100°C; t: 6 h		
	Ultrasound (US)			
	Calcination	T: 550°C; t: 2 h		
Experimental results showed that the modified coal fly ash catalyst could improve biodiesel yields under ultrasound assisting system.			Waste cooking oil transesterification: Catalyst loading/FAME • 4.97 wt%/95.57 wt% FAME	
KX-CFA (A-CFA) CFA-derived zeolite KX	CFA calcination	T: 850°C; t: 2 h	• Diffraction peaks: crystalline quartz and mullite (XRD); spherical CFA particles, CFA, octahedral crystals ion exchanged zeolite KX (FESEM); mesoporous zeolitic material: N_2 ads/des type II isotherm; S_{BET}: 735.8 m^2/g	[120]
	CFA acidification	HCl; T: 80°C; t: 1.5 h;		
	Alkali fusion	NaOH:CFA = 1:1–1:2; T: 400–600°C; t: 1 h;		
	Hydrothermal	T: 90–120°C; t: 4–24 h		

Catalyst type	Catalyst synthesis		CFA and catalyst characteristics	Ref.
(A-CFA or B/CFA)	Method(s)	Conditions	Catalyst efficiency	
	Ion exchange	1.0 M CH_3COOK		
	Calcination	T: 500 °C; t: 2 h		
The effective utilization of fly ash for zeolite KX synthesis and its use as a catalyst for transesterification would improve ecological balance and helps in value addition.			Soybean oil transesterification: Catalyst loading/FAME • 3.0 wt%/81.2 wt% FAME	
Animal bone/CFA (B/CFA) Impregnated CFA-based catalyst	CFA drying Animal bones calcination	T: 105°C; t: overnight; T: 900°C; t: 2 h;	• CFA chemical composition (wt%): 56.6 SiO_2, 23.2 Al_2O_3, 5.8 Fe_2O_3, and 7.9 CaO (AAS); overall crystalline phases: quartz, mullite (CFA), dicalcium silicate (Ca_2SiO_4), hydroxyapatite ($Ca_5(PO_4)_3OH$), β-tricalcium phosphate, and CaO (XRD); surface morphology: cenospheres (CFA), rod like crystalline particles (impregnated fly ash catalysts C10, C20, and C30) (SEM); basicity: 5.1–17.4 mmoles HCl/g; mesoporous solids: N_2 ads/des type III isotherm; S_{BET} (m^2/g): 1.7 CFA, 100 CABP, 11.3 C10, 7.1 C20, 4.2 C30	[121]
	Impregnation (wet)	T: 70°C; t: 4 h; L: S = 10: 1; pH:12.1; AT: 24 h		
	Calcination	T: 900°C; t: 2 h		
Animal bones (calcium enriched waste materials) impregnated in fly ash might be a potential source of catalyst in biodiesel production.			Mustard oil transesterification: Catalyst loading/FAME • 10 wt%/90.4 wt% FAME	
Kaliophilite (A-CFA) CFA-derived kaliophilite catalyst	Geopolymer synthesis	Alkali activator (KOH in potassium water glass); T: 80°C; t: 24 h	• Amorphous aluminosilicate, quartz and mullite crystals (CFBFA), amorphous geopolymer, and $KAlSiO_4$ (as-synthesized kaliophilite catalyst) (XRD); irregular CFBFA particles (30 μm), dense structure (geopolymer), prismatic crystals (~1 μm) (kaliophilite) (SEM); medium-strength basic sites (K-O ion pairs) and high strength basic sites (surface O^{2-} ion) (TPD-CO_2); mesoporous catalyst: N_2 ads/des type IV isotherm; S_{BET}: 3.49 (m^2/g)	[122]
	Hydrothermal	Geopolymer monolith:50 ml KOH; T: 180°C; t: 24 h		
	Drying kaliophilite	T: 105°C; t: 12 h		
Circulating fluidized bed fly ash (CFBFA) was used to synthesize kaliophilite catalyst via a facile and low-energy two-step process: fabrication of amorphous CFBFA geopolymer and hydrothermal transformation of CFBFA based geopolymer into kaliophilite. This catalyst affords three benefits: high value-added reutilization of CFBFA industrial by-products, low-energy synthesis of kaliophilite, and low-cost production of biodiesel.			Canola oil transesterification: Catalyst loading/FAME • 5.0 wt%/99.2 wt% FAME	
FA/Na-X (A-CFA) FA-derived zeolite Na-X	Hydrothermal	—	• Low Si/Al ratio preferentially result in zeolite FA/Na-X (XRD); faujasite phase irregular crystals (unique morphology) (SEM); S_{BET} (m^2/g): 320 (FA/Na-X), 257 (FA/Na-X)	[116]
	Ion exchange	L:S = 10:1; 1.0 M CH_3COOK; T: 60–70°C; t: 24 h		
	Calcination	T: 500 °C; t: 2 h		
Fly ash transformed into a zeolite Na-X phase and exchanged with K proved to be suitable for use as a catalyst in biodiesel synthesis under less rigorous conditions.			Sunflower oil transesterification: Catalyst loading (FA/K-X)/FAME • 3 wt%/85.5 wt% FAME	

Catalyst type	Catalyst synthesis		CFA and catalyst characteristics	Ref.
(A-CFA or B/CFA)	Method(s)	Conditions	Catalyst efficiency	
KNO_3/CFA (B/CFA) Impregnated CFA-based catalyst	Impregnation (wet)	KNO_3 aq. stock; solution; L:S = 1:1; KNO_3:FA = 1:1	• Crystalline phases: α-quartz, hematite, mullite (CFA), KNO_3 (KNO_3/CFA catalyst) surface morphology: cenospheres (CFA), potassium impregnated fly ash spherical particles aggregates (>10 μm) (SEM); N_2 ads/des type III isotherm; S_{BET}: 0.55 (m^2/g)	[123, 124]
	Calcination	T: 500–700°C; t: 5 h		
Fly ash loaded with KNO_3 was used as a solid base catalyst in the transesterification of sunflower oil to methyl esters to make a meaningful utilization of fly ash.			Sunflower oil transesterification: Catalyst loading/FAME • 5 wt%/87.5 wt% FAME	
K-Zeolite (A-CFA) CFA-derived K-Zeolite	CFA drying CFA calcination	T: 80°C; t: overnight; T: 900°C; t: 3 h;	• Main crystalline phases: hexagonal quartz (SiO_2), orthorhombic mullite crystalline phase ($3Al_2O_3{\cdot}2SiO_2$) (CFA), K-Zeolite ≡ K-CHA zeolite (potassium type zeolite) (hydrothermally activated XRD pattern); prism-like crystals zeolite crystals (SEM micrographs); mesoporous solids: N_2 ads/des type IV isotherm; S_{BET} (m^2/g): 2.1 coal fly ash, 24.7 K-Zeolite	[125]
	Hydrothermal	5 M KOH (aq. stock solution); CFA:KOH = 1:4; T: 160°C; t: 8 h		
	Drying product	T: 80°C; t: overnight		
	Calcination	T: 450°C; t: 4 h		
The obtained K-Zeolite can be used in biodiesel industry. Utilization of biodiesel by-product glycerol is of great importance for sustainability of biodiesel industry. Conversion of glycerol to value-added chemicals increases the profitability of biodiesel production.			Glycerol transesterification: Catalyst loading/Glycerol carbonate • 4 wt%/96.0 wt% Glycerol carbonate	
CaO/Fly ash (B/CFA) CaO/Fly ash catalyst Impregnated Fly ash-based catalyst	Starting material	(a) 50 wt% Fly ash (b) 50 wt% Ca $(NO)_3{\cdot}$ $4H_2O$ $Ca(NO)_3{\cdot}4H_2O$ = CaO_P	• Crystalline phases: quartz (SiO_2), calcium oxide (CaO), dicalcium silicate (Ca_2SiO_4), and calcium hydroxide $Ca(OH)_2$ (XRD); basicity: $H\_ < 8.2$ (FA), $H\_ > 9.3$ (C1, C2 and C3); S_{BET} (m^2/g): 24.3 C2 (800°C), 909.8 C2 (850°C)	[126]
	Impregnation (wet)	—		
	Calcination	T: 800, 850 and 900°C		
	CaO/FA catalyst	CaO_p:FA = 70:30 (C1);		
		CaO_p:FA = 80:20 (C2);		
		CaO_p:FA = 90:10 (C3)		
Palm fly ash supported calcium oxide (CaO) catalyst was prepared through impregnation method and used in transesterification from off-grade palm oil for biodiesel manufacturing. The efficiency of CaO/Fly ash is affected by its basic strength.			Palm oil transesterification: Catalyst loading/FAME • 6 wt%/71.77 wt% FAME	
CFA-HT (A-CFA) FA-hydrotalcite catalyst	Alkali fusion	FZ	• XRD patterns consistent with hydrotalcite materials; basicity (mmoles HCl/g): 36.6 FZ-HT, 28.8 C-HT and 12.4 F-HT; surface morphology: cenospheres (Fly ash), octahedral crystals (FZ), platelet-like morphology (C-HT), platelet-like structures of HT (F-HR and FZ-HT); mesoporous solids: N_2 ads/	[127]
	Hydrothermal			
	Coprecipitation	C-HT		
		F-HT		
		FZ-HT		
	Calcination	T: 500 °C; t: 6 h		

Catalyst type	Catalyst synthesis		CFA and catalyst characteristics	Ref.
(A-CFA or B/CFA)	Method(s)	Conditions	Catalyst efficiency	
			des type III isotherm (Fly ash), type II isotherm (FZ), type IV isotherm (F-HT), type IV isotherm (FZ-HT); S_{BET} (m^2/g): 1.7 Fly ash, 323.2 FZ, 32.9 C-HT, 39.6 F-HT, and 476.6 FZ-HT	
Mg-Al hydrotalcite-like catalysts were prepared from fly ash and fly ash-based zeolite by coprecipitation method. The activity of prepared catalyst was estimated in mustard oil transesterification. The FAME yield tends to increase with increasing BET surface area.			Mustard oil transesterification: Catalyst loading/FAME • 7 wt%/93.4 wt% FAME	
***A. granosa* and *P. undulata*/CFA (B/CFA)** CaO/CFA catalyst Impregnated CFA-based catalyst	Drying starting material	(a) Shells T: 110°C; t: 6 h (b) Fly ash T: 105°C; t: 10 h	• Identified phases: SiO_2 (crystalline phase), Al_2O_3 amorphous phase (Fly ash), Ca_2SiO_4 dicalcium silicate (calcined impregnated catalyst) (XRD); smaller morphology size of particles: *P. undulata* shell less than 75 μm, *A. granosa* shell less than 85 μm, fly ash supported CaO catalyst had the size less than 30 μm	[128]
	Impregnation (wet)	—		
	Calcination	T: 800°C; t: 3 h		
Fly ash supported CaO catalyst derived from waste mollusk shell of *Anadara granosa* and *Paphia undulata* was used for palm oil transesterification. This catalyst could gain the yield of FAME of 92 and 94 wt% for *A. granosa* and *P. undulata* shell, respectively.			Palm oil transesterification: Catalyst loading/FAME • 6 wt%/94.0 wt% FAME	
Sulfated fly ash (SFA) (A-CFA) SFA-sulfated fly ash catalyst	Sulfonation	—	• Crystalline phases: quartz, mullite, hematite, lime; thermally stable up to 550°C; acid sites: 0.401 mmol/g (NH_3-TPD), basic sites: 0.197 mmol/g (CO_2-TPD); SO_4^{2-} groups is confirmed by FTIR analysis; surface morphology: homogeneous distribution of small spherical pores on FA surface, large connected spherical pores on SFA catalyst surface; SFA crystallite size: 16.8 nm; S_{BET} (m^2/g): 38.3	[129]
The goal is on the fly ash utilization for the development of sulfated fly ash (SFA) catalyst synthesis under solvent-free conditions. The use of SFA catalyst has been found to be advantageous in biodiesel synthesis from feedstock with high free fatty acids content.			Maize acid oil esterification: Catalyst loading/FAME • 5 wt%/98.3 wt% FAME	
CFA-Zeolite X (A-CFA) CFA-derived Zeolite X	Alkali fusion	T: 450–600°C; t: 1–2 h; CFA:NaON = 1:1–1:2.5	• Identified crystalline phases: Pure single-phase zeolites X and A under following conditions: FA: NaOH = 1:1.2, crystallization time 1 h (Zeolite X), 12 h (Zeolite A), fusion temperature 550°C, crystallization temperature 110°C, and calcination temperature 800°C; cations exchange Zeolite A (highest value); S_{BET} (m^2/g): 167.4 (Zeolite X), 24.1 (Zeolite A)	[130]
	Hydrothermal	10–30 wt% $NaAlO_2$; L:S = 10:1; T: 90–120°C; t: 24 h; AT = 12–16 h		
	Ion exchange	1 M CH_3COOK; L:S = 10:1; T: 60°C; t: 24 h		
	Calcination	T: 500 °C; t: 2 h		

Catalyst type	Catalyst synthesis		CFA and catalyst characteristics	Ref.
(A-CFA or B/CFA)	Method(s)	Conditions	Catalyst efficiency	
Different types of single-phase zeolites (Zeolite X and Zeolite A) with high cations exchange capacity were synthesized from alkali fusion followed by hydrothermal treatment of coal fly ash as source material. Coal fly ash was used successfully for production of biodiesel in mustard oil transesterification with suitable calorific value (37.5 MJ/kg).			Mustard oil transesterification: Catalyst loading/FAME • 5 wt%/84.6 wt% FAME	

Table 4.
Synthesis of CFA based heterogeneous catalyst for biodiesel production.

of zeolite A. In this study, the K-exchanged form of zeolite X exhibits suitable catalytic properties during the production of biodiesel from soybean oil.
From previous and similar investigations, it is obvious, that the activity of the zeolite-based catalyst can be improved by impregnation of mainly alkaline metals, such as potassium. The content of some alkali metals (Na and K) in biodiesel fuel is regulated by EN 14214, (max. concentration 5 ppm). Due to high leaching affinity, such catalysts are unsuitable. In order to obtain an active and stable catalytic form, previously treated CFA can be modified with different CaO-based active catalytic components. Volli et al. [121] investigated the utilization of CFA by impregnation of calcium from animal bones in order to synthesize catalysts for biodiesel production from mustard oil. The highest catalytic activity (TAG conversion of 90.4%) is achieved by catalyst with 10 wt% loaded animals bones powder on CFA. However, further increasing of animal bone powder loading on CFA leads to decreasing of catalytic activity. Waste materials such as eggshells are efficient as a high calcium source, which can be used as biodiesel catalysts. Carbonate eggshell form can be converted into active oxide form using simple synthesis methods (thermal activation and modification techniques).

4. Biodiesel synthesis over CFA based catalyst

In recent years, most studies have shown that CFA can successfully catalyze transesterification of various oily feedstock in order to produce biodiesel. Pure CFA is practically inactive, but in the modified form it could exhibit high catalytic activity. In **Table 5** are shown different catalysts and their catalytic performance for biodiesel production from various feedstock. Xiang et al. [122] investigated alkali activated CFA modified by sodium sulfate under hydrothermal conditions, whereby transesterification reaction was carried out under microwave [23] and ultrasound [122] conditions. The high catalytic activity was achieved for short reaction time and it was shown that catalyst could be used even eight times without any loss of catalytic activity. Other modification methods are based on CFA conversion into zeolite or hydrotalcite, impregnation of alkali or alkali earth metals or ion exchange of previously mentioned zeolites.

CaO from chicken eggshell supported on CFA exhibits the highest catalytic activity (97.0%) and stability (16 reaction cycles) in the transesterification reaction. Volli et al. [124] prepared CaO from animal bones supported on fly ash and tested in biodiesel production. The satisfactory conversion (90.4%) was achieved after 6 h, whereas the catalyst suffered negligible loss of activity when tested for 5 cycles of reuse. On the other hand, Bhandari et al. [120] and Volli and Purkait [130] used potassium ion exchanged fly ash zeolite for biodiesel production, where the prepared catalyst gave yield of 81.2% and conversion of 84.6% for 8 and 7 h,

Catalyst	Feedstock	Reaction condition				C or Y	RC	Refs.
		T	MOR	CC	t			
Modified^{MW} CFA	WCO	66.2	9.67	3.99	0.1	94.9 (C)	8	[23]
Modified^{US} CFA	WCO	—	10.71	4.97	0.03	95.6 (C)	8	[119]
KX-CFA	Soybean oil	65	6	3	8	81.2 (C)	—	[120]
Eggshell/CFA	Soybean oil	70	6.9	1	5	97.0 (C)	16	[118]
Animal bone/CFA	Mustard oil	65	5.5	10	6	90.4 (C)	5	[121]
Sodalite	Soybean oil	65	12	4	2	95.5 (C)	—	[117]
Kaliophilite	Canola oil	85	15	5	6	99.2 (C)	4	[122]
FA/Na-X	Sunflower oil	65	6	3	8	83.5 (Y)	3	[116]
KNO_3/CFA	Sunflower oil	160	15	15	5	86.1 (C)	—	[123]
KNO_3/CFA	Sunflower oil	120	15	5	8	81.5 (C)	1	[124]
K-Zeolite	Glycerol	75	3	4	1.5	90.2 (C)	5	[125]
CaO/Fly ash	Palm oil	70	6	6	3	71.7 (C)	—	[126]
CFA-HT	Mustard oil	65	12	7	6	93.4 (Y)	—	[127]
A. granosa/CFA	Palm oil					92.0 (Y)	3	[128]
P. undulata/CFA	Palm oil					94.0 (Y)		[128]
Sulfated fly ash (SFA)	Maize acid oil	125	15	5	3	98.3 (C)	3	[129]
CFA-Zeolite X	Mustard oil	65	12	5	7	84.6 (C)	3	[130]

T = reaction temperature (°C), MOR = methanol/oil molar ratio, CC = catalyst concentration (wt%), t = reaction time (h), C or Y = conversion or yield (%), RC = reaction cycle, WCO = waste cooking oil.

Table 5.
Biodiesel synthesis over CFA based catalysts.

respectively. Except for zeolites, active catalytic form or adequate catalytic support can be achieved by conversion of CFA into layered double hydroxides known as hydrotalcites [127]. High biodiesel yield can be obtained by using such materials under mild conditions. Lathiya et al. [129] synthesized a sulfated fly ash catalyst, which exhibits high catalytic activity, but in comparison with other presented catalytic systems, such activity can be achieved under more rigorous conditions, which is a feature of acid heterogeneously catalyzed biodiesel production.

5. Conclusion

This review reports a brief overview of the developments of various heterogeneous catalysts derived from industrial and biological waste materials as an efficient solid base catalyst for biodiesel production. As one of the few renewable energy fuel cost-effective options that can be recycled, low-cost biodiesel generation brings with it economic as well as social and environmental benefits. The fundamentals of methanolysis, the role of various process parameters and factors affecting biodiesel production from different feedstock are highlighted to guide future research and development on this topic. The development of heterogeneous fly ash-based catalysts supported with alkaline and alkaline earth metal (oxides, hydroxides, salts) gained a great awareness due to the wide availability of alkaline/alkaline earth metal-rich waste materials and their corresponding high catalytic activity in the methanolysis of triacylglycerol oils.

Acknowledgements

This work was supported by the Ministry of Education, Science and Technological Development of the Republic of Serbia within the framework of the project III 45001.

Author details

Miroslav Stanković[1]*, Stefan Pavlović[1], Dalibor Marinković[1], Marina Tišma[2], Margarita Gabrovska[3] and Dimitrinka Nikolova[3]

1 Department of Catalysis and Chemical Engineering, Institute of Chemistry, Technology and Metallurgy, National Institute, University of Belgrade, Belgrade, Serbia

2 Faculty of Food Technology, Josip Juraj Strossmayer University of Osijek, Osijek, Croatia

3 Institute of Catalysis, Bulgarian Academy of Sciences, Sofia, Bulgaria

*Address all correspondence to: mikastan@nanosys.ihtm.bg.ac.rs

References

[1] Marinković D, Stanković M, Veličković A, Avramović J, Miladinović M, Stamenković O, et al. Calcium oxide as a promising heterogeneous catalyst for biodiesel production: Current state and perspectives. Renewable and Sustainable Energy Reviews. 2016;**56**: 1387-1408. DOI: 10.1016/j.rser.2015.12.007

[2] Marinković D, Avramović A, Stanković M, Stamenković O, Jovanović D, Veljković V. Synthesis and characterization of spherically-shaped $CaO/\gamma\text{-}Al_2O_3$ catalyst and its application in biodiesel production. Energy Conversion and Management. 2017;**144**: 399-413. DOI: 10.1016/j.enconman.2017.04.079

[3] Shan R, Lu L, Shi Y, Yuan H, Shi J. Catalysts from renewable resources for biodiesel production. Energy Conversion and Management. 2018;**178**: 277-289. DOI: 10.1016/j.enconman.2018.10.032

[4] Gebremariam SN, Marchetti JM. Biodiesel production through sulfuric acid catalyzed transesterification of acidic oil: Techno economic feasibility of different process alternatives. Energy Conversion and Management. 2018;**174**: 639-648. DOI: 10.1016/j.enconman.2018.08.078

[5] Wang YT, Fang Z, Zhang F. Esterification of oleic acid to biodiesel catalyzed by a highly acidic carbonaceous catalyst. Catalysis Today. 2019;**319**:172-181. DOI: 10.1016/j.cattod.2018.06.041

[6] Lu W, Alam MA, Wu C, Wang Z, Wei H. Enhanced deacidification of acidic oil catalyzed by sulfonated granular activated carbon using microwave irradiation for biodiesel production. Chemical Engineering and Processing Process Intensification. 2019; **135**:168-174. DOI: 10.1016/j.cep.2018.10.019

[7] Wang YT, Fang Z, Yang XX, Yang YT, Luo J, Xu K, et al. One-step production of biodiesel from Jatropha oils with high acid value at low temperature by magnetic acid-base amphoteric nanoparticles. Chemical Engineering Journal. 2018;**348**:929-939. DOI: 10.1016/j.cej.2018.05.039

[8] Jeon Y, Chi WS, Hwang J, Kim DH, Kim JH, Shul YG. Core-shell nanostructured heteropoly acid-functionalized metal-organic frameworks: Bifunctional heterogeneous catalyst for efficient biodiesel production. Applied Catalysis B: Environmental. 2019;**242**:51-59. DOI: 10.1016/j.apcatb.2018.09.071

[9] Farooq M, Ramli A, Subbarao D. Biodiesel production from waste cooking oil using bifunctional heterogeneous solid catalysts. Journal of Cleaner Production. 2013;**59**:131-140. DOI: 10.1016/j.jclepro.2013.06.015

[10] Mansir N, Teo SH, Rashid U, Saiman MI, Tan YP, Alsultan GA, et al. Modified waste egg shell derived bifunctional catalyst for biodiesel production from high FFA waste cooking oil. A review. Renewable and Sustainable Energy Reviews. 2018;**82**:3645-3655. DOI: 10.1016/j.rser.2017.10.098

[11] Andrade TA, Martín M, Errico M, Christensen KV. Biodiesel production catalyzed by liquid and immobilized enzymes: Optimization and economic analysis. Chemical Engineering Research and Design. 2019;**141**:1-14. DOI: 10.1016/j.cherd.2018.10.026

[12] Adewale P, Vithanage LN, Christopher L. Optimization of enzyme-catalyzed biodiesel production from crude tall oil using Taguchi method. Energy Conversion and Management.

2017;**154**:81-91. DOI: 10.1016/j.enconman.2017.10.045

[13] Guldhe A, Singh B, Mutanda T, Permaul K, Bux F. Advances in synthesis of biodiesel via enzyme catalysis: Novel and sustainable approaches. Renewable and Sustainable Energy Reviews. 2015;**41**:1447-1464. DOI: 10.1016/j.rser.2014.09.035

[14] Jeon KW, Shim JO, Cho JW, Jang WJ, Na HS, Kim HM, et al. Synthesis and characterization of Pt-, Pd-, and Ru-promoted Ni-$Ce_{0.6}Zr_{0.4}O_2$ catalysts for efficient biodiesel production by deoxygenation of oleic acid. Fuel. 2019;**236**:928-936. DOI: 10.1016/j.fuel.2018.09.078

[15] Viêgas CV, Hachemi I, Freitas SP, Arvela PM, Aho A, Hemming J, et al. A route to produce renewable diesel from algae: Synthesis and characterization of biodiesel via in situ transesterification of *Chlorella alga* and its catalytic deoxygenation to renewable diesel. Fuel. 2015;**155**:144-154. DOI: 10.1016/j.fuel.2015.03.064

[16] Farobie O, Leow ZYM, Samanmulya T, Matsumura Y. New insights in biodiesel production using supercritical 1-propanol. Energy Conversion and Management. 2016;**124**:212-218. DOI: 10.1016/j.enconman.2016.07.021

[17] Aboelazayem O, Gadalla M, Saha B. Valorisation of high acid value waste cooking oil into biodiesel using supercritical methanolysis: Experimental assessment and statistical optimisation on typical Egyptian feedstock. Energy. 2018;**162**:408-420. DOI: 10.1016/j.energy.2018.07.194

[18] Korkut I, Bayramoglu M. Selection of catalyst and reaction conditions for ultrasound assisted biodiesel production from canola oil. Renewable Energy. 2018;**116**:543-551. DOI: 10.1016/j.renene.2017.10.010

[19] Korkut I, Bayramoglu M. Ultrasound assisted biodiesel production in presence of dolomite catalyst. Fuel. 2016;**180**:624-629. DOI: 10.1016/j.fuel.2016.04.101

[20] Tan SX, Lim S, Ong HC, Pang YL. State of the art review on development of ultrasound-assisted catalytic transesterification process for biodiesel production. Fuel. 2019;**235**:886-907. DOI: 10.1016/j.fuel.2018.08.021

[21] Jin LJ, Wen CY. Production of biodiesel by transesterification of Jatropha oil with microwave heating. Journal of the Taiwan Institute of Chemical Engineers. 2017;**75**:43-50. DOI: 10.1016/j.jtice.2017.03.034

[22] Hong IK, Jeon H, Kim H, Lee SB. Preparation of waste cooking oil based biodiesel using microwave irradiation energy. Journal of Industrial and Engineering Chemistry. 2016;**42**:107-112. DOI: 10.1016/j.jiec.2016.07.035

[23] Xiang Y, Xiang Y, Wang L. Microwave radiation improves biodiesel yields from waste cooking oil in the presence of modified coal fly ash. Journal of Taibah University of Science. 2017;**11**:1019-1029. DOI: 10.1016/j.jtusci.2017.05.006

[24] Lam MK, Lee KT, Mohamed AR. Homogeneous, heterogeneous and enzymatic catalysis for transesterification of high free fatty acid oil (waste cooking oil) to biodiesel: A review. Biotechnology Advances. 2010;**28**:500-518. DOI: 10.1016/j.biotechadv.2010.03.002

[25] Hindryawati N, Maniam GP, Karim MR, Chong KF. Transesterification of used cooking oil over alkali metal (Li, Na, K) supported rice husk silica as potential solid base catalyst. Engineering Science and Technology, an International Journal. 2014;**17**:95-103. DOI: 10.1016/j.jestch.2014.04.002

[26] Su M, Yang R, Li M. Biodiesel production from hempseed oil using alkaline earth metal oxides supporting copper oxide as bi-functional catalysts for transesterification and selective hydrogenation. Fuel. 2013;**103**:398-407. DOI: 10.1016/j.fuel.2012.07.009

[27] Arzamendi G, Arguiñarena E, Campo I, Zabala S, Gandía LM. Alkaline and alkaline-earth metals compounds as catalysts for the methanolysis of sunflower oil. Catalysis Today. 2008;**1**: 305-315. DOI: 10.1016/j.cattod.2007.11.029

[28] Salinas D, Sepúlveda C, Escalona N, Fierro JLG, Pecch G. Sol-gel La_2O_3-ZrO_2 mixed oxide catalysts for biodiesel production. Journal of Energy Chemistry. 2018;**27**:562-572. DOI: 10.1016/j.jechem.2017.11.003

[29] Guldhe A, Moura VRC, Singh P, Rawat I, Moura ME, Sharma Y, et al. Conversion of microalgal lipids to biodiesel using chromium-aluminum mixed oxide as a heterogeneous solid acid catalyst. Renewable Energy. 2017; **105**:175-182. DOI: 10.1016/j.renene.2016.12.053

[30] Fan M, Liu Y, Zhang P, Jiang P. Blocky shapes Ca-Mg mixed oxides as a water-resistant catalyst for effective synthesis of biodiesel by transesterification. Fuel Processing Technology. 2016;**149**:163-168. DOI: 10.1016/j.fuproc.2016.03.029

[31] Navajas A, Campo I, Moral A, Echave J, Sanz O, Montes M, et al. Outstanding performance of rehydrated Mg-Al hydrotalcites as heterogeneous methanolysis catalysts for the synthesis of biodiesel. Fuel. 2018;**211**:173-181. DOI: 10.1016/j.fuel.2017.09.061

[32] Dias APS, Bernardo J, Felizardo P, Correia MJN. Biodiesel production over thermal activated cerium modified Mg-Al hydrotalcites. Energy. 2012;**41**:344-353. DOI: 10.1016/j.energy.2012.03.005

[33] Navajas A, Campo I, Arzamendi G, Hernández WY, Bobadilla LF, Centeno MA, et al. Synthesis of biodiesel from the methanolysis of sunflower oil using PURAL® Mg-Al hydrotalcites as catalyst precursors. Applied Catalysis B: Environmental. 2010;**100**:299-309. DOI: 10.1016/j.apcatb.2010.08.006

[34] Reyero I, Velasco I, Sanz O, Montes M, Arzamendi G, Gandía LM. Structured catalysts based on Mg-Al hydrotalcite for the synthesis of biodiesel. Catalysis Today. 2013;**216**: 211-219. DOI: 10.1016/j.cattod.2013.04.022

[35] Alismaeel ZT, Abbas AS, Albayati TM, Doyle AM. Biodiesel from batch and continuous oleic acid esterification using zeolite catalysts. Fuel. 2018;**234**:170-176. DOI: 10.1016/j.fuel.2018.07.025

[36] Jammal NA, Hamamre ZA, Alnaief M. Manufacturing of zeolite based catalyst from zeolite tuft for biodiesel production from waste sunflower oil. Renewable Energy. 2016; **93**:449-459. DOI: 10.1016/j.renene.2016.03.018

[37] Wang YY, Chen BH. High-silica zeolite beta as a heterogeneous catalyst in transesterification of triolein for biodiesel production. Catalysis Today. 2016;**278**:335-343. DOI: 10.1016/j.cattod.2016.03.012

[38] Wang T, Xu Y, He Z, Zhou M, Yu W, Shi B, et al. Fabrication of sulphonated hollow porous nanospheres and their remarkably improved catalytic performance for biodiesel synthesis. Reactive and Functional Polymers. 2018;**132**:98-103. DOI: 10.1016/j.reactfunctpolym.2018.09.014

[39] Jaya N, Selvan BK, Vennison SJ. Synthesis of biodiesel from pongamia oil using heterogeneous ion-exchange resin catalyst. Ecotoxicology and

Environmental Safety. 2015;**121**:3-9. DOI: 10.1016/j.ecoenv.2015.07.035

[40] Kitakawa NS, Kitakawa NS, Ihara T, Nakashima K, Yonemoto T. Production of high quality biodiesel from waste acid oil obtained during edible oil refining using ion-exchange resin catalysts. Fuel. 2015;**139**:11-17. DOI: 10.1016/j.fuel.2014.08.024

[41] Deboni TM, Hirata GAM, Shimamoto GG, Tubino M, Meirelles AJA. Deacidification and ethyl biodiesel production from acid soybean oil using a strong anion exchange resin. Chemical Engineering Journal. 2018;**333**: 686-696. DOI: 10.1016/j.cej.2017.09.107

[42] Monge JA, Bakkali BE, Trautwein G, Reinoso S. Zirconia-supported tungstophosphoric heteropolyacid as heterogeneous acid catalyst for biodiesel production. Applied Catalysis B: Environmental. 2018;**224**:194-203. DOI: 10.1016/j.apcatb.2017.10.066

[43] Xiang HX, Ke CK, Wei Y, Liu HC, Li L, Hao WP, et al. Amino acid-functionalized heteropolyacids as efficient and recyclable catalysts for esterification of palmitic acid to biodiesel. Fuel. 2016;**165**:115-122. DOI: 10.1016/j.fuel.2015.10.027

[44] Vassilev SV, Tascón JMD. Methods for characterization of inorganic and mineral matter in coal: A critical overview. Energy & Fuels. 2003;**17**: 271-281. DOI: 10.1021/ef020113z

[45] Vejahati F, Xu Z, Gupta R. Trace elements in coal: Associations with coal and minerals and their behavior during coal utilization—A review. Fuel. 2010; **89**:904-911. DOI: 10.1016/j.fuel.2009.06.013

[46] World Energy Council (WEC). World Energy Resources Report [Internet]. 2016. Available from: https://www.worldenergy.org/assets/images/imported/2016/10/World-Energy-Resources-Full-report-2016.10.03.pdf [Accessed: 04 August 2019]

[47] Melikogly M. Clean coal technologies: A global to local review for Turkey. Energy Strategy Reviews. 2018; **22**:313-319. DOI: 10.1016/j.esr.2018.10.011

[48] BP Statistical Review of World Energy/68th edition [Internet]. 2019. Available from: https://www.bp.com/content/dam/bp/business-sites/en/global/corporate/pdfs/energy-economics/statistical-review/bp-stats-review-2019-coal.pdf [Accessed: 08 August 2019]

[49] IEA. Clean Coal Centre (CCC TCP) [Internet]. 2017. Available from: https://www.iea.org/tcp/fossilfuels/ccc/ [Accessed: 29 September 2019]

[50] Tang X, Snowden S, McLellan BC, Höök M. Clean coal use in China: Challenges and policy implications. Energy Policy. 2015;**87**:517-523. DOI: 10.1016/j.enpol.2015.09.041

[51] Mishra MK, Khare N, Agrawal AB. Scenario analysis of the CO_2 emissions reduction potential through clean coal technology in India's power sector: 2014–2050. Energy Strategy Reviews. 2015;**7**: 29-38. DOI: 10.1016/j.esr.2015.03.001

[52] Bhatt A, Priyadarshini S, Mohanakrishnan AA, Abri A, Sattler M, Techapaphawit S. Physical, chemical, and geotechnical properties of coal fly ash: A global review. Case Studies in Construction Materials. 2019;**11**:1-11. DOI: 10.1016/j.cscm.2019.e00263

[53] Hall ML, Livingston WR. Fly ash quality, past, present and future, and the effect of ash on the development of novel products. Journal of Chemical Technology and Biotechnology. 2002;**77**: 234-239. DOI: 10.1002/jctb.538

[54] Yao ZT, Ji XS, Sarker PK, Tang JH, Ge LQ, Xia MS, et al. A comprehensive

review on the applications of coal fly ash. Earth-Science Reviews. 2015;**141**: 105-121. DOI: 10.1016/j.earscirev.2014.11.016

[55] Thompson RL, Bank T, Montross S, Roth E, Howard B, Verba C, et al. Analysis of rare earth elements in coal fly ash using laser ablation inductively coupled plasma mass spectrometry and scanning electron microscopy. Spectrochimica Acta Part B. 2018;**143**: 1-11. DOI: 10.1016/j.sab.2018.02.009

[56] Blissett RS, Rowson NA. A review of the multi-component utilization of coal fly ash. Fuel. 2012;**97**:1-23. DOI: 10.1016/j.fuel.2012.03.024

[57] Vassilev SV, Vassileva CG. Methods for characterization of composition of fly ashes from coal-fired power stations: A critical overview. Energy & Fuels. 2005;**19**:1084-1098. DOI: 10.1021/ef049694d

[58] Asl SMH, Javadian H, Khavarpour M, Belviso C, Taghavi M, Maghsudi M. Porous adsorbents derived from coal fly ash as cost-effective and environmentally-friendly sources of aluminosilicate for sequestration of aqueous and gaseous pollutants: A review. Journal of Cleaner Production. 2019;**208**:1131-1147. DOI: 10.1016/j.clepro.2018.10.186

[59] Mondal S, Ghar A, Satpati AK, Sinharoy P, Singh DK, Sharma JN, et al. Recovery of rare earth elements from coal fly ash using TEHDGA impregnated resin. Hydrometallurgy. 2019;**185**:93-101. DOI: 10.1016/j.hydromet.2019.02.005

[60] Jegadeesan G, Al-Abed SR, Pinto P. Influence of trace metal distribution on its leachability from coal fly ash. Fuel. 2008;**87**:1887-1893. DOI: 10.1016/j.fuel.2007.12.007

[61] BP Energy Outlook: 2018 Edition [Internet]. 2018. Available from: https://www.bp.com/content/dam/bp/en/corporate/pdf/energy-economics/energy-outlook/bp-energy-outlook-2018.pdf [Accessed: 20 October 2019]

[62] Ram LC, Masto RE. Fly ash for soil amelioration: A review on the influence of ash blending with inorganic and organic amendments. Earth-Science Reviews. 2014;**128**:52-74. DOI: 10.1016/j.earscirev.2013.10.003

[63] Wesche K, editor. Fly Ash in Concrete: Properties and Performance. 1st ed. London: CRC Press (Taylor & Francis Group); 1991. DOI: 10.1201/9781482267051. 356p

[64] Vichaphund S, Aht-Ong D, Sricharoenchaikul V, Atong D. Characteristic of fly ash derived-zeolite and its catalytic performance for fast pyrolysis of Jatropha waste. Environmental Technology. 2014;**35**: 2254-2261. DOI: 10.1080/09593330.2014.900118

[65] Franus W, Wiatros-Motyka MM, Wdowin M. Coal fly ash as a resource for rare earth elements. Environmental Science and Pollution Research. 2015;**22**: 9464-9474. DOI: 10.1007/s11356-015-4111-9

[66] Cokca E, Yilmaz Z. Use of rubber and bentonite added fly ash as a liner material. Waste Management. 2004;**24**:153-164. DOI: 10.1016/j.wasman.2003.10.004

[67] Moreno N, Querol X, Andrés JM, Stanton K, Towler M, Nugteren H, et al. Physico-chemical characteristics of European pulverized coal combustion fly ashes. Fuel. 2005;**84**:1351-1363. DOI: 10.1016/j.fuel.2004.06.038

[68] Chindaprasirt P, De Silva P, Sagoe-Crentsil K, Hanjitsuwan S. Effect of SiO_2 and Al_2O_3 on the setting and hardening of high calcium fly ash-based geopolymer systems. Journal of Materials Science. 2012;**47**:4876-4883. DOI: 10.1007/s10853-012-6353-y

[69] Terzić A, Radojević Z, Miličić LJ, Pavlović LJ, Aćimović Z. Leaching of the potentially toxic pollutants from composites based on waste raw material. Chemical Industry and Chemical Engineering Quarterly. 2012;**18**:373-383. DOI: 10.2298/CICEQ111128013T

[70] Lee SH, Kim HJ, Sakai E, Daimon M. Effect of particle size distribution of fly-ash-cement system on the fluidity of cement pastes. Cement and Concrete Research. 2003;**33**: 763-768. DOI: 10.1016/S0008-8846(02) 01054-2

[71] Koukouzas NK, Zeng R, Perdikatsis V, Xu W, Kakaras EK. Mineralogy and geochemistry of Greek and Chinese coal fly ash. Fuel. 2006;**85**: 2301-2309. DOI: 10.1016/j.fuel. 2006.02.019

[72] Takada T, Hashimoto I, Tsutsumi K, Shibata Y, Yamamuro S, Kamada T, et al. Utilization of coal ash from fluidized-bed combustion boilers as road base material. Resources, Conservation and Recycling. 1995;**14**: 69-77. DOI: 10.1016/S0921-3449(95) 80001-8

[73] Lanyerstorfer C. Pre-processing of coal combustion fly ash by classification for enrichment or rare earth elements. Energy Reports. 2018;**4**:660-663. DOI: 10.1016/j.egry.2018.10.010

[74] Hower JC, Dai S, Seredin VV, Zhao L, Kostova IJ, Silva LFO, et al. Note on the occurrence of yttrium and rare earth elements in coal combustion products. Coal Combustion and Gasification Products. 2013;**5**:39-47. DOI: 10.4177/CCGD-D-13-00001-1

[75] Birik D, White JC. Rare earth elements, in bituminous coals and underclays of the Sydney Basin, Nova Scotia: Element sites, distribution, minerology. International Journal of Coal Geology. 1991;**19**:219-251. DOI: 10.1016/0166-5162(91)90022-B

[76] Pan J, Zhou C, Tang M, Cao S, Liu C, Zhang N, et al. Study on the modes of occurrence of rare earth elements in coal fly ash by statistics and a sequential chemical extraction procedure. Fuel. 2019;**237**:555-565. DOI: 10.1016/j.fuel.2018.09.139

[77] Fiket Z, Medunić G, Turk MF, Kniewald G. Rare earth elements in superhigh-organic sulfur Rusa coal ash (Croatia). International Journal of Coal Geology. 2018;**194**:1-10. DOI: 10.1016/j. coal.2018.05.002

[78] Georgakopoulos A, Filippidis A, Fournaraki AK, Turiel JLF, Llorenes JF, Mousty F. Leachability of major and trace elements of fly ash from Ptolemais Power Station, Northern Greece. Energy Sources. 2002;**24**:103-113. DOI: 10.1080/00908310252774426

[79] Kashiwakura S, Kumagai Y, Kubo H, Wagatsuma K. Dissolution of rare earth elements from coal fly ash particles in a dilute H_2SO_4 solvent. Open Journal of Physical Chemistry. 2013;**3**: 69-75. DOI: 10.4236/ojpc.2013.32009

[80] Tuan LQ, Thenepalli T, Chilakala R, Vu HHT, Ahn JW, Kim J. Leaching characteristics of low concentration rare earth elements in Korean (Samcheok) CFBC bottom ash samples. Sustainability. 2019;**11**:1-11. DOI: 10.3390/su11092562

[81] Seredin VV. Rare earth element-bearing coals from the Russian Far East deposits. International Journal of Coal Geology. 1996;**30**:101-129. DOI: 10.1016/0166-5162(95)00039-9

[82] Wagner NJ, Matiane A. Rare earth elements in select Main Karoo Basin (South Africa) coal and coal ash samples. International Journal of Coal Geology. 2018;**196**:82-92. DOI: 10.1016/ j.coal.2018.06.020

[83] Folgueras MB, Alonso M, Fernandez FJ. Coal and sewage sludge

ashes as sources of rare earth elements. Fuel. 2018;**192**:128-139. DOI: 10.1016/j.fuel.2016.12.019

[84] Huang Z, Fan M, Tian H. Rare earth elements of fly ash from Wyoming's Powder River Basin Coal. Journal of Rare Earths. 2019;**38**:1-8. DOI: 10.1016/j.jre.2019.05.004

[85] Flues M, Sato IM, Scapin MA, Cotrim MEB, Camargo IMC. Toxic element mobility in coal and ashes of Figueira Coal Power Plant, Brazil. Fuel. 2013;**103**:430-436. DOI: 10.1016/j.fuel.2012.09.045

[86] Silva LFO, DaBoit K, Sampaio CH, Jasper A, Andrade ML, Kostova IJ, et al. The occurrence of hazardous volatile elements and nanoparticles in Bulgarian coal fly ashes and the effect on human health exposure. Science of the Total Environment. 2012;**416**:513-526. DOI: 10.1016/j.scitotenv.2011.11.012

[87] Goodarzi F. Characteristics and composition of fly ash from Canadian coal/fired power plants. Fuel. 2006;**85**:1418-1427. DOI: 10.1016/j.fuel.2005.11.022

[88] Tang Q, Liu G, Zhou C, Zhang H, Sun R. Distribution of environmentally sensitive elements in residential soils near a coal/fired power plant: Potential risks to ecology and children's health. Chemosphere. 2013;**93**:2473-2479. DOI: 10.1016/j.chemosphere.2013.09.015

[89] Medunić G, Kuharić Z, Krivohvalek A, Fiket Z, Radjenović A, Gödel K, et al. Geochemistry of Croatian superhigh-organic-sulphur Rasa coal, imported low-S coal, and bottom ash: Their Se and trace metal fingerprints in seawater, clover, foliage, and mushroom specimens. International Journal of Oil Gas and Coal Technology. 2018;**18**:3-24. DOI: 10.1504/IJOGCT.2018.10006334

[90] Kurda R, Silvestre JD, Brito J. Toxicity and environmental and economic performance of fly ash and recycled concrete aggregates use in concrete: A review. Heliyon. 2018;**4**:1-45. DOI: 10.1016/j.heliyon.2018.e00611

[91] Petrotou A, Skordas K, Papastergios G, Filippidis A. Factors affecting the distribution of potentially toxic elements in surface soils around and industrialized area of northwestern Greece. Environment and Earth Science. 2012;**65**:823-833. DOI: 10.1007/s12665-011-1127-4

[92] Sushil S, Batra VS. Analysis of fly ash heavy metal content and disposal in three thermal power plants in India. Fuel. 2006;**85**:2676-2679. DOI: 10.1016/j.fuel.2006.04.031

[93] Kim Y, Kim K, Jeong GY. Study of detailed geochemistry of hazardous elements in weathered coal ashes. Fuel. 2017;**193**:343-350. DOI: 10.1016/j.fuel.2016.12.080

[94] Ćujić M, Dragović S, Djordjević M, Dragović R, Gajić B. Environmental assessment of heavy metals around the largest coal fired power plant in Serbia. Catena. 2016;**139**:44-52. DOI: 10.1016/j.catena.2015.12.001

[95] Keegan TJ, Farago ME, Thornton I, Hong B, Colvile RN, Pesch B, et al. Dispersion of As and selected heavy metals around a coal-burning power station in central Slovakia. Science of the Total Environment. 2006;**358**:61-71. DOI: 10.1016/j.scitotenv.2005.03.020

[96] Ayanda OS, Fatoki OS, Adekola FA, Ximba BJ. Characterization of fly ash generated from Matla Power Station im Mpumalanga, South Africa. Journal of Chemistry. 2012;**9**:1788-1795

[97] Penilla RP, Bustos AG, Elizalde SG. Immobilization of Cs, Cd, Pb and Cr by synthetic zeolites from Spanish low-calcium coal fly ash. Fuel. 2006;**85**:823-832. DOI: 10.1016/j.fuel.2005.08.022

[98] Baba A, Kaya A. Leaching characteristics of solid waste from thermal power plants of western Turkey and comparison of toxicity methodologies. Journal of Environmental Management. 2004;**73**: 199-207. DOI: 10.1016/j.jenvman. 2004.06.005

[99] Hower JC, Robertson JD. Chemistry and petrology of fly ash derived from the co-combustion of western United States coal and tire-derived fuel. Fuel Processing Technology. 2004;**85**:359-377. DOI: 10.1016/j.fuproc.2003.05.03

[100] Manz OE. Coal fly ash: A retrospective and future look. Fuel. 1999;**78**:133-136. DOI: 10.1016/ S0016-2361(98)00148-3

[101] Vassilev SV, Vassileva CG. A new approach for the combined chemical and mineral classification of the inorganic matter in coal. 1. Chemical and mineral classification systems. Fuel. 2009;**88**:235-245. DOI: 10.1016/j. fuel.2008.09.006

[102] Gollakota ARK, Volli V, Shu C-M. Progressive utilization prospects of coal fly ash: A review. Science of the Total Environment. 2019;**672**:951-989. DOI: 10.1016/j.scitotenv.2019.03.337

[103] Gajić G, Djudjević L, Kostić O, Jarić S, Mitrović M, Pavlović P. Ecological potential of plants for phytoremediation and ecorestoration of fly ash deposits and mine wastes. Frontiers in Environmental Science. 2018;**6**:1-24. DOI: 10.3389/fenvs. 2018.00124

[104] Tiwari MK, Bajpai S, Dewangan UK, Tamrakar RK. Suitability of leaching test methods for fly ash and slag: A review. Journal of Radiation Research and Applied Science. 2015;**8**: 523-537. DOI: 10.1016/j.jrras.2015.06.003

[105] Ahmaruzzaman M. A review on the utilization of fly ash. Progress in Energy and Combustion Science. 2010;**36**: 327-363. DOI: 10.1016/j.jpecs.2009.11.003

[106] Abdalqader AF, Jin F, Al-Tabbaa A. Development of greener alkali-activated cement: Utilization of sodium carbonate for activating slag and fly ash mixtures. Journal of Cleaner Production. 2016;**113**:66-75. DOI: 10.1016/j. clepro.2015.12.010

[107] Huang X, Zhao H, Zhang G, Li J, Yang Y, Ji P. Potential of removing Cd (II) and Pb(II) from contaminated water using a newly modified fly ash. Chemosphere. 2020;**242**:125148. DOI: 10.1016/j.chemosphere.2019.125148

[108] Cardoso AM, Paprocki A, Ferret LS, Azevedo CMN, Pires M. Synthesis of zeolite Na-P1 under mild conditions using Brazilian coal fly ash and its applications in wastewater treatment. Fuel. 2015;**139**:59-67. DOI: 10.1016/j.fuel.2014.08.016

[109] Mushtaq F, Zahid M, Bhatti IA, Nasir S, Hussain T. Possible applications of coal fly ash in wastewater treatment. Journal of Environmental Management. 2019;**240**:27-46. DOI: 10.1016/j. envman.2019.03.054

[110] Belviso C. State-of-the-art applications of fly ash from coal and biomass: A focus on zeolite synthesis processes and issues. Progress in Energy and Combustion Science. 2018;**65**: 109-135. DOI: 10.1016/j.pecs.2017. 10.004

[111] Tauanov Z, Shah D, Inglezakis V, Jamwal PK. Hydrothermal synthesis of zeolite production from coal fly ash: A heuristic approach and its optimization for system identification of conversion. Journal of Cleaner Production. 2018;**182**: 616-623. DOI: 10.1016/j. clepro.2018.02.047

[112] Iqbal A, Sattar H, Haider R, Munir S. Synthesis and characterization of pure phase zeolite 4A from coal fly

ash. Journal of Cleaner Production. 2019;**219**:258-267. DOI: 10.1016/j.clepro.2019.02.066

[113] Feng W, Wan Z, Daniels J, Li Z, Xiao G, Yu J, et al. Synthesis of high quality zeolites from coal fly ash: Mobility of hazardous elements and environmental applications. Journal of Cleaner Production. 2018;**202**:390-400. DOI: 10.1016/j.clepro.2018.08.140

[114] Fukasawa T, Horigome A, Tsu T, Karisma AD, Maeda N, Huang A-N, et al. Utilization of incineration fly ash from biomass power plants for zeolite synthesis from coal fly ash by hydrothermal treatment. Fuel Processing Technology. 2017;**167**:92-98. DOI: 10.1016/j.fuproc.2017.06.023

[115] Ojumu TV, Du Plessis PW, Petrik LF. Synthesis of zeolite A from coal fly ash using ultrasonic treatment—A replacement for fusion step. Ultrasonics Sonochemistry. 2016;**31**: 342-349. DOI: 10.1016/j.ultsonch.2016.01.016

[116] Babajide O, Musyoka N, Petrik L, Ameer F. Novel zeolite Na-X synthesized from fly ash as a heterogeneous catalyst in biodiesel production. Catalysis Today. 2012;**190**: 54-60. DOI: 10.1016/j.cattod.2012.04.044

[117] Manique MC, Lacerda LV, Alves AK, Bergmann CP. Biodiesel production using coal fly ash-derived sodalite as a heterogeneous catalyst. Fuel. 2017;**190**:268-273. DOI: 10.1016/j.fuel.2016.11.016

[118] Chakraborty R, Bepari S, Banerjee A. Transesterification of soybean oil catalyzed by fly ash and egg shell derived solid catalyst. Chemical Engineering Journal. 2010;**165**:798-805. DOI: 10.1016/j.cej.2010.10.01

[119] Xiang Y, Wang L, Jiao Y. Ultrasound strengthened biodiesel production from waste cooking oil using modified coal fly ash as catalyst. Journal of Environmental Chemical Engineering. 2016;**4**:818-824. DOI: 10.1016/j.jece.201512.031

[120] Bhandari R, Volli V, Purkait MK. Preparation and characterization of fly ash based mesoporous catalyst for transesterification of soybean oil. Journal of Environmental Chemical Engineering. 2015;**3**:906-914. DOI: 10.1016/j.jece.2015.04.008

[121] Volli V, Purkait MK, Shu CM. Preparation and characterization of animal bone powder impregnated fly ash catalyst for transesterification. Science of the Total Environment. 2019; **669**:314-321. DOI: 10.1016/j.scitotenv.2019.03.080

[122] He PY, Zhang YJ, Chen H, Han ZC, Liu LC. Low-energy synthesis of kaliophilite catalyst from circulating fluidized bed fly ash for biodiesel production. Fuel. 2019;**257**:1-10. DOI: 10.1016/j.fuel.2019.116041

[123] Babajide O, Musyoka N, Petrik L, Ameer F. Use of coal fly ash as a catalyst in the production of biodiesel. Petroleum and Coal. 2010;**52**:261-272

[124] Kotwal MS, Niphadkar PS, Deshpande SS, Bokade VV, Joshi PN. Transesterification of sunflower oil catalyzed by flyash-based solid catalysts. Fuel. 2009;**88**:1773-1778. DOI: 10.1016/j.fuel.2009.04.004

[125] Algoufi YT, Hameed BH. Synthesis of glycerol carbonate by transesterification of glycerol with dimethyl carbonate over K-zeolite derived from coal fly ash. Fuel Processing Technology. 2014;**126**:5-11. DOI: 10.1016/j.fuproc.2014.04.004

[126] Helwani Z, Fatra W, Saputra E, Maulana R. Preparation of CaO/Fly ash as a catalyst inhibitor for transesterification process of palm oil in

biodiesel production. IOP Conference Series: Materials Science and Engineering. 2018;**334**:1-10

[127] Volli V, Purkait MK. Preparation and characterization of hydrotalcite-like materials from flyash for transesterification. Clean Technologies and Environmental Policy. 2016;**18**: 529-540. DOI: 10.1007/s10098-015-1036-4

[128] Hadiyanto H, Lestari SP, Abdullah A, Widayat W, Sutanto H. The development of fly ash-supported CaO derived from mollusk shell of *Anadara granosa* and *Pahia undulate* as heterogeneous CaO catalyst in biodiesel synthesis. International Journal of Energy and Environmental Engineering. 2016;7:297-305. DOI: 10.1007/s40095-016-0212-6

[129] Lathiya DR, Bhatt DV, Maheria KC. Sulfated fly-ash catalyzed biodiesel production from maize acid feedstock: A comparative study of Taguchi and Box-Behnken design. ChemistrySelect. 2019;**4**:4392-4397. DOI: 10.1002/slct.201803916

[130] Volli V, Purkait MK. Selective preparation of zeolite X and A from fly ash and its use as catalyst for biodiesel production. Journal of Hazardous Materials. 2015;**297**:101-111. DOI: 10.1016/j.hazmat.2015.04.066

Section 3

Biofuels and Hydrogen Systems

Chapter 11

Biomass Carbonization

Mahmoud Amer and Ahmed Elwardany

Abstract

Carbonization is the art of reinventing the waste biomass into a carbon−/energy-rich charcoal. It redefines the principles of renewable energy and power generation. Char is produced by a pyrolysis process in which the biomass is heated in an inert atmosphere to high temperatures until absorbed volatiles are expelled thus enriching its heating value and energy content. Carbonization itself is an old process that is being used till now, but the renewed interest in it especially with biomass is because it opens new doors for commercial and scientific applications. The carbon can be extracted from the produced char to form the precious graphite and graphene. This chapter provides a general overview about slow pyrolysis processes including carbonization and the torrefaction process which is a mild carbonization process. The characterization of different biomass species and their effect on the carbonization process and the final product will be also discussed. Different carbonization processes and methodologies which vary in the process parameters will be addressed, and the most promising ones will be highlighted. An important addition to this chapter is the general design parameters, methodologies, and factors that must be taken into consideration when designing carbonization reactors for lab and industrial designs.

Keywords: biomass, carbonization, torrefaction, slow pyrolysis, biochar, charcoal

1. Introduction

The depletion and rising cost of the fossil fuels along with the rising greenhouse effect created energy and environmental concerns. This urged researches to look for a proper alternative for fossil fuels thus appeared the increasing interest in renewable energy sources. Renewable energy is defined as the energy obtained from naturally repetitive and persistent flows of energy occurring in the local environment [1]. This can be achieved without demolishing the current infrastructure for power generation by finding alternative fuels that can work on the fossil fuels burners currently in place. These alternatives can be produced from biomass because they are the only alternative renewable source of carbon. Biomass is a term for organic waste that stems from plants and produced by photosynthesis making them all green energy sources. During their formation, biomasses already used the exact same amount of carbon dioxide they will later emit as exhaust, so they are usually referred to as a carbon neutral fuel when it is burnt [2–4].

Waste biomass encompasses a wide range of materials, and their availability is advantageous. The wastes usually have a trivial value and could be considered as a pollutant to the environment due to their wrongful disposal methods. There exist many sources of waste biomass either from agriculture or industrial sectors as waste from agricultural processes or frozen vegetables/fruits factories, respectively.

The usage of these different kinds of biomasses has positive economic and environmental impacts and can be called a truly renewable source, because it is a by-product of a different primary production which adds a real additional income. There exist numerous pathways to upgrade the biomass into biofuel either by thermochemical or biochemical methods [5, 6]. In this chapter we are concerned with the thermochemical conversion.

Biomass thermochemical conversion is performed by heat application to start the decomposition of organic materials by cracking the hydrocarbonaceous bonds to transform it into a richer form of energy. Thermochemical conversion processes include combustion, pyrolysis, carbonization, co-firing, gasification, and liquefaction [7]. Pyrolysis is considered as the starting point of all thermochemical conversion technologies because it involves all chemical reactions to form solid "char," liquid "bio-oil," and gas "biogas" carried out under oxygen-free or oxygen-limiting conditions. The gases and the bio-oil are from the volatile fraction of biomass, while the char is mostly the fixed carbon component [8].

The proportions of these pyrolysis products depend on the pyrolysis method (fast, slow, or conventional) and reaction parameters. In pyrolysis, biomass is heated in the absence of air.

Pyrolysis can be classified according to the applied heating rate during the process into fast and slow pyrolysis. In fast pyrolysis, the heating rate is very high, and the residence time is lower; these parameters increase the bio-oil yield fraction. On the other hand, slow pyrolysis has lower heating rates that result in higher solid yields or char; hence, our focus will be on it. Slow pyrolysis is even further classified into carbonization as well as torrefaction. This classification depends on the operating temperature of the process and the residence time. Carbonization process utilizes higher operating temperatures and longer residence times than torrefaction. Carbonization aims to produce a highly carbonaceous product, while torrefaction could be a pretreatment process for further processing [9]. The product from any of the two slow pyrolysis processes is called "char," while charcoal is reserved for the carbonization products. Charcoal refers to the highly carbonaceous product that is intended to be used as a fuel. Furthermore, charcoal can be used in the smelting and sintering processes as a reductant in the metallurgical industry [10]. Char can also be further processed to get activated carbon which is used as an adsorbent [11]. Furthermore, char has a good potential to be used as a soil fertilizer, hence termed "biochar" [12].

Carbonization is the oldest known thermochemical process that allowed humans to convert wood into charcoal, which is the first biofuel that has been used by humans to step out of the Stone Age by refining ores into metals. Carbonization was performed in the early ages by gathering the wood into a cone-shaped pile, covering it with earth, slowly combusting the wood, and allowing for the water content and volatile substances to exit from a central chimney, turning the wood into coal.

This chapter will focus on the slow pyrolysis especially carbonization process for biomass conversion. A brief discussion will be provided for the carbonization products and their characterization techniques. Finally, an overview of carbonization reactor design along with their production techniques concludes the chapter.

2. Carbonization vs. torrefaction

Sometimes carbonization process is confused with other thermochemical processes like torrefaction. Both carbonization and torrefaction are two branches from the pyrolysis process. They both include heating the biomass at specified temperature for a specific residence time until its properties are upgraded for better use either for fuel, fertilizing, or adsorption applications. And even they have the same general

	Torrefaction	Carbonization
Temperature (°C)	200-300	>300
Residence time	< 2 hours	>2 hours and could reach days
Operating pressure	Atmospheric	Up to 10 atmospheres
Environment	Inert or oxygen limited	Inert or oxygen limited
Heating rate	Low	Low
Main outcome	High energy density	High fixed carbon
Moisture content (%)	<5	<5
Fixed carbon (%)	30	85
Volatiles (%)	65	10
Bio-oil yield (%)	5	30
Biogas yield (%)	15	35
Solid yield (%)	80	35
Hydrophobicity	Hydrophobic	Hydrophobic

Table 1.
Comparison between torrefaction and carbonization [4, 9, 13].

definition which is given for pyrolysis. The main objective of the carbonization process is to obtain the maximum attainable fixed carbon, but this comes on the account of the product which will suffer low energy yield, energy density and hydrocarbon content. On the other hand, torrefaction aims at maximizing the energy density and energy yield, but this will not provide high carbon content. So, based on the product motivation of both processes, they differ in their process parameter settings in terms of temperature range and residence time which produce either torrefied or carbonized products. Carbonization temperature and time are much higher than of torrefaction. **Table 1** shows the main differences between the carbonization and torrefaction processes. It can be seen in **Table 1** that the environment for both torrefaction and carbonization can be either inert or oxygen limited. Sometimes even small traces of oxygen could provide better results especially with carbonization more than torrefaction. This limited amount partially combusts small fraction of the biomass which delivers enough internal heat to successfully complete either the carbonization process. This also provides more homogeneous product due to better heat distribution across the biomass [9, 13]. Carbonization removes almost all the volatiles in the biomass. Torrefaction retains most of the volatiles except for the very light ones that get easily cracked at low temperatures.

3. Biomass carbonization process

The production of solid biomass fuel has been practiced for thousands of years through carbonization providing early people with charcoal for heat purposes and iron ore extraction. The carbonization process occurs slowly heating the biomass to high temperatures exceeding 400°C and for several hours. The products from this process are charcoal when it is used as fuel, biochar when used as fertilizer or soil amendments, biocoke for metal extraction, and finally activated carbon when regular charcoal is upgraded for adsorption and purification purposes. These products are further explained later in Section 5.

Figure 1 shows the complete carbonization process with all its stages until its intended products are reached. It also shows how the biomass should look in each stage. A brief explanation of each process is provided indicating key features of each stage. The heat source for this complete process should be by a means of renewable energy like initially the burning part of the raw biomass or by solar heating. Electrical heating is only viable in research but inapplicable and not cost-effective for industrial scale.

The different process regions and heating stages indicated in both **Figures 1** and **2** are further explained as follows.

3.1 Predrying

This is the first step of the carbonization process where the biomass is heated from the room temperature to the drying temperature which is approximately

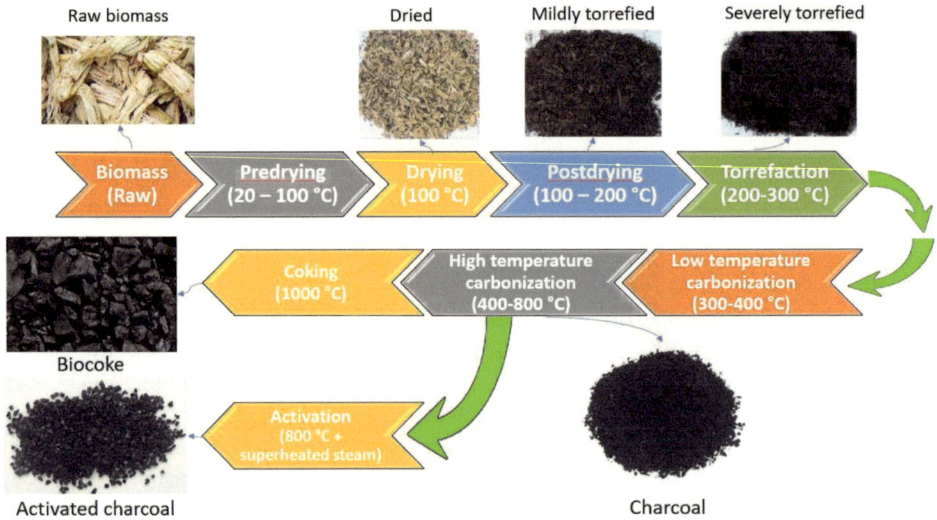

Figure 1.
Complete carbonization process including its products.

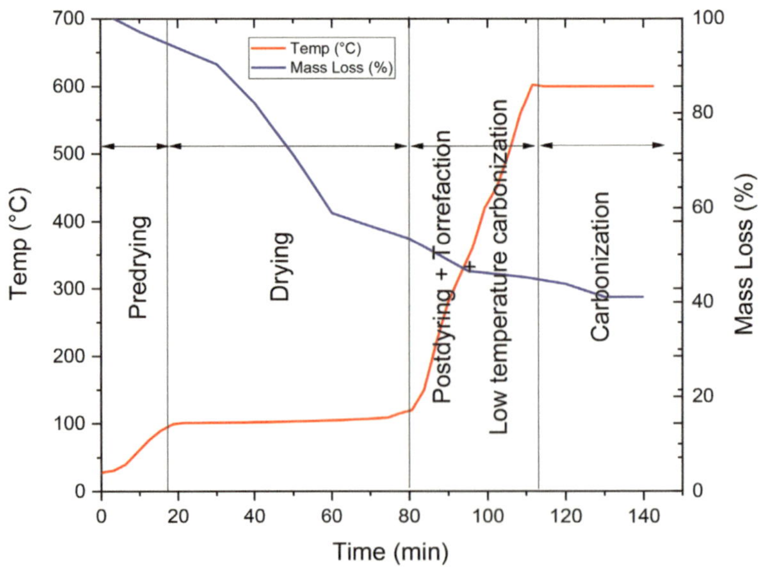

Figure 2.
Temperature and mass loss profiles for carbonization process.

100°C. During this stage, the heat is mostly transferred by conduction from the reactor's walls to the biomass. This stage usually has a steep temperature rise. The heat supplied to the biomass accounts mainly for increasing the moisture temperature to the evaporation temperature.

3.2 Drying

This step is the highest step in energy consumption because of the high latent heat of water especially for biomasses with high initial moisture content like water hyacinth and sewage sludge. All heat supplied during this stage is just to convert the moisture content in the biomass from the liquid to the vapor phase. The biomass's temperature is almost constant during this stage regardless of the reactor's outer temperature because evaporation occurs at constant temperature. In order to reduce the heat loss during this stage, sun drying should be employed for the raw biomass before its introduction to the reactor.

3.3 Postdrying

After drying is complete, the heat energy added to the biomass starts to increase its temperature. This stage ranges between 100 and 200°C. During this stage no significant decomposition starts within the biomass. Only the water molecules bound inside the fibers and very light volatiles are expelled during this stage. This stage does not require high heat and is usually quick.

3.4 Torrefaction

Torrefaction occurs between 200 and 300°C. During this stage the decomposition starts. The first constituent of the biomass to start decomposing is the hemicellulose. By the end of this stage, almost all hemicellulose is decomposed, and small fractions of the cellulose start decomposing. If the heating process is stopped at this stage to keep the temperature constant, a torrefied product with high energy density and low fixed carbon content is obtained.

3.5 Low-temperature carbonization

This stage occurs between 300 and 400°C. During this stage, both hemicellulose and cellulose completely decompose, and the lignin starts to degrade. The biomass' structure keeps breaking and decomposing to produce slightly richer fixed carbon content compared to torrefaction. However, this results in lower energy density.

3.6 High-temperature carbonization

If heating continues over 400°C, a very-high-carbon-rich charcoal is obtained. The degree of lignin degradation depends on the final temperature. During this stage, very high fixed carbon contents are obtained.

3.7 Coking

If the purpose of the carbonization is to use the product in metal ore extraction, then the heating is increased until the temperature exceeds 1000°C. This produces biocoke which is extremely carbon rich.

3.8 Activation

If the purpose is to produce activated carbon for adsorption, the heating continues until 800°C in superheated steam to remove the tar. The product of this stage has a very large pore volume making it most suited for adsorption and purification applications. This is done by heating the already made charcoal to 800°C in a superheated steam atmosphere. This superheated steam performs two major functions during the activation process. The first is to isolate the charcoal from the oxidative environment ensuring that it does not burn. The second is to remove the tarry residue which is blocking the finely structured pores inside the charcoal. After this steam heating, the product must be cooled in a non-oxidative environment then stored in an airtight container. This activation process makes the inner pores in the charcoal more accessible for adsorption. It also increases the pore surface area which gives the activated charcoal excellent adsorption capacity.

3.9 Cooling

This is the final step in the process line. If the product is removed from the reactor exactly after the carbonization ends, it will ignite, because the reactor's final temperature is much higher than the ignition temperature of the produced charcoal. So, it must be cooled first before it is exposed to the oxidative atmosphere. Usually the extracted heat during the cooling stage is recovered in the form of hot air to be used for biomass predrying before it enters the reactor. This could drastically reduce the very high input heat during the drying stage and increase the overall efficiency of the system.

Figure 2 shows the typical temperature and mass loss profiles for the biomass during the stages of the carbonization process. The mass loss during the heating process depends on the temperature and the biomass's constituents. At temperatures less than 100°C, the mass loss is for surface moisture and light volatiles, 100–120°C for loss of water, 120–200°C for volatile compounds, 200–300°C for hemicellulose, 275–350°C for cellulose, and 160–900°C for lignin [9, 14, 15].

Before discussing the specifics of the different products from the carbonization process, we need first to understand how to characterize these products in the next section.

4. Characterization of the biomass and the carbonized solids

The use of any biomass for conversion to energy will be affected by its characteristics. These initial characteristics will change after being carbonized. This information influences and determines the overall efficiency of the conversion process as well as the overall economics. These characteristics need to be understood in detail before any judgment is made on the carbonized products.

These following mentioned properties are of the greatest importance for characterization.

4.1 Proximate analysis

Proximate analysis gives the general composition of the biomass in terms of gross components such as moisture content (MC), volatile matter (VM), ash (ASH), and fixed carbon (FC). This proximate analysis is performed by thermogravimetric analysis (TGA). The TGA is performed in an inert atmosphere, usually nitrogen, with an initial temperature heating rate of 12°C/min from the room temperature to 105°C with a holding time of 15 min to remove the moisture content. Then the sample is heated to 900°C with a heating rate of 20°C/min to perform the pyrolysis; then a holding time of 15 min is established at 900°C to ensure all volatiles are removed. After that the temperature is reduced to 850°C then pure oxygen or air fills the test chamber instead of the nitrogen to burn the remaining sample. The remaining amount after the burning is the ash content. The fixed carbon is obtained by subtracting the ash content from the already existing amount before oxygen introduction [4, 16]. The abovementioned temperature profile is shown in **Figure 3** [4].

Figure 4 shows a sample for a TGA analysis and how the proximate analysis can be obtained from it. Each one of the four results is briefly explained below. **Table 2** presents some proximate and ultimate analyses results for selected raw biomasses.

4.1.1 Moisture content (MC)

Biomasses are always characterized with high moisture content as all plants depend heavily on water to grow. The total moisture content of some biomass can reach over 90% such as water hyacinth [17]. MC is undesirable during biomass carbonization because water drains very-high-energy amounts to evaporate and this energy is non-recoverable. Thus, sun drying is very important as it does not consume any power and it makes it easier to finish the drying at the carbonization plant. MC is also an important factor to assess the cost of transportation for the raw biomass. The MC for pyrolyzed or carbonized biomasses should be negligible.

4.1.2 Volatile matter (VM)

VM emerges from the biomass devolatilization which is the cracking of the heavy hydrocarbons in the biomass into lighter volatiles. These volatiles are in the gaseous phase during the pyrolysis process. If these volatiles are captured and cooled, part of them will condensate to become bio-oil, and the remaining will stay as no condensable gases which are called biogas. VM percentage varies according to the TGA process itself as VM depends highly on the heating rate [9]. As the heating rate increases, VM decreases as the time for devolatilization is reduced.

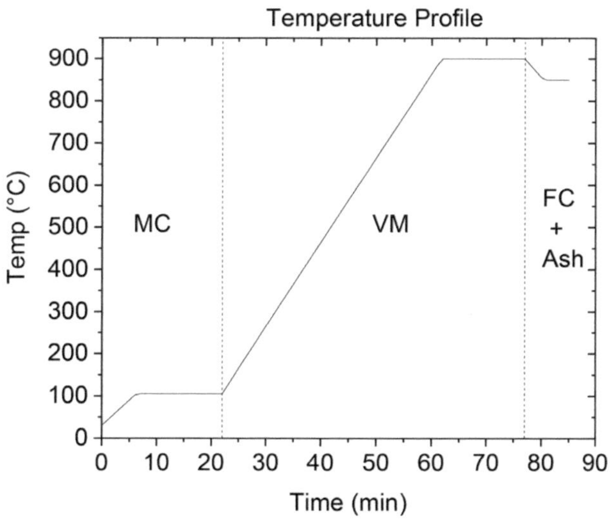

Figure 3.
The temperature profile of TGA test to obtain the proximate analysis [4].

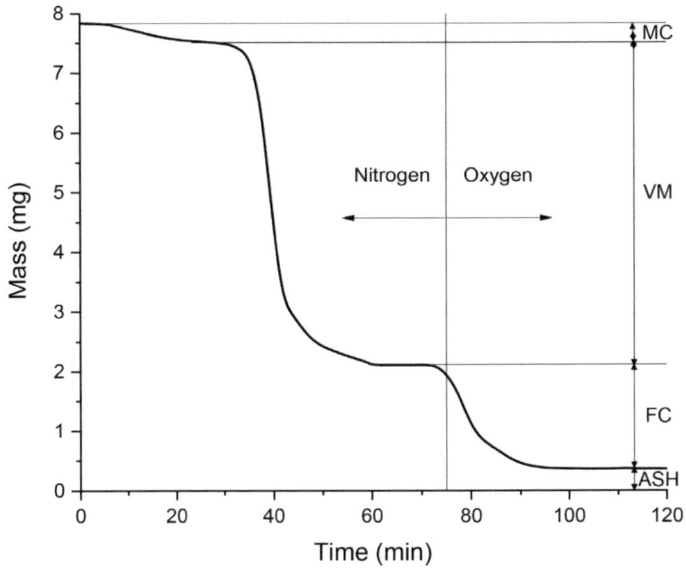

Figure 4.
TGA analysis sample for cotton stalks raw sample [4].

Biomass	Wood	Wheat straw	Rice husk	Rice straw	Lignite coal	Bituminous coal
MC (%)	20	16	8.2	8.5	34	11
VM (%)	82	59	69.5	64.7	29	35
FC (%)	17	21	12.5	8.9	31	45
Ash (%)	1	4	9.8	17.7	6	9
C (%)	51.6	48.5	50.7	52.8	56.4	73.1
H (%)	6.3	5.5	6.9	7.3	4.2	5.5
O (%)	41.5	45.6	41.4	39.1	18.4	8.7
N (%)	0	0.3	1	0.65	1.6	1.4
S (%)	0.1	0.1	—	0.11	—	1.7
Heating value (Mj/kg)	18.6	17.3	21.9	22	26.8	34

Table 2.
Proximate and ultimate analyses results for some raw biomasses [2, 4].

4.1.3 ASH content (ASH)

The ash obtained from biomass conversion does not necessarily come entirely from the biomass itself. During biomass collection and handling, some dirt, rock residue, other impurities, or small pieces of trash are often collected and mixed with the original biomass. All these impurities constitute the major inorganic component of the biomass feedstock which finally remain as ash. So, it is highly recommended that the samples used for characterization and analysis should be washed first and screened to ensure that the collected samples represent the studied biomass. These screening or cleaning processes will not be applicable in large-scale industries as they will cost time and money to swift through the huge biomass bulk. So, these

Biomass		Proximate analysis (%wt)						
		Wet basis				Dry basis		
		MC	VM	Ash	FC	VM	Ash	FC
Rice husk	Raw	8.2	69.5	9.8	12.4	75.4	10.6	14
	Dried	2.7	73.3	10.4	13.5	75.3	10.6	14.1

Table 3.
Comparison between the wet and dry basis proximate analysis results.

impurities have to be accepted as a drawback in large-scale industrial plants as they will slightly reduce the quality of the carbonized product.

4.1.4 Fixed carbon (FC)

The fixed carbon (FC) is determined from Eq. (1).

$$\%FC = 100 - (\%Ash + \%VM + \%MC) \tag{1}$$

FC represents the solid carbon in the biomass that remains in the char after devolatilization.

4.1.5 Proximate analysis calculation criteria

The proximate analysis can either be calculated based on dry basis (db) which excludes the moisture content or wet basis (wb) which includes the moisture content. Usually the results obtained from the TGA analysis are based on wet basis as they show the moisture content. However, it is not fair to consider the moisture content in judging in which biomass is better than the other because the drying efficiency can relatively affect all proximate analysis parameters. **Table 3** shows that wet analysis cannot be a good judge on the biomass' quality due to different moisture contents of the original sample and after converting to dry basis, both results are the same.

In order to calculate the proximate analysis based on the wet or dry basis, Eqs. (2, 3) are used, respectively.

$$\%property = property/(MC + VM + FC + Ash) * 100\% \tag{2}$$

$$\%property = property/(VM + FC + Ash) * 100\% \tag{3}$$

where property refers to MC, VM, FC, and Ash for wet basis and refers only to VM, FC, and Ash for dry basis.

4.2 Elemental composition

The elemental composition is obtained by performing elemental or ultimate analysis which is sometimes called CHNSO analysis. This analysis results in the composition of the hydrocarbon fuel in terms of its basic elements. The results are based on dry ash-free (DAF) analysis. This means that the moisture content (MC) and the inorganic materials (ASH) are disregarded from the results. The MC and ASH are obtained from the abovementioned proximate analysis in Section 4.1. Some sample elemental analysis results for some raw biomasses are presented in **Table 2**. A typical ultimate analysis result is given in the following form shown in Eq. (4).

$$C + H + N + S + O + MC + ASH = 100\%, \quad (4)$$

where C, H, N, S, and O are the mass percentages of carbon, hydrogen, nitrogen, sulfur, and oxygen, respectively, in the fuel.

Generally, in any biomass you will find major percentages of C, H, and O and minor percentages of N. However, S is not usually found in most of the biomasses which is a plus when their carbonized products are combusted, as they will not generate the corrosive sulfuric acid (H_2SO_4) or produce harmful SO_2 emissions. The moisture content is basically water which has the chemical form of H_2O; the hydrogen and oxygen in this water aren't included in the H and O in Eq. (4). However, this water is expressed as a general term in MC. Thus, the hydrogen and oxygen in the elemental analysis represent only the amounts in the organic components of the fuel itself.

From the ultimate analysis results, we can get the atomic ratios (H/C) and (O/C). H/C and O/C are often represented as Y-X plot which is called "van Krevelen" diagram as shown in **Figure 5** for some general classifications of different hydrocarbon fuels. This figure shows that biomass has very high relative amounts of oxygen and hydrogen, which results in relatively low heating values. The idea from pyrolysis is to move on the van Krevelen diagram from the biomass area to the coal area. The higher the pyrolysis temperature, the closer the produced charcoal approaches coal in the lower left corner of the van Krevelen diagram. To reach anthracite, pyrolysis temperature and residence time should be greatly increased which is done during carbonization.

4.3 Heating value (HV)

The heating value is defined as the amount of released heat by unit mass or unit volume for a fuel initially at 25°C when it is combusted, and the products have been cooled to 25°C again. If this heat includes the latent heat of water vaporization (i.e., the water vapor in the products is condensed), then it is called higher heating value (HHV), else, it is called lower heating value (LHV) [9]. HV could be obtained either experimentally using a calorimeter or numerically using empirical equations. These empirical equations can either depend on the results from proximate analysis, elemental analysis, or both combined in a single Equation [18]. One of the most

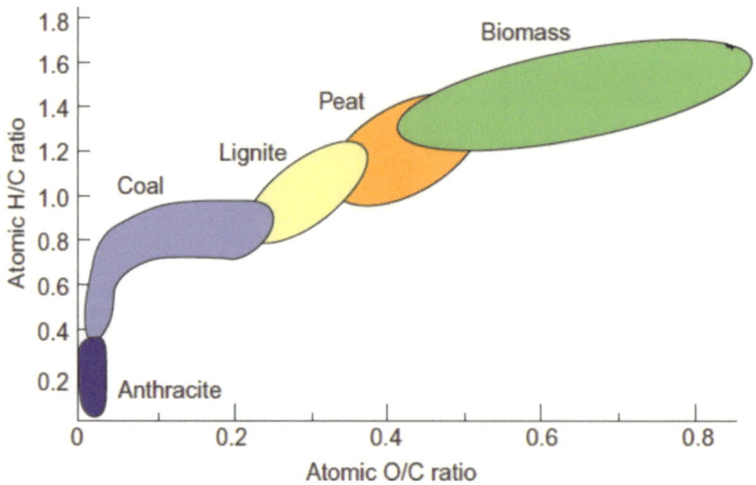

Figure 5.
Van Krevelen diagram, adapted from Basu [9].

used correlations to empirically calculate the HHV from the elemental analysis is provided by Ozyuguran et al. [19] which is shown in Eq. (5). This equation is mostly used because it has the highest value of regression coefficient (R^2) of 0.9441 when compared to experimental results, takes all five parameters (C, H, N, S, O) into consideration, can be used with all biomasses, and has a mean absolute error of 0.3178%, an average absolute error of 1.6978%, an average bias error of 0.0549%, and a root mean square deviation of 0.4230 [19]. The correlation is as follows:

$$HHV\left(\frac{MJ}{kg}\right) = -4.9140 + 0.2611\,[\%N] + 0.4114\,[\%C] + 0.6114\,[\%H] - 0.02097\,[\%O] + 0.3888\,[\%S] \quad (5)$$

where $\%N$, $\%C$, $\%H$, $\%O$, and $\%S$ are the mass percentages of the elemental nitrogen, carbon, hydrogen, oxygen, and sulfur based on dry ash-free ultimate analysis.

For more information about the numerical equations to calculate either the higher or the lower heating value, refer to the article of Ozyuguran et al. [19] and Vargas-Moreno et al. [18].

4.4 Solid and energy yields and enhancement factor

The solid yield (SY) defines the percentage remaining mass fraction of the original biomass after carbonization as shown in Eq. (6). The solid yield could either be defined based on dry basis (db) or wet basis (wb) if the original biomass mass is taken after or before drying, respectively. As biomass contains high water levels, it would be unjust to use the SY on wb because it will give very low numbers, so it is usually taken based on dry basis. Sometimes even when the raw biomass might contain a large amount of impurities and ashes, the SY is taken based on dry ash-free (DAF) basis. The enhancement factor (EF) represents the percentage increase in the heating value due to the carbonization process which is shown in Eq. (7); it is usually calculated based on the higher heating value.

Neither solid yield nor enhancement factor can singularly define the effectiveness of the carbonization or pyrolysis process. The reason is that when the pyrolysis conditions are more severe (i.e., higher temperature and residence time), the remaining mass will become lower. However, the product's higher heating value will be higher. So, in order to correctly judge the performance of the pyrolysis process, both factors should be taken into consideration. The result of this is the energy yield (EY) which is the multiplication of both solid yield and the enhancement factor as shown in Eq. (8).

The solid yield (SY), enhancement factor (EF) of HHV, and energy yield (EY) should be always calculated for all pyrolyzed biomasses to monitor the overall effectiveness and efficiency of the conversion process [20, 21].

$$SY(\%) = \frac{M_{carbonized}}{M_{raw}} * 100 \quad (6)$$

$$EF = \frac{HHV_{carbonized}}{HHV_{raw}} \quad (7)$$

$$EY(\%) = \frac{M_{carbonized}}{M_{raw}} * \frac{HHV_{carbonized}}{HHV_{raw}} * 100 = SY * EF \quad (8)$$

where M_{raw} and $M_{carbonized}$ represent the masses of the raw and torrefied samples, respectively.

4.5 Grindability

The fibrous and plastic nature of raw biomass makes it very hard to handle and transport especially if pneumatic transport is used. They could clog the system and will need large pneumatic channels to pass through them. This creates a process difficulty. Also, if biomass is required to replace coal or be used with coal in co-combustion systems, it must be finely ground and pulverized into very small and fine particles which is very hard for raw biomass and will cost a lot. The grindability is evaluated by calculating the Hardgrove Grindability Index (HGI) [22]. The HGI is measured by a small ball milling machine where a standard mass of 50 g is ground for a given time subject to a known ball force. The resulting product is sieved to measure amounts dropping below 75 μm which is compared against some standards to find the HGI value. The higher the HGI index, the lower the power required to grind the sample [9]. Esteban and Carrasco performed a study to compare the power required to grind one ton of raw wood biomass and coal to a fineness of d_{50} of approximately 500 μm. They found that biomass would require from 130 to 170 kWh where coal would need less than 20% of that power which is in the range of 7 to 36 kWh [23]. Pyrolysis, either torrefaction or carbonization, completely breaks down the biomass' cell structure due to the high heat applied. The treatment temperature is the most influential parameter for grinding. This makes the carbonized products the most brittle and least fibrous which can be even ground by hand. Generally speaking, carbonization highly increases the HGI and thus the grindability of the treated biomass making it fit to be used with regular coal pulverizers, pneumatic systems, and burners allowing for carbonized biomass to replace the fossil coal.

4.6 Hydrophobicity

Raw biomass is hygroscopic due to their fibrous nature which means that it can easily absorb moisture even after drying. This means that if raw biomass is required to be used as a source of energy, it must be dried exactly prior to its burning as it cannot be kept dry during storage. As mentioned above in Section 4.1.1, moisture is a bad thing when using and handling biomass. Biomass consists of three main components, hemicellulose, cellulose, and lignin; their moisture uptake is in the same order from higher to lower [24, 25]. As hemicellulose decomposes in the range of 220–300°C, 275–350°C for cellulose, and 160–900°C for lignin [9, 14, 15]. This means that all pyrolysis treatments which start from 200°C result in almost complete breakdown of the hemicellulose, converting the treated biomass from hygroscopic into hydrophobic nature. This means that carbonized biomass is the most hydrophobic. This eases their handling, storage, and transportation. The hydrophobicity can be determined by the immersion test. In this test, a small sample weight of approximately 0.5 g with particle size less than 1 mm is immersed in deionized water at room temperature in a sintered glass filter for 2 hours. The sample is then air-dried for 1 hour. After that its moisture content is estimated, and the moisture uptake can be identified [22].

5. Carbonization products

5.1 Charcoal (fuel)

Charcoal is the name given to the carbonization product when the main intent of its use is fuel. It is the oldest carbonized fuel ever used by humans. Nowadays charcoal is produced and used as smokeless fuel source or for barbeques. Charcoal is produced at relatively low carbonization temperatures up to 500°C but at a very slow heating rate which takes a very long time. According to Antal and Mochidzuki [26], the fixed carbon content of charcoal should have less than 30% volatile matter and less than 5% ash content. These values correspond to charcoal calorific values between 28 and 33 MJ/kg.

5.2 Activated charcoal

Activated charcoal is an updated version of the regular charcoal with very high pore number and surface area. This means that it cannot be obtained directly from the carbonization process and it needs further processing to obtain it. The activation process is a two-step process. First the charcoal is produced from regular carbonization. The second step is performed by removing the tar from the inner pores of the charcoal by means of a superheated steam. The activation process is further explained in Section 3.8. The activated charcoal is a valuable product used in the chemical and environmental industries in adsorption. It is also used to remove pollutants from fluids. The pore surface area required for the activated charcoal is specified by the physical structures using the Brunauer–Emmett–Teller (BET) surface area technique. The commercial grades of activated carbons require a specific surface area between 500 and 1500 m^2/g. This is a much better upgrade from the carbonized biochar which has a maximum BET surface area of 500 m^2/g [27].

5.3 Biochar (fertilizer)

Biochar is the name given to charcoal when its intention is to be used as a fertilizer. To produce biochar, carbonization occurs at high temperatures more than 500°C. Biochar is known for its carbon sequestration potential and soil amendment properties. Biochar is considered as a greenhouse gas reduction route because it increases the carbon content stably retained in the soil. Due to the high thermal stresses applied on the biochar during its formation, it becomes very tough to degrade with time. It can remain in its form for hundreds of years and retaining this carbon in the soil ensuring its fertility for this time [28]. The International Biochar Initiative (IBI) provides a standardized definition for the biochar's characteristics that is most suitable for soil amendment [29].

Sometimes when there is a need for new cultivation area, the first place to look for is the forest remains. These remains are burnt which results in forming biochar which has high carbon content and enhances the soil and improves its fertility. However, if these forest remains are large, this burning could contribute to the overall greenhouse effect because the burning is not controlled and is performed in open air so most of the carbon is burnt and only small parts of the total carbon in biomass is retained as stable solid char in the soil. The higher the degree of carbonization, the higher the fixed carbon amount and the better is the property of biochar [30].

5.4 Biocoke

For metal ore extraction, even regular fossil coal is not enough. The coal is carbonized or coked at very high temperatures to produce the coke. So, biomass can be severely carbonized to produce the biocoke which is a substitute for conventional coke. This severe carbonization process is called coking. Coking temperature can exceed 1000°C. Biocoke should have a volatile matter content less than 15% and ash content between 0.5 and 5% [26]. During metal extraction process, biocoke is heated with metallic ores with oxides or sulfides. The carbon in the biocoke combines with oxygen and sulfur which facilitates metal extraction. Biocoke has been used for extraction of iron from iron ore during the dawn of metallurgical industries. Steel industry is one of the highest contributors in CO_2 emissions due to their heavy carbon and coke usage. This motivated the use of biocoke in blast furnace to reduce heavy carbon dioxide emissions from the iron and steel industries. Sometimes, the biocoke is pulverized and injected into the blast furnaces.

5.5 Advantages of the carbonization products

Carbonization products have a large number of benefits which can be obtained through the recycling of the biomass waste as follows:

- Renewable solid fuel source
- Carbon-negative process
- Reduces dependability on carbon positive fossil fuels
- Improves the fertility of the soil and subsequently the crop production yield
- Less chemical fertilizers required
- Soil remediation
- Improves water quality
- Counteracts land desertification by increasing its fertility
- Improves water and nutrition retention in the soil

6. Carbonization reactors

There exist hundreds of simple and complicated carbonization reactors with different technologies such as kilns, retorts, and converters. Kiln describes the simplest traditional char-making equipment from wooden logs. Retorts and converters are the industrial reactors that can produce char as well as capture some volatiles and bio-oil. The difference between retorts and converters is the size of the biomass used. Retorts are used for long and thick wood logs, while converters are used with small particles of biomass [31]. These reactors can be characterized according to many classifications as follows [32]:

- The required final product
- Mode of operation (batch or continuous)
- Heat transfer method (direct, indirect, or microwave)
- Input heat source (electric, gas, or biomass combustion)
- Biomass loading method

- Operating pressure
- Construction material (soil, brick, concrete, or steel)
- Mobility (stationary, portable)

In this section, the used characterization method is the feeding system of the biomass into the reactor which includes batch and continuous systems.

6.1 Batch reactors

These reactors are the simplest and easiest ones to build and operate. However, they have some influential drawbacks. The most important disadvantage is the inhomogeneous quality of charcoal due to varying temperature gradients inside the reactor, uneven gas circulation, partial combustion, and uneven particle sizes. Thus, to have a homogeneous product, a good temperature control must be employed to improve the quality of the final product.

6.1.1 Kilns

Kiln is one of the simplest and primitive designs in biomass slow pyrolysis carbonization reactors that usually use wood. Kilns do not need external heat source to heat up; they are built in a way that supplies the required heat internally from partially burning the wood. **Figure 6** shows a primitive construction for the mound kiln where the wood logs are stacked on the ground and an earth covering is built over it, leaving small openings at the bottom for air supply. The stacking and the small openings limit the oxygen supply to the wood just enough to burn some wood to provide heat for carbonization. Since the kiln is well insulated, all the generated heat is retained inside it, and that helps slow down the thermal degradation of the wood into charcoal [9]. The central stack is added to the kiln to remove vapors and volatiles from the pile during the carbonization process and to provide access to the pile for the initial ignition of the wood. Once a small fire is established inside the kiln, the carbonization process begins. The color and density of the flue gases are used to judge the process maturity. White smoke indicates the moisture removal, and it is observed early in the beginning of the process. Yellow smoke indicates the initiation of the pyrolysis, and the kiln's temperature reached 250°C at least. Blue smoke indicates the near completion of the carbonization process. The

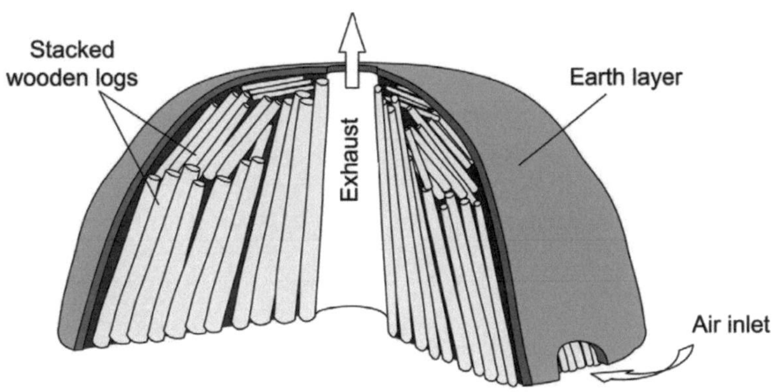

Figure 6.
Cross-sectional view of the mound kiln, adapted from Ronsse et al. [13].

carbonization time depends on the batch size and can reach a maximum of 1 month for very large kilns exceeding 28 tons per batch.

A modification to earth kilns has been made using steel and brick kilns where instead of covering the stack with earth material, the kiln is made from steel or bricks. This provides much better temperature distribution and production homogeneity. Using bricks improves the thermal insulation and the distributed air infiltration inside the kiln [33].

6.1.2 Retorts

The major difference between kilns and retorts is how the pyrolysis heat is generated which is shown in **Figure 7**. As mentioned in the previous section, heat is generated internally in the kilns. In retorts, heat is generated from an external heat source then transferred to the biomass by different means which are:

- Direct heating (the hot gases pass through the reactor and come in contact with the biomass)
- Indirect heating (the heat from the hot gases are transferred to the biomass through the retort's wall)

In both direct and indirect heating, the heat can be generated through different routes which are:

- Burning some raw biomass to generate the main heat which is aided with and after burner to burn the volatiles released from the carbonization process as shown in **Figure 7(b)**.

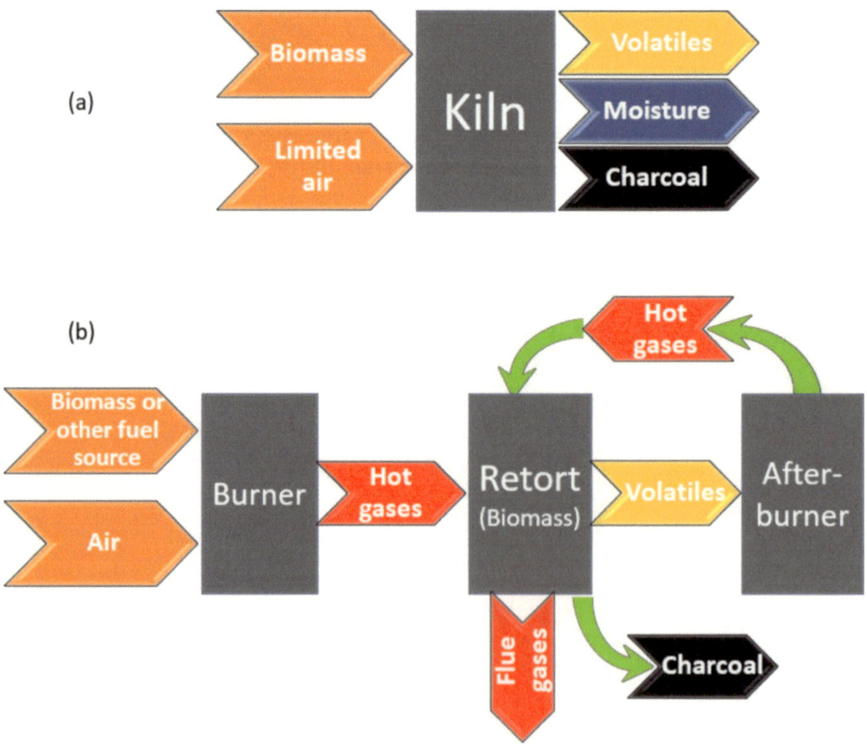

Figure 7.
Comparison between the process sequence of (a) kilns and (b) retorts.

- Burning some external fuel like gas for the main heating aided with burning the volatiles.

- Burning biomass or another external fuel only for the main burner without utilizing the volatiles. But this is the least efficient way and will not be cost-effective.

The advantage of retorts over kilns is that there is no partial combustion for the biomass inside the reactor which increases the output char yield. However, in the case of direct retort heating, partial combustion could still occur. This could occur if the combustion in the burner or the after burner is lean. This will result in excess air in the hot gases which will come into contact with the biomass leading to partial combustion. However, this partial combustion is much less than what happens in the kilns because in kilns it is how heat is generated, while in retorts it's a by-product of direct heating. Also, in the direct heating, sample size must be controlled because small sizes would increase the pressure drop of the hot gases across the retort.

6.2 Continuous reactors

Industrial applications require continuous supply of the carbonization product required so they cannot always have the privilege of working with batch reactors. So continuous reactors had to be developed. Almost all continuous reactors incorporate a sort of mixing mechanism. This greatly increases the product homogeneity and char yield and reduces the carbonization residence time. These are all advantages from the industrial need's point of view. However, continuous systems are more complicated than batch systems as they incorporate many moving heated parts, conveyors, and motors. The speed of these reactors must be optimized according to the required outcome to control the residence time and ensure that the biomass does not exit prematurely. There exist numerous designs for the continuous systems, but we will only consider two of the most used ones: the rotary drum and rotary screw carbonizers. In order for the process to be continuous, the combustion source should be continuous as well. That's why all these reactors utilize the volatiles during the carbonization process to heat the reactor. At the beginning of the process or if the volatiles cannot produce the required heat, there will be an extra biomass screw feeder directly to the burner [34]. All these continuous systems will require an external source for electricity to drive the electrical motors controlling the biomass feeder, rotating drum, afterburner's air blower, and the cooler.

6.2.1 Rotary drum

Rotary drums are basically cylindrical retorts mounted on geared electrical motor as shown in **Figure 8**. It is obvious from the figure that the drum is inclined with the direction of the flow. This is to ensure the flow continuity by gravity as well as good mixing. The degree of inclination and the drum's rotating speed control the flow speed as well as the residence time. The flow control here is not exact as it will differ according to the biomass' shape and physical structure.

6.2.2 Rotary screw

Figure 9 shows the basic construction of the screw carbonizer. It is similar to the rotary drum type except that it does not require being inclined like the drum type,

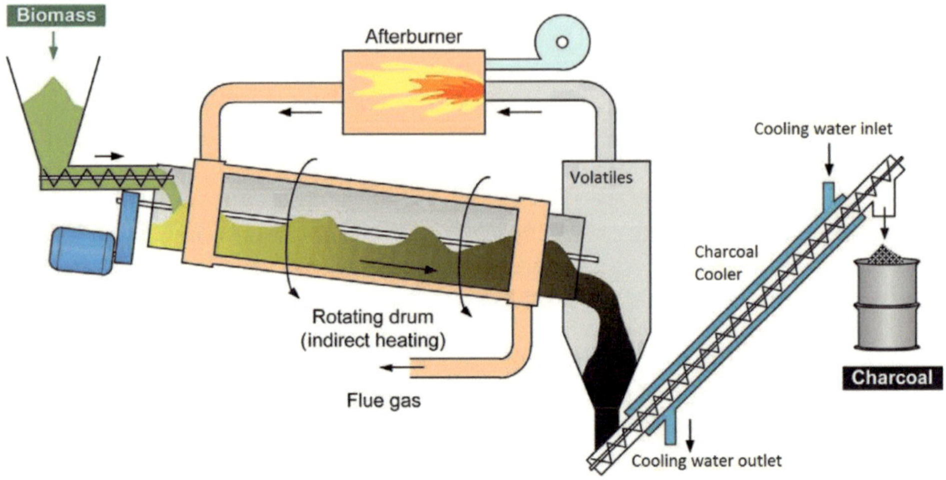

Figure 8.
Rotary indirectly heated drum adapted from Ronsse [13].

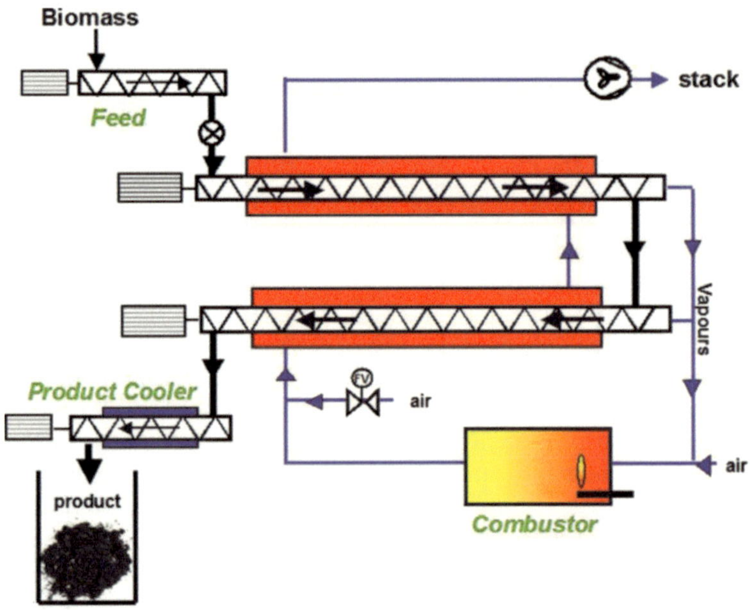

Figure 9.
Rotary indirectly heated screw carbonizer.

and the biomass is transported through the carbonization zone via the rotating helical screw. The screw movement controls the exact flow speed and the residence time. As the volume between two screw teeth is constant, an exact correlation between volumetric charcoal flow rate and the rotating speed can be obtained. So, simply by manipulating the screw's rotational speed, the flow can be adjusted. This reactor can accommodate variable sized biomasses because of the high mechanical power of the driving screw. This system is prone to clogging problems which might interrupt the operation if the biomass is too lumpy. The use of mechanical screw force to move the biomass is advantageous as this mechanical force aids in breaking large biomass lumps into fine powder.

DOI: http://dx.doi.org/10.5772/intechopen.90480

7. Conclusion

Biomass is a renewable, sustainable, and carbon neutral source of solid fuel, soil fertilizer, water and air purifier, and metal ore extractor. Carbonization, which is a slow pyrolysis process, can produce these varied products. Carbonization process is a thermochemical process that has been performed for hundreds of years to upgrade the raw biomass into useful solid product. The process conditions depend on the required product. Both charcoal and biochar are produced by the same process with heating up to 500°C; however they differ in their usage which is either as fuel or soil amendment and fertilizer, respectively. Biocoke production process is the same as for charcoal and biochar except that the heating continues up to and might be exceeding 1000°C. Activated carbon is made from charcoal which is post treated with superheated steam to remove the tarry residue and improving its surface area and pore volume making it suitable for adsorption applications. Several characterization techniques should be performed on both the raw biomass and its carbonized products to judge the carbonization system's effectiveness.

Kilns are mostly used when the manual labor is experienced and just needs basic knowledge of performing this carbonization process. Retorts, although they are still simple, are more complicated than kilns and harder to manufacture. Both kilns and retorts are batch systems. Continuous reactors are gaining raised interest from the industry due to their time effectiveness, automated process, homogeneous product, and continuous product supply. Two of the most common continuous systems are the rotary drum and rotary screw systems. They are almost the same except in the biomass transportation mechanism. Rotary screw is more controllable than rotary drum. As the process parameters depend on the raw biomass characteristics, the reactors must be adjusted and optimized with different operating conditions for each biomass used. These continuous systems enabled the overall system size to be reduced opening new pathways to making them portable and easily transported according to the biomass availability.

Conflict of interest

The authors declare no conflict of interest.

Abbreviation

TGA	Thermogravimetric analysis
MC	Moisture content
VM	Volatile matter
FC	Fixed carbon
HV	Heating value
HHV	Higher heating value
LHV	Lower heating value
SY	Solid yield
EY	Energy yield
EF	Enhancement factor
DAF	Dry ash-free basis
db	Dry basis
wb	Wet basis
BET	Brunauer–Emmett–Teller
HGI	Hardgrove Grindability Index

Author details

Mahmoud Amer[1,2] and Ahmed Elwardany[1,2]*

1 Fuels and Combustion Engines Lab, Energy Resources Engineering Department, School of Energy Resources, Environment, Chemical and Petrochemical Engineering, Egypt-Japan University of Science and Technology (E-JUST), Egypt

2 Mechanical Engineering Department, Faculty of Engineering, Alexandria University, Alexandria, Egypt

*Address all correspondence to: ahmed.elwardany@ejust.edu.eg

References

[1] Twidell J, Weir T. Renewable Energy Resources. Routledge: Taylor Francis; 2015

[2] McKendry P. Energy production from biomass (part 1): Overview of biomass. Bioresource Technology. 2002; **83**:37-46

[3] Saidur R, Abdelaziz E, Demirbas A, Hossain M, Mekhilef S. A review on biomass as a fuel for boilers. Renewable and Sustainable Energy Reviews. 2011; **15**:2262-2289

[4] Amer M, Nour M, Ahmed M, Ookawara S, Nada S, Elwardany A. The effect of microwave drying pretreatment on dry torrefaction of agricultural biomasses. Bioresource Technology. 2019;**286**:121400

[5] McKendry P. Energy production from biomass (part 2): Conversion technologies. Bioresource Technology. 2002;**83**:47-54

[6] Zhang L, Xu CC, Champagne P. Overview of recent advances in thermo-chemical conversion of biomass. Energy Conversion and Management. 2010;**51**: 969-982

[7] Patel M, Zhang X, Kumar A. Techno-economic and life cycle assessment on lignocellulosic biomass thermochemical conversion technologies: A review. Renewable and Sustainable Energy Reviews. 2016;**53**:1486-1499

[8] Jameel H, Keshwani DR. Thermochemical conversion of biomass to power and fuels. In: Biomass to Renewable Energy Processes. Taylor Francis; 2017. pp. 375-422

[9] Basu P. Biomass Gasification, Pyrolysis and Torrefaction: Practical Design and Theory. Academic Press, Elsevier; 2018

[10] Purwanto H, Zakiyuddin AM, Rozhan AN, Mohamad AS, Salleh HM. Effect of charcoal derived from oil palm empty fruit bunch on the sinter characteristics of low grade iron ore. Journal of Cleaner Production. 2018; **200**:954-959

[11] Benedetti V, Patuzzi F, Baratieri M. Characterization of char from biomass gasification and its similarities with activated carbon in adsorption applications. Applied Energy. 2018;**227**: 92-99

[12] López-Cano I, Cayuela M, Sánchez-García M, Sánchez-Monedero M. Suitability of different agricultural and urban organic wastes as feedstocks for the production of biochar—Part 2: Agronomical evaluation as soil amendment. Sustainability. 2018;**10**:2077

[13] Ronsse F, Nachenius RW, Prins W. Carbonization of biomass. In: Recent Advances in Thermo-Chemical Conversion of Biomass. Elsevier; 2015. pp. 293-324

[14] Chen W-H, Lu K-M, Tsai C-M. An experimental analysis on property and structure variations of agricultural wastes undergoing torrefaction. Applied Energy. 2012;**100**:318-325

[15] Zhao C, Jiang E, Chen A. Volatile production from pyrolysis of cellulose, hemicellulose and lignin. Journal of the Energy Institute. 2017;**90**:902-913

[16] De S, Agarwal AK, Moholkar V, Thallada B. Coal and Biomass Gasification. Springer; 2018

[17] Batanouny K, El-Fiky A. The water hyacinth (Eichhornia crassipes Solms) in the Nile system, Egypt. Aquatic Botany. 1975;**1**:243-252

[18] Vargas-Moreno J, Callejón-Ferre A, Pérez-Alonso J, Velázquez-Martí B.

A review of the mathematical models for predicting the heating value of biomass materials. Renewable and Sustainable Energy Reviews. 2012;**16**: 3065-3083

[19] Ozyuguran A, Akturk A, Yaman S. Optimal use of condensed parameters of ultimate analysis to predict the calorific value of biomass. Fuel. 2018;**214**: 640-646

[20] Chen W-H, Peng J, Bi XT. A state-of-the-art review of biomass torrefaction, densification and applications. Renewable and Sustainable Energy Reviews. 2015;**44**:847-866

[21] Kai X, Meng Y, Yang T, Li B, Xing W. Effect of torrefaction on rice straw physicochemical characteristics and particulate matter emission behavior during combustion. Bioresource Technology. 2019;**278**:1-8

[22] Kanwal S, Chaudhry N, Munir S, Sana H. Effect of torrefaction conditions on the physicochemical characterization of agricultural waste (sugarcane bagasse). Waste Management. 2019;**88**: 280-290

[23] Esteban LS, Carrasco JE. Evaluation of different strategies for pulverization of forest biomasses. Powder Technology. 2006;**166**:139-151

[24] Pejic BM, Kostic MM, Skundric PD, Praskalo JZ. The effects of hemicelluloses and lignin removal on water uptake behavior of hemp fibers. Bioresource Technology. 2008;**99**: 7152-7159

[25] Li H, Liu X, Legros R, Bi XT, Lim C, Sokhansanj S. Torrefaction of sawdust in a fluidized bed reactor. Bioresource Technology. 2012;**103**:453-458

[26] Antal MJ, Mochidzuki K, Paredes LS. Flash carbonization of biomass. Industrial & Engineering Chemistry Research. 2003;**42**: 3690-3699

[27] Major J, Steiner C, Downie A, Lehmann J. Biochar effects on nutrient leaching. In: Biochar for environmental management. Routledge: Taylor Francis; 2012. pp. 303-320

[28] Zimmerman AR. Abiotic and microbial oxidation of laboratory-produced black carbon (biochar). Environmental Science & Technology. 2010;**44**:1295-1301

[29] International Biochar Initiative (IBI). Available from: https://biochar-international.org/characterizationstandard/ [Accessed: 23 September 2019]

[30] Crombie K, Mašek O, Sohi SP, Brownsort P, Cross A. The effect of pyrolysis conditions on biochar stability as determined by three methods. GCB Bioenergy. 2013;**5**:122-131

[31] Garcia-Nunez J, Pelaez-Samaniego M, Garcia-Perez M, Fonts I, Abrego J, Westerhof R, et al. Historical developments of pyrolysis reactors: A review. Energy & Fuels. 2017;**31**: 5751-5775

[32] Boateng AA, Garcia-Perez M, Mašek OE, Brown R, del Campo B. Biochar production technology. In: Biochar for Environmental Management. Routledge: Taylor Francis; 2015. pp. 95-120

[33] Lehmann J, Czimczik C, Laird D, Sohi S. Stability of biochar in soil. Biochar for Environmental Management: Science and Technology. Earthscan; 2009:183-206

[34] Duku MH, Gu S, Hagan EB. Biochar production potential in Ghana—A review. Renewable and Sustainable Energy Reviews. 2011;**15**:3539-3551

Chapter 12

Aviation Fuels and Biofuels

Jonathan Baumi, Caroline Milani Bertosse and Carmen Luisa Barbosa Guedes

Abstract

Aviation industry consumes about 177 billion liters of kerosene, moving more than 25,000 aircraft and 6 billion passengers. To achieve that, civil aviation in 2015 generated about 781 million tons of CO_2 corresponding to 2% anthropogenic emissions of this greenhouse gas, and all required energy is derived from fossil sources. To reduce the environmental impact and to create alternative energy sources to bring energy security, it is of great importance to increase researching and development, so that it becomes viable to produce biokerosene. This chapter aims to present some varieties of biomass and its derivatives being studied as raw materials for new aviation fuels such as ethanol, butanol, fatty acid methyl esters, and fusel oil.

Keywords: biomass, jet fuel, avgas

1. Introduction

Since the beginning of the first civilization, human being has been seeking for progress and better quality of life, developing new technologies and new materials with the aim of making the tasks simpler. The first technology appeared thousands of years ago, using utensils made from chipped stones and evolved in quality and quantity mainly after the industrial revolution in the nineteenth century, to the present day with a great range of consumer goods such as electronics, textiles, automobiles, food, cosmetics, and furniture, among others. To keep and make possible these great advances of mankind, even greater quantities of energy, a essencial necessity, after the industrial revolution had grown rapidly requiring the use of new sources of energy [1, 2].

With the invention of internal combustion engines in the nineteenth century, and the commercialization and popularization of automobiles in the twentieth century, there arose a great and growing need for fuels, which for many years was supplied only with petroleum products, a natural nonrenewable resource. The impacts caused by the oil exploration and the combustion of its derivatives have become serious environmental problems causing a great increase in the emission of greenhouse gases and pollutants, being the main CO_2, SOx, and NOx [3, 4]. Currently, the demand for renewable energy sources is increasing, aiming at reducing anthropogenic CO_2 emissions and reducing society's dependence on oil, and increasing interest in biofuels from biomass.

In the 1970s, Brazil became a pioneer in public policies for the use of biofuels in the energy matrix with the implementation of laws that made mandatory the addition of 10% ethanol in gasoline in 1973 and the launch of the Proálcool program in 1975, which aimed at partial replacement of gasoline by ethanol in the vehicle fleet, a measure taken in response to the first major oil crisis in the world. The first national biodiesel production program, Pro-Oil, was jointly launched in Brazil, which due to few stimuli for the production of vegetable oils and the fall of oil prices in the 1980s, was unsuccessful, ending in 1986 [5, 6].

The world aviation industry basically uses two types of fuel with distinct characteristics, known as aviation gasoline (AVGAS) and aviation kerosene (QAV) or jet fuel, which consumes around 177 billion liters of fuel worldwide, moving more than 25,000 aircraft. Almost all of this volume comes from petroleum derivatives, a nonrenewable source that handled more than 6 billion passengers [7]. In 2015, civil aviation generated about 781 million tonnes of CO_2 corresponding to 2% of all anthropogenic greenhouse gas emissions according to the Air Transport Action Group (ATAG). Already in 2009, the European Aviation Commission (EAC) determined that emissions from civil aviation should be reduced by 20% by the year 2020, with the use of renewable sources being a highly promising alternative for this goal to be met [8]. Due to the fact that the main oil producers are located in politically unstable countries politically unstable and prices of oil and its derivatives are highly volatile, the need for alternative sources of energy goes beyond the environmental issue.

1.1 Aviation gasoline (AVGAS)

Aviation gasoline known worldwide as AVGAS is a fossil fuel that distills in the range between 30 and 170°C containing gasoline-derived and diesel compounds. This fuel consists mainly of isoparaffins with five to nine carbons, and also, aromatics in small amounts. The main component is alkylate, which is basically a mixture of isooctane (2,2,4-trimethyl pentane) with olefins. Isooctane is a branched chain alkane, octane isomer, which is standard 100 on the octane scale [9].

Aviation gasoline is used exclusively in small airplanes that have spark ignition engines such as aircraft used in agricultural aviation, small commercial aviation, private aviation, pilot training, and in experimental and sporting aircraft. The compound with molecular formula $Pb(C_2H_5)_4$ called tetraethyl lead or TEL is an additive for aviation gasoline that increases the octane rating but it is toxic and releases lead particles into the air [10]. In several parts of the world, this additive is prohibited from being added to the gasoline of land vehicles, however, it is still used in aviation gasoline, with the current technology, it is only possible to achieve an octane rating of more than 100 with the current economic viability through the addition of tetraethyl lead in aviation gasoline in accordance with the standards of the American Society for Testing and Materials—ASTM D3341 or D5059. The most used low lead aviation gasoline is known internationally as AVGAS 100 LL that presents differentiated properties, performance requirements from other gasolines. Throughout the process of handling, transporting, and storing of aviation gasoline, special equipment to the product is used, and the system is periodically inspected to ensure that it is thoroughly clean and free of any possibility of contamination.

The specification of aviation gasoline is practically the same throughout the world, including high calorific value (43.5 MJ kg^{-1}) and low freezing point (−58°C). Constant product monitoring includes daily draining of the storage tank and supply units, filtering of the product prior to filling and periodic inspection of the respective filters. The classification of aviation gasoline is given according to its resistance to detonation in Octane Motor (MON) units according to the ASTM D2700 test. Due to some varieties of aviation gasoline marketed worldwide, the identification of tubing and tanks for refueling was standardized, thus avoiding inadequate fuel supply (**Table 1**).

Type	Color	Lead content (g L^{-1})	Feature
80	Red	0.14	No longer marketed
82 UL	Purple	0.0	Low performance engines
100	Green	1.12	Has been replaced by AVGAS 100 LL
100 LL	Blue	0.56	Worldwide standard

Table 1.
Apparent color and anti-knock power of aviation gasoline [11].

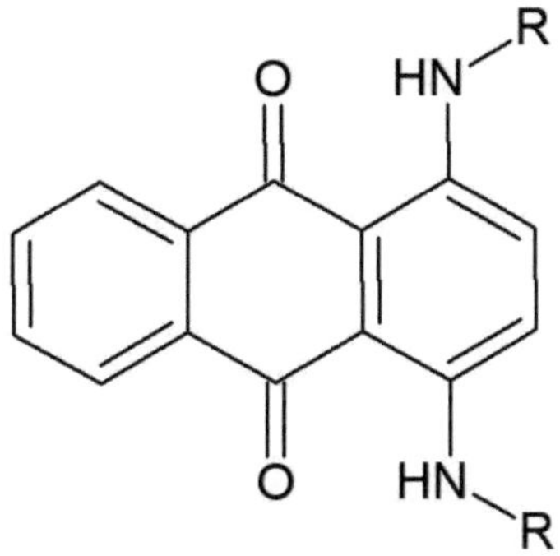

Figure 1.
Blue dye for AVGAS.

AVGAS 100 LL, corresponding to octane 100 and low level of tetraethyl lead is identified with the blue color due to the addition of dye based on 1,4-dialkylamino-anthraquinone (**Figure 1**).

The combustion of a tetraethyl lead fuel generates emissions of both lead and lead oxides, which can accumulate and cause serious damage to the engine. To avoid this accumulation, sequestrants or chelants are used that react with lead compounds and form more volatile species. The most used additives are 1,2-dibromoethane and 1,2-dichloroethane [12]. Despite the significant decrease, it is estimated that the effects of lead to human health cause losses in the US economy of approximately US $ 1.06 billion annually due to public health impacts [13].

The challenges of developing an alternative to lead as an additive in aviation gasoline involve maximum compatibility with existing engines and high reliability. When comparing an airplane engine and a car engine, we have to consider that the operating regime is quite different, in aviation, 80–90% of the time the engine is in high energy demand, and in cars, only 10% on average. Aviation gasoline consists of different types of hydrocarbons that have a higher flash point than that of vehicular gasoline, which starts to release flammable vapor at −42.8°C to optimize its ignition characteristics necessary for the engines [14].

The process of homologation of new additives and fuels is time consuming and costly, especially when compared to vehicular fuels. Despite this, there is not yet an additive that can leave aviation fuels with high octane, since even the use of oxygenated additives have their disadvantages, especially in aviation fuels: increased fuel consumption and increased production of nitrogen oxides.

1.2 Aviation kerosene

Aviation kerosene, also known as jet fuel or by the acronym QAV, is a petroleum source fuel that is in the distillation range between 150 and 300°C and is used in a medium and large turbine type engine. It is produced by fractionation through distillation at atmospheric pressure, followed by treatments, and is suitable for the generation of energy by combustion in aircraft gas turbine engines. It must remain liquid and homogeneous until the zone of combustion of the aircraft, and have calorific power as high as possible. The specification of aviation kerosene, Jet A-1, is performed in each country to be compatible with the Aviation Fuel Quality Requirements for Operated Systems (AFQRJOS). In order to guarantee the quality of the product until delivery to the final consumer, the Quality Assurance System (QMS) covers the whole distribution chain from the refinery through maritime and land terminals, bases, airport storing, until the supply of the aircraft.

When 1 kg of kerosene is burned, an average of 14 g of NOx; 0.8 g of SOx; 3.15 kg of CO_2; 4 g of CO; and 0.6 g of hydrocarbons are generated, compounds that have a high environmental impact mainly in relation to greenhouse effect and acid rain [15]. In world, commercial aviation is used Jet A1, regulated in Brazil by ANP resolution n ° 37; in the US by the Federal Aviation Administration (FAA) through ASTM D1655 [16, 17]. Jet A kerosene is only used in the USA and differs from Jet A1 at the point of freezing, which is −40°C, higher than −47°C (**Table 2**).

The main fuels used in military aviation are JP-4 and JP-8, with JP-4 kerosene falling out of use since the 1990s due to safety concerns. This fuel is regulated worldwide by the U.S. Military Specification MIL-PRF-5624S and DEF STAN 91-88. JP-8 kerosene is similar to Jet A1 kerosene with the difference of having anti-corrosion additives, dispersants, antifreeze agents, and antioxidants defined by MIL-DTL-83133, DEF STAN 91-87. There is also JP-8 + 100 kerosene containing additives that increase heat resistance at 37°C (100°F) to 218°C over regular JP-8 [18, 19].

In military aviation, there is also kerosene JP5 (European F44), JP7 (US only), JP8 (F34 European), which are chemically similar to the Jet A1 differentiating with respect to antifreeze and antioxidant additives [20] are used. The basic composition of Jet A and Jet A1 kerosene are described in **Table 3**.

In addition to Jet A and Jet A1 kerosene, Jet B type kerosene is the most volatile, thus handling is more dangerous. It has a freezing point below −47°C and is used only in extremely cold regions, such as Canada and Russia. It is mainly composed by hydrocarbons of 5–15 carbon atoms, being actually a mixture of gasoline with kerosene.

In Russia, there is TS-1 kerosene, standardized by the GOST 10227, which differs from Jet A-1 by the freezing point below −57°C and a flash point of 28°C, being lower than 40°C of Jet A1. In addition to Russia, there is also China, which has five

Regulatory agency	Country	Standard/resolution	Jet fuel name
Agência Nacional do Petróleo, Gás Natural e Biocombustíveis (ANP)	Brazil	Resolution n°37	Jet A1
Federal Aviation Administration (FA)	USA	ASTM D1655/ASTM 6615	Jet A e Jet A1/ Jet B
Transport Canada Civil Aviation (TCCA)	Canada	CAN/CGSB-3.23 CAN/ CGSB-3.22	Jet A/A1/Jet B
Civil Aviation Authority (CAA)	UK	DefStan 91–91	Jet A1
European Aviation Safety Agency (EASA)	EU	AFQRJOS	Jet A1
Federal Air Transport Agency (FATA)	Russia	GOST 10227/GOST R 52050	TS-1/Jet A1
Civil Aviation Administration of China (CAAC)	China	GB 6537	No 3

Table 2.
Main regulatory agencies and technical standards for assessing the quality of civil aviation kerosene in the world.

Compound	Formula	Type
n-Octane	C_8H_{18}	n-paraffin
2-Methylheptane	C_8H_{18}	Isoparaffin
1-Methyl-1-ethylcyclopentane	C_8H_{14}	Cycloparaffin
Ethyl-cyclohexane	C_8H_{16}	Cycloparaffin
o-Xylene	C_8H_{10}	Aromatic
p-Xylene	C_8H_{10}	Aromatic
Cis-Decalin	$C_{10}H_{18}$	Cycloparaffin
Tetralin	$C_{10}H_{12}$	Aromatic
Naftalene	$C_{10}H_8$	Aromatic
n-Dodecane	$C_{12}H_{26}$	n-paraffin
2-Methylundecane	$C_{12}H_{26}$	Isoparaffin
1-Ethylnaftalene	$C_{12}H_{12}$	Aromatic
n-Hexylbenzene	$C_{12}H_{18}$	Aromatic
n-Hexadecane	$C_{16}H_{34}$	n-paraffin
2-Methylpentadecane	$C_{16}H_{34}$	Isoparaffin
n-Decylbenzene	$C_{16}H_{26}$	Aromatic

Table 3.
Jet A and Jet A1 chemical composition [21].

aviation kerosene standards: No. 1 and No. 2, with a flash point near 28°C and freezing point below −60 and −50°C, respectively; No. 3 which is similar to Jet A1; No. 4 which is similar to Jet B; and No. 5 is a kerosene similar to No. 3 but with a high flash point. Currently, practically all kerosene sold in China is No. 3 [22].

The National Agency for Petroleum, Natural Gas and Biofuels—ANP in Brazil, through Resolution 37 of December 2009, specified the technical standard of aviation kerosene, as well as the technical standards for quality control. The quality standards of this resolution are based on the international standards specified in ASTM 1655 and DefStan 91-91. Due to the fact that each country determines its aviation kerosene quality standards, a list of quality requirements for the worldwide commercialization of this fuel, known as Aviation Fuel Quality Requirements for Jointly Operated System (AFQRJOS), has been created. AFQRJOS is based on standards D1655 and DefStan 91-91 and serves as the standard of quality for the main suppliers in the world market as Agip; BP; Chevron Texaco; Exon Mobil; Kuwait Petroleum; Shell; Statoil and Total. For the commercialization of aviation kerosene, rigid physicochemical standards and characteristics are applied, having to deal with approximately 30 tests, a number higher than bioethanol (15 tests) and biodiesel (18 tests).

Aviation kerosene must have a high calorific value (42.8 MJ/kg) coupled with a low specific gravity (0.775–0.820 g mL^{-1}) due to the issue of total aircraft weight, energy efficiency, and flight autonomy [23]. Another important property is the freezing point, due to the working conditions, the fuel should not solidify or form crystals at temperatures below −40°C for Jet A and −47°C for Jet A1 [24]. The moisture content should be low enough to avoid the growth of microorganisms and to reduce corrosivity [25]. The desirable properties of kerosene make the development of biokerosene quite complicated and it is a major obstacle in the search for new alternatives and processes. **Table 4** lists the most commonly used quality technical standards worldwide and the standards in Brazil.

	Unit	ASTM 1655-4a	Def Stan 91-91	ANP n° 37
Density a 15°C	g mL^{-1}	0.775–0.840	0.775–0.840	0.771–0.836* (20°C)
Viscosity a −20°C	mm^2 s^{-1}	8.0 (max)	8.0 (max)	8.0 (max)
Acid value	mg KOH g^{-1}	0.100	0.0012	0.015
Flash point	°C	38 (min)	38 (min)	38 or 40
Heat of combustion	MJ kg^{-1}	42.8 (min)	42.8 (min)	42.8 (min)
Freezing point	°C	−47	−47	−47
Sulfur	%	0.3	0.3	0.3
Aromatics	%	25	25	25
Smoke point	Mm	25 (min)	25 (min)	25
JFTOT Delta P (260°C)	mmHg	25	25	25
Conductivity	pS m^{-1}	50–450	50–600	50–600
Maximum boiling point	°C	300 (max)	300 (max)	300 (max)

Table 4.
Jet A-1 specifications.

Observing ASTM D1655, Def Stan 91-91 and ANP n° 37, it can be concluded that there is a convergence in, practically, all the parameters, differing only in electrical conductivity and acidity. As aviation gasoline, aviation kerosene-containing tubing and refueling tanks are identified by a black color-coded adhesive with an auxiliary color to indicate the type of kerosene (Jet A, Jet A1, or Jet B).

In order to meet the demanding quality parameters of aviation kerosene, compounds known as additives are used, with the objective of improving specific physicochemical properties:

Static charge dissipator: kerosene produced in refineries has low electrical conductivity and this can cause static electricity to accumulate by moving fluid in tanks and filters. To avoid this accumulation a compound known as Stadis 450, which is allowed by ASTM 1655, ANP n ° 37 and DefStan 91-91 in contents up to 5.0 mg/L is added and its use is optional.

Metal deactivator: the presence of metals in the fuel can catalyze oxidation reactions and directly impact thermal stability. The most commonly found metals are Copper and Zinc. The most commonly used additive is N,N-disalicylidene-1,2-propanediamine, which has a chelating action and is used in contents up to 5.7 mg/L [26].

Antioxidants: the formation and presence of peroxides may cause deterioration in the quality of the fuel and may form gums and particulate matter. The use of antioxidant additive is mandatory by ANP n ° 37 and DefStan 91-91 in contents of 17–24 mg L^{-1} and optional by ASTM D1655. The most commonly used compound for this purpose is butylated hydroxytoluene (BHT).

Lubricity enhancer: it has the function of reducing the corrosive effect of the fuel as well as assisting in the anti-wear effect. Its use is not mandatory in kerosene for civil aviation although it is permitted in the Brazilian, US, European Union, and UK legislations and is more commonly used in military fuels (**Figures 2** and **3**).

Icing inhibitors: it is used because of the presence of water in the kerosene, dissolved water can form crystals that can cause clogging of the fuel filter. It is considered optional and is rarely used for civil aviation and most used in military applications and the most commonly used additive is diethylene glycol methyl ether (DiEGME) (**Figure 4**) that may be used as a biocide [27].

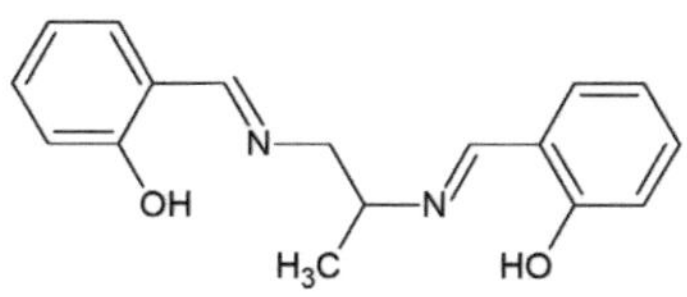

Figure 2.
N, N-*disalicylidene-1,2-propanediamine.*

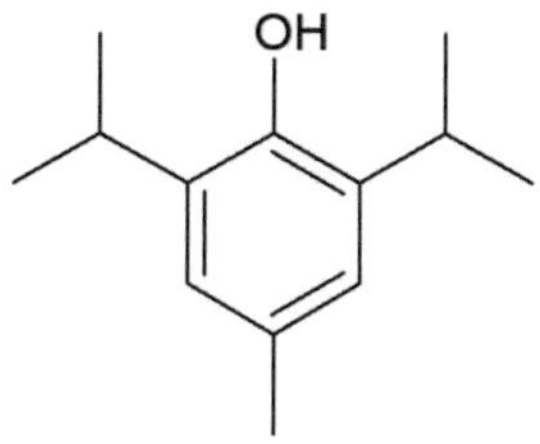

Figure 3.
BHT chemical structure.

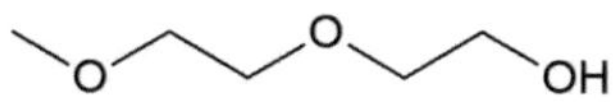

Figure 4.
Diethylene gylcol methyl ether.

Additive type (chemical or brand name)	AFQRJOS Jet A-1	ASTM (Jet A)	DEF STAN 91-91 (Jet A-1)	DEF STAN 91-86 (F-44)	DEF STAN 91-88 (F-40)	IATA (Jet A-1)	CAN-GSB 3.24 (F-34/ F-44)	GE D 50TF2	P&W SB, No. 2016
Anti-oxidant	R	O	R	R	R	R	O/R	O	R
2,6-Ditertiary-butyl phenol	•	•	•	•	•	•	•	•	•
2,6-Ditertiary-butyl-2-methyl phenol	•	•	•	•	•	•	•	•	•
2,4-Dimethyl-6-tertiary-butyl phenol	•	•	•	•	•	•	•	•	•
Mix 75%(min) 2,6-ditertiary-butyl phenol	•	•	•	•	•	•	•	•	•
Mix 25%(max) tertiary and tritertiary butyl phenols	•	•	•	•	•	•	•	•	•
Mix 72%(min) 2,4-dimethyl-6-tertiary-butyl phenol	•	•	•	•	•	•	•	•	•
Mix 28%(max) methyl and dimethyl tertiary-butyl phenols	•	•	•	•	•	•	•	•	•

Additive type (chemical or brand name)	AFQRJOS Jet A-1	ASTM (Jet A)	DEF STAN 91-91 (Jet A-1)	DEF STAN 91-86 (F-44)	DEF STAN 91-88 (F-40)	IATA (Jet A-1)	CAN-GSB 3.24 (F-34/ F-44)	GE D 50TF2	P&W SB, No. 2016
Mix 55%(min) 2,4-dimethyl-6-tertiary-butyl phenol	•	•	•	•	•	•	•	•	•
15%(min) 2,6-ditertiary-butyl-4-methyl phenol	•	•	•	•	•	•	•	•	•
30%(max) methyl and dimethyl tertiary-butyl phenols	•	•	•	•	•	•	•	•	•
Static dissipator additive	R	O	R		R	R	R	O	R
Stadis 450		•	•		•	•	•	•	•
Sigbol									•
Anti-icing additive	A	A	O	R	R	A	O/A	O	O
Ethylene glycol monomethyl ether							•		•
Ethylene glycol monomethyl ether and Methyl alcohol									•
Diethylene glycol monomethyl ether	•	•	•	•	•	•	•	•	•
Corrosion inhibitors		A	O	R	R	A	A	O	O
Apollo PRI-19		•	•	•	•		•	•	•
Hitec 580		•	•	•	•	•	•	•	•
Nalco 5403		•	•	•	•	•	•	•	•
DCI-4[a]		•	•	•	•	•	•	•	•
DCI-6[a]			•	•	•	•	•	•	•
Nalco 5405						•	•	•	•
Spec-Aid 8Q22						•	•		•
Unicor J						•	•		•
Tolad 351						•	•		•
Tolad 4410			•	•	•	•	•		•
RPS-613						•	•		•
Hitec 515									•
Tolad 245								•	
Mobilad F-800								•	
PWA-536									•
Metal deactivator	O	O	O	O	O	O	O	O	O

Additive type (chemical or brand name)	**AFQRJOS Jet A-1**	**ASTM (Jet A)**	**DEF STAN 91-91 (Jet A-1)**	**DEF STAN 91-86 (F-44)**	**DEF STAN 91-88 (F-40)**	**IATA (Jet A-1)**	**CAN-GSB 3.24 (F-34/ F-44)**	**GE D 50TF2**	**P&W SB, No. 2016**
N,N′-Disalicylidene-1,2-propanediamine	•	•	•	•	•	•	•	•	•
N,N′-Disalicylidene-1,2-cyclohexanediamine								•	
Thermal stability additive		A						O	O
Spec-Aid 8Q462								•	•
AeroShell Performance Additive 101								•	•
Turboline FS100C								•	•
Turboline FS100								•	•
JFA-5									•
Leak detection additive		O	O				O	O	O
Tracer A		•	•				•	•	•
Biocide		E				O		O	O
Biobor JF						•		•	•
Kathon FP 1.5						•		•	•

O—Optional: the additive may be added by the fuel manufacturer to the extent permitted by specification without consulting customers. The supplier may be required to declare its presence; A—Agreement: purchasing authorities may require that an additive be used to the extent permitted by specification. If the fuel supplier desires to add it, he must secure agreement of the customer; E—Engine Manufacturer's Agreement: specification authorities may require agreement by engine manufacturers; R—Required: the additive must be introduced at the level specified to meet a specific handling requirement. The point of addition is not necessarily into refinery production (R for anti-oxidant treatment refers to hydrotreated fuel).

Table 5.
Additives for aviation kerosene.

Table 5 lists the main additives required for jet fuel, including the ones mentioned before.

Between 2003 and 2008, the price of aviation kerosene rose 462%, reaching almost US$ 4 per gallon [28] and the economic crisis in 2009 caused a sharp fall reaching close to US$ 1 per gallon. The quotation in October 2016 is US$ 1.47 per gallon [29], data that shows the unpredictability of this input. The Brazilian market follows the external market oscillations, with large variations in the price of this input, which are quite pronounced varying 47% between the years 2008 and 2009, 25% between 2014 and 2016, and comparing 2009–2014, the increase reached 75%. **Table 6** shows the average consumer prices in 11 largest Brazilian capitals between 2008 and 2016.

In the civil aviation industry, the main impact factor in operating costs is the fuel, which corresponds to approximately 40% of the total costs followed by aircraft 20% and operating expenses with 17% (**Table 7**). Therefore, it is extremely important to increase the supply of aviation kerosene, which can be contemplated with new processes and new sources of raw material, and consequently the reduction of its selling price.

Average prices 2008–2016 (without taxes)									
Year	**2008**	**2009**	**2010**	**2011**	**2012**	**2013**	**2014**	**2015**	**2016**
Price (US$/L)	0.44	0.29	0.33	0.41	0.48	0.52	0.54	0.46	0.41

Table 6.
Jet A1 kerosene prices in Brazil between 2008 and 2016 [30].

Category	2009	2010	2011	2012	2013	2014
Crew	12%	12%	13%	11%	10%	10%
Fuel	30%	33%	36%	39%	37%	37%
Depreciation of flight equipment	4%	4%	4%	4%	4%	3%
Aircraft rental, maintenance, and insurance	20%	15%	13%	14%	17%	20%
Airport fees	2%	2%	2%	2%	3%	3%
Air navigation rates	4%	3%	3%	3%	3%	3%
Indirect costs	9%	9%	9%	9%	9%	9%
Operational expenses of public air services	21%	21%	19%	17%	18%	17%

Table 7.
Fight expenses (2009–2014) [31].

2. Biofuels for aviation

The intention of using biofuels in the aviation industry is relatively recent, as research projects have grown significantly after 2010 (**Figure 5**).

Currently, the use of biofuels in aviation suffers from the high production costs that are decreasing and still amount to two to four times the price of fuel of fossil origin, a difference that has already reached 30 times in previous years [33].

During Rio + 20, Azul Linhas Aéreas used sugarcane-derived kerosene supplied by Amyris in an Embraer E-195 for a demonstration flight. Other Brazilian airlines flying with biofuels: GOL Linhas Aéreas, which used biokerosene from corn oil and fry oil in a Boeing 737-800, and LATAM Linhas Aéreas in 2010 using a fuel derived from jatropha oil [34].

Sgouridis and co-workers estimate that 0.5% of the world's jet fuel used in 2009 came from renewable sources and projects, which in 2024, the share will be between 15.5 and 30.5%, depending on public policies and investments and there will be a transport capacity increased by 140% with 20% lower emissions compared to 2004 [35].

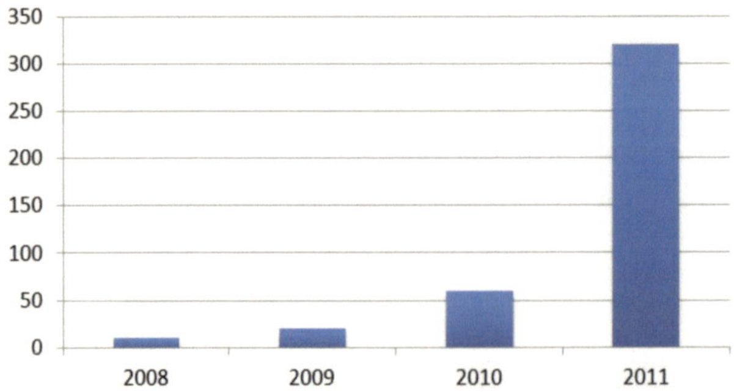

Figure 5.
Number of initiatives for biokeresene production [32].

Production process	Maximum blending
Synthetic paraffinic kerosene (SPK) Hydroprocessed esters and fatty acids (HEFA)	Up to 50%
Synthesized iso-paraffinic (SIP)	Up to 10%

Table 8.
Allowed biokerosene blends [38].

The production of aviation kerosene may be based on the liquefaction and gasification of biomass as well as biomolecules such as alcohols and sugars. The standards DEF STAN 91-91 and ASTM 7566 allow the insertion of paraffins from the Fischer-Tropsch process from renewable sources as semi-synthetic fuel in mixtures of up to 50% with petroleum kerosene.

The Fischer-Tropsch process is based on the conversion of biomass (coal, natural gas, lignocellulosic feedstock, sugars, and vegetable oils) into synthesis gas (syngas), consisting of H_2 and CO, which are then polymerized in paraffins, olefins, and compounds carbonylates [36]. Currently, the production of synthetic paraffinic kerosene known as Synthetic Parafinic Kerosene (SPK) is through the use of Fischer-Tropsch (FT) process or hydrotreatment of vegetable oils. We can cite the South African SASOL that produces kerosene from natural gas and coal as well as the North American Syntroleum [37].

In Brazil, ANP resolution n° 63 and in the USA, ASTM D7566 regulates the use of kerosene from biomass with mixtures ranging up to 10% or up to 50%, varying according to the raw material, and the production process (**Table 8**).

2.1 Why use biofuels?

- Reduction of greenhouse gases, mainly the reduced emission of CO_2. Biofuels have a favorable balance of CO_2, depending on the starting biomass and the route used.

- Reduction of atmospheric pollution, such as reduction of SOx, NOx, and CO. The burning of biofuels is generally more efficient and clean than that of petroleum products.

- Increased energy security. The variability of energy sources generates market stability and reduces the risk of scarcity.

- Development of the bioenergy industry, generating jobs, and new possibilities of intellectual capital.

- Decreased dependence on oil, which is a limited natural resource.

We currently have two main bioenergy production chains in Brazil and in the world: the ethanol and biodiesel industry (fatty acid esters). The use and modification has been intensively studied in the partial replacement of both aviation gasoline and aviation kerosene.

2.1.1 Ethanol

The sugar and ethanol industry in Brazil was introduced in the seventeenth century and became one of the main sources of bioenergy inserted in the national energy matrix in the twentieth century. In the 1970s, Brazil became a pioneer in public policies for the use of biofuels to implement laws that made it mandatory to add 10% ethanol in gasoline in 1973 and the launch of the Próalcool program in 1975 that aimed to partially replace gasoline by ethanol in the vehicle fleet, a measure taken in response to the first major global oil crisis.

According to the Brazilian Ministry of Agriculture, Livestock, and Supply [39], sugarcane production in 2016 was 631.8 million tons and national sugar production was 35.4 million tons, while the ethanol production was 28.526.000 m^3 in that year.

Of this total, 57.1% refers to hydrous ethanol fuel 16.296.000 thousand m^3 [40] and the production of anhydrous ethanol, which is mixed with gasoline A to form gasoline C, recorded an increase of 1.9% in relation to the previous year, totaling 12.230.000 thousand m^3.

Ethanol has been present in civil aviation for more than 10 years with the launch by EMBRAER of the Ipanema agricultural aircraft (EMB-202A) that was designed to be moved to 100% of this biofuel [41]. This aircraft was an adaptation of the aviation-powered gasoline engine (AVGAS 100LL). Two years after the homologation of Ipanema DCTA introduced a flex engine in the T25 model, a less radical and more technologically advanced concept [42].

The use of ethanol replacing aviation gasoline has similar effects to the replacement of vehicular gasoline, such as an increase in average engine power of 7% and a reduction in autonomy of 25–40% due to its lower calorific value (28 MJ kg^{-1} vs. 43.5 MJ kg^{-1}) with higher economic benefits related to the higher cost of aviation gasoline, on average US$ 0.80 per liter, and environmental benefits as a renewable fuel, to reduce CO_2 emissions by up to 63% and with total absence of lead and sulfur [43]. Another advantage that must be mentioned is the tolerance for the presence of water in the fuel, since in airplanes, it is of utmost importance to check the presence of humidity to avoid problems in flight being a standard procedure in this branch of civil aviation.

Due to its large proportions, the sugar and alcohol industry generates several and abundant residues such as vinasse, bagasse, straw, and fusel oil, the last three having an important role in the bioenergy industry and have a great potential for application in civil aviation, either by the use of pure ethanol or by new biofuels developed through this waste.

2.2 Fatty acid methyl esters

Fatty acid methyl esters (FAME) are mainly used as biodiesel, a renewable biofuel obtained through the transesterification reaction of triacylglycerols using methyl alcohol in the presence of a catalyst producing long chain fatty acids esters and glycerin as a co-product (**Figure 6**) [44, 45]. The catalysis used may be acidic, enzymatic, or alkaline, with the most commonly used catalysts being the homogeneous alkalines [46]. These include potassium hydroxide (KOH), sodium hydroxide (NaOH), and sodium methoxide ($CH_3O^-Na^+$).

The main raw materials for the production of fatty acid esters are vegetable oils and animal tallow, which are mainly made up of triacylglycerols. Among the plant sources are the oilseeds, such as the palm (*Elaeis guineensis*), sunflower (*Helianthus annuus*), *Jatropha curcas*, and mainly soybean (*Glycine max*), which corresponds to 68.6% of the Brazilian national production and between animal sources is the bovine tallow with a contribution of 17.3% [47].

Blends of methyl esters of jatropha fatty acids and frying oil have been studied, blends up to 20% v/v with Jet A1 kerosene may be suitable in accordance with ASTM D1655, limited by the fact that these esters have high viscosity (13.02 $mm^2\ s^{-1}$ at 50% blend) and also a very far cloud point (−18°C) from desired temperature for Jet A1 kerosene [21].

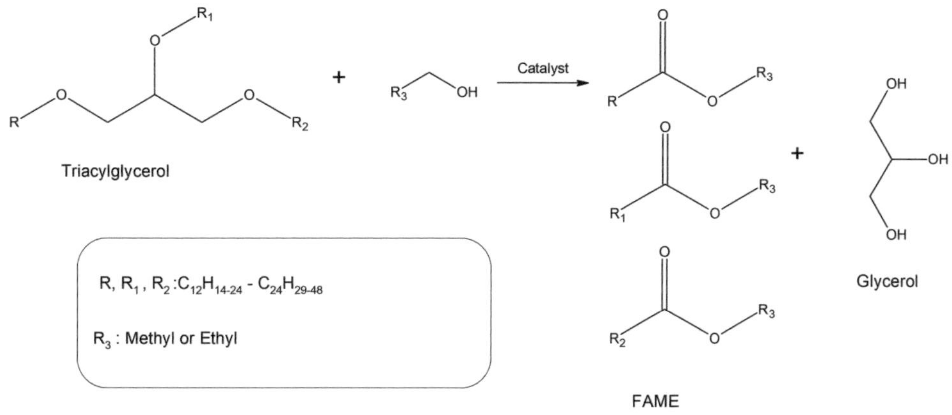

Figure 6.
Transesterification reaction.

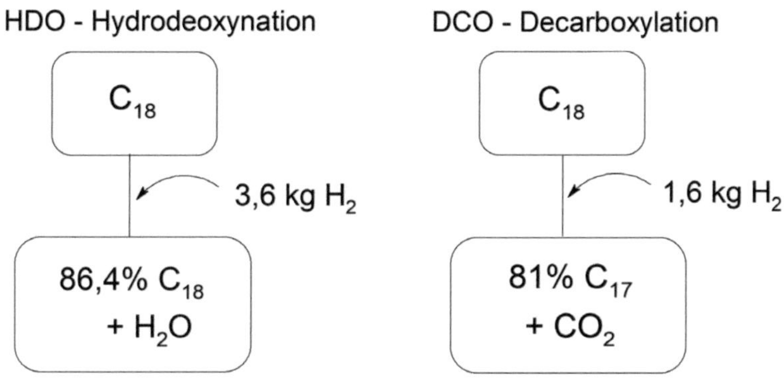

Figure 7.
HEFA for n-paraffin obtention [52].

In addition to conventional sources of fatty acids, we have microalgae, photosynthesizing organisms that have a great potential for lipid production. Studies with *Nannochloris sp., Botryococcus braunii, Chlorella sp.,* and *Scenedesmus sp.* showed great potential for production of lipids and hydrocarbons of high unsaturation, which is unsuitable for application as biodiesel but can serve as an intermediary input in the production of biokerosene [48, 49].

It is considered that the use of fatty acid esters as a partial substitute for Jet A1 kerosene is quite complicated due to very different physicochemical properties, mainly relative to low calorific power, high kinematic viscosity, high metal content, problems with reactions of hydrolysis, and very high freezing point in addition to the possibility of increased oxidative degradation due to the unsaturations of the vegetable oils.

Although their use as esters has not been shown to be consistent, these products open up a great potential as a feedstock for kerosene synthesis by Hydroprocessing of Esters and Fat Acids (HEFA) [50], a route already recognized by ASTM D7566 to be a viable route, mainly due to the great availability and established market [51].

The HEFA process consists basically in the reaction of vegetable oils or biodisel with hydrogen to remove the carbonyl groups by reducing or forming CO_2, causing loss of a chain carbon, and the formation of n-paraffins, known as Hydrogenated Vegetable Oil (HVO) (**Figure 7**).

The n-paraffins formed in the process have properties suitable for diesel, with high cetane number and high viscosity, which do not yet have properties desired for

Jet A1 kerosene. To improve fuel quality, n-paraffins are subjected to a hydroisomerization (HIS) process for the formation of isoparaffins, a class of compounds highly desired for this purpose.

3. Potential raw materials for aviation biofuels production

As there is an increasing demand for aviation fuels and the difficulties encountered to use methyl esters of fatty acids (FAME), there is a necessity to look for alternative raw materials.

3.1 Fusel oil

Fusel oil is one of the co-products of fermentation for the production of fuel ethanol composed of the less volatile fraction of the distillation [53]. The term "fusel" is derived from German and means "lower"; the composition of that co-product may vary according to the substrate used and fermentative conditions, as well as its yield, which may be 1–11 L for each 1000 L of ethanol produced [53, 54].

The composition of the fusel oil and the contents of the components are shown in **Table 9**.

Table 9 shows two major components together, the isoamyl and isobutyl alcohols (**Figure 8**), contributing up to 65% of the fusel oil according to Pérez et al. [53] and up to 85% according to Patil et al. [55].

The isoamyl alcohol derived from fermentative media is a metabolite generated in the decomposition of an amino acid, isoleucine, mainly by the action of the aminotransferases enzymes and pyruvate decarboxylases, and alternatively

Compound	Content (%)
Isoamyl alcohol	38–85
Isobutyl alcohol	15–20
Methyl alcohol	0.5–2.7
n-Propyl alcohol	0.6–1.4
Ethyl formate	>0.1
2-Butanol	>0.1
Methyl benzoate	>0.1
1-Butanol	>0.2
Hexyl acetate	>0.1
n-Octanol	>0.1
n-Decanol	>0.1
Methyl acetate	>0.1
Amyl acetate	>0.1
3-Pentanol	>0.1
Acetaldehyde	>0.1
Propylaldehyde	>0.01
Amyl alcohol	>0.1

Table 9.
Fusel oil composition.

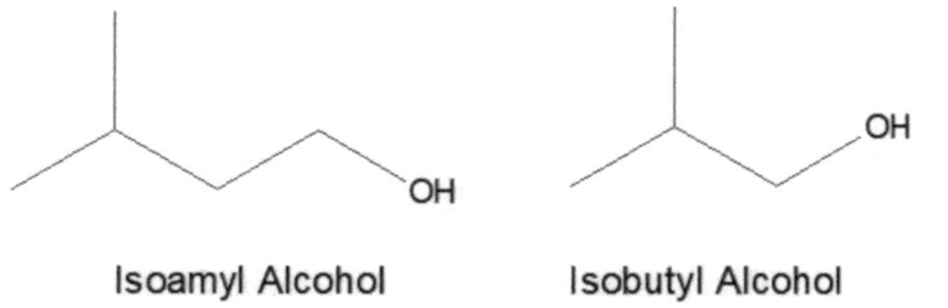

Figure 8.
Representation of the molecular structures of the components with the highest concentration in the fusel oil.

also by routes involving a-ketoacid dehydrogenase and acylCoA hydrolase.
In the last decade, fusel oil has been studied for several applications such as synthesis of lubricants by esterification with oleic acid with heavier fractions, obtaining inputs for the pharmaceutical industry, and also in the perfumery industry [56, 57].

Isoamyl alcohol was studied as a total and partial substituent of the vehicle gasoline in 50% fraction, with an increase of up to 25% in the CO emission, with a lower combustion efficiency and a possible shortening of the engine life, indicating that its use as fuel is practically unfeasible in Otto cycle engines [58].

Although its use in pure form is not suitable for use as a fuel, this residue may be a source of branched carbon chain inputs as a starting material for the synthesis of new biofuels as well as for oligomerization reactions in the synthesis of isoparaffins, which are known by producing high-octane fuels to produce kerosene, aviation gasoline, diesel, and lubricating oils with routes similar to those applied to ethanol and n-butanol [59].

3.2 n-Butanol

In contrast to ethanol, n-butanol does not yet have a significant participation in the energy matrix due to its application as a solvent and in the production of resins and to the fact that some countries including Brazil are not self-sufficient in this input. It is considered as a second generation biofuel with higher energy properties and better mixing with paraffinic fuels compared to ethanol.

Butanol can be produced by means of fermentative raw materials of amylaceas and saccharines, obtained by the route acetone-butanol-ethanol (ABE), which is the second most used fermentation process worldwide only losing to the fermentation of ethanol [60, 61]. The most commonly used bacteria for butanol production are *Clostridium acetobutylicum, Clostridium beijerinckii, Clostridium saccharoperbutylacetonicum,* and *Clostridium saccharobutylicum* using various raw materials such as cane molasses, corn husk, cassava flour, bagasse, straw, and sugar cane vinasse [62].

Mixtures of n-butanol and Jet A kerosene were studied and showed a decrease of 5000 rpm in turbine rotation due to its low calorific power (33.08 MJ/kg) compared to Jet A-1 (43.28 MJ/kg), but which had the advantage of significantly reducing NOx and CO emissions by up to 50 and 35%, respectively [63].

Other obstacles in its use are the kinematic viscosity at −20°C of 12.84 $mm^2 s^{-1}$ well above 8.00 $mm^2 s^{-1}$ allowed in the current standards, which can be bypassed in mixtures of up to 20% with Jet A1 kerosene and its high cloud point [64].

The fact that n-butanol has a hydroxyl group makes the fuel more hygroscopic and also denser than a paraffinic kerosene, impairing its energy efficiency. **Figure 9** shows routes for the production of aviation kerosene starting from ethanol and n-butanol known as alcohol to jet (AtJ) [65].

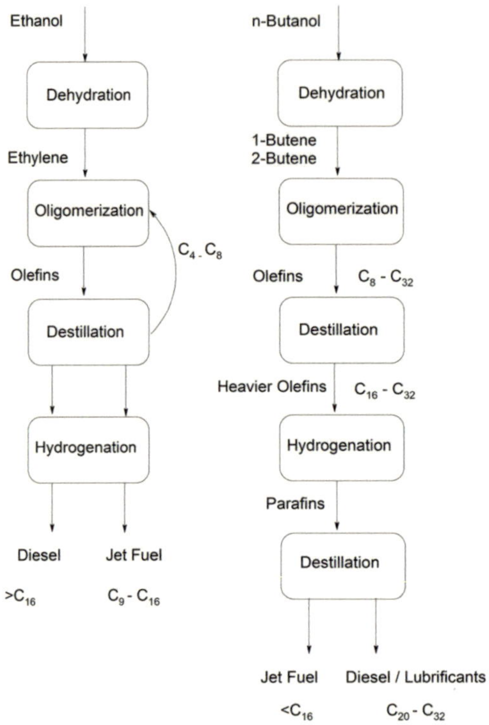

Figure 9.
Routes of obtaining aviation kerosene by alcohol polymerization.

Author details

Jonathan Baumi, Caroline Milani Bertosse and Carmen Luisa Barbosa Guedes*
Departamento de Química, LAFLURPE, Universidade Estadual de Londrina (UEL), Centro de Ciências Exatas (CCE), Londrina, PR, Brazil

*Address all correspondence to: carmen@uel.br

References

[1] Veraszto EV, da Silva D, Miranda NA, Simon FO. Tecnologia: Buscando uma definição para o conceito. Prisma.com. 2008;7:60-85

[2] Farias LM, Sellitto MA. Uso da energia ao longo da história: evolução e perspectivas futuras. Revista Liberato. 2011;**12**:1-106

[3] Coronado CR, Carvalho Junior JAC, Quispe CA, Sotomonte CR. Ecological efficiency in glycerol combustion. Applied Thermal Engineering. 2014;**63**:97-104. DOI: 10.1016/j.applthermaleng.2013.11.004

[4] Demirbas A. Progress and recent trends in biodiesel fuels. Energy Conversion and Management. 2008;**50**:14-34. DOI: 10.1016/j.enconman.2008.09.001

[5] Pousa GPAG, Santos ALF, Suarez PAZ. History and policy of biodiesel in Brazil. Energy Policy. 2007;**35**:5393-5398

[6] Stattman SL, Hospes O, Mol APJ. Governing biofuels in Brazil: A comparison of ethanol and biodiesel policies. Energy Policy. 2013;**61**:22-30. DOI: 10.1016/j.enpol.2013.06.005

[7] Cantarella H, Nassar AM, Cortez LAB, Baldassin Junior R. Potential feedstock for renewable aviation fuel in Brazil. In: Environmental Development. 15th ed. New York: Elsevier BV; 2015. pp. 52-63. DOI: 10.1016/j.envdev.2015.05.004

[8] Chiaramonti D, Prussi M, Buffi M, Tacconi D. Sustainable bio kerosene: Process routes and industrial demonstration activities in aviation biofuels. Applied Energy. 2014;**136**:767-774. DOI: 10.1016/j.apenergy.2014.08.065

[9] Petrobrás. Querosene de aviação: Informações Técnicas [Internet]. 2014. Available from: http://sites.petrobras.com.br/minisite/assistenciatecnica/public/downloads/QAV-Informa%C3%A7%C3%B5es-T%C3%A9cnicas-v.1.3-29.pdf [Accessed: 28 May 2018]

[10] Farah MA. Petróleo e seus derivados: definição, constituição, aplicação, especificações, características de qualidade. 1st ed. Rio de Janeiro: LTC; 2012. p. 226

[11] American Society For Testing and Materials. D910: Standard Specification for Leaded Aviation Gasolines. West Conshohocken, PA: ASTM International; 2017

[12] Seyferth D. The rise and fall of Tetraethyllead. 2. Organometallics. 2003;**22**:5154-5178. DOI: 10.1021/om030621b

[13] Wolfe PJ, Giang A, Ashok A, Selim NE, Barrett SRH. Costs of IQ loss from leaded aviation gasoline emissions. Environmental Science & Technology. 2016;**50**:9026-9033

[14] Esler D. Getting the Lead Out: The Future of Avgas [Internet]. 2015. Available from: http://aviationweek.com/bca/getting-lead-out-future-avgas [Accessed: 28 October 2017]

[15] Rocha AC, Martins MPP, Santos CB. Estimativa das emissões de dióxido de carbono proveniente da aviação civil regular em um dia específico. In: IV Workshop de Mudanças Climáticas e Recursos Hídricos do Estado de Pernambuco e do I Workshop Internacional sobre Mudanças Climáticas e Biodiversidade (IV WMCRHPE); 29-31 October 2012, Recife, Brazil

[16] Agência Nacional de Petróleo (ANP). Gás Natural e Biocombustíveis Resolução n° 37 DOU 2.12.2009 [Internet]. 2012. Available from:

http://www.anp.gov.br/SITE/acao/download/?id=24408 [Accessed: 28 October 2016]

[17] American Society For Testing and Materials. D1655: Standard Specification for Aviation Turbine Fuel. West Conshohocken, PA: ASTM International; 2017

[18] Arkoudeas P, Kalligeros S, Zannikos F, Anastopoulos G, Karonis D, Korres D. Study of using JP-8 aviation fuel and biodiesel in CI engines. Energy Conversion and Management. 2003;**44**:1013-1025

[19] Schmitigal J, Tebbe J. JP-8 and other Military Fuels [Internet]. 2011. Available from: www.dtic.mil/cgi-bin/GetTRDoc?AD=ADA554221 [Accessed: 26 February 2018]

[20] Yilmaz N, Atmanli A. Sustainable alternative fuels in aviation. Energy. 2017;**40**:1378-1386

[21] Baroutian S, Aroua MK, Raman AAA, Shafie A, Ismail RA, Hamdan H. Blended aviation biofuel from esterified Jatropha curcas and waste vegetable oils. Journal of the Taiwan Institute of Chemical Engineers. 2013;**44**:911. DOI: 10.1016/j.jtice.2013.02.007

[22] SHEEL Global. Civil Jet Fuel [Internet]. 2018. Available from: https://www.shell.com/business-customers/aviation/aviation-fuel/civil-jet-fuel-grades.html [Accessed: 05 June 2018]

[23] Mohammad M, Kandaramath TH, Yaakoba Z, Sharmac YC, Sopiana K. Overview on the production of paraffin based-biofuels via catalytic hydrodeoxygenation. Renewable and Sustainable Energy Reviews. 2013;**22**:121-132. DOI: 10.1016/j.rser.2013.01.026

[24] Hari TK, Yaakob Z, Binitha NN. Aviation biofuel from renewable resources: Routes, opportunities and challenges. Renewable and Sustainable Energy Reviews. 2015;**42**:1234-1244. DOI: 10.1016/j.rser.2014.10.095

[25] Gaylarde CC, Bento FM, Kelley J. Microbial contamination of stored hydrocarbon fuels and its control. Revista de Microbiologia. 1999;**30**. DOI: 10.1590/S0001-37141999000100001

[26] Waynick JA. The development and use of metal deactivators in the petroleum industry: A review. Energy & Fuels. 2001;**15**:1325-1340. DOI: 10.1021/ef010113j

[27] Rabaev M, Shapira D, Geva J, Fass R, Sivan A. Effect of the fuel system icing inhibitor diethylene glycol monomethyl ether on the biodegradation of jet fuel in soil. International Biodeterioration & Biodegradation. 2009;**63**:311-315

[28] Llamas A, García-martínez MJ, Al-lal AM, Canoira L, Lapuerta M. Biokerosene from coconut and palm kernel oils: Production and properties of their blends with fossil kerosene. Fuel. 2012;**102**:483-490. DOI: 10.1016/j.fuel.2012.06.108

[29] IATA. Fuel price Analysis [Internet]. 2016. Available from: http://www.iata.org/publications/economics/fuel-monitor/Pages/price-analysis.aspx [Accessed: 28 January 2018]

[30] Agência Nacional de Petróleo (ANP). Gás Natural e Biocombustíveis Resolução n° 37 DOU 2.12.2009 [Internet]. 2018. Available from: http://www.anp.gov.br [Accessed: 28 October 2017]

[31] Agência Nacional de Aviação Civil (ANAC). Anuário do Transporte Aéreo [Internet]. 2016. Available from: http://www.anac.gov.br/assuntos/dados-e-estatisticas/anuario/Anuario2016.zip [Accessed: 10 March 2018]

[32] International Civil Aviation Organization (ICAO). A summary of

research and perspectives. In: The ICAO Workshop of Aviation and Alternative Fuels; Montréal, 10-12 February; 2009

[33] Kousoulidou M, Lonza L. Biofuels in aviation: Fuel demand and CO_2 emissions evolution in Europe toward 2030. Transportation Research Part D: Transport and Environment. 2016;**46**:166-181. DOI: 10.1016/j.trd.2016.03.018

[34] Cortez LAB, Nigro FEB, Nassar AM, Cantarella H, Horta Nogueira LA, Moraes MAFD, et al. Flightpath to Aviation Biofuels in Brazil: Action Plan. 1st ed. Campinas: Unicamp; 2013. 56 p

[35] Sgouridis S, Bonnefoy PA, Hansman RJ. Air transportation in a carbon constrained world: Long-term dynamics of policies and strategies for mitigating the carbon footprint of commercial aviation. Transportation Research Part A: Policy and Practice. 2011;**45**:1077-1091. DOI: 10.1016/j.tra.2010.03.019

[36] Atsonios K, Kougioumtzis MA, Panopoulos KD, Kakaras E. Alternative thermochemical routes for aviation biofuels via alcohols synthesis: Process modeling, techno-economic assessment and comparison. Applied Energy. 2015;**138**:346. DOI: 10.1016/j.apenergy.2014.10.056

[37] Renewable Energy Group (REG). [Internet]. 2016. Available from: http://www.regfuel.com/technologies/fischer-tropsch-gas-to-liquid [Accessed: 27 October 2016]

[38] Agência Nacional de Petróleo (ANP). Gás Natural e Biocombustíveis Resolução N° 63 DOU 05/12/2014 [Internet]. 2014. Available from: https://www.legisweb.com.br/legislacao/?id=278197 [Accessed: 08 June 2018]

[39] MAPA. Produção Brasileira de Cana-de-açúcar, Açúcar e Etanol [Internet]. 2018. Available from: http://www.agricultura.gov.br/assuntos/sustentabilidade/agroenergia/arquivos-producao/PRODUOBRASILEIRADECANADEACARACAREETANOLatualizadoem19062018.pdf. [Accessed: 28 January 2018]

[40] Brasil. Balanço Energético Nacional. Empresa de Pesquisa Energética (EPE) [Internet]. 2016. Available from: https://ben.epe.gov.br/downloads/Relatorio_Final_BEN_2017.pdf [Accessed: 28 January 2018]

[41] União da Indústria de Cana-de-açúcar (UNICA) [Internet]. 2014. Available from: http://www.unica.com.br/noticia/1183323992035374661/movido-a-etanol-por-cento2C-aviao-agricola-ipanema-completa-10-anos-por-cento0D-por-cento0A/ [Accessed: 08 June 2018]

[42] Silveira V. Mistura em pleno vôo. Scientific American Brasil. 2006;**44**:30-34

[43] Stump D. Sustentabilidade e Tecnologias de Baixo Carbono no Brasil [Internet]. 2011. Available from: https://www.usp.br/mudarfuturo/cms/?cat=53 [Accessed: 08 June 2018]

[44] Knothe G, Gerpen JV, Krahl J, Ramos LP. Manual do Biodiesel. 1st ed. São Paulo: Blucher; 2006. 352 p

[45] Leung DYC, Wu X, Leung MKH. A review on biodiesel production using catalyzed transesterification. Applied Energy. 2009;**87**:1083-1095. DOI: 10.1016/j.apenergy.2009.10.006

[46] Lôbo IP, Ferreira SLC, Cruz RS. Biodiesel: Parâmetros de qualidade e métodos analíticos. Química Nova. 2009;**32**:1596-1608. DOI: 10.1590/S0100-40422009000600044

[47] De Boni LAB. Tratamento da glicerina bruta e subprodutos obtidos da reação de transesterificação de

sebo bovino utilizada para a produção de biodiesel [thesis]. Universidade Luterana do Brasil; 2008

[48] Stepan E, Enascutaa CE, Oprescua EE, Radua ERA, Galana AM, Vasilievicia G, et al. Intermediates for synthetic paraffinic kerosene from microalgae. Fuel. 2016;**172**:29-36. DOI: 10.1016/j.fuel.2016.01.027

[49] Gouveia L, Marques AE, Silva TL, Reis A. *Neochloris oleabundans* UTEX #1185: A suitable renewable lipid source for biofuel production. Journal of Industrial Microbiology & Biotechnology. 2009;**36**:821. DOI: 10.1007/s10295-009-0559-2

[50] Schuchardt U, Franco TT, de Melo JCP, Cortez LAB. Sustainable aviation fuels for Brazil. Biofuels, Bioproducts and Biorefining. 2014;**8**:151-154. DOI: 10.1002/bbb.1481

[51] American Society For Testing and Materials. D7566: Standard Specification for Aviation Turbine Fuel Containing Synthesized Hydrocarbons. West Conshohocken, PA: ASTM International; 2017

[52] Starck L, Pidol L, Jeuland N, Chapus T, Bogers P, Bauldreay J. Production of hydroprocessed esters and fatty acids (HEFA)—Optimisation of process yield. Oil & Gas Science and Technology – Revue D'ifp Energies Nouvelles. 2014;**71**:10-23. DOI: 10.2516/ogst/2014007

[53] Pérez ER, Cardoso DR, Franco DW. Análise dos álcoois, ésteres e compostos carbonílicos em amostras de óleo fúsel. Química Nova. 2001;**24**:10-12. DOI: 10.1590/s0100-40422001000100003

[54] Patil AG, Koolwal SM, Butala HD. Fusel Oil: Composition, removal and potential utilization. International Sugar Journal. 2002;**104**:51-58

[55] Leyva F, Orjuela A, Gil I, Vargas J, Rodríguez G. Vapor–liquid equilibrium of isoamyl alcohol+isoamyl propionate and propionic acid+isoamyl propionate systems at 50.00, 101.33 and 150.00 kPa. Fluid Phase Equilibria. 2013;**356**:56-62. DOI: 10.1016/j.fluid.2013.07.024

[56] Özgülsün A, Karaosmanôglu F, Tüter M. Esterification reaction of oleic acid with a fusel oil fraction for production of lubricating oil. Journal of the American Oil Chemists' Society. 2000;**77**:105-109. DOI: 10.1007/s11746-000-0017-5

[57] Liaw W, Ho S, Wang J, Hu OY, Li J. Determination of morphine by high-performance liquid chromatography with electrochemical detection: Application to human and rabbit pharmacokinetic studies. Journal of Chromatography B: Biomedical Sciences and Applications. 1998;**714**:237-245. DOI: 10.1016/s0378-4347(98)00230-8

[58] Solmaz H. Combustion, performance and emission characteristics of fusel oil in a spark ignition engine. Fuel Processing Technology. 2015;**133**:20-28. DOI: 10.1016/j.fuproc.2015.01.010

[59] Brooks KP, Snowden-Swan IJ, Jones SB, Butcher MG, GSJ L, Anderson DM, et al. Biofuels for Aviation. 1st ed. Massachusetts: Academic Press; 2016. pp. 109-150. DOI: 10.1016/B978-0-12-804568-8.00006-8

[60] Al-Shorgani NKN, Shukorb H, Abdeshahiand P, Nazira MYM, Kalilb MS, Hamida AA, et al. Mint: Process optimization of butanol production by *Clostridium saccharoperbutylacetonicum* N1-4 (ATCC 13564) using palm oil mill effluent in acetone–butanol–ethanol fermentation. Biocatalysis and Agricultural Biotechnology. 2015;**4**:244. DOI: 10.1016/j.bcab.2015.02.004

[61] Dürre P. Biobutanol: An attractive biofuel. Biotechnology Journal. 2007;**2**:1525-1534. DOI: 10.1002/biot.200700168

[62] Li HG, Maa XX, Zhanga QH, Luob W, Wua YQ, Lia XH. Enhanced butanol production by solvent tolerance *Clostridium acetobutylicum* SE25 from cassava flour in a fibrous bed bioreactor. Bioresource Technology. 2016;**221**:412

[63] Mendez CJ, Parthasarathy RN, Gollahalli SR. Performance and emission characteristics of butanol/jet A blends in a gas turbine engine. Applied Energy. 2014;**118**:135-140. DOI: 10.1016/j.apenergy.2013.12.011

[64] Chuck CJ, Donnelly J. The compatibility of potential bioderived fuels with Jet A-1 aviation kerosene. Applied Energy. 2014;**118**:83-91. DOI: 10.1016/j.apenergy.2013.12.019

[65] Wang W, Tao L. Bio-jet fuel conversion technologies. Renewable and Sustainable Energy Reviews. 2016;**53**:801-822. DOI: 10.1016/j.rser.2015.09.016

Chapter 13

Hydrogen Technologies for Mobility and Stationary Applications: Hydrogen Production, Storage and Infrastructure Development

Martin Khzouz and Evangelos I. Gkanas

Abstract

The present chapter focuses on hydrogen technologies for both stationary and mobility/transportation applications. Hydrogen production from sustainable resources for the generation of pure and low cost hydrogen is described in the chapter. Several potential hydrogen production techniques are introduced and analyzed. The challenges and the advantages of each production method will be discussed. Furthermore, the chapter will introduce hydrogen infrastructure development for mobility applications and will discuss hydrogen storage challenges. Hydrogen production for fuel cell technologies requires an improvement regarding sustainability of the hydrogen supply and an improvement regarding decentralized hydrogen production. Moreover, hydrogen economy as far requires a large scale and long term storage solution to meet the increasing demand.

Keywords: hydrogen production, hydrogen storage, hydrogen infrastructure, hydrogen applications, hydrogen economy

1. Introduction

The establishment of hydrogen economy for mobility applications led to a new movement for energy sector. The ultimate target is to reduce carbon dioxide emission by achieving a transition towards green transport system. There are three main challenges for the global adoption of a hydrogen economy; the production of hydrogen, the storage of hydrogen and the further development of the hydrogen fuel cells. Currently, hydrogen is produced from non-renewable a resource which forms a short term solution. The energy supply sector requires new, reliable and innovative technologies which support green energy supply and achieves the required economy targets. Hydrogen is an energy carrier as it can be extracted from renewable and non-renewable resources which can be converted to electrical power and water by using fuel cells. Hydrogen-powered fuel cells are electrochemical cells that convert chemical energy of hydrogen into electricity via redox reactions. The fuel cells are able to deliver a renewable energy at the point of use with no emission which makes a solution to the problem of greenhouse gases emissions.

	Natural gas	Gasoline	Hydrogen
Energy by weight	1.2 X > gasoline	43 MJ/kg	1.2 X > gasoline
Energy by volume	1.5 X < gasoline	120 MJ/gallon	1.5 X < gasoline
Boiling point	−162°C	Liquid at room temp	−252.87°C
Buoyancy relative to air	2 X lighter	3.75 X heavier	14 X lighter
Odor	Mercaptan	Yes	Odorless
Toxicity	Some	High	None
Color	No	Yes	No

Table 1.
Physical properties of natural gas, gasoline compared to hydrogen.

Fuel cells have been portrayed as humanity's solution to the global energy challenge and the answer to the negative environmental impacts from the prevalent energy production methods of today. The current chapter looks at hydrogen as a clean and efficient energy carrier for the future and fuel cells as the power generators, as well as the realistic steps needed to take advantage of the benefits that have been seen in the fuel cells today.

The advantages of hydrogen as fuel over other fuel sources, the several feed stocks available and current methods for hydrogen extraction are discussed, indicating the most commercially-feasible hydrogen production path for fuel cell applications.

While moving from the fossil fuel era towards a greener fuel future, the use of hydrogen as energy carrier is highlighted by a lot of advantages [1]. One of the major advantages that play a major role for hydrogen's fuel is the high specific energy (Lower heating value per unit mass); as hydrogen is lighter than gasoline or diesel for the same amount of energy (**Table 1**). Moreover, hydrogen has the highest utilization efficiency as compared to the fossil fuels, **Table 1** explains physical properties if hydrogen compared with natural gas and gasoline. It was reported

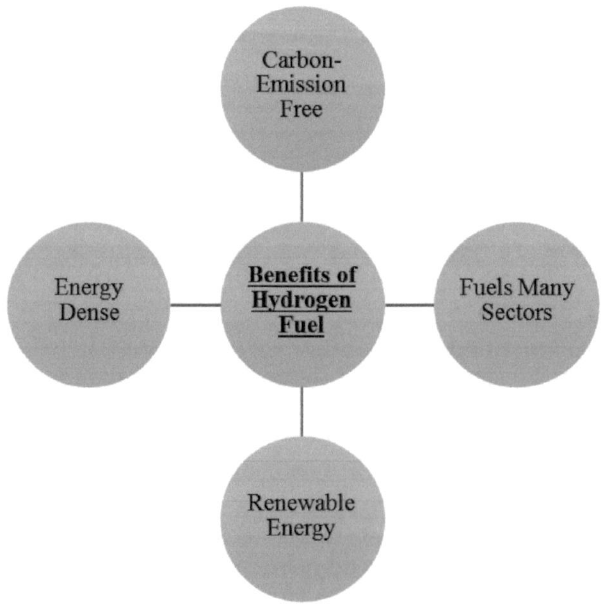

Figure 1.
Benefits of hydrogen energy carrier.

that hydrogen is 39% more efficient than fossil fuels [2]. This means high output energy per unit of mass. Secondly, hydrogen is a clean source of energy, as the product of hydrogen oxidation is only water and heat, independently on the process the oxidation occurs in; whether using it in a combustion engine or a fuel cell, or the type of energy it was converted to; thermal, electrical or mechanical. Thirdly, hydrogen can be produced from different raw materials, and then stored both as compressed, liquefied and via solid state reactions to be used in all types of applications. Finally, hydrogen is a stable non-corrosive element, however, safety aspects needs to be considered. **Figure 1** explains benefits of hydrogen energy carrier.

2. Sources of hydrogen production

Hydrogen is an energy carrier with high potential for green and sustainable applications, as it is carbon-emission free, energy dense, can supply many energy sectors and can be produced form renewable sources. Hydrogen is currently used at many industrial sectors; mainly at chemical industries and refineries. Although, the last two decades, there has been an increased interest from the automotive industry for a low pollution propulsion system. Thus, a carbon emission-free hydrogen method of producing hydrogen is required for automobile applications. In general, as renewable hydrogen, can be defined the hydrogen that is formed from solar, tidal, wind, hydro and supplied by sustainable feedstock. However, sustainable methods are expensive to produce hydrogen and require several technological advances for mass production. **Figure 2** shows that majority of hydrogen used today is produced from fossil fuels. Almost 95% of produced hydrogen is used at the site where it is produced [3]. Renewable hydrogen does not yet contribute to hydrogen market as the cost of renewable power still uncompetitive.

Hydrogen can be produced in large-scale centralized production plants or in small scale distributed production plants, decentralized generation. Currently the hydrogen production via methane reforming in centralized plants is cheaper than the equivalent production in decentralized plants. However, hydrogen distribution (hydrogen delivery infrastructure) and storage technical barriers will make hydrogen production in decentralized plants more feasible as the demand on hydrogen increases. At this early stage of hydrogen energy technology, small scale production is suitable to fulfill the low hydrogen demand and avoid infrastructural unavailability issue. In the long term however, large scale production will be a basic

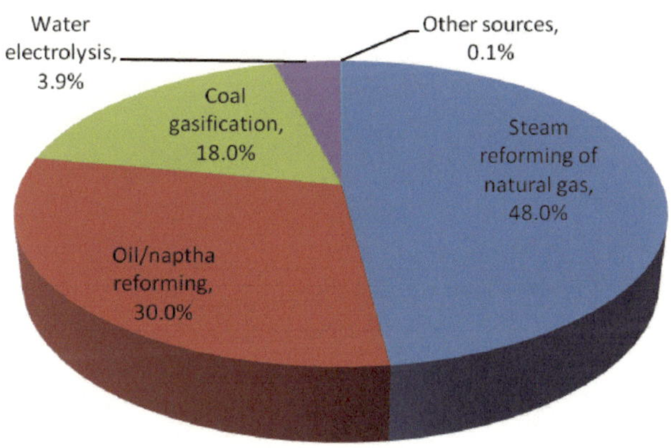

Figure 2.
Global hydrogen production from main sources [6].

requirement to meet the future expected demand, and solutions must be found for the distribution challenge.

The world annual hydrogen production is around 368 trillion m^3 [4]. Almost 50%of this amount is produced by steam reforming for natural gas [5]. Hydrogen can be produced in various processes, such as, thermo-chemical, electrochemical, photochemical, photo-catalytic and photo-electrochemical processes. The feed stocks for hydrogen and the production technologies are shown in **Figure 2**. The hydrogen energy feedstock and production routs are explained in **Figure 3**. The main three routes for hydrogen production are; reforming from hydrocarbons, electrolysis and thermal decomposition or cracking.

2.1 Hydrogen from fossil fuels

Hydrogen at commercial usage is obtained from various hydrocarbon feedstocks such as; methane and fossil fuels which can be oil, gas and coal. The extraction method to produce hydrogen is a well-developed technology at large scale production such as steam reformation, fuel oxidation and gasification [7]. The fossil fuel recourses are main contributor to carbon dioxide emissions and the resources for fossil fuels are depleting which makes hydrogen produced from feedstock non-sustainable. Fossil fuels including coal and natural gas are currently the most common way to produce hydrogen. Steam reforming is one example of reforming technology; oxygen is also used instead of steam. Reforming techniques are named depending on the oxidation agent use; this include, steam reforming for light hydrocarbons (methane), partial oxidation for heavy hydrocarbons such as coal and oil and mixed technology known as auto-thermal reforming. After reforming process, the reforming gas mixture must be cleaned to obtain purity of hydrogen required for the application.

2.2 Hydrogen from biomass

One of the emerging technologies for a renewable and sustainable hydrogen feedstock is to use biomass gasification. This method allows the production of

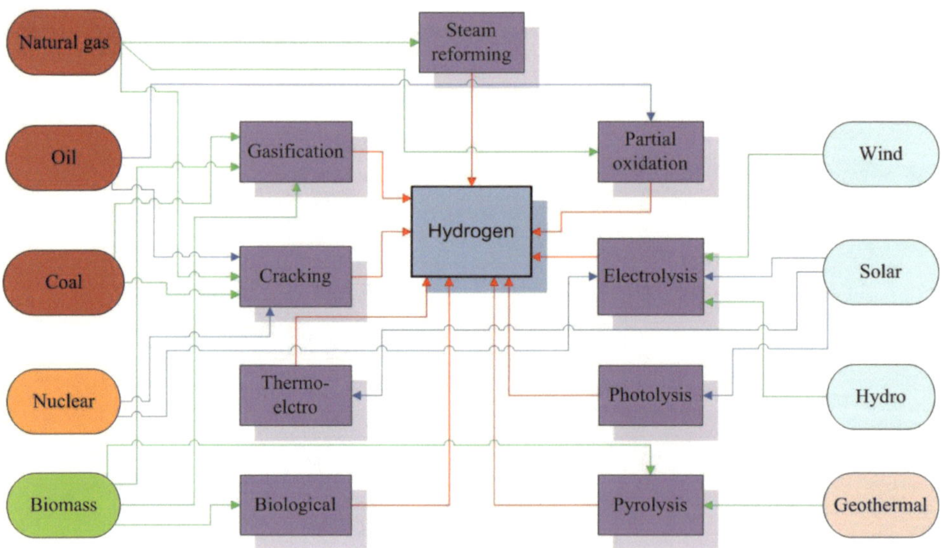

Figure 3.
Hydrogen energy production routes and hydrogen extraction methods [5].

hydrogen with low carbon dioxide emissions. The most popular biomass resources for producing hydrogen are often supplied from agriculture or solid municipal waste [8]. Hydrogen can be extracted from biomass fuels by pyrolysis, biological reactions and gasification [9]. The process used depends on the biomass feedstock. The production via biomass feedstock the most popular method for hydrogen production due to the availability of its gasfires. Further studies are needed to improve catalysts and microorganisms used for pyrolysis and biological reactions to increase the efficiency and hydrogen production rate. Eqs. (1) and (2) are explaining the main mechanism for the pyrolysis (Eq. (1)) and gasification (Eq. (2)).

High temperature gasification is also used to produce hydrogen. The process allows hydrogen production without producing CO_2. The light hydrocarbons are heated in plasma up to 1600°C and converted to carbon and hydrogen. This is also known as high temperature plasma arc gasification.

Biological fermentation process is also used to produce hydrogen using bacterial or algae decomposition of water. The production process can be performed using photo biological at sunlight or at dark conditions.

$$\text{Biomass} \xrightarrow{\text{pyrolysis}} H_2 + CO_2 + CO + \text{Hydrocarbon gases} \tag{1}$$

$$\text{Biomass} \xrightarrow{\text{gasification}} H_2 + CO_2 + CO + N_2 \tag{2}$$

2.3 Hydrogen from nuclear sources

Mainly, Hydrogen production from nuclear energy can be achieved by using thermochemical processes. Excess heat is generated in a nuclear reactor. Thermal energy will split molecules by applying high temperature. With no catalyst, a steam of 2500°C will dissociate into hydrogen and oxygen. High temperature is considered a problem for material design therefore some catalyst is considered to split water down at 900°C [10].

The by-product heat that is produced during this process is used for hydrogen production method. Several reactor technologies are used to produce hydrogen and the production method is considered economically attractive [11]. The process itself is low carbon dioxide emission. Nuclear generated electricity does not produce CO_2 emissions. There are four methods to use nuclear energy to produce hydrogen; electrolysis, high temperature steam electrolysis, thermochemical water splitting and steam reforming with heat input from a nuclear reactor. The waste heat is utilized in thermochemical water splitting and high temperature electrolysis which both methods are efficient compared to low temperature electrolysis.

In summary, all methods of hydrogen production might be used with nuclear power as a source of electricity and heat to produce hydrogen. The problem with hydrogen production facility at nuclear power plant that hydrogen production facility can be treated as heat sink and flammable substance is in the system which makes a potential hazard for safe operation of a nuclear facility as well as hydrogen safe handling [10].

2.4 Hydrogen from water

Electrolysis process is used to split water electrochemically to hydrogen and oxygen. The hydrogen production from water is considered green upon the electrical supply green routes [12]. The electricity can come from different sources and the electricity source determines whether any CO_2 is produced to generate hydrogen.

For example, wind energy produces no CO_2 but fossil fuel does. Photo-electrolysis is also considered a green hydrogen route as the process depends on sunlight to power the process [13].

The electrolysis process is performed using an electrolyser; there are three types of electrolyser; alkaline electrolyser, proton exchange membrane electrolyser and solid oxide electrolyser. All consist of two electrodes, water and an electrolyte which is a substance containing free ions that make the substance electrically conductive. The name of electrolyser depends on electrolyte used. Alkaline and proton exchange membrane electrolyser are defined as low temperature electrolyser (less than 100°C) and solid oxide electrolyser is classified as high temperature (900°C). High purity of hydrogen is produced using water electrolysis and the standard potential of a water electrolysis cell is 1.23 V at 25°C.

3. Non-renewable hydrogen production

Currently, this method is the most economical viable for hydrogen production. The current hydrogen production economy is dependent on fossil fuel resources and almost 96% of hydrogen produced is generated from hydrocarbon fuel [14]. Mostly, the hydrogen produced from non-renewable resources is performed in large central chemical plants. The centralized production at large scale chemical plants is used for mass production then hydrogen is transported. Large steam reformers are one of the examples of mass hydrogen production where hydrogen is transported to the location of use via pipelines, road or by ship. For hydrogen produced from non-renewable is a manufactured fuel from fossil fuel including natural gas and coal, methanol and water where electricity is supplied from non-renewable resources.

3.1 Steam-methane reforming (SMR)

Natural gas is used at most common way to produce hydrogen by steam reforming. The process is used for light hydrocarbons. Natural gas is composed of methane (CH_4). Steam reforming is performed at two stages Eq. (3) and Eq. (4) where methane is mixed with water steam at high temperature and heat is absorbed (Eq. (3)), the reaction itself is endothermic and requires a catalyst. The required temperature to perform the reaction is around 850°C. The second step of the process is the water gas shift reaction (Eq. (4)) and this involves a heat release (exothermic reaction). The combination of both reactions results a net of heat addition to the process. The total actual efficiency of the process is rarely above 80% [15]. The process itself is a demanding thermal process where methane undergoes a steam reformation or partial oxidation. The partial oxidation uses oxygen rather than steam as oxidizing agent to perform reaction as this process can maintain a high temperature where it is producing heat [16]. Combination of steam and oxygen for performing the reaction is also used in hydrogen production and the process is known as auto-thermal reforming. **Table 2** compares various reforming methods.

$$CH_4 + H_2O \rightarrow CO + 3H_2; \Delta H = +206\ kJ/mol \tag{3}$$

$$CO + H_2O \rightarrow H_2 + CO_2 \Delta H = -41\ kJ/mol \tag{4}$$

The process of steam methane reforming is performed as a two stage reaction. The first reaction is an endothermic catalytic reaction where methane is reformed using steam to syngas under the existence of catalyst. The process requires steam at

DOI: http://dx.doi.org/10.5772/intechopen.91676

	Steam reforming	Partial oxidation	Auto-thermal reforming
Steam availability	Steam is part of process	Partially burns a fuel, no steam is required	Both steam and air are required
Chemical reaction	$C_nH_mO_z + (n-z)H_2O \rightarrow$ $nCO + \left(n + \frac{m}{2} - z\right)H_2$	$C_nH_mO_z + \frac{1}{2}(n-z)O_2 \rightarrow$ $nCO + \frac{m}{2}H_2$	$C_nH_mO_z + yO_2 + 2\left(n - y - \frac{z}{2}\right)H_2O \rightarrow$ $nCO_2 + 2\left(n - y - \frac{z}{2} + \frac{m}{4}\right)H_2$
Maximum percent purity of H2 in the products mix (vol%)	$100 \times \left(\frac{2n+0.5m-z}{3n+0.5m-z}\right)$ combined steam reforming and water gas shift reaction	$100 \times \left(\frac{n+0.5m}{2n+0.5m}\right)$ combined partial oxidation and water gas shift reaction	$100 \times \left(\frac{2n-2y-z+0.5m}{3n-2y-z+0.5m}\right)$ auto-thermal reforming as shown in equation above
Maximum H2 yield (mol H2 per mol CnHmOz)	$2n + 0.5m - z$, combined steam reforming and water gas shift reaction	$n + 0.5m$, combined partial oxidation and water gas shift reaction	$2n - 2y - z + 0.5m$, auto-thermal reforming as shown in equation above
Heat requirments	Largely endothermic, consume a lot of heat, disadvantage	Exothermic, produce heat, advantage	Must be controlled under slightly exothermic reaction, start-up requires heat, advantage
Start-up period for reformer application	Long start-up period, since steam is required, disadvantage	Short start-up period, since no steam is required, advantage	Short start-up period, steady state conditions must be controlled, advantage

Table 2.
Comparison of different hydrocarbon reforming methods [20].

temperature of 700–1100°C and relatively high pressure range (20–30 bar) [17]. The outcome of the first stage reaction is syngas which is a mixture of hydrogen and carbon monoxide. The second stage requires converting carbon-monoxide to further produce hydrogen. The process is known as water gas shift reaction and it's an exothermic reaction which can occur at two steps, high temperature water gas shift reaction at 350°C followed by a low temperature water gas shift reaction at 200°C [18].

The SMR process has an efficiency of 60–75% and it is well established technology for hydrogen production [19]. One of the main advantages of using MSR for hydrogen production is the product purification using pressure swing adsorption that can produce hydrogen purity greater than 99%. The drawback of such method is that an endothermic process requires heat which is mainly provided by burning natural gas. The SMR produces high yield of hydrogen (50 wt% of CH_4) but with some impurities and the high demand for energy requirements will increase the production cost.

Hydrogen via SMR can be produced at both centralized and decentralized facilities. In the case of centralized production, liquefied or gaseous hydrogen is distributed to the area of application via pipes or tank trailers. For decentralized production, hydrogen is produced and stored at the location of usage. Due to the low hydrogen demand, a design with a capacity of 100 Nm^3/h with 65% efficiency (LHV) is commercially available [19]. This process of hydrogen production is currently the most favorable method for both centralized and decentralized production in already developed countries with piped methane infrastructure available and the process' efficiency is relatively high compared to the other methods. However, one drawback for this system is that CO_2 is produced via exothermic water gas shift reaction between CO and excess steam, and this adds the need for gas purification process.

3.2 Methanol steam reforming

Methanol (CH_3OH) is also used for hydrogen production purposes where the reaction temperature required to perform the process (200–300°C) is much less as compared to methane steam reforming [21]. The reformation process produces up to 75% of hydrogen concentration and maximum yield of 18.75 wt% of CH_3OH. The advantage of methanol steam reforming is that methanol is at liquid form which makes this fuel suitable for stationary hydrogen applications where reformation temperature is relatively low. In addition, methanol can be produced from renewable resources [22]; it has low sulfur content and high hydrogen to carbon ratio. The process requires use of methanol where pure methanol is expensive and the hydrogen produced has some impurities which requires further purification. Methanol is currently produced from coal and natural gas. The total world demand for methanol is about 41 million metric tons [22]. Methanol is also produced from bio-based resources as landfill gas, pig manure and paper mill black liquor. The viable method of producing methanol is using hydrocarbon fossil fuels [22]. The methanol reforming is an attractive method as the absence of C-C bonds in methanol allows the reaction to take place at low temperature. Because of low energy chemical bonds, reforming can be granted at faster start up and low material cost. Copper based catalyst is used at methanol steam reforming reaction (Eq.5). In real reaction, reformate gases usually contain CO, CO_2, H_2, H_2O and CH_3O [23]. **Table 3** compares both methane and methanol fuels in hydrogen production.

$$CH_3OH + H_2O \rightarrow CO_2 + 3H_2; \Delta H = +49.5\, kJ/mol \tag{5}$$

Feature	Methane	Methanol
Maximum percent purity of H_2 in the products mix (vol%)	80%	75%
Maximum H_2 yield	4 mol/mol of CH_4 (or 50 wt% of CH_4)	3 mol/mol of CH_3OH (or 18.75 wt% of CH_3OH)
Catalyst composition	Nickel based	Copper based
Selectivity	Hydrogen and CO	Hydrogen and CO_2
Size	Large (three purification units and gas cleaning unit)	Small (one purification unit)
Phase	Gas	Liquid
Temperature	Up to 800°C	Up to 350°C
H_2 produced	High	Low
CO produced	High	Low
Material	Expansive reactor material	Cheaper material as low temperature design

Table 3.
Methanol and methane fuels comparison for hydrogen production [20].

3.3 Electrolysis of water

Supplying electrical current to split water into hydrogen and oxygen is known as water electrolysis. The process efficiency is determined by electrical and thermal behaviors of the electrochemical reaction. The electrolysis process produces high purity hydrogen 99.999% with an efficiency of 70% [24, 25]. One major issue with the electrolysis process is that the electrodes tend to degrade overtime causing an increase of resistance, forcing the performance to slow down. The life span of the electrolyser is short and whole reaction process is slow in comparison to the large scale reforming process [26]. Currently, small scale electrolysers are used to produce hydrogen; however, hydrogen produced via this method is strongly affected by high electricity prices. The main aim for water electrolysis is to use the surplus electricity generated from renewable resources such as, wind, solar, hydropower, geothermal so it can be used in all different applications when stored in hydrogen. Renewable electrical generators need to be further developed to achieve energy sustainability.

$$H_2O_{(l)} \xrightarrow{electrolysis} H_{2_{(g)}} + \frac{1}{2}O_{2_{(g)}} \Delta H^\circ = +285.8\ \text{kJ/mol} \tag{6}$$

3.4 Coal gasification

Gasification is a process where raw materials such as coal, biofuel, biomass and other carbonaceous materials reacts at high temperature under controlled amount of steam and oxygen to produce hydrogen and carbon monoxide.

Economically, coal is attractive feed stock for hydrogen production by gasification due to the stability of coal prices compared to natural gas [27]. However, the complexity and high cost of the gasification process makes it a less favorable hydrogen source. The complexity of the process comes from the need of feedstock pre-treatment, multi-stage cleaning of raw syngas and CO conversion. This makes this process only feasible for large scale (centralized) hydrogen production. Coal to hydrogen conversion can achieve 52% of low heating value [28]. The major problem facing coal gasification process is removing CO_2 from the gas shift reaction, which decreases the overall efficiency and increases hydrogen production cost.

$$Coal + H_2O \rightarrow H_2 + CO + impurities \tag{7}$$

4. Renewable sources and methods

The hydrogen routes where the energy source does not contribute to greenhouse gases are defined renewable. Renewable sources of energy such as solar energy, wind energy, geothermal, tidal, and hydro and biomass energy are the most popular for power generation as well as for hydrogen production. The future potential for hydrogen production will depend on solar energy, wind energy and biomass where it can supply electricity for potential hydrogen production method.

Renewable resources for hydrogen production will include biomass, ethanol, wind farms, tidal energy, solar and hydroelectric. The method of production and whether the hydrogen has been produced without emitting CO_2 is known as green hydrogen. Greener hydrogen routes such as from electrolysis are playing a major role in hydrogen for mobility applications. Decentralized hydrogen production depends on renewable resources where on-demand hydrogen requirements are delivered to end user such as using personal electrolyser for home use.

4.1 Solar energy

Solar energy is used in photovoltaic systems to generate electricity. It is also used on concentrated collectors to collect thermal energy. The solar energy can meet the global energy requirements, if it stored and terraformed into another energy carrier forms. The disadvantage of long sunlight hour's availability makes this technology less attractive for hydrogen production. The cost of implementation of solar capturing system and the space availability when employed at residential area are also drawbacks of such technology [29]. The solar energy can produce thermal energy which can be used to supply energy for hydrogen production methods such as thermolysis process, thermochemical process and steam reformation process. It also can produce photon energy which can supply energy for electrolysis using photovoltaics and photo-electrolysis [30]. Solar photovoltaic has low temperature specification and it is assumed direct solar collection type to produce hydrogen and oxygen using electrolysis and photo-electrolysis. It also can be used to produce hydrogen by photo-biolysis process. The solar thermal has high temperature specification where steam reforming; thermolysis and cracking methods can be used to produce hydrogen from above mentioned routes.

4.2 Solar photovoltaic (PV)

The solar panels are used to generate electricity connected to the main supply grid. Demonstration of such system for hydrogen generation on a small scale has been performed at research level [31]. The system requires special current control unit to perform electrolysis process. The system operation will depend on amount of exposed sunlight where additional electricity from the grid required performing the process. Depending on photovoltaic hydrogen power system alone is a challenging technology and the hydrogen production efficiency is still under research. The PV performance depends on temperature and amount of exposed sun where DC-DC converter requires advanced monitoring system to supply enough power supply for electrolyser.

4.3 Photon-electrolysis

The photon-electrolysis process has low efficiency output [32] that makes this technology not a suitable solution for massive hydrogen production. The technology is under research and development and currently the process itself is not economically attractive.

4.4 Solar steam-methane reforming

The solar steam-methane reforming system applies direct or indirect solar heat to generate the required high temperature demand to fulfill the requirements for the chemical reaction. The thermal energy is harvested at power plant for such technology where heat transfer using a fluid is maintained to supply heat to catalytic reactor tubes. Syngas will be produced in a closed loop reaction to increase the efficiency and to maintain the high temperature requirements of the reaction. For indirectly heated reforming process, the working fluid transfers the heat energy from the solar receiver to the catalyst container at a closed loop system using primary and secondary heat transfer system. The directly heated reformer contains imbedded catalyst container where the thermal energy source will increase the catalyst temperature. This method does not require a secondary loop system, whereas the cost of design can be reduced. The challenges for such hydrogen production method are the requirement for constant solar radiation to maintain the chemical process and the cost for system development [33]. **Table 4** compares most popular hydrogen production methods and **Table 5** summaries current commercial hydrogen production methods.

5. Hydrogen storage

The future of hydrogen as an energy carrier greatly depends on the method at which hydrogen can be stored (**Table 6**). The fact that hydrogen has a low boiling point and a very low density both in gaseous (0.08988 g/L @ STP) and liquid (70.99 g/L) forms makes hydrogen storage complicated and costly [5]. Hydrogen can be stored via changing physical conditions such as its temperature, pressure or phase, or by changing chemical conditions of various solid and liquid compounds. The latter method includes metal hydrides, complex hydrides, carbon nanotubes, Metal Organic Frameworks (MOFs), 2D materials as well as light hydrocarbons as intermediate hydrogen carriers [34].

Process	Feedstock required	Efficiency	Current use	Renewable or non-renewable
Steam reforming	Hydrocarbons	60–75%	Commercially	Non-renewable
Partial oxidation	Hydrocarbons	60–75%	Commercially	Non-renewable
Electrolysis	Water and electricity	70%	Commercially	Non-renewable
Solar steam Reforming	Hydrocarbons	60–75%	Commercially	Renewable
Photo – electrolysis of water	Water and solar	12.4%	Under Research	Renewable
Thermo – water splitting	Water and heat	30%	Under Research	Non-renewable
Photolysis	Water and solar	0.5%	Under Research	Renewable

Table 4.
Hydrogen production process comparison.

	Steam methane reforming	Methanol steam reforming	Electrolysis of water
Advantages	• Widely accepted method • A high efficiency of 60–75% • Reliable and stable method • Can be used for various industries	• Requires lower temperatures than methane SR to work • Produces high efficiency reactions	• It is an on-demand fuel type • Can be used for various industries • Produces high hydrogen purities and efficiency rate
Disadvantages	• Carbon capture and storage • High operating and maintenance costs • Non-renewable method, releases greenhouse gases	• Hydrogen impurities • Feedstock required is costly	• The electrodes are prone to degrading over time requiring replacement • Electrolysis solution requires replacing once used up
Areas for Improvement	• Production of more efficient catalysts and reduce costs • Develop a viable renewable method of SMR for future use	• To develop on the system integrating a renewable aspect to it • To provide more popularity to the method	• Photo-electrolysis does not yield a high efficiency rate • Further research is required for a faster rate of reaction

Table 5.
Most popular hydrogen production methods comparison.

Parameter	Unit	2017 goal	Ultimate
System fill time (5 kg)	min	3.3	2.5
Cycle life	cycles	1500	1500
Operating temperature	°C	40/60	40/60
System volumetric capacity	kg H_2/L system	0.040	0.070
System gravimetric capacity	kg H_2/kg system	0.055	0.075

Table 6.
Summary on the targets for hydrogen storage systems set by the United States Department of Energy for on-board vehicular applications in year 2017 [35].

Chemical storage methods in high volumetric energy capacity with low cost for on-board hydrogen application are currently under research and development. Nowadays, commercial hydrogen is stored as a gas in high pressure cylinders (350 bar and 700 bar) or as liquid hydrogen at 20 K in cryogenic tanks. The process of hydrogen pressurization and liquefaction consumes energy which adds an extra cost to the hydrogen delivery stage.

Hydrogen storage technologies (**Table** 7) are investigated and mainly hydrogen is stored in three categories; compressed gas storage, cryogenic (liquid) storage and solid storage. The alternatives solutions for high pressure storage are known as; liquefaction, absorption by chemical compounds, metal hydrides, chemical hydrides and carbon nanotube. Hydrogen is good as a gravimetric energy carrier which means it has the most energy of any fuel in proportion to its mass. The problem with hydrogen is its volumetric energy is the worst. When hydrogen is compressed or liquefied, the volumetric capacity increases. The target is to increase both volumetric and gravimetric capacities.

	Viable capacity (kg)	Market	Commercial availability	Volumetric/ gravimetric ratios
Steel cylinder	1–100	Delivered tanks, bundled on trucks or stationary for mass storage	Yes	Medium to high volume. Low weight
Carbon fiber compressed gas tank	0.1–5	In demonstration fuel-cell cars and on fuel-cell and internal combustion engine busses	Yes	Medium to high volume. Medium weight
Carbon fiber-wrapped aluminum gas tank	0.1–100	Cars and busses	Yes	Medium to high volume Low to medium weight
Cryogenic	0.1–500	Stationary storage, tank trucks,	Yes	High volume. High weight
Metal hydride	0.01–1	Small fuel cells, small vehicles	Yes	Medium to high weight Medium to high volume
Chemical hydride	0.01–1	Small applications	No	Medium to high weight Medium to high volume
Single-walled nanotubes	0.001–100	All markets, but too expensive now	No	High weight. High volume

Table 7.
Current hydrogen storage technologies [36].

5.1 Compressed gas storage

Commercially, two standard pressures (35 and 70 MPa) are used for automotive applications. A high compressive cost and a safety factor for storage medium are some of the main challenges for such storage technology. Currently, there are four types of hydrogen vessels [37, 38], **Figure 4**. Type I; the vessel operates at low pressures, such vessels are the heaviest type of storage vessel. However, single material tank makes those vessels the cheapest to manufacture. Type II vessel is a composite thick tank wrapped with fiber resin. The tank can resist high internal pressure; the tank itself is heavy as metallic wall is thick. Type III is a composite vessel consisting of a thick or thin metallic wall fully wrapped with a fiber resin. The tank provides extra internal pressures resistance. Type IV vessel is a polymeric fully wrapped with fiber resin. It is the lightest storage vessel with a highest pressure resistance. In summary, Type I is the cost effective tank but their weight is the heaviest. Type IV storage tank can hold up to 700 bars with the lightest weight but the cost is two folds of Type I. **Table 8** explains the cost, weight and operation pressure of each tank type.

5.2 Liquefaction of hydrogen

Cryogenic tanks are used to store hydrogen at liquefied form. Hydrogen is at liquid form at 20 K (−253°C) which has a high cost and it requires energy to be maintained at liquid phase. The liquid form is more desirable than gaseous storage. It has higher energy storage density than gaseous form which gives liquefied hydrogen a content advantage. In general, the cryogenic tanks are of cylindrical

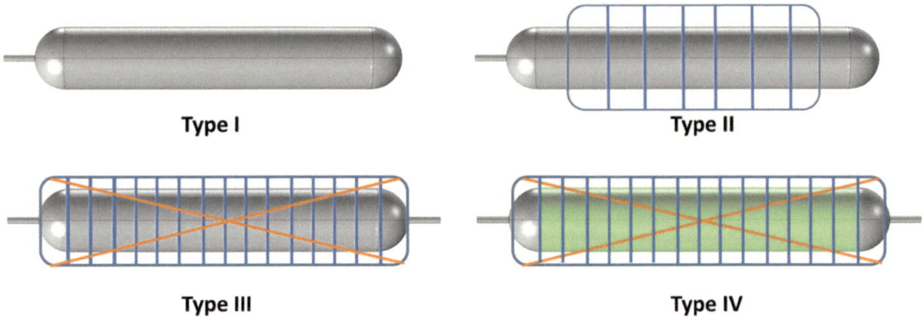

Figure 4.
Hydrogen compressed gas storage, vessel types [37].

	Type I	Type II	Type III and IV
Cost ($/liter)	5	5.5	11
Weight (lb/liter)	3.0	0.9	0.75 to 1
Maximum Pressure (bar)	200	263	300 to 700

Table 8.
Manufacture cost, weight and maximum pressure of four types pressure vessels [39].

	Compressed hydrogen 200 bar	Compressed hydrogen 350 bar	Compressed hydrogen 700 bar	Liquefied Hydrogen 2	Single wall nano tube
Storing energy (kJ/kg)	10,300	12,264	14,883	42,600	15,998
Volumetric energy content (MJ/m3)	714	2492	3599	3999	2159
Energy content (MJ/kg)	1.05	8.04	7.20	16.81	3.6

Table 9.
Relevant parameters for possible five hydrogen storage forms [41].

shape, as it is easier to manufacture and some of them are of spherical shape. A multilayer material for insulation heat is a current development priority for liquid hydrogen. Hydrogen boil-off or hydrogen loss is a challenge for hydrogen storage [40]. Some multilayers design and polymeric films are used to avoid energy loss.

Table 9 describes five possible hydrogen storage forms, compressed hydrogen at 200, 350 and 700 bar, liquefied form and nanotubes. **Table 9** explains the advantage of storing hydrogen at liquid form. The weight is three times less than compressed gas and it has higher energy storage content.

5.3 Metal hydrides

Hydrogen can be stored in solid state form by either physisorption or by chemisorption. Metal hydrides are chemical compounds that allow hydrogen to be stored at high density and under moderate temperature and pressure conditions [42].

Safety issues of gaseous storage	Safety issues of liquid hydrogen storage	Safety issues of metal hydrides storage
Loss of containment	Loss of containment	Pyrophoric materials
Materials-blistering	Boil off	Explosion and loss of containment
Heating effect during filling	Ice formation	Toxicity and heat

Table 10.
Some potential safety issues of hydrogen storage methods.

The hydrogen storage at such method is safer in comparison to the high pressure gas and liquid hydrogen storage. Metals hydrides exist within metals or alloys, they have physical properties which make them suitable for hydrogen storage such as; low absorption, high volumetric energy density, oxidation resistivity and some of them benefits of good reversibility [35]. The drawback of such storage method can be related to the cost of metal hydrides, metal hydrides being heavy and the reaction kinetic to release hydrogen is slow [43]. The safety issues of hydrogen storage in gaseous, liquefied and solid forms are stated in **Table 10**.

6. Hydrogen energy and fuel cells today

Nowadays, hydrogen energy and fuel cells are still considered as potential candidates to replace fossil fuels, but yet, several steps are necessary to be done towards the commercialization for real life applications. However, an increasing number of companies, big and small alike are investing in the field as they see the future potential for power generation in different applications. It worth mentioning that fuel cell research in its majority is focused on either PEMFC or SOFC, and with time these two types are becoming competitive technologies for some applications.

Hydrogen energy and fuel cell work at the moment can be divided into three major areas; research and development, materials and components supply, and resources training. On the top of that, governmental support and setting laws with incentives for clean energy suppliers and users is raising the awareness of the public and encourages more investments in the field. Some investors in the field of hydrogen energy and fuel cells identified their interest on a particular application, and they have been trying to produce prototypes for products that can be a step towards being a commercial product such as; Transport, Grid support, Domestic CHP, Auxiliary power generation, Portable applications, Educational demonstrations and Testing and measurement systems [44]. Other investors found their interest in identifying a specific component from the fuel cell system and became suppliers to the others working on the system integration and developing prototypes. The market for such companies is already there; however, the demand is quite small at the moment. Most of these investors either hold some research on their own or with collaboration with academia to develop their component or produce new alternatives to compete in a continually advancing field. Components can be one part of the fuel cell or hydrogen production device, hydrogen storage devices, system controllers, and testing and measurement equipment's.

At the moment, PEMFCs and SOFCs are holding the lions share in terms of R&D and commercialization. Together, PEMFC and SOFC cover the whole power generation rate spectrum for all different applications: PEMFC is suitable for mobile, portable, and transport with power rates up to 100 kW, while SOFC is suitable for medium to high power generation in hundred MWs. See **Table 11**.

Application	Power rate	Fuel cell needed
Portable application	Up to 500 W	PEMFC, DMFC
Domestic applications	5–10 kW	PEMFC, SOFC
Transport application	50–100 kW	PEMFC
Central power generation	100 kW–100 MW	SOFC, MCFC

Table 11.
Fuel cell applications.

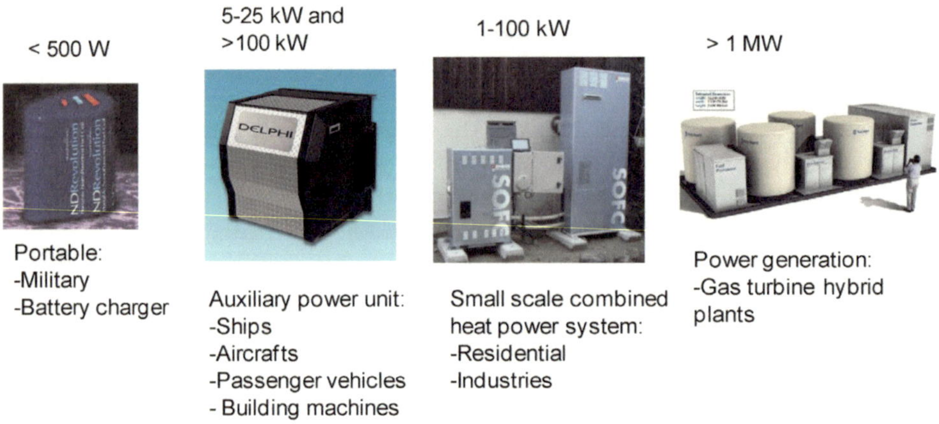

Figure 5.
SOFC applications.

Different types of fuel cells are used depending upon their applications. The operation temperature is the main issue when fuel cell is selected. The most popular fuel cells are solid oxide fuel cell and polymer electrolyte fuel cell. The solid oxide fuel cell designed to operate at high temperatures 700–1000°C. The fuel quality can be a mixture of hydrogen and carbon contained fuel. The main application is stationary power generation and auxiliary power units as shown in **Figure 5**. Polymer electrolyte fuel cell is a low temperature operation fuel cell (below 90- up to 180°C). The low temperature gives a faster start up but this mean a high purity of

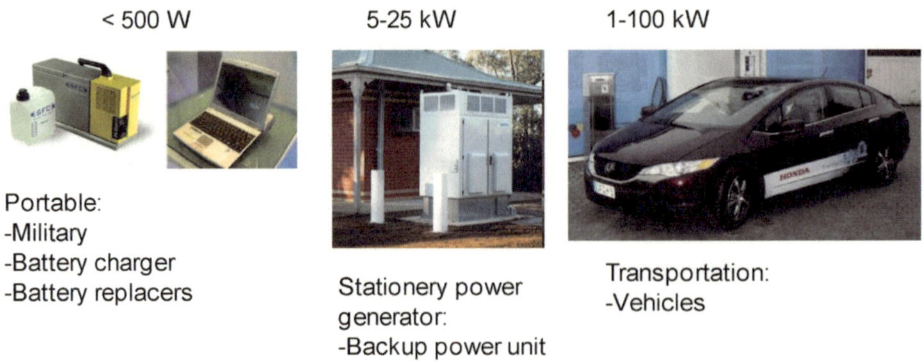

Figure 6.
PEMFC applications.

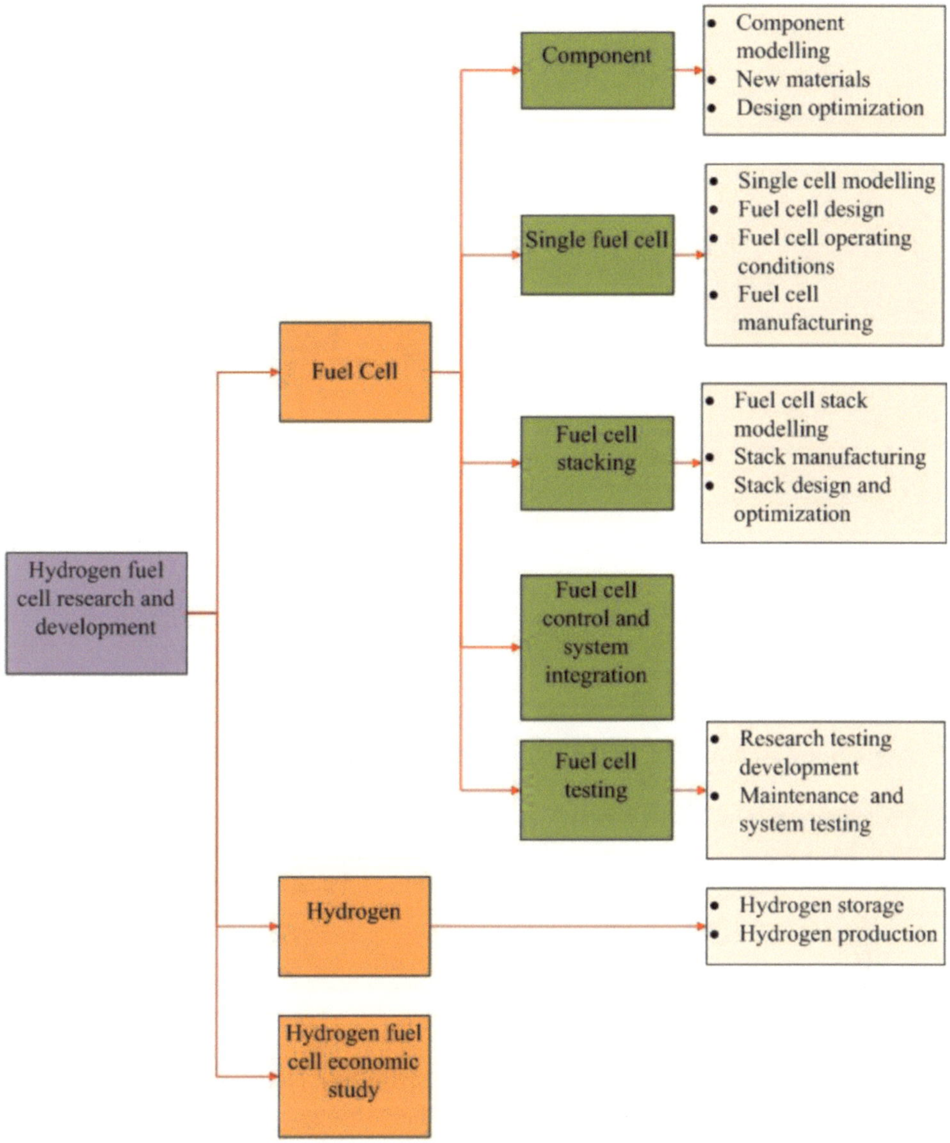

Figure 7.
Hydrogen fuel cell technology developments.

hydrogen must be supplied. The main applications for such fuel cell a portable power generation and fuel cell vehicle are shown in **Figure 6**.

Fuel cells have the advantages of zero end use emissions, no moving parts, quite operation and high efficiency compared to the combustion engine. In addition, different types of fuel cells can provide power generation within various operating conditions that can meet the needs of various applications. However, there are challenges facing fuel cells application (**Figure 7**); these challenges include: hydrogen fuel availability, fuel cell reliability, cost of fuel cell and hydrogen, fuel cell operation durability and wide public acceptance for hydrogen technology [45, 46].

Hydrogen economy has many benefits but it still facing many obstacles and challenges that limit hydrogen of being independent economy. **Figure 8** shows the key limitations.

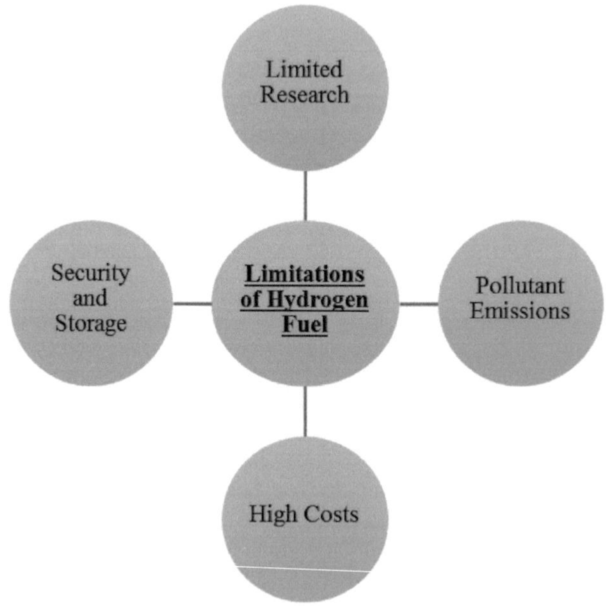

Figure 8.
Limitation of hydrogen fuel cell technology.

7. Steps needed for fuel cell commercialization

To achieve commercialization prospects (**Figure 9**), several technical-technological barriers must be identified. The most important barriers to be addressed are: cost of technology, fuel cell system integration and customer requirements [46].

8. Conclusions

The chapter addressed the main hydrogen production methods and introduced various future green hydrogen routes and opportunities. Hydrogen production for fuel cell technologies requires an improvement regarding sustainability of the hydrogen supply and an improvement regarding decentralized hydrogen production. Electrolysis for on-demand applications can be a potential route when it is coupled with green power supply. Hydrogen can be used for load balancing if stored properly, especially when integrated with renewable energy system. Moreover, hydrogen economy as far requires a large scale and long term storage solution to meet the increasing demand. Hydrogen storage is facing several challenges, such as high pressure storage increase with the size of container; cryogenic storage has a problem with longer and large quantity storage. Solid state storage requires a scale up in case of being used for stationary purposes improve the thermodynamic properties of the materials for storage/release at moderate temperature and pressure.

DOI: http://dx.doi.org/10.5772/intechopen.91676

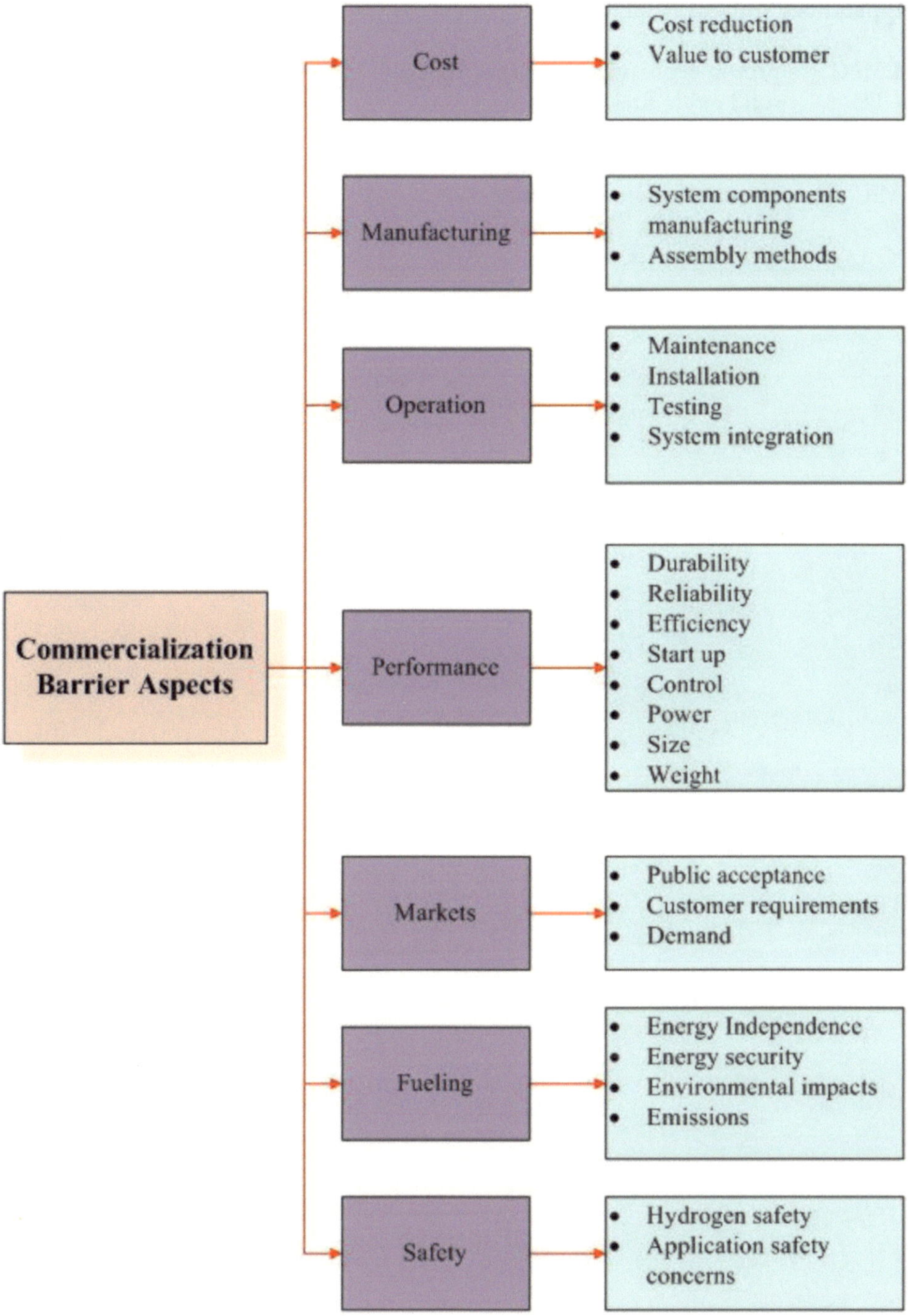

Figure 9.
Commercialization aspects of hydrogen fuel cell technology.

Appendices and nomenclature

PEMFC	proton exchange membrane fuel cell
SOFC	solid oxide fuel cell
DMFC	direct methanol fuel cell
MCFC	molten carbonate fuel cell
SMR	steam methane reforming
PV	photovoltaic
DC	direct current

Author details

Martin Khzouz[1*] and Evangelos I. Gkanas[2]

1 Department of Systems Engineering, Military Technological College, Muscat, Oman

2 Hydrogen for Mobility Lab, School of Mechanical, Automotive and Aerospace Engineering, Coventry University, Coventry, UK

*Address all correspondence to: marcin.khzouz@gmail.com

References

[1] Balat M. Hydrogen in fueled systems and the significance of hydrogen in vehicular transportation. Energy Sources, Part B: Economics, Planning, and Policy. 2007;**2**:49-61

[2] Staffell I, Scamman D, Velazquez Abad A, Balcombe P, Dodds PE, Ekins P, et al. The role of hydrogen and fuel cells in the global energy system. Energy & Environmental Science. 2019;**12**:463-491

[3] Available from: http://www.airproducts.com/Products/Gases/Hydrogen.aspx

[4] Lemus RG, Martínez Duart JM. Updated hydrogen production costs and parities for conventional and renewable technologies. International Journal of Hydrogen Energy. 2010;**35**:3929-3936

[5] Balat M. Potential importance of hydrogen as a future solution to environmental and transportation problems. International Journal of Hydrogen Energy. 2008;**33**:4013-4029

[6] Ewan BCR, Allen RWK. A figure of merit assessment of the routes to hydrogen. International Journal of Hydrogen Energy. 2005;**30**:809-819

[7] Hydrogen Production Energy Conversion Efficiencies. 2020 [updated 2020; cited 8 May 2020]. Available from: https://h2tools.org/hyarc/hydrogen-data/hydrogen-production-energy-conversion-efficiencies

[8] Xuan J, Leung MKH, Leung DYC, Ni M. A review of biomass-derived fuel processors for fuel cell systems. Renewable and Sustainable Energy Reviews. 2009;**13**:1301-1313

[9] Demirbas A, Arin G. Hydrogen from biomass via pyrolysis: Relationships between yield of hydrogen and temperature. Energy Sources. 2004;**26**:1061-1069

[10] Verfondern K. Nuclear energy for hydrogen production. Energy Technology. Forschungszentrum Jülich; 2007;**58**

[11] El-Emam RS, Khamis I. Advances in nuclear hydrogen production: Results from an IAEA international collaborative research project. International Journal of Hydrogen Energy. 2019;**44**:19080-19088

[12] Bhandari R, Trudewind CA, Zapp P. Life cycle assessment of hydrogen production via electrolysis – A review. Journal of Cleaner Production. 2014;**85**:151-163

[13] Shakya BD, Aye L, Musgrave P. Technical feasibility and financial analysis of hybrid wind–photovoltaic system with hydrogen storage for Cooma. International Journal of Hydrogen Energy. 2005;**30**:9-20

[14] Energy USDO. Available from: https://afdc.energy.gov/fuels/hydrogen_production.html

[15] Brown F. A comparative study of fuels for on-board hydrogen production for fuel-cell-powered automobiles. International Journal of Hydrogen Energy. 2001;**26**:381-397

[16] Milner-Elkharouf L, Khzouz M, Steinberger-Wilckens R. Catalyst development for Indirect Internal Reforming (IIR) of methane by partial oxidation. International Journal of Hydrogen Energy. 2020;**45**(8):5285-5296

[17] Twigg MV. Catalyst Handbook. London: Wolfe Publishing Ltd; 1989

[18] Broda M, Manovic V, Imtiaz Q, Kierzkowska AM, Anthony EJ, Müller CR. High-purity hydrogen via the sorption-enhanced steam methane reforming reaction over a synthetic CaO-based sorbent and a Ni catalyst.

Environmental Science & Technology. 2013;**47**:6007-6014

[19] Ramsden T, Ruth M, Diakov V, Laffen M, Timbario TA. Hydrogen pathways: Updated cost, well-to-wheels energy use, and emissions for the current technology status of ten hydrogen production, delivery, and distribution scenarios: National Renewable Energy Laboratory (NREL). Golden, CO (United States); 2013

[20] Khzouz M. The Development and Characterization of Ni-Cu/Al_2O_3 Catalyst for Hydrogen Production Via Multi-fuel Reforming. Birmingham, UK: University of Birmingham; 2014

[21] Arzamendi G, Diéguez PM, Montes M, Centeno MA, Odriozola JA, Gandía LM. Integration of methanol steam reforming and combustion in a microchannel reactor for H2 production: A CFD simulation study. Catalysis Today. 2009;**143**:25-31

[22] Palo DR, Dagle RA, Holladay JD. Methanol steam reforming for hydrogen production. Chemical Reviews. 2007; **107**:3992-4021

[23] Khzouz M, Gkanas EI, Du S, Wood J. Catalytic performance of Ni-Cu/Al_2O_3 for effective syngas production by methanol steam reforming. Fuel. 2018;**232**:672-683

[24] de Fátima Palhares DDA, Vieira LGM, Damasceno JJR. Hydrogen production by a low-cost electrolyzer developed through the combination of alkaline water electrolysis and solar energy use. International Journal of Hydrogen Energy. 2018;**43**: 4265-4275

[25] Fe C, Kaddami M, Mikou M. Effect of operating parameters on hydrogen production by electrolysis of water. International Journal of Hydrogen Energy. 2017;**42**:25550-25557

[26] Ursua A, Gandia LM, Sanchis P. Hydrogen production from water electrolysis: Current status and future trends. Proceedings of the IEEE. 2012; **100**:410-426

[27] Li G, Zhang K, Yang B, Liu F, Weng Y, Liu Z, et al. Life cycle analysis of a coal to hydrogen process based on ash agglomerating fluidized bed gasification. Energy. 2019;**174**:638-646

[28] Al-Zareer M, Dincer I, Rosen MA. Analysis and assessment of a hydrogen production plant consisting of coal gasification, thermochemical water decomposition and hydrogen compression systems. Energy Conversion and Management. 2018;**157**: 600-618

[29] Kabir E, Kumar P, Kumar S, Adelodun AA, Kim K-H. Solar energy: Potential and future prospects. Renewable and Sustainable Energy Reviews. 2018;**82**:894-900

[30] Dincer I, Joshi A. Solar Based Hydrogen Production Systems. New York: Springer-Verlag; 2013

[31] Hollmuller P, Joubert J-M, Lachal B, Yvon K. Evaluation of a 5 kWp photovoltaic hydrogen production and storage installation for a residential home in Switzerland. International Journal of Hydrogen Energy. 2000;**25**: 97-109

[32] Jia J, Seitz LC, Benck JD, Huo Y, Chen Y, Ng JWD, et al. Solar water splitting by photovoltaic-electrolysis with a solar-to-hydrogen efficiency over 30%. Nature Communications. 2016;**7**: 13237

[33] Wang H, Hao Y. Thermodynamic study of solar thermochemical methane steam reforming with alternating H_2 and CO_2 permeation membranes reactors. Energy Procedia. 2017;**105**: 1980-1985

[34] Gkanas EI, Khzouz M. Study on the hydrogenation of an mm-based AB5-intermetallic for sustainable building applications. International Journal of Energy Research. 2019;**43**(14):8033-8048

[35] Rusman NAA, Dahari M. A review on the current progress of metal hydrides material for solid-state hydrogen storage applications. International Journal of Hydrogen Energy. 2016;**41**:12108-12126

[36] Partnership USD. Hydrogen Storage Tech Team Roadmap. Available from: https://www.energy.gov/sites/prod/files/2017/08/f36/hstt_roadmap_July2017.pdf2017

[37] Barthélémy H. Hydrogen storage – Industrial prospectives. International Journal of Hydrogen Energy. 2012;**37**: 17364-17372

[38] Barthelemy H, Weber M, Barbier F. Hydrogen storage: Recent improvements and industrial perspectives. International Journal of Hydrogen Energy. 2017;**42**:7254-7262

[39] Markets T. Available from: https://www.compositesworld.com/articles/the-markets-pressure-vessels

[40] Mital SK, Gyekenyesi JZ, Arnold SM, Sullivan R, Manderscheid JM, Murthy PLN. Review of current state of the art and key design issues with potential solutions for liquid hydrogen cryogenic storage tank structures for aircraft applications. Cleveland, OH, USA: National Aeronautics and Space Administration; 2006

[41] Di Profio P, Arca S, Rossi F, Filipponi M. Comparison of hydrogen hydrates with existing hydrogen storage technologies: Energetic and economic evaluations. International Journal of Hydrogen Energy. 2009;**34**:9173-9180

[42] Gkanas E, Khzouz M. Metal Hydride Hydrogen Compression Systems – Materials, Applications and Numerical Analysis. John Wiley & Sons; 2018

[43] Khzouz M, Gkanas EI, Girella A, Statheros T, Milanese C. Sustainable hydrogen production via LiH hydrolysis for unmanned air vehicle (UAV) applications. International Journal of Hydrogen Energy. 2020;**45**(8):5384-5394

[44] Veziroğlu TN, Şahin S. 21st Century's energy: Hydrogen energy system. Energy Conversion and Management. 2008;**49**:1820-1831

[45] Wang J, Wang H, Fan Y. Techno-economic challenges of fuel cell commercialization. Engineering. 2018;**4**: 352-360

[46] Eichman J, Brouwer J, Samuelsen S. Exploration and prioritization of fuel cell commercialization barriers for use in the development of a fuel cell roadmap for California. Journal of Fuel Cell Science and Technology. 2010;**7**(5): 051017-1/12. DOI: 10.1115/1.4000689

Section 4

Solar Energy

Chapter 14

The PV/Wind System for Sustainable Development and Power Generation with Real Dynamic Input Datasets in the Distribution Power Systems

Emel Bakmaz, Kemal Aygul, Burak Esenboga, Tugce Demirdelen and Mehmet Tumay

Abstract

Rapid population growth and industrialization in developing countries cause an increase in demand for energy. In order to meet this energy demand, two types of resources are used: renewable energy and nonrenewable energy. Nonrenewable sources, also called fossil fuels, cause environmental problems in serious and dangerous dimensions. For this reason, it is a necessity to find alternatives. It is a renewable energy source that can be used as an alternative to fossil fuels. This chapter deals with power control of a PV/wind system for power generation with dynamic input dataset. The main contribution of this chapter is that it is the first time to use real data from PV/wind system and observe the system reliability with real-time simulation results. The proposed system consists of doubly fed induction-based wind generator, rotor-side converter (RSC), grid-side converter (GSC), solar arrays, DC-DC converter and grid-side converter, and grid and dynamic loads. The aim of the proposed strategy is to use wind and solar energies with maximum efficiency by simulating the real condition of wind and insolation with input datasets. The modeling and the validation of the operation of the system and its controllers are done by using PSCAD/EMTDC.

Keywords: energy demand, renewable energy, PV/wind system, power generation, dynamic input, residential neighborhood

1. Introduction

Energy demand increases with industrialization, population growth, and technological development day by day, so the concept of energy has gained much attention, and sustainability of energy resources is one of the most significant issues and problems in the world today. The rapid depletion of energy resources, the unconscious use of resources such as petroleum, coal, and nuclear energy that are not self-renewing, and the pollution from these resources in the environment and

IntechOpen

in the atmosphere have led people to use renewable energy resources. To meet the world's ever-increasing energy needs, scientists have done a lot of work to popularize the use of renewable energy sources in all areas and have developed projects like solar batteries to use the sun and projects to use windmills that make use of wind power. In this chapter, real data is obtained from solar panels and wind turbines in Adana, located in the Mediterranean region, and then used in simulation modeling and analysis. Solar and wind energy potential maps are shown for Adana in **Figure 1**.

Looking at recent studies in the literature, it seems that studies on the subject have increased in recent years. González, Riba, and Rius [1] presented a methodology for optimization of life cycle cost of hybrid renewable energy systems (HRES) based on wind, solar photovoltaic (PV), and biomass power. It is stated that HRES are alternative to enhance renewable energy production worldwide. Jain, Karampuri, and Somasekhar [2] proposed an induction motor pumping system which is fed from solar PV. They integrated the variable frequency control, sample-averaged zero-sequence elimination, and the maximum power point tracking (MPPT) algorithms in order to keep the system stable under various environmental situations. Long, Eghlimi, and Zhang [3] offered an optimization model to optimize the hourly load dispatch and the structure of a PV/wind hybrid system. They have stated that the stable operation of the hybrid power system is a more dependent wind power system than PV power system. Patra, Ray, and Mohanty [4] examined a stand-alone wind-diesel-PV hybrid system and found that the main issues are in compensating the reactive power and regulating the transient as a stable. Authors showed by simulations that transient voltage balance case in UPFC is more functional than that of STATCOM and SVC and the efficiency of UPFC with fuzzy sliding mode is enhanced compared to PI and fuzzy PI controller. Parida and

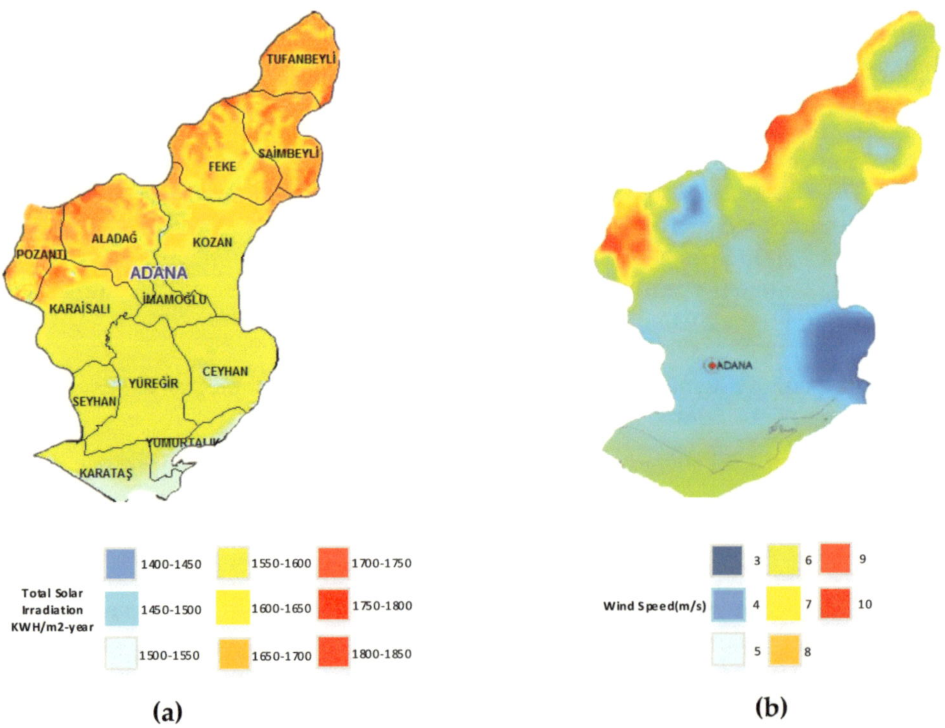

Figure 1.
Adana energy potential map. (a) Adana solar energy potential map and (b) wind energy potential map.

Chatterjee [5] proposed a strategy for controlling rotor current of doubly fed induction generator (DFIG) in wind-solar hybrid system to improve generation capacity of WECS in the system. Al-Masri and Ehsani [6] performed a study about economic aspects of connecting a wind-PV hybrid system into the Jordanian power system. Shen C. and Shen Y. [7] proposed an output filter integrated to a novel dual-input photovoltaic-wind converter (DIPWC), and operation principles are summarized at this study. Solar photovoltaic power and wind turbine energy are operated at the same time and individually. 1-kW prototype power converter is built and measured to verify the correctness of the DIPWC. Also, it is mentioned that key part of the DIPWC is the inductor at the output. Cho, Chun, and Hong [8] proposed an algorithm for size optimization of a hybrid system that consists of battery, solar PV, wind generator, and diesel generator. For the performance evaluation of the proposed method, MATLAB is used. Kalla et al. [9] proposed adaptive sliding mode control algorithm for a micro-grid (MG) consisting of hydropower and wind and solar PV generation systems in order to control micro-grid voltage and frequency and also improve power quality of micro-grid.

Palmer et al. [10] studied the impact of solar radiation on the geographical distribution of solar installations. They observed that variations on irradiance may be reduced by the aggregate of the installations in the neighborhood. Ekstrom et al. [11] presented a methodology for evaluating the volatility of a system which contains both PV and wind energy generation systems. A year scale scheduling method based on interval optimization for hydro-PV-wind hybrid system is proposed by Liu, Tan, and Jiang [12]. The method is implemented on a real case study in Yunnan. Luna et al. [13] modeled and designed an energy management system and integrated it to a micro-grid. Shanthi, Uma, and Keerthana [14] proposed a power transfer scheme by minimizing the number of converters used for PV/wind system connected to the grid. Authors stated that the proposed system is capable of simplifying the process of integration of DC systems into AC system. In [15] a micro-grid which consists of renewable sources, battery, and energy management system for mentioned elements is proposed. In [16] wind-PV hybrid system which is capable to operate in on-grid and off-grid modes and developed a power management strategy for the mentioned system is proposed. Askarzadeh [17] proposed a system that integrates solar PV and wind and tidal energies with the battery storage system and proposed crow search algorithm (CSA) for the optimization of the proposed system. Kant, Jain, and Singh [18] integrated diesel-wind-PV sources with BESS as a stand-alone micro-grid. Rezkallah et al. [19] conducted a study to deal with the design process and application of active power control (APC) to reduce the number of sensors at the wind-PV-battery energy combination hybrid autonomous power generation system. Modeling, stability analysis, control design, and parameters of this hybrid combination of power generation system are presented.

Tiwari, Singh, and Goel [20] conducted a study about autonomous wind-solar hybrid energy system. Theoretical and experimental validation of autonomous wind-solar hybrid system consisting of a wind generator with doubly fed induction generator and solar PV array is also done. The new system is modeled in MATLAB and under negative conditions such as varying load, unstable wind speeds, and solar irradiation, and stator windings of DFIG are balanced with low total harmonic distortion (THD). The prototype of the system is developed using 3.7 kW DFIG and 5 kW solar array simulator to validate the simulation results. A study by Einan, Torkaman, and Pourgholi [21] conducts a new methodology or technique for isolated micro-grid including battery energy storage systems. To confirm the performance, fuzzy controller and cuckoo optimization algorithm are checked against particle swarm optimization (PSO) algorithm and genetic algorithm. The optimized fuzzy-cuckoo controller relatively differs from the other methods, and results showed that the fuzzy-cuckoo controller has better performance than other methods.

Ji et al. [22] presented the benefits of the wind and PV combined energy storage and transmission dual power generation system. To analyze the benefits of the wind-PV combined system, a multi-angle evaluation index system of the wind/PV/ energy storage is designed. Mendoza, Sumper, and Arellano [23] conducted a case study for reducing the fossil fuel consumption by generating electricity from renewable energy sources. Two systems are selected from 12 system designs. The technique used in this study can be implemented to the SIDS or other small islands in order to plan island electricity systems that will reach low emission targets in their electricity generation.

The effect of PV and variable speed wind power plants on the frequency response of the US Eastern Interconnection and Texas Interconnection is investigated by Liu, You, and Liu [24]. Controlling the frequency of the power system is also investigated. Validated system models are simulated and compared with realistic renewable information. In addition, inertia and governor control are introduced. Results show that the governor and inertia control reduced the effect of the increasing renewable penetration levels on the frequency response of the US Eastern Interconnection and Texas Interconnection and, also, show that optimal solution has higher reliability and fewer greenhouse gas emission.

The study by Hussain et al. [25] presents the optimization method named iterative filter selection approach to design the hybrid renewable energy system with maximum reliability and minimum project cost. The system includes battery storage, wind turbine, photovoltaic arrays, AC load, and dump load. As iterative filter selection approach is performed, minimization of surplus power is considered in the design process. Moreover, the proposed method is compared with some other techniques in the literature. Results show that duration analysis or computational time shortens and dump load size tolerance is shorted out for better results. Akram, Khalid, and Shafiq [26] conducted capacity optimization and development of the hybrid energy storage system resources such as wind-solar and battery-supercapacitor in a grid-connected MG. Optimization problems are formulated and solved to decrease computational time and complexity of the system. Optimization method is based on a few factors about the MG system. This method is tested using real power demand, solar irradiation, and wind speed data.

Akram, Khalid, and Shafiq [27] presented two iterative algorithms to determine appropriate sizes of renewable energy resources and energy storage for economic, reliable, and efficient operation. These algorithms named as source sizing algorithms and battery sizing algorithms are used to avoid over- and undersizing. Also, forced outage rates of wind turbine and PV and the utilization factor of battery energy storage system are considered, and the effectiveness of the proposed approach is depicted thanks to MATLAB simulation results. Gonzales et al. [28] presented a method for analyzing the stability of weak power system. It is emphasized that the main effects on weak power systems are decreasing voltage regulation and decreasing primary frequency regulation. PV power plants and wind farms proposed and validated using the real system in Bolivia due to this reason. Khaled, Eltamaly, and Beroual [29] implemented PSO to ensure the optimal power flow. Authors also introduced a study about optimal power flow of a power system integrated with distributed wind and PV as renewable distributed generation elements. The purpose of optimal power flow researches is to optimize certain objectives by adjusting power system variables. Modified PSO is used in simulation to achieve the global minimum more accurately and faster than other methods. Results show that the cost of generation and operation can be reduced by using renewable distributed generation.

This chapter presents the power control of a PV/wind system for power generation with dynamic input dataset. One of the most significant and main advantages of this chapter is that it is the first time to use real data from PV/wind system and observe the system reliability with real-time simulation results. The proposed system consists of doubly fed induction-based wind generator, rotor-side converter, grid-side converter, solar arrays, DC-DC converter, and grid and dynamic loads. The aim of the proposed strategy is to use wind and solar energy with maximum efficiency by simulating the real condition of wind and insolation with input datasets. The modeling and the validation of the operation of the system and its controllers are done by using PSCAD/EMTDC. To confirm the control, design, and the operation of the proposed system, the detailed review is done, and the simulation results are shown and indicated.

Due to the restrictions among existing literature, the aim of this chapter is as below:

1. To deal with power control of a PV/wind system for power generation

2. To use real data from PV/wind system and observe the system reliability with real-time simulation results

3. To acquire simulation results and theoretical analysis from a real industrial network model in PSCAD

4. To consist of doubly fed induction-based wind generator, RSC, GSC, solar arrays, DC-DC converter grid, and dynamic loads

5. To use wind and solar energies with maximum efficiency by simulating the real condition of wind and insolation with input datasets

6. To confirm the control system and the operation of the proposed system

This chapter primarily demonstrates the aforementioned six objectives of the proposed system.

2. Design of the proposed system

The proposed system consists of two parts: PV and wind system. These systems are examined in detail.

2.1 Solar photovoltaic system

2.1.1 Solar photovoltaic panel

The solar energy is converted into DC electrical power by a photovoltaic (PV) system. The parameters of an equivalent circuit of a PV are the open-circuit voltage (V_{oc}), the short-circuit current (I_{sc}), the maximum voltage (V_{max}), and the maximum power point (MPP). The I–V stands for Current-Voltage and P-V curves are shaped by these parameters. The PV cell represented by a current source is connected parallel with a diode. The current, I_g, varies with the solar radiation level that the surface is exposed to. The current, I_d, which flows through the diode, is the main reason for the nonlinear I–V characteristic of the PV cell (**Figure 2**).

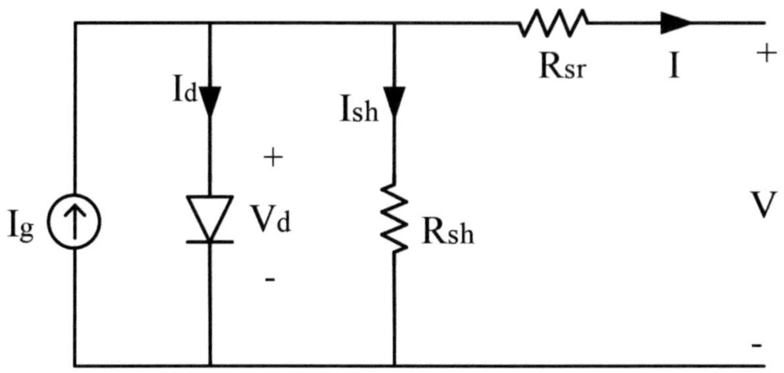

Figure 2.
Equivalent circuit of solar PV cell.

When we apply Kirchhoff's current law to the circuit,

$$I = I_g - I_d - I_{sh} \tag{1}$$

The substitution of the related expressions for the shunt branch current I_{sh} and diode current I_d in Eq. (1) yields the following:

$$I = I_g - I_0 \left[\exp\left(\frac{V + IR_{sr}}{\frac{nkT_c}{q}} \right) - 1 \right] - \left(\frac{V + IR_{sr}}{R_{sh}} \right) \tag{2}$$

I_g stands for the photocurrent, and it is the function of solar radiation on the cell and cell temperature:

$$I_g = I_{scR} \frac{G}{G_R} [1 + \alpha_T (T_c - T_{cR})] \tag{3}$$

The current I_0 in Eq. (2) is called the saturation current, a function of cell temperature, and given by

$$I_0 = I_{0R} \left(\frac{T_c}{T_{cR}} \right) \exp\left(\frac{1}{T_{cR}} - \frac{1}{T_c} \right) \frac{qe_g}{n_k} \tag{4}$$

A group of cells can be connected together either in a series or parallel combination. By connecting solar cells, PV modules can be created. Similarly, by connecting PV modules together, a PV array can be created. The output power of such PV systems can be calculated by the following equation:

$$P_{pv\,(t)} = H_t(t) x PVA x \mu_c(t) \tag{5}$$

$\mu_c(t)$ in Eq. (6) is the hourly efficiency of the PV system which can be obtained in terms of the cell temperature:

$$\mu_c(t) = \mu_{cr}[1 - \beta_t x (T_c(t) - T_{cr})] \tag{6}$$

$$T_c(t) = T_a + \lambda H(t) \tag{7}$$

$$\text{PVA} = \frac{1}{8760} \sum_{t=1}^{8760} \frac{P_{L,av}(t) F_s}{H_t \eta_c(t) V_f} \tag{8}$$

A PV panel generally consists of tempered glass, ethylene-vinyl acetate (EVA) layers, solar cells, and back-sheet materials shown in **Figure 3**. Tempered glass has been treated by heat or chemicals to increase its strength. Glass can be double or single due to increasing strength, but triple glass is not recommended because of reducing efficiency and low thermal transmittance. EVA is the adhesive material that includes ethylene, vinyl, and acetate chemicals, and its layers provide excellent thermal protection for solar cells. EVA layers provide to not pass too much of the sunshine trying to reach the solar cells. Back-sheet contributes to mechanical balance, durability for harsh weather, and high-performance properties for the PV module.

Figure 3.
PV panel layout.

2.1.2 DC-DC boost converter

The input voltage of the step-up (boost) converter is always less than the output voltage. **Figure 4** shows the working principle of the DC-DC boost converter. **Figure 4(a)** shows the on state. This state is included to store power in the inductor, charge the inductor, which increases the current through the inductor.

Figure 4(b) shows the off state. This state is included to discharge power from the inductor into the load and the capacitor and decrease current through the inductor.

2.1.3 Three-phase inverter

Another component is the three-phase inverter shown in **Figure 5**. Inverters are generally used for high-power applications. A six-transistor and six-diode

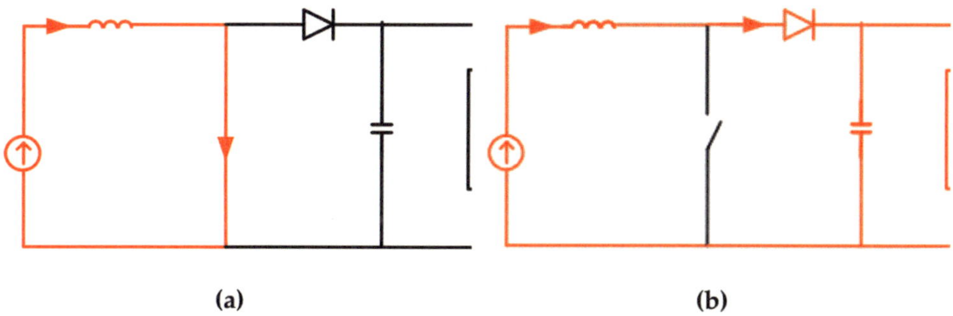

Figure 4.
(a) On-state step-up converter and (b) off-state step-up converter.

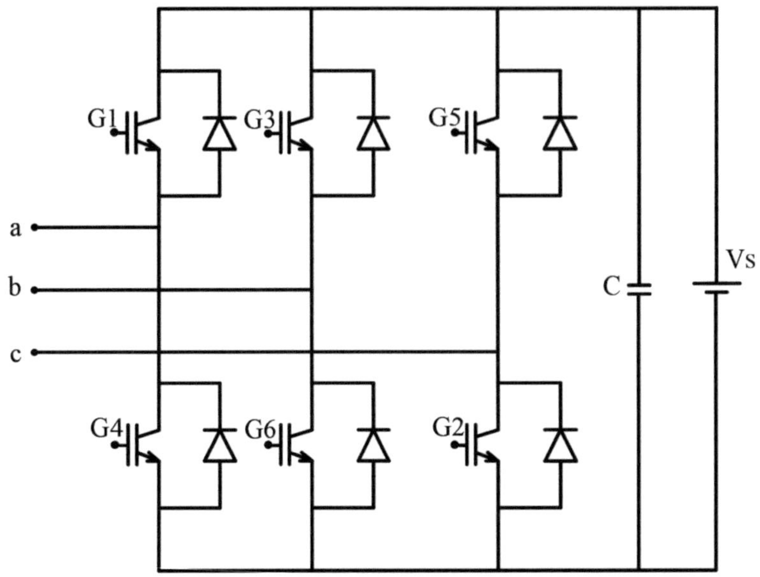

Figure 5.
Three-phase inverter.

Switch states for three-phase voltage source inverter					
State	**State no**	**Switch states**	**V_{ab}**	**V_{bc}**	**V_{ca}**
G5, G6, G1 on, others off	1	101	Vs	−Vs	0
G6, G1, G2 on, others off	2	100	Vs	0	−Vs
G1, G2, G3 on, others off	3	110	0	Vs	−Vs
G2, G3, G4 on, others off	4	010	−Vs	Vs	0
G3, G4, G5 on, others off	5	011	−Vs	0	Vs
G4, G5, G6 on, others off	6	001	0	−Vs	Vs
G1, G3, G5 on, others off	7	111	0	0	0
G2, G4, G6 on, others off	8	000	0	0	0

Table 1.
Switch states for three-phase voltage source inverter.

configuration is generally used for a three-phase output shown in the figure below. Two types of control signals can be applied to the transistors: 180° conduction and 120° conduction. The 180° conduction provides for better usage of the switches and is the preferred method. The switch states for a three-phase voltage source inverter (VSI) are shown in **Table 1**.

2.2 Wind system

2.2.1 Wind turbine

Wind turbine technology is developed day by day, and several models are designed to increase the power capability of the wind turbines. The mechanical power of the wind turbine is related to its power coefficient C_p and wind velocity v and represented by

$$P_{mech} = \frac{1}{2} C_p \rho \pi R^2 v^3 \tag{9}$$

The air density is represented by ρ, and the radius of the turbine propeller, given by Eq. (9), is represented by R. The power coefficient can be described as the ratio of the electric power produced by a wind turbine divided by the total amount of wind power flowing into the turbine blades at a certain wind speed. Every wind turbine has a unique power coefficient value. Power coefficient is broadly defined as a function of the tip/speed/ratio λ, and the equation is given by the value w corresponding to the rotational speed of the wind turbine:

$$\lambda = \frac{wR}{v} \tag{10}$$

When λ gets a specific value, it ensures a maximum of C_p. Turbine rotational speed value provides to capture the maximum mechanical power attainable from the wind for a specific wind velocity. Thus, turbine speed is followed precisely. When the power coefficient is assumed to be at the optimum for a particular wind turbine, the tip/speed/ratio values producing the maximum mechanical power of the wind turbine reach their optimum value. This operation is effective only if the nominal wind speed with the existing wind power does not exceed the rated capacity of the generator.

2.2.2 Doubly fed induction generator

The back-to-back converter is a bidirectional frequency converter. This means that it must manage power flow in both directions. The RSC and GSC and a common DC bus combine to form this back-to-back converter. Park's model is a special model frequently used for induction generators. The study analysis is easier when the rotor variables are referred to the stator side. A rotating reference frame is more preferable, but in this study, a static stator-oriented reference frame is more appropriate. Also, linear magnetic circuits are not negligible. Park's model can be expressed by the equations below:

$$v_s = R_s\, i_s + \frac{d}{dt}\phi_s \tag{11}$$

$$v_r = R_r i_r + \frac{d}{dt}\phi_r - jw\phi_r \tag{12}$$

For this model, v is the voltage, i is the current, R is the resistance, ϕ is the magnetic flux, and w is the rotor electrical speed. The subscripts s and r indicate stator and rotor quantities.

Stator voltage depends on the grid because of the connection of the stator to the grid. The machine is controlled by the rotor voltage v_r because rotor voltage is adjusted by the converter. The stator and rotor fluxes are given by

$$\phi_s = L_s i_s + L_m i_r \tag{13}$$

$$\phi_r = L_r i_r + L_m i_s \tag{14}$$

L_s and L_r are expressed as the stator and rotor inductance. L_m is the magnetizing inductance. Rotor voltage is the relatively important variable. This voltage can be calculated from Eqs. (15) and (16):

$$v_{r0} = \frac{L_m}{L_s}\left(\frac{d}{dt} - jw\right)\phi_s, \sigma = 1 - \frac{L_m^2}{L_s L_r} \tag{15}$$

where σ is the leakage factor and σL_r is the rotor transient inductance. Eqs. (12) and (15) yield

$$v_r = \frac{L_m}{L_s}\left(\frac{d}{d_t} - jw\right)\phi_s + \left(R_r + \sigma L_r\left(\frac{d}{dt} - jw\right)\right)^r i_r \tag{16}$$

v_r is the rotor voltage presented in Eq. (16). If the rotor is an open-circuit state, it means that i_r is equal to 0 and one term of the rotor voltage is v_{r0} given by Eq. (17).

$$v_{r0} = \frac{L_m}{L_s}\left(\frac{d}{dt} - jw\right)\phi_s \tag{17}$$

Another term of the rotor voltage (v_r) exists when the current (i_r) flows in the rotor. Voltage drop existing on rotor resistance (R_r) and the rotor transient inductance (σL_r) cause rotor voltage.

2.2.3 Rotor-side and grid-side converters

The main task of the RSC is to manage the reactive power and the active power of the generator. The fluctuations that affect stator reactive power and the electromagnetic torque are also removed by the RSC. The main task of the GSC is to maintain DC-link voltage stability against all possible directions and magnitudes of the rotor power. In addition, the GSC is responsible to ensure operation with zero reactive power, in other words, with a unity power factor. That is to say, the GSC does not exchange reactive power; it only exchanges active power with the grid. As a result, it can be said that the reactive power exchange is only carried out by the stator.

3. Control of the proposed system

The circuit topology and the controller of the proposed system can be seen in **Figure 6**. There are two main parts to the controller of the proposed system: the PV system and wind system controllers.

3.1 Maximum power point tracking (MPPT)

The terminal voltage of the photovoltaic system is generally adjusted according to the MPP voltage by an incremental conductance method. MPP is obtained from the instantaneous conductance of the photovoltaic module in comparison to the incremental conductance of the solar photovoltaic module. While instantaneous conductance is the current divided by the voltage, incremental conductance is the difference in current divided by the difference in voltage:

$$\frac{\Delta I}{\Delta V} = \frac{I}{V} \tag{18}$$

$$\frac{\Delta I}{\Delta V} > -\frac{I}{V} \tag{19}$$

$$\frac{\Delta I}{\Delta V} < -\frac{I}{V} \tag{20}$$

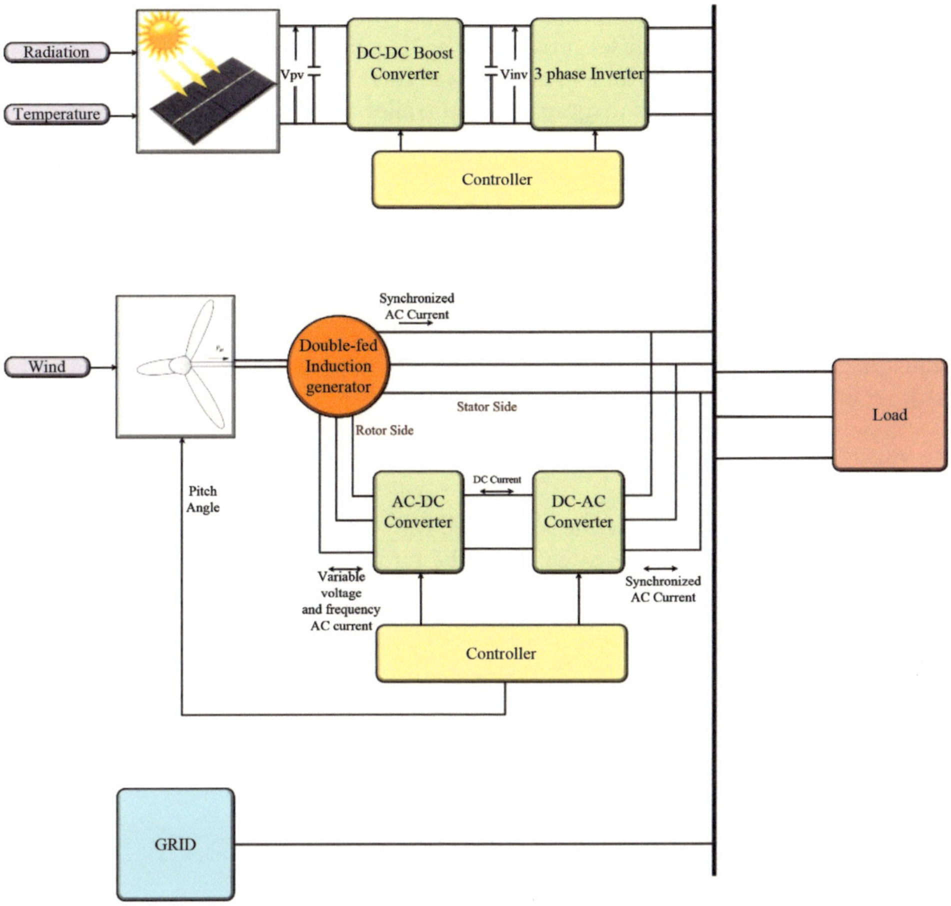

Figure 6.
Proposed system and controller.

Eq. (18) shows that the PV module is operating at the MPP. Equation (19) shows the left-hand region of the MPP on the P-V curve, and Eq. (20) shows the right-hand region of the MPP on the P-V curve. The slope of the P-V curve is equal to zero at MPP and is given by Eq. (21):

$$\frac{dP}{dV} = 0 \tag{21}$$

And also

$$\frac{dP}{dV} = I\left(\frac{dV}{dV}\right) + V\left(\frac{dI}{dV}\right) \tag{22}$$

$$\frac{dP}{dV} = I + V\left(\frac{dI}{dV}\right) \tag{23}$$

By using Eq. (21) and Eq. (23), Eq. (24) is derived and it has small permitted errors

$$\left|I + V\left(\frac{dI}{dV}\right)\right| = e \tag{24}$$

The incremental method is used to find the MPP of the PV module. It helps to reduce power loss and system cost and improve the PV efficiency. It has the great advantage of being a fast power tracking process. However, if the solar irradiation is low due to a small current differentiation, it will be unbalanced, but it has more accuracy, higher efficiency, and a faster power tracking process than other methods. The detailed block diagram and controller are shown in **Figures** 7 and **8**.

3.2 DC-DC boost converter controller

There are different types of controllers for different industries. Proportional-integral (PI) controllers are appropriate for various application areas. By tuning the parameters of these controllers, the desired output can be produced. This is a primary necessity in the implementation of these controllers. For this reason, the adjustment of the parameters of the controller has to be done with high speed and high accuracy shown in **Figure 9**.

3.3 Three-phase inverter controller

The difference between the measured DC-link value of the inverter and the reference DC-link voltage was calculated. The error signal obtained by the conventional PI controller was used for the comparison of reference PI signals with carrier signals to generate the gate signals shown in **Figure 10**.

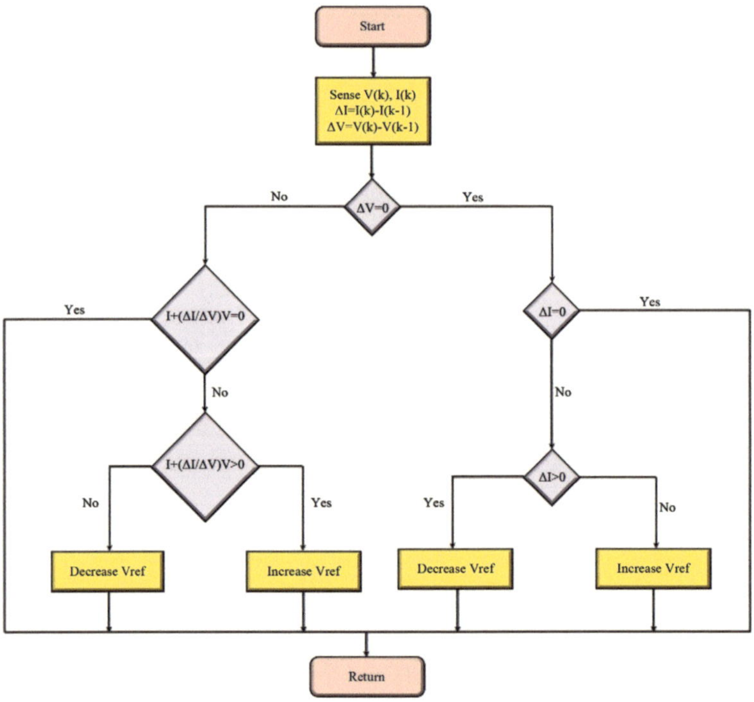

Figure 7.
Incremental method block diagram.

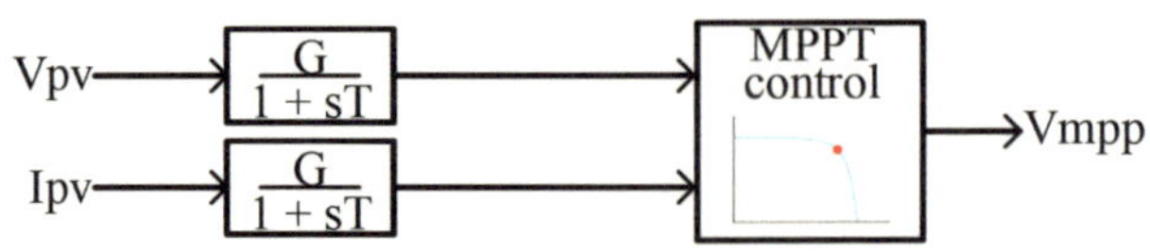

Figure 8.
MPPT controller.

3.4 Wind system controller

DFIG is named as "doubly fed" because two electrical parts of the machine called as stator and rotor are connected to electrical sources, as shown in **Figure 11**. There is a direct connection between the electric grid and the stator circuit; however, rotor windings are linked to an AC/DC/AC converter over slip rings. The converter consists of two types of converters: the RSC (C_{rotor}) and the GSC (C_{grid}). In **Figure 11**, the capacitor performs as the DC voltage source. A coupling inductor, L, exists between the GSC and the grid used to connect to C_{grid}. The induction generator converts the rotational power of the wind turbine into electrical power and this generated power is transferred to the electrical grid. In addition, pitch angle commands are used to regulate the generated power of the wind turbine, the DC busbar voltage, and also the reactive power—or the voltage at the grid terminals. Pitch angle control is also enabled to limit the maximum output power to protect the generator against a blast of wind.

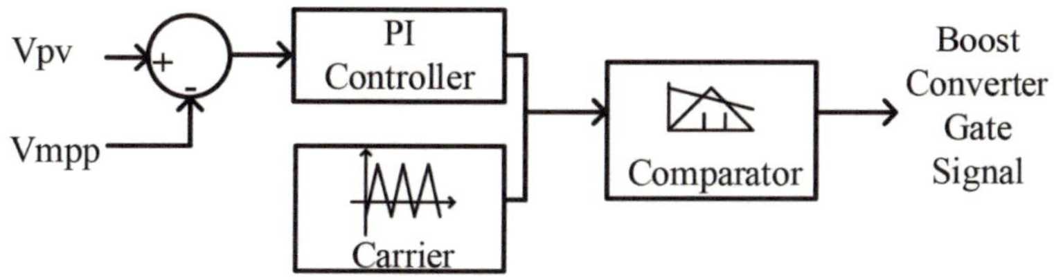

Figure 9.
DC-DC boost converter controller.

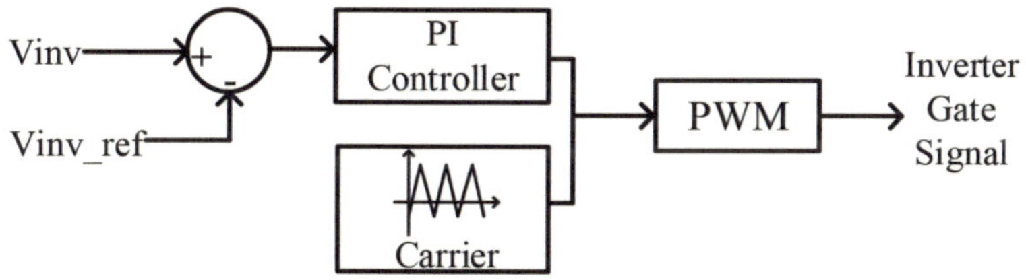

Figure 10.
Three-phase inverter controller.

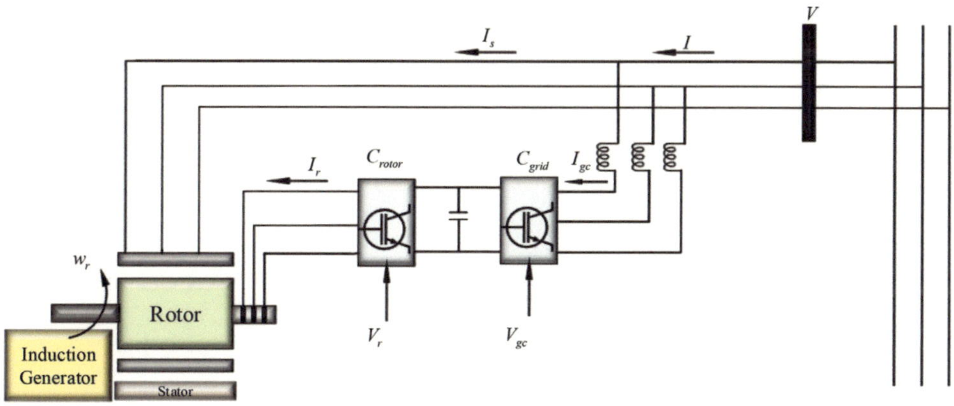

Figure 11.
DFIG-based wind system.

The torque and speed of the DFIG shown in **Figure 11** are controlled by the RSC. The RSC primarily aims to control the rotor currents and thus deliver the desired torque at the shaft of the machine. The RSC regulates the output power of the wind turbine thanks to the torque controller. The real generated power appearing at the grid side of the wind turbine is added to both the mechanical and electrical power losses and is compared to the reference power. The rotor speed error is reduced by using a proportional-integral controller. The reference rotor current added in the rotor windings by the RSC controls the electromagnetic torque T_e. The actual I_{qr} component of the positive sequence current is compared to I_{qr_ref}, and the error is minimized to zero by using a proportional-integral controller. The output of the controller for regulating the current is the voltage V_{qr} generated by C_{rotor} (**Figure 12**).

The GSC provides the flowing control of the real and reactive power from the turbine system to the grid, as shown in **Figure 13**. DC bus capacitor voltage is regulated by the grid-side converter. The control system consists of two control loops. A DC voltage controller forms the outer regulation loop. The current controller uses the output of the DC voltage controller as the reference current. A current controller and feed-forward terms which predict C_{grid}—the output voltage—form the inner regulation loop. This loop regulates the magnitude and phase of the GSC-generated voltage (V_{gc}) from the I_{dgc_ref} generated by the DC voltage controller and indicated by I_{q_ref} reference.

Pitch angle control is necessary to protect against network faults shown in **Figure 14**. When the network experiences a fault, the generator continues to generate power; therefore the DC-link capacitance will be overcharged. This means

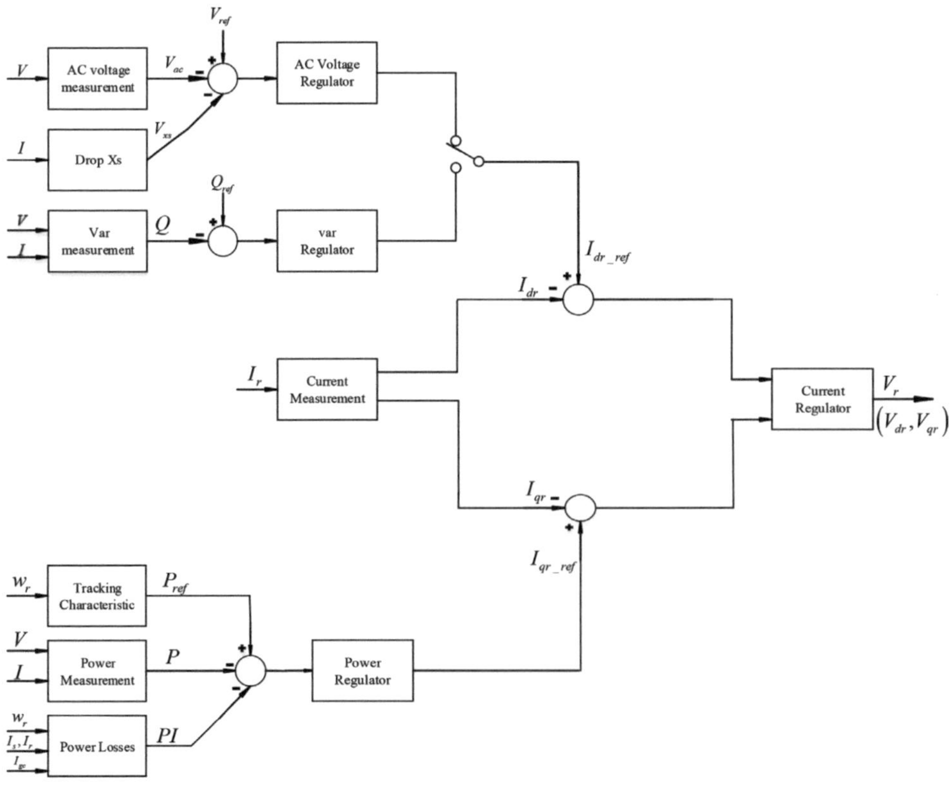

Figure 12.
RSC controller.

that there is no control over the blade speed and the torque. Pitch angle control is often preferred to braking mechanisms.

In addition to these measures, the reactive current flowing in the RSC controls the reactive power at the grid terminals. The wind turbine has two regulation

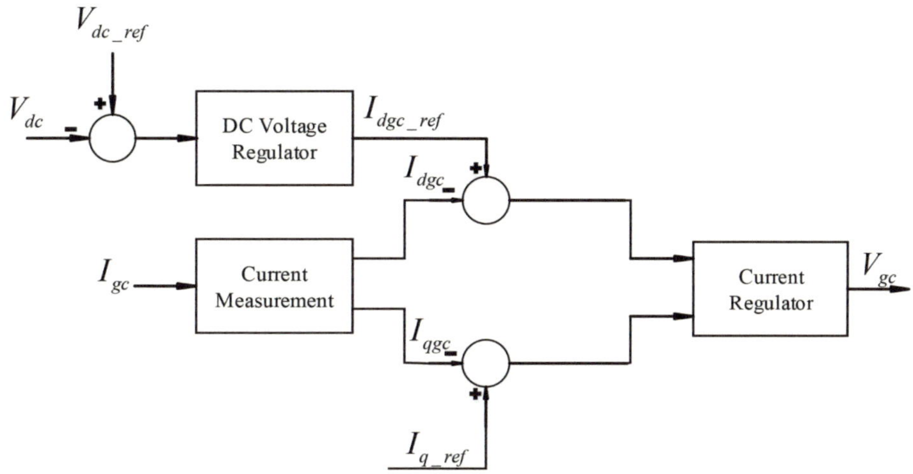

Figure 13.
GSC controller.

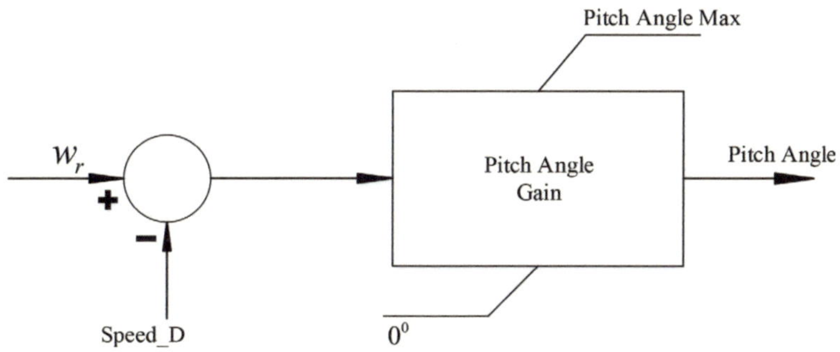

Figure 14.
Pitch angle controller.

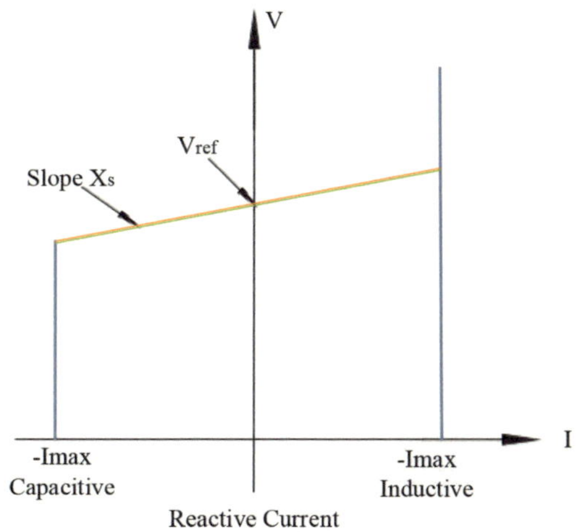

Figure 15.
I-V characteristic curve.

modes. The first of these is the voltage regulation mode; it fulfills the following I-V characteristic shown in **Figure 15**.

When the reactive current reaches the maximum current values, the voltage is optimized by the reference voltage V_{ref}, and the I-V characteristic has a specified slope shown in **Figure 15**. Another mode is the VAR regulation mode; the VAR controller provides constant reactive power at the grid terminals.

As a result, doubly fed induction generators are components that generate electricity in wind turbines. While the power generated by the wind turbine is converted into electrical power, they adjust the amplitude and frequency of their output voltages to maintain them at a constant value and also control the power factor thanks to their control mechanism and structure.

4. Results and discussion

The combined PV/wind system with real dynamic input datasets was validated by a simulation program called PSCAD/EMTDC. The simulation aims to evaluate the accuracy and efficacy of the control strategy used for implementation in power distribution systems. The related parameters of the proposed system are shown in **Table 2**. The nominal frequency of the power grid is 50 Hz for this simulation.

In this chapter, the configuration includes a variable speed-controlled wind turbine and a DFIG. This topology is used for tracking the efficient use of wind

Parameters	Value
480 kW PV array	
Series-connected PV modules	20
Parallel-connected PV modules	16
Series-connected PV cells	108
Parallel-connected PV cell strings	12
Reference solar radiation	1000 W/m^2
Reference PV panel temperature	25°C
PV cell	
Effective area	0.01 m^2
Series resistance	0.02 ohm
Shunt resistance	1000 ohm
Ideality factor of the diode	1.5
Bandgap energy	1.103 eV
Saturation current	10–9 A
Short circuit current	2.5 A
Wind energy conversion system	
Cut-in speed	4 m/s
Rated speed	10.5 m/s
Cut-out speed	25 m/s
Rated power	5.1 MVA

Table 2.
Parameters of proposed system.

power. The power coefficient, *Cp*, which is a function of wind speed/machine speed is selected to be 1.2 p.u. (per unit). For this reason, the variations in wind speed require variation in machine speed in order to keep the system operating at maximum *Cp*. In the *Cp* implementation used here, the nominal power of the turbine is obtained when rotating at 1.2 p.u. mechanical speed. As described earlier, RSC and GSC were used in this topology. The GSC is in charge of controlling the DC-link voltage, and the reactive power control is normally set to inject zero Q (reactive power of the system) for this converter.

As seen from **Figure 16**, the cut-in speed is 4 m/s. The turbine has an oscillation in its power rate and voltage level after start-up. It can be because of the synchronization point with the grid. A start-up PI controller is used to set up the voltage on the stator terminals before synchronizing. The rated speed is 11 m/s. Beyond the rated speed, output power is almost constant even if the wind speed accelerates. As the wind speed increases above the rated speed, the mechanical force on the wind turbine increases. After a certain point, this force may damage the turbine. Because

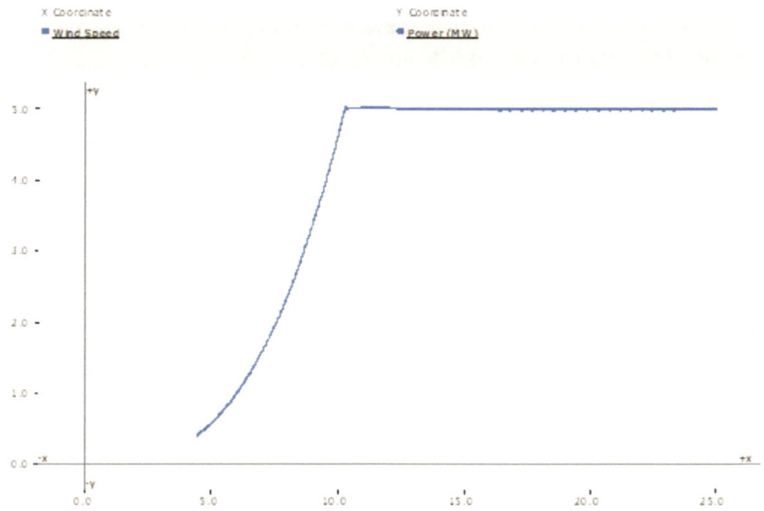

Figure 16.
The curve of wind speed versus output power.

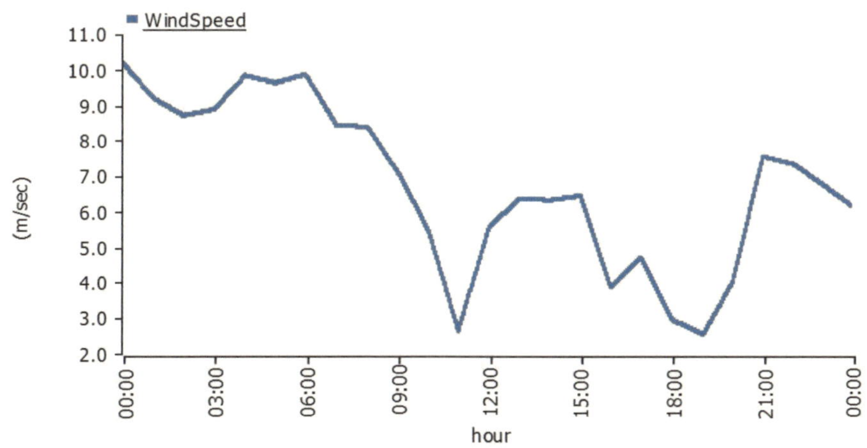

Figure 17.
Wind speed versus hour for 1 day.

of this, the braking system to stop the rotor beyond a certain wind speed—which is called the cut-out speed—is reached. The cut-out speed is almost 25 m/s.

The data included in **Figure 17** includes wind speed in units of m/sec over a 24-hour period.

The solar irradiance in units of W/m^2 is represented by a 24-hour time scale in **Figure 18**. As shown above, the solar power varies over the day, and it has its peak value at noon. This is the cell temperature (in units of degrees centigrade) graphic in **Figure 18**, which is also important for power generation from the solar energy source.

This is the cell temperature in units of centigrade degree (°C) graphic in **Figure 18** which is also important for the power generation from the solar energy source.

The active power flow seen from the point of common coupling (PCC) during a day is shown in **Figure 19**. **Figures 18–20** show that PV plant generates power only when solar radiation is available (in this case between 6 a.m. and 4 p.m.), and the output of wind power generation system is proportional to wind speed. The graph of the load which is real consumption data of a neighborhood and the graph of active power flow to the electrical grid are also seen at **Figure 19**. When the electric power demand of the loads is greater than the total produced power by wind power generation system and the PV power generation system, the lacking power is

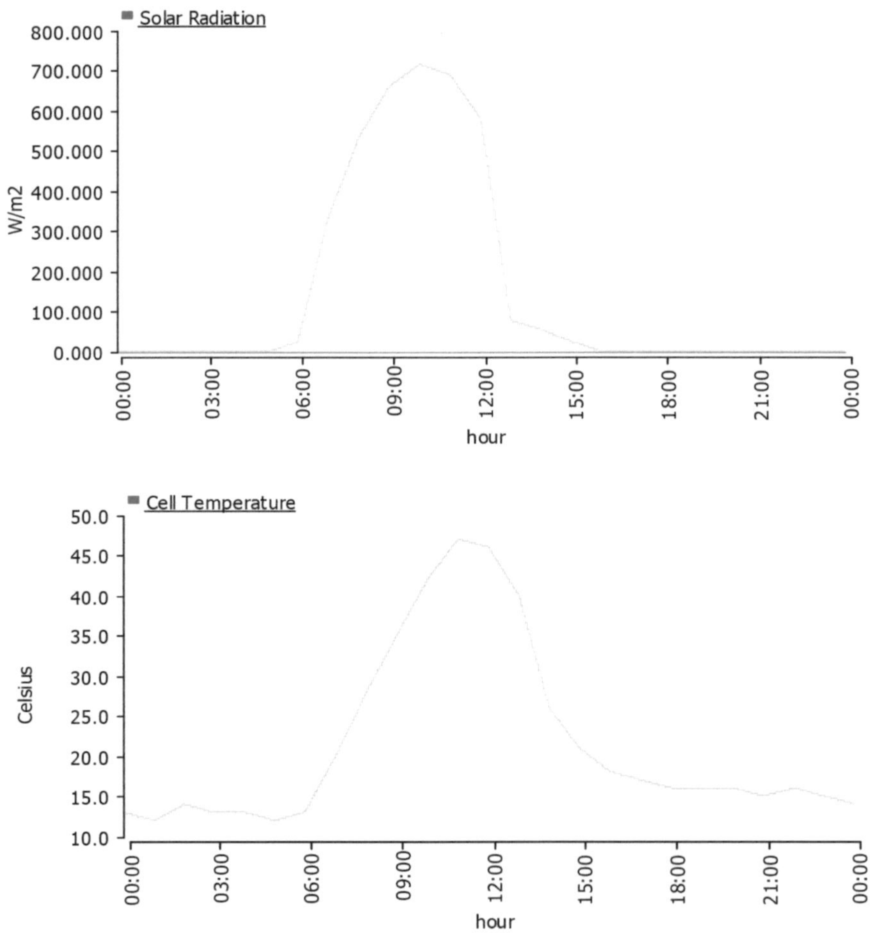

Figure 18.
Solar radiation and cell temperature versus hour for 1 day.

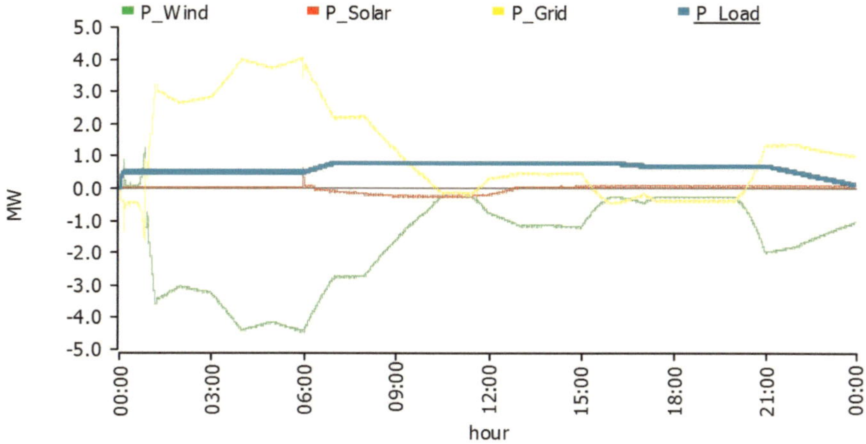

Figure 19.
Active power flow versus hour for 1 day.

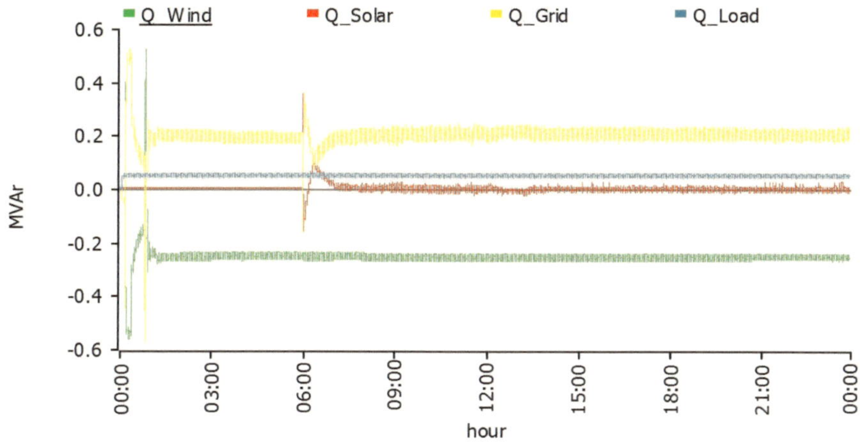

Figure 20.
Reactive power flow versus hour for 1 day.

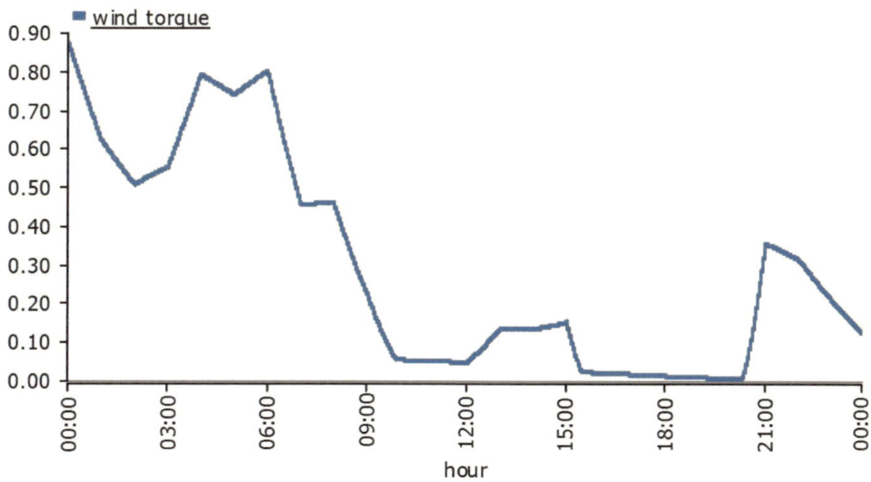

Figure 21.
Wind torque versus hour for 1 day.

provided from the electrical grid. Contrary, when the power demand of the loads is less than the total produced power by wind power generation system and PV power generation system, the remaining power is transferred to the electrical grid.

The reactive power flow seen from the PCC is shown in **Figure 20**. It is wind torque graphic in **Figure 21** which changes according to wind speed.

5. Conclusion

In this chapter, attention is mainly focused on the power control of a combined PV/wind system for power generation. The goal is to use real data from both the combined PV/wind systems and observe the system reliability with real-time simulation results. In this simulated PV/wind system, it is engineered to reach the maximum energy available from real conditions of wind speed and solar irradiance using dynamic input datasets. The simulation results illustrate and verify the operation and control of the proposed system. The chapter showed that the use of renewable energy resources provides higher energy potential, cost efficiency, operating and maintenance costs, an effective solution to energy demand, an attractive investment alternative, and environmental sustainability. Moreover, the period of financial return is shortened thanks to a combined PV/wind system, compared to stand-alone PV or wind systems. In addition, the sun and wind energy complete each other when the weather conditions are continuously changing. Thereby, energy production is provided all the time, and the system reliability is also consolidated by preventing sudden electrical loss. This chapter is a great example for further studies in terms of applicability because a real PV/wind system can be established, observed, and supported by performing simulation studies.

Author details

Emel Bakmaz[1], Kemal Aygul[1], Burak Esenboga[2], Tugce Demirdelen[2*] and Mehmet Tumay[2]

1 Department of Electrical and Electronics Engineering, Cukurova University, Adana, Turkey

2 Department of Electrical and Electronics Engineering, Adana Alparslan Turkes Science and Technology University, Adana, Turkey

*Address all correspondence to: tdemirdelen@atu.edu.tr

References

[1] González A, Riba J-R, Rius A. Optimal sizing of a hybrid grid-connected photovoltaic–wind–biomass power system. Sustainability. 2015;**7**: 12787-12806

[2] Jain S, Karampuri R, Somasekhar VT. An integrated control algorithm for a single-stage PV pumping system using an open-end winding induction motor. IEEE Transactions on Industrial Electronics. 2016;**63**:956-965

[3] Long H, Eghlimi M, Zhang Z. Configuration optimization and analysis of a large scale PV/wind system. IEEE Transactions on Sustainable Energy. 2017;**8**:84-93

[4] Patra S, Ray PK, Mohanty A. Robust fuzzy-sliding mode based UPFC controller for transient stability analysis in autonomous wind-diesel-PV hybrid system. IET Generation Transmission and Distribution. 2016;**10**: 1248-1257

[5] Parida A, Chatterjee D. Model-based loss minimisation scheme for wind solar hybrid generation system using (grid-connected) doubly fed induction generator. IET Electric Power Applications. 2016;**10**: 548-559

[6] Al-Masri HM, Ehsani M. Feasibility investigation of a hybrid on-grid wind photovoltaic retrofitting system. IEEE Transactions on Industry Applications. 2016;**52**:1979-1988

[7] Shen C-L, Shen Y-S. Output filter design for a novel dual-input PV-wind power converter by energy balance principle. Applied Sciences. 2016;**6**:263

[8] Cho J-H, Chun M-G, Hong W-P. Structure optimization of stand-alone renewable power systems based on multi object function. Energies. 2016;**9**: 649

[9] Kalla UK, Singh B, Murthy SS, Jain C, Kant K. Adaptive sliding mode control of standalone single-phase microgrid using hydro, wind, and solar PV array-based generation. IEEE Transactions on Smart Grid. 2018;**9**:6806-6814

[10] Palmer D, Koubli E, Cole I, Betts T, Gottschalg R. Comparison of solar radiation and PV generation variability: System dispersion in the UK. IET Renewable Power Generation. 2017;**11**: 550-557

[11] Ekstrom J, Koivisto M, Mellin I, Millar RJ, Lehtonen M. A statistical model for hourly large-scale wind and photovoltaic generation in new locations. IEEE Transactions on Sustainable Energy. 2017;**8**:1383-1393

[12] Liu Y, Tan S, Jiang C. Interval optimal scheduling of hydro-PV-wind hybrid system considering firm generation coordination. IET Renewable Power Generation. 2017;**11**:63-72

[13] Luna AC, Diaz NL, Graells M, Vasquez JC, Guerrero JM. Mixed-integer-linear-programming-based energy management system for hybrid pv-wind-battery microgrids: Modeling, design, and experimental verification. IEEE Transactions on Power Electronics. 2017;**32**:2769-2783

[14] Shanthi P, Uma G, Keerthana MS. Effective power transfer scheme for a grid connected hybrid wind/ photovoltaic system. IET Renewable Power Generation. 2017;**11**:1005-1017

[15] Merabet A, Tawfique Ahmed K, Ibrahim H, Beguenane R, Ghias AMYM. Energy management and control system for laboratory scale microgrid based wind-PV-battery. IEEE Transactions on Sustainable Energy. 2017;**8**:145-154

[16] Basaran K, Cetin NS, Borekci S. Energy management for on-grid and

off-grid wind/PV and battery hybrid systems. IET Renewable Power Generation. 2017;**11**:642-649

[17] Askarzadeh A. Electrical power generation by an optimised autonomous PV/wind/tidal/battery system. IET Renewable Power Generation. 2017;**11**: 152-164

[18] Kant K, Jain C, Singh B. A hybrid diesel-wind PV-based energy generation system with brushless generators. IEEE Transactions on Industrial Informatics. 2017;**13**:1714-1722

[19] Rezkallah M, Hamadi A, Chandra A, Singh B. Design and Implementation of active power control with improved P& O method for wind-PV-battery-based standalone generation system. IEEE Transactions on Industrial Electronics. 2018;**65**:5590-5600

[20] Tiwari SK, Singh B, Goel PK. Design and control of autonomous wind–solar system with DFIG feeding 3-phase 4-wire loads. IEEE Transactions on Industry Applications. 2018;**54**: 1119-1127

[21] Einan M, Torkaman H, Pourgholi M. Optimized fuzzy-cuckoo controller for active power control of battery energy storage system, photovoltaic, fuel cell and wind turbine in an isolated micro-grid. Batteries. 2017;**3**:23

[22] Ji H, Niu D, Wu M, Yao D. Comprehensive benefit evaluation of the wind-PV-ES and transmission hybrid power system consideration of system functionality and proportionality. Sustainability. 2017; **9**:65

[23] Mendoza-Vizcaino J, Sumper A, Galceran-Arellano SPV. Wind and storage integration on small islands for the fulfilment of the 50-50 renewable electricity generation target. Sustainability. 2017;**9**:905

[24] Liu Y, You S, Liu Y. Study of wind and PV frequency control in U.S. power grids—EI and TI case studies. IEEE Power and Energy Technology Systems Journal. 2017;**4**:65-73

[25] Hussain S, Al-ammari R, Iqbal A, Jafar M, Padmanaban S. Optimisation of hybrid renewable energy system using iterative filter selection approach. IET Renewable Power Generation. 2017;**11**: 1440-1445

[26] Akram U, Khalid M, Shafiq S. An innovative hybrid wind-solar and battery-supercapacitor microgrid system—Development and optimization. IEEE Access. 2017;**5**: 25897-25912

[27] Akram U, Khalid M, Shafiq S. Optimal sizing of a wind/solar/battery hybrid grid-connected microgrid system. IET Renewable Power Generation. 2018;**12**:72-80

[28] Gonzales Arispe JC, Ponce Vasquez AB. Wind and PV farms integration within power systems using static and dynamic simulations. IEEE Latin America Transactions. 2018;**16**:148-154

[29] Khaled U, Eltamaly AM, Beroual A. Optimal power flow using particle swarm optimization of renewable hybrid distributed generation. Energies. 2017;**10**:1013

Chapter 15

Recovery of Photovoltaic Module Heat Using Thermoelectric Effect

Felix A. Farret and Emanuel A. Vieira

Abstract

The growing demand for renewable energy sources, in particular for solar technologies, requires more detailed studies to increase power and efficiency. Among them, thermoelectric energy conversion is a well-known technology used for decades including solar thermal generators (STEG), radioisotope thermoelectric generators (RTG), automotive thermoelectric generators (ATG) and thermoelectric generators (TEG). This chapter aims to demonstrate that the thermoelectric effect (Seebeck effect) can be used to harness the thermal energy retained in photovoltaic panels to increase their overall efficiency with its direct conversion into electrical energy and vice versa. It is also observed that solar radiation can be converted directly into electric energy, as in photovoltaic modules, or yet can be converted directly into electricity, as in thermoelectric modules. It is emphasised that although the energy conversion by thermoelectric effect still has low electrical efficiency, this source is characterised by a high degree of reliability, low maintenance, appreciable durability and absence of moving parts, and it allows generating electric energy through recovery of the thermal energy from several industrial processes. At the end of this chapter is presented a case study related to the thermal energy absorbed by a polycrystalline photovoltaic module to illustrate their increased efficiency and power in thermoelectric-photovoltaic cogeneration.

Keywords: photovoltaic-thermoelectric cogeneration, thermal energy, thermoelectric effect, renewable energy sources, solar energy

1. Introduction

In this chapter, the thermoelectric effect is used to exploit the thermal energy accumulated in the operation of photovoltaic panels, reversing it into electric energy. For this, a brief presentation of the thermoelectric effect is made to obtain an equivalent circuit of a thermoelectric module and its associations (series and parallel).

Traditionally, the term thermoelectric effect or thermoelectricity involves three effects that can be identified separately: the Seebeck effect, the Peltier effect and the Thomson effect. **Figure 1** illustrates these three thermoelectric effects.

From the discovery of Alexandro Volta related to the production of electricity by the mere contact between different metals, the researches on thermoelectricity took

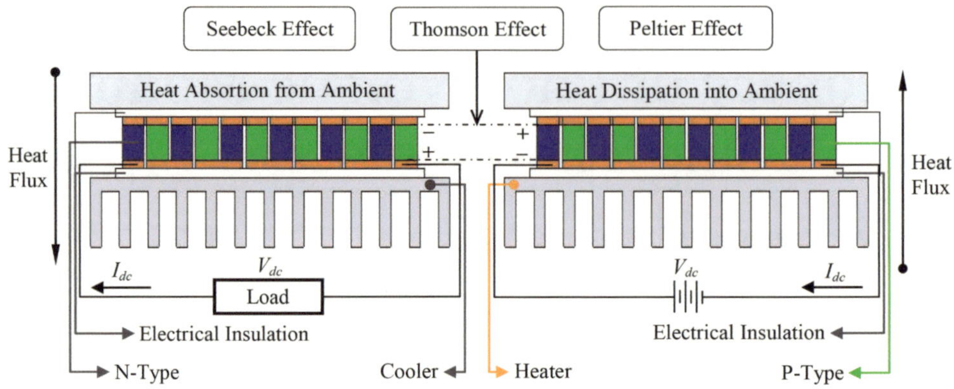

Figure 1.
Thermoelectric effect.

impulse and began looking for the origin of the thermoelectric effects [1]. These effects can be reunited as follows:

- Seebeck effect (1822) explaining that an emf is generated by a pair of different conductors whose ends are not at the same temperature with respect to their junction.

- Peltier effect (1834) which refers to the release or absorption of heat at the junction of two different materials when an electric current flows through from one conductor to the other.

- Joule effect (1840) that relates the production of heat at a given time whenever the material is carrying electric charges.

- Thomson effect (1856) refers to the production or absorption of heat when an electric current passes through a circuit made of a single material and under a temperature difference across its length.

The material properties listed above are used to explain how the thermoelectric effect is able to produce a direct conversion of a temperature difference into an electrical voltage and vice versa. On the left side of **Figure 1**, the thermal Seebeck effect appears as a temperature gradient on the faces of a thermoelectric module coupled to a pair of semiconductors, whereby an electric current flows through the N-P direction of the junction. The Peltier effect represented on the right side of **Figure 1** consists of the production of a temperature gradient across the junction between two conductors (or semiconductors) of different materials in a closed loop when subjected to an electric voltage. On the other hand, thermoelectric generators can act as heat engines, being less bulky and having no moving parts [2].

In principle, the association of the Peltier-Seebeck and Thomson effects may be thermodynamically reversible, whereas the Joule effect is not reversible. This nonreversibility of the Joule effect causes many authors to consider the existence of only two thermoelectric effects.

2. Thermoelectric materials

The application of semiconductor materials in the construction of thermocouples in some special cases has contributed significantly for electricity generation, refrigeration and heating, as well as being of interest to theoretical physics based on the modern science of semiconductors. The energy applications of these thermoelements are mainly in thermoelectric coolers, industrial and domestic heating devices, thermoelectrically generated sound, ultrasonic generators, vacuum thermoelements, space applications in satellite power, marine applications in buoys and headlights, as well as in medicine applications of cardiac pacemaker feeding.

For the sake of completeness, the principle of direct conversion of thermal energy into electricity is also widely used today in temperature measurements by means of thermocouples or thermo-junctions based on metals or metal alloys. In this particular, **Table 1** shows the Seebeck coefficient for some metal alloys and semiconductors used to make thermoelectric modules on a commercial scale.

The economical and practical applications of thermoelectric materials depend on the characteristics of the available thermoelements, their efficiency, temperatures involved in the operation, stability of operation and costs involving raw materials in the final preparation. As a rule, it should be noted that the thermal cycle of a thermoelectric generation differs from others because it consists essentially of a solid-state phenomenon where the thermal energy is converted directly into electric energy. This direct conversion has made possible the construction of simple devices such as transducers or power sources, which have practically no maintenance and no moving parts, which makes them highly interesting in exact sciences.

The thermoelectric materials can be made of conductors, semiconductors and insulators. In a more careful classification, it is necessary to take into account the microscopic characteristics referring to the electron behaviour in the valence layer of the material by action of an electric field [3]. A thermoelectric material with high electrical conductivity, high Seebeck coefficient and low thermal conductivity can be considered a good thermoelectric material [4].

It should be said at this stage that the efficiency of a thermoelectric device depends both on the operating temperature of the materials and on the characteristics of these materials, which can be expressed by the figure of merit given in Eq. (1).

Semiconductor	Seebeck coefficient (μV/°C)
Bi_2Te_3 (P-type)	−230
$Bi_{2x}Sb_xTe_3$ (P-type)	300
Sb_2Te_3 (P-type)	185
PbTe	−180
$Pb_3Ge_{39}Se_{58}$	1670
$Pb_6Ge_{36}Se_{58}$	1410
$Pb_9Ge_{33}Se_{58}$	−1360
$Pb_{13}Ge_{29}Se_{58}$	−1710
$Pb_{15}Ge_{37}Se_{58}$	−1990
$SnBi_4Te_7$	120
$SnBi_3Sb_1Te_7$	151
$SnBi_{2.5}Sb_{1.5}Te_7$	110
$SnBi_2Sb_2Te_7$	90
$PbBi_4Te_7$	−53

Table 1.
Seebeck coefficients for metal alloys and semiconductors.

$$Z = \frac{\alpha^2}{\sigma\kappa} \tag{1}$$

where:
α is the Seebeck coefficient.
σ is the electric conductivity.
κ is the thermal conductivity.
Z is the figure of merit.

Eq. (2) relates the figure of merit to temperature. However, the figure of merit does not have a constant behaviour as the temperature varies [1], but it can also be expressed in a dimensionless form $Z\overline{T}$ [4]:

$$Z\overline{T} = \left(\frac{\alpha^2}{\sigma\kappa}\right)\overline{T} \tag{2}$$

where:
$\overline{T}$ is the average temperature between the faces (hot and cold).
$Z\overline{T}$ is the figure of merit dependent on temperature.

3. Thermoelectric module

A "modern" thermoelectric converter essentially consists of a series of thermoelectric semiconductors of N-type and P-type materials alternately connected. It is noticeable that these semiconductor elements are electrically connected in series and thermally in parallel by a series of metal contacts placed at each two ceramic plates (**Figure 2a**).

A typical thermoelectric device consists of two ceramic substrates (alumina) (**Figure 2a**), which serves both as base and electrical insulation, for thermoelements (e.g. bismuth telluride) electrically connected in series and thermally in parallel as in the case of ceramics [1]:

$$T_{\max} = 0.5Z \cdot T \tag{3}$$

Conventional thermoelectric devices have various specifications according to the type of application. The dimensions commonly range from 3 to 5 mm in thickness and from 56 to 60 mm in lateral length. The faces exposed to the outside have a square geometry. The maximum rate of heat pumping varies from 1 to 125 W [5].

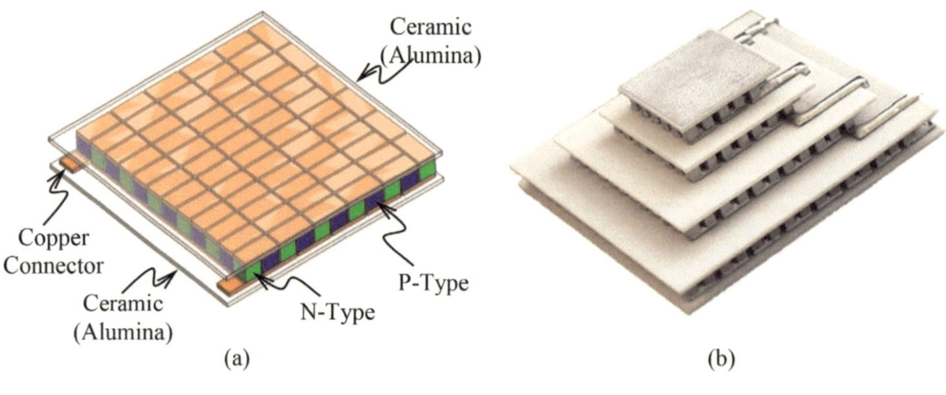

Figure 2.
Thermoelectric module: (a) single TEG and (b) multistage TEG.

The maximum temperature difference between the hot and cold sides can reach 70° C. Typical thermoelectric modules contain from 3 to 127 thermocouples.

There are multistage (cascade) thermoelectric devices designed to meet the requirements of large temperature differentials (ΔT can reach 130°C). The lowest achievable temperature is <−100°C for multistage devices [5].

There are thermoelectric modules made up of several layers known as multistage thermoelectric modules (see **Figure 2b**). The maximum temperature difference obtained between the interfaces of each layer in a multistage thermoelectric module is given by Eq. (3).

4. Equivalent circuit of thermoelectric modules

Section 1 explained how the Seebeck effect could generate an electrical voltage from a temperature difference between the junction of two distinct semiconductor materials and the ends of them. Eq. (4) states that the higher the temperature differences between the thermoelectric junctions, the greater the electric potential generated [6]:

$$V = N_j \int_{T_1}^{T_2} [\alpha_2(T) - \alpha_1(T)]dt \tag{4}$$

where:

N_j is the number of junctions for identical thermoelectric pairs.

T_1 and T_2 are the temperatures at the extremes of the material pair.

α_1 and α_2 are the Seebeck coefficients of the semiconductor materials.

The equivalent electrical circuit of a thermoelectric module can be seen in **Figure 3a**, where the output voltage V_i across the terminals of a no-load module is given by Eq. (5). As this circuit is not connected to any load, it does not circulate electrical current through it:

$$V_i = \alpha \Delta T \tag{5}$$

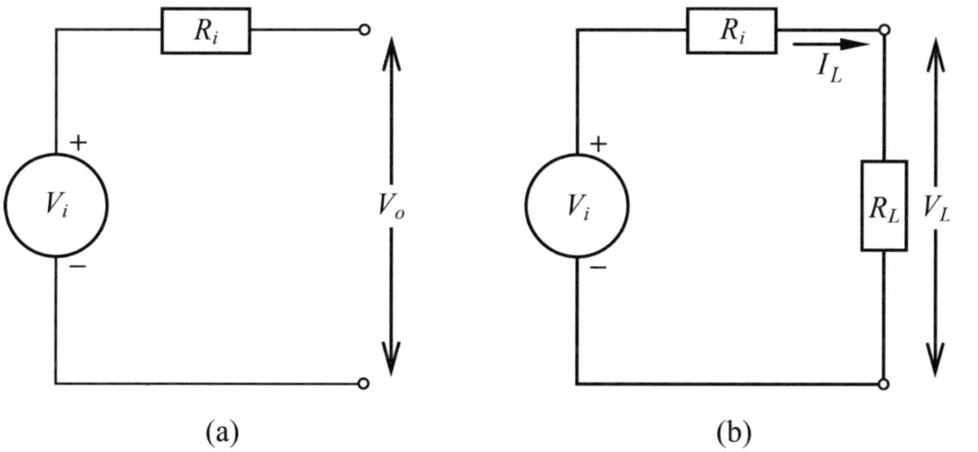

Figure 3.
Equivalent electric circuit: (a) open circuit and (b) circuit with load.

where:
V_i is the input source voltage.
α is the absolute coefficient of Seebeck.
ΔT is the temperature difference between faces of the thermoelectric module.

When a load R_L is connected across the circuit terminals of **Figure 3b**, a current I_L will flow through it according Eq. (6):

$$I_L = \frac{V_i}{R_i + R_L} \tag{6}$$

It is shown in **Figure 3b** that the load voltage V_L can be defined as the total voltage generated minus the internal voltage drops in the module, as in Eq. (7) [7]:

$$V_L = V_i - I_L R_i \tag{7}$$

Eq. (7) can then be rewritten in terms of temperature difference and the Seebeck coefficient by replacing I_L given by Eq. (6) and V_i given by Eq. (5). The load voltage V_L can be rewritten as a function of the temperature and the Seebeck coefficient as in Eq. (8):

$$V_L = \alpha\, \Delta T - \frac{\alpha\, \Delta T}{R_i + R_L} R_i \tag{8}$$

The power generated by a thermoelectric module is a function of the generated voltage as given by Eq. (9):

$$P_L = \frac{V_i^2 R_L}{(R_i + R_L)^2} \tag{9}$$

The power dissipated in the load can also be expressed as a function of the temperature difference across the faces of the thermoelectric module, according to Eq. (10):

$$P_L = \frac{(\alpha^2 \Delta T^2) R_L}{(R_i + R_L)^2} \tag{10}$$

Mathematically, the maximum power in the load $P_{L,\max}$ is obtained by deriving the expression of electric power in relation to R_L and equalling to zero according to Eq. (11):

$$\frac{V_i^2 (R_i + R_L) - 2\, R_L V_i^2}{(R_i + R_L)^2} = 0 \tag{11}$$

Based on the identity of Eq. (11), it is concluded that $R_L = R_i$, both nonzero resistors. The internal resistance R_i of each thermoelectric module is obtained by the algebraic sum of the internal thermal resistances of the metallic material of the junction with the sum of resistances of the N-P junctions and with the thermal resistance of the physical contacts [8].

By making $R_i = R_L = R_{\max,pot}$ and replacing everything in Eq. (10), the maximum power is given as a function of temperature by Eq. (12) [9]:

$$P_{L,\max} = \frac{\alpha^2 \Delta T^2}{4\, R_{\max,pot}} \tag{12}$$

It should be noted that the performance of a thermoelectric module can be characterised by the Seebeck coefficient, the internal resistance and the thermal conductance [7]. The thermal conductance of the thermoelectric module is given as a function of the thermal conductance of each P-N junction. As the P-N elements have equal volumes, the thermal conductance will be the same for both semiconductors, and it is expressed by Eq. (13) [7]:

$$K = 2\frac{kA_j}{l} \tag{13}$$

where:
A_j is the area of each thermoelectric junction.
κ is the thermal conductivity.
l is the thermoelectric module length.

The heat absorbed $\dot{Q}_h$ and the heat removed $\dot{Q}_c$ are expressed by Eqs. (14) and (15) [7]:

$$\dot{Q}_h = \alpha\, I_L T_h + K(T_h - T_c) - \frac{1}{2} I_L^2 R_i \tag{14}$$

$$\dot{Q}_c = \alpha\, I_L T_c + K(T_h - T_c) - \frac{1}{2} I_L^2 R_i \tag{15}$$

where:
T_h is the temperature of the hot face of the thermoelectric pair.
T_c is the temperature of the cold face of the thermoelectric.
$\alpha\, I_L T_h$ is the heat rate of the thermoelectric module.
$\alpha\, I_L T_c$ is the heat rate of the thermoelectric module.
$\frac{1}{2} I_L^2 R_i$ is the power dissipated.
$K(T_h - T_c)$ is the heat pumping between two thermal reservoirs.

The efficiency of the thermoelectric module can be obtained from the ratio of the electric output power P_L and the heat received in the thermoelectric module according to Eq. (16):

$$\eta_{TEG} = \frac{P_L}{\dot{Q}_q} \tag{16}$$

According to [10], the heat transferred through the module determines the thermoelectric efficiency. Therefore, the thickness of the module is the parameter that most affects the efficiency which is expressed as in Eq. (17):

$$\eta_{TEG} = \frac{I_L^2 R_L w_t}{\kappa\, A_{TEG} \Delta T} \tag{17}$$

where:
w_t is the thickness of the thermoelectric module.
A_{TEG} is the face area of the thermoelectric module.

5. Association of thermoelectric modules

The arrangement of connections between thermoelectric modules resembles that of the photovoltaic modules, which can be arranged in series or in parallel. If the objective is to increase the electrical voltage across the thermoelectric modules, the series connection must be used. If the aim is to increase the current through the modules, they must be connected in parallel. A third option is a mixed series-parallel arrangement that provides an increase in power either by the increase in voltage or current or both. In this way, the load current generated by the association of the thermoelectric modules is represented by Eq. (18) [10]:

$$I_L = \frac{N_s\, \alpha\, \Delta T}{\frac{N_s R_i}{N_P} + R_L} \tag{18}$$

where:

N_s and N_p are, respectively, the number thermoelectric modules in series and parallel.

The voltage generated across the load by an arrangement of thermoelectric modules is given by Eq. (19) [10]:

$$V_L = R_L \left(\frac{N_s\, \alpha\, \Delta T}{\frac{N_s R_i}{N_P} + R_L} \right) \tag{19}$$

The maximum thermoelectric power occurs when the internal resistance R_i is equal to the load resistance R_L, and it is given by Eq. (20) [10]:

$$P_{L,\max} = \frac{N_T [\alpha\, (T_h - T_c)]^2}{4\, R_i} \tag{20}$$

where:

N_T is the total number of modules.

With Eq. (21) it is possible to determine the heat input in the thermoelectric generator in watts [10]:

$$\dot{Q}_w = N_T \left[\frac{\alpha\, T_h I_L}{N_T} - 0.5 R_i \left(\frac{I_L}{N_P} \right)^2 + k(T_h - T_c) \right] \tag{21}$$

where:

N_T is the total number of thermoelectric modules.

The efficiency of the thermoelectric generator is given by Eq. (22), which relates the maximum power generated with the array of modules and the amount of heat that had been absorbed by the array [10]:

$$\eta_{TEG} = \frac{P_{L,\max}}{\dot{Q}_w} \tag{22}$$

6. Layer temperatures in the photovoltaic modules

The determination of the temperature in a photovoltaic module is of great importance for thermoelectric generation. For this, a Ross model was developed to simplify the determination of temperature in a photovoltaic cell located between the layers of the photovoltaic modules caused by the ambient temperature and solar radiation as shown in **Figure 4**.

Ross determined the temperature coefficients for photovoltaic cells according to the residential installation form (**Table 2**) [11, 12].

From the Ross observation, the temperature of the photovoltaic cell can be determined with Eq. (23) [11, 12]:

$$T_{Cell} = T_{amb}K_S G_i \tag{23}$$

where:

T_{amb} is the ambient temperature.

K_S is the coefficient of Ross.

G_i is solar radiation.

The mechanism of heat transfer between layers that cause the temperature difference between the photovoltaic cell and the polyvinyl fluoride layer, which commercial trade mark is Tedlar®, occurs by conduction and can be expressed by Eq. (24) [13]:

$$T_{Cell} - T_{Tedlar} = \frac{\dot{Q}}{\sum R_{themal}} \tag{24}$$

where:

$\sum R_{thermal}$ is the sum of the thermal resistances of the photovoltaic cell.

$\dot{Q}$ is the useful energy of solar radiation.

T_{Tedlar} is the temperature of Tedlar®.

T_{cell} is the temperature of the photovoltaic cell.

Table 3 lists the thickness of the photovoltaic module layers with the thermal conductivities of each constituent. The thermal resistance between the Glass-Tedlar® layers is determined then by Eq. (25) [13]:

$$\sum R_{themal} = \frac{d_{Glass}}{k_{Glass}} + \frac{d_{EVA}}{k_{EVA}} + \frac{d_{Tedlar}}{k_{Tedlar}} \tag{25}$$

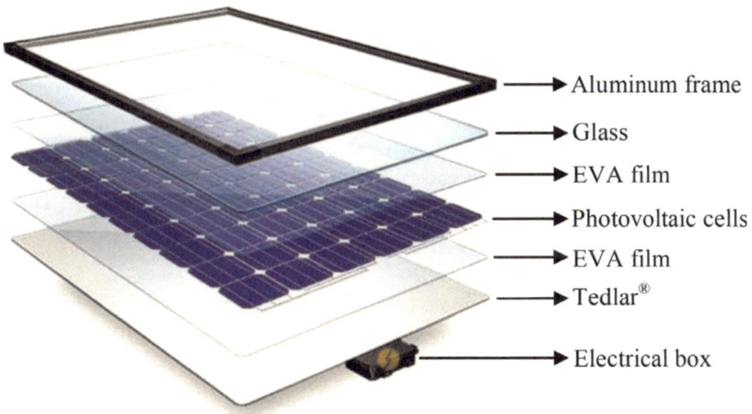

Figure 4.
Layers of a photovoltaic module.

Type of PV installation	Ross coefficient
Integrated into the roof	0.058
Small distance to the roof (<10 cm)	0.036
Great distance to the roof (>10 cm)	0.027
Free	0.020

Table 2.
Ross coefficients, K_S.

Description	Values
Layer thickness of EVA, d_{EVA}	0.5 mm
Layer thickness of Glass, d_{Glass}	0.4 mm
Layer thickness of Tedlar®, d_{Tedlar}	0.15 mm
Thermal conductivity of EVA, k_{EVA}	0.34 W/m^2 K
Thermal conductivity of Glass, k_{Glass}	1.0 W/m^2 K
Thermal conductivity of Tedlar®, k_{Tedlar}	0.167 W/m^2 K

Table 3.
PV layer thicknesses and thermal conductivity of some elements [6].

where:
d_{Glass} is the thickness of the glass.
d_{EVA} is the thickness of the ethylene vinyl acetate (EVA).
d_{Tedlar} is the thickness of the Tedlar®
k_{Glass} is the thermal conductivity of the glass.
k_{EVA} is the thermal conductivity of the ethylene vinyl acetate (EVA).
k_{Tedlar} is the thermal conductivity of the Tedlar®.

By knowing the approximate temperature of the photovoltaic cell and isolating the term Tedlar® in Eq. (24), the temperature at the lower surface of the Tedlar® layer is estimated by Eq. (26) [13]:

$$T_{Tedlar} = T_{Cell} - \left[\frac{\dot{Q}}{\frac{d_{Glass}}{k_{Glass}} + \frac{d_{EVA}}{k_{EVA}} + \frac{d_{Tedlar}}{k_{Tedlar}}} \right] \tag{26}$$

7. Temperature of the absorption plate and its occupation in the thermoelectric area

Much of the radiation on photovoltaic panels is absorbed as heat flowing through the constituent layers. To calculate the temperature of the absorption plate, the parameters shown in **Table 4** take into account the dimensions of the photovoltaic plate. The internal temperature of the plate is determined by Eq. (27) [13, 14].

The temperature on the hot side of the thermoelectric module is considered to be the internal temperature of the absorption plate given by Eq. (28):

$$T_{\text{int}, board} = T_{Tedlar} - \frac{L_A}{\kappa_A A_A} \tag{27}$$

where:
L_A is the thickness of the absorption plate.
k_A is the thermal conductivity of the absorption plate.

Definition	Values
Thickness, L_A	0.005 m
Thermal conductivity, k_A	237 W/m^2°C^{-1}
Absorption plate area, A_A	0.0643 m^2

Table 4.
Characteristics of the absorption plate.

A_A is the area of the plate that absorbs heat.
T_{Tedlar} is the temperature of Tedlar®.
$T_{\text{int}, board}$ is the internal temperature of the absorption plate.

The area occupied by thermoelectric modules is a limiting factor for determining the size of the thermoelectric conversion. To estimate the area that the thermoelectric system will occupy, one can use Eq. (28):

$$A_T = \mathrm{w}\,\ell\,N_T \tag{28}$$

where:
A_T is the total area of the TEG module.
w is the width of the TEG.
ℓ is the length of the TEG.

8. Determination of the power obtained from a TEG

The power generated by a TEG depends on the internal temperature in the thermoelectric absorption plate, which is an essential parameter for determining the thermoelectric energy generated. Therefore, using Eq. (18) for the considered parameters, the load current can be expressed by Eq. (29):

$$I_{R_L} = \frac{N_s\alpha(T_{\text{int}, board} - T_{amb})}{\frac{N_sR_i}{N_P} + R_L} \tag{29}$$

Reordering Eq. (19) for the data parameters, the load voltage is finally determined by Eq. (30):

$$V_c = R_L\left[\frac{N_s\alpha(T_{\text{int}, board} - T_{amb})}{\frac{N_sR_i}{N_P} + R_L}\right] \tag{30}$$

The maximum power of the thermoelectric generation given by Eq. (31) occurs when the internal resistance R_i is equal to the load resistance R_L:

$$P_{L,\max} = \frac{N_T[\alpha(T_{\text{int}, board} - T_{amb})]^2}{4\,R_L} \tag{31}$$

9. Case study

In this case study, the steps used in the laboratory of CEESP-UFSM to determine the production capacity of a small photovoltaic-thermoelectric cogeneration assembly are shown. The temperature estimation used as input parameter the data of the automatic meteorological station of the National Institute of Meteorology (INMET) for the city of Santa Maria-RS, Brazil, in the period of January 17, 2018–January 23, 2018. The same methodological procedure of this case study can be used to any other location by taking into account the historical temperature data of the place.

In the simulations performed for this section, the reference photovoltaic module is the YL010P-17B 1/13 from the manufacturer YINGLI SOLAR with the characteristics described in **Table 5**. As a prototype, the parameters of this TEG module using the bismuth telluride thermoelectric element are listed in **Table 6**. From the data of solar radiation, ambient temperature and wind speed, it was possible to estimate

Description	Values
Rated voltage	17.1 V
Rated current	0.59 A
Open circuit voltage	24.8 V
Short circuit current	0.65 A
Peak power	$10 \pm 5\%$ W_p
Conversion efficiency	11.1%
External module area	0.09 m^2

Table 5.
Data of the photovoltaic modules, YL010P-17B 1/13.

Description	Values
Maximum voltage	15.4 V
Maximum current	10.5 A
Internal resistance	1.24 Ω
Maximum temperature variation	67°C
Seebeck coefficient	0.02875 V/K
Thermoelectric junctions	127 pairs
Width, w	40 mm
Length, ℓ	40 mm
Thickness	3.9 mm

Table 6.
Parameters of the thermoelectric module.

with Eq. (23) the temperature of the photovoltaic cell. The temperature of the Tedlar® layer is given by Eq. (26) with the thermal flow directed to the thermoelectric generator. With this, the load power generated by the thermoelectric modules is determined with Eq. (31) with the modules subjected to a positive temperature gradient, i.e. the temperature of the Tedlar® layer is higher than the ambient temperature.

The thermoelectric microgenerator used in this case study is composed of three parts, absorption plate, thermoelectric modules and heat sink, as shown in **Figure 5** considering the dimensions and the internal temperature of the photovoltaic plate. A thin layer of thermal paste was used to improve the heat transfer between the photovoltaic module and the constituent layers of the thermoelectric microgenerator. The electrical connections of the thermoelectric microgenerator are shown in **Figure 6** where the thermoelectric modules are arranged in series in a total of six modules which in turn are connected in parallel with the photovoltaic module.

Figure 7 is the record of temperatures measured in the week between January 17, 2018, and January 23, 2018. The blue line data represents the temperature of the photovoltaic cell, and the red line represents the ambient temperature. Notice that there is a significant difference between the ambient temperature and the temperature of the photovoltaic cell. With this, heat reduces the efficiency of the photovoltaic cells but has a reasonable potential for thermoelectric cogeneration. The internal temperature of the absorption plate is the same as the hot surface of the

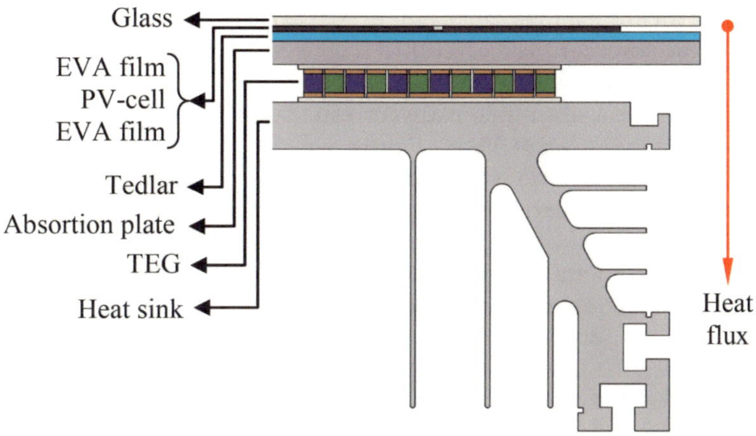

Figure 5.
Layer distribution of the PV-TEG system.

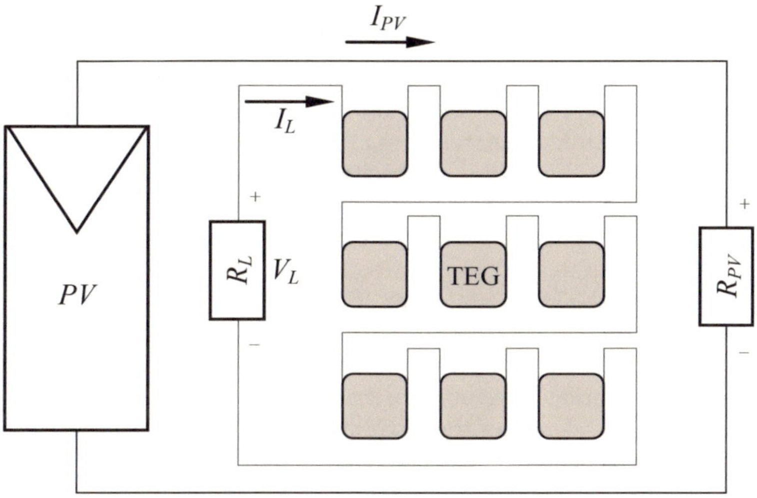

Figure 6.
General diagram of PV-TEG cogeneration.

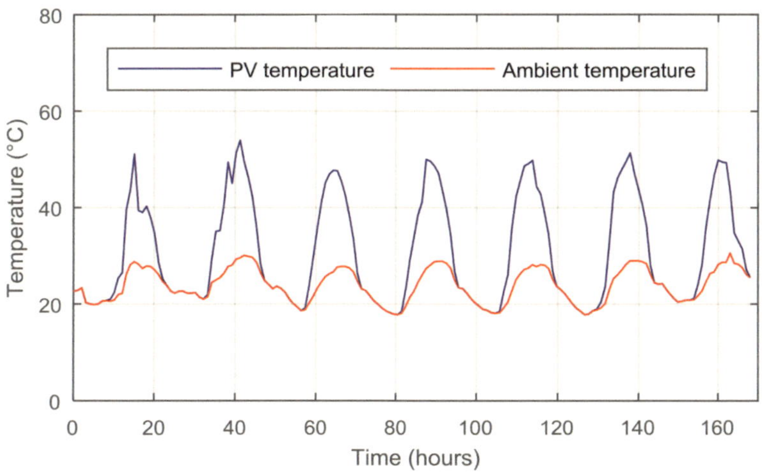

Figure 7.
Temperatures of the photovoltaic cell and the environment.

thermoelectric modules and is calculated by Eq. (27). The average temperature estimated for the absorber plate is 0.0002°C which is lower than the temperature of the Tedlar® layer given by Eq. (26).

The area of the heat-absorbing plate corresponds to 71.44% of the area of the photovoltaic panel, which is 0.09 m^2. **Figure 8** shows the temperature variation in the photovoltaic cell, the temperature of the absorber plate and the ambient temperature. In this figure, the theoretical temperature of the absorber plate is slightly less than the temperature of the photovoltaic cell and is also significantly higher than the ambient temperature. The greater is the difference between these parameters, the greater is the thermoelectric generation as specified by Eq. (31). The most important parameter for increasing thermoelectric generation is the temperature difference between the hot side and the cold side of the module. The average temperature difference between the TEG surfaces in the practical tests was 23.88°C for the given period. This estimation considered that the heat sink temperature was maintained at 5°C above room temperature.

Figure 9 is an estimate of the thermoelectric generation calculated for the period from January 17, 2018, to January 23, 2018.

The total thermoelectric generation estimated with the data used in this section is 96.86 W for a 1-week period accounting for 71 hours of electricity generation. The temperature variation between the absorber plate and the TEG ambient temperature was always positive. During this period, the estimated average of hourly energy production in the TEG microgenerator was 1.37 Wh.

Figure 10 shows the theoretical electricity either in the photovoltaic module or in the TEG module during the period considered in this section. Soon after, the power of the thermoelectric generator was estimated using Eqs. (23)–(31). The total theoretical energy generated in the photovoltaic module was 479 W, and the total energy generated in the TEG module was 96.86 W.

It is important to mention at this point that both generation profiles shown in **Figure 10** have the same shape a part of their scales. This is due to the fact that there is a proportional correlation between incident solar irradiance and the bottom surface of the photovoltaic module where the TEGs are connected to the absorber plate.

Table 7 shows the theoretical electricity generated by the cogeneration system described in this case study. This shows that after coupling the TEG module to the photovoltaic panel, the total electric energy generated had an increase of 20.2% when compared to the photovoltaic system without cogeneration.

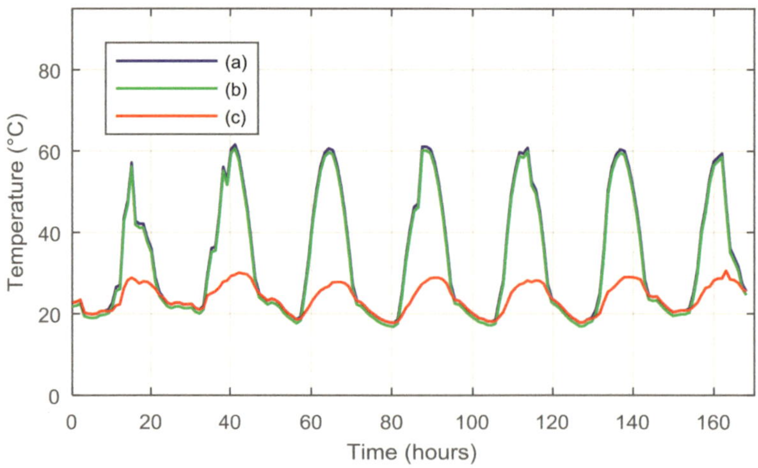

Figure 8.
Temperatures of (a) PV-cell, (b) inner face of the absorption plate, and (c) ambient.

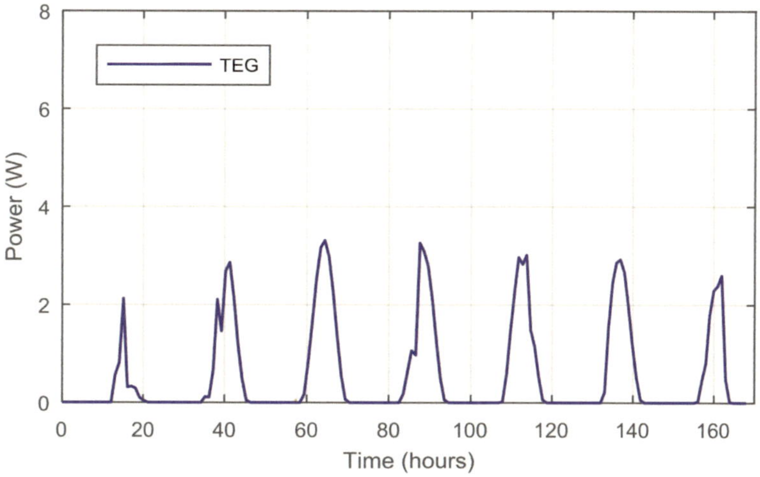

Figure 9.
Estimation of the TEG energy.

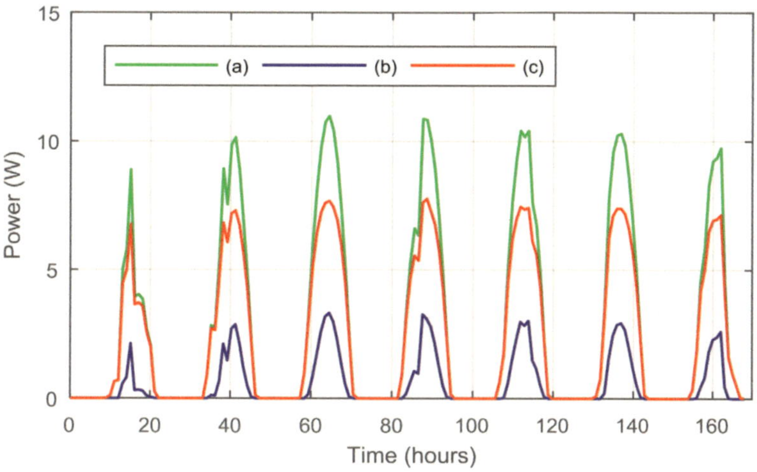

Figure 10.
Energy generation profiles: (a) PV module with cogeneration, (b) TEG module, and (c) PV module without cogeneration.

Electric power	
PV module without cogeneration	388.14 W
PV module with cogeneration	479 W
TEG module	96.86 W
PV + TEG cogeneration	575.86 W

Table 7.
Total theoretical electric power along a week.

10. Conclusion

This chapter analyses the use of thermoelectric generators in photovoltaic cogeneration to exploit the residual thermal energy in a direct, renewable, viable, efficient and sustainable way. It shows that the capture of the residual energy by the Seebeck effect for cogeneration of electric energy can be useful and efficient in several fields of human activities concerning its easy application, low weight, reduced size and simplicity of operation.

A thermoelectric generator operates noninvasively, indirectly taking advantage of the thermal energy of other systems where it is inserted without negative interference in the process. Another characteristic of the thermoelectric generators is their modularity, which allows the expansion of cogeneration to match certain output power or to change their way in association with other thermoelectric modules.

Even if the thermoelectric generation is still limited to low temperature gradients ($\Delta T < 25°C$), it can be seen that this technology will have benefits when associated with photovoltaic systems. This is due to the increase of the heat transfer from photovoltaic modules to the environment, at the same time, generating an additional amount of electric energy in the TEG module without increasing any further conversion area exposed to the sun.

The heat transfer technique used in the case study described in this chapter shows that it is possible to exploit the residual thermal energy of a photovoltaic module to improve its own performance. This has been suggested as a practical way to improve thermoelectric cogeneration (Seebeck effect) with photovoltaic modules (PV-TEG) and to increase energy efficiency of other associated hybrid systems.

The possibilities of thermoelectric-photovoltaic cogeneration are still greatly reduced by the limitations imposed by their low temperature gradients and efficiency. An improvement can be expected, for example, with a significant increase in the number of interconnected thermoelectric modules. Another point is that there is a limitation in the contact area between the thermocouples and the bottom surface of the photovoltaic module. Of course, this would increase the assembly cost and complexity of the generator, which could only be justified in particular cases that would not allow any other alternatives.

Author details

Felix A. Farret* and Emanuel A. Vieira
Center of Excellence in Energy and Power Systems (CEESP), Federal University of Santa Maria (UFSM), Santa Maria, Brazil

*Address all correspondence to: fafarret@gmail.com

References

[1] Vieira E. Increasing the Efficiency of Photovoltaic Modules with Utilization of Thermal Energy for Thermoelectric Generation of Solid State [master thesis]. Santa Maria, RS: Post-Graduation Program in Electrical Engineering; 2018. p. 128

[2] Farret FA, Simões GM. Integration of Renewable Sources of Energy. 2nd ed. Hoboken, New Jersey, USA: John Wiley and Sons Ltd; 2018. LCCN: 2017007716

[3] Radhakrishnan R. “A Review of Thermoelectric, Handbook Macro to Nano,” D.M. Rowe (editor). Materials and Manufacturing Processes. 2008; **23**(6):626-627. DOI: 101080/10426910802135819

[4] Ismail B, Ahmed W. Thermoelectric power generation using waste-heat energy as an alternative green technology. Recent Patents on Electrical Engineering. 2009;**2**(1):27-39

[5] Riffat SB, Ma X. Thermoelectrics: A review of present and potential applications. Applied Thermal Engineering. 2003;**23**(8):913-935. DOI: 101016/s1359-4311(03)00012-7

[6] Daud MMM, Nor NBM, Ibrahim T. Novel hybrid photovoltaic and thermoelectric panel. In: 2012 IEEE International Power Engineering and Optimization Conference, PEOCO 2012: Conference Proceedings; 2012. pp. 269-274

[7] Hsiao YY, Chang WC, Chen SL. A mathematic model of thermoelectric module with applications on waste heat recovery from automobile engine. Energy. 2010;**35**(3):1447-1454. DOI: 10.1016/j.energy.2009.11.030

[8] Carvalho C.A.R. of Feasibility Study of the Use of Heat Exhaust for Generation of Electric Energy in Automobiles [master thesis]. Taubaté University; 2012. p. 68

[9] Phillips SS. Characterizing the thermal efficiency of thermoelectric modules. Bachelor of Science, Massachusetts Institute of Technology. Massachusetts: Cambridge; 2009:1-9

[10] Yodovard P, Khedari J, Hirunlabh J. The potential of waste heat thermoelectric power generation from diesel cycle and gas turbine cogeneration plants. Energy Sources. 2001;**23**(3):213-224. DOI: 101080/00908310151133889

[11] Ross RGJ. Interface design considerations for terrestrial solar cell modules. In: Photovoltaic Specialist Conference 1976; 1976. pp. 801-806

[12] Ross RGJ. Flat-plate photovoltaic array interface design optimization. In: Proc. 14th IEEE Photovoltaic Specialist Conference 1980; 1980. pp. 1126-1132

[13] Krauter S, Preiss A. Comparison of module temperature measurement methods. In: Conference Record of the IEEE Photovoltaic Specialists; 26617(4): 2009. pp. 333-338. DOI: 10.1109/pvsc.2009.5411669

[14] Inropera FP, DeWitt DP, Bergman TL, Lavine AS. Fundamentals of heat and mass transfer. In: Water. Vol. 6th. 2002. p. 997. ISBN: 9780471457282

Chapter 16

Renewable Energy Application for Solar Air Conditioning

Rubeena Kousar, Muzaffar Ali, Nadeem Ahmed Sheikh, Faik Hamad and Muhammad Kamal Amjad

Abstract

This chapter presents an overview of various solar air conditioning technologies such as solar PV, absorption, desiccant, and adsorption cooling systems. It includes feasibility and comparative analysis of numerous standalone and hybrid configurations of solar cooling systems, which were investigated in past. In addition, recent developments in use of solar energy as a regeneration source to dehumidify desiccant wheel in different applications are also discussed. Details of system technologies and climate-based performance comparison in terms of various performance factors, for example, COP_{th}, Q_{latent}, $Q_{sensible}$, COP_{solar}, SF, PES, and $\eta_{collector}$ for solar-assisted configurations are highlighted. It is observed that hybridization of solar solid desiccant system results more efficient and cost-effective cooling system as latent and sensible loads are treated independently, especially when regeneration process of desiccant wheel is integrated with solar energy. This review will help to explore further improvements in solar-assisted cooling systems.

Keywords: cooling technologies, solar air conditioning, hybrid desiccant, solar collectors, separate load handling

1. Introduction

Earth has varying climates and environmental conditions depending upon the location and the time of the year. Air conditioning is meant to change the environmental conditions of a space by regulating its humidity, temperature, distribution, and cleanliness [1]. Whereas there are many objectives of developing the heating, ventilation, and air conditioning (HVAC) systems, the ultimate objective is to provide human comfort against extreme weather conditions. Various studies in literature report the fact that human performance is affected by extreme weather conditions. For example, Gagge et al. [2] studied subjects at different temperature ranges (12–48°C) and compared their physical response while concluding that the environmental conditions had drastic effects on the performance of human beings. Decreased performance could be resulted in humid and hot environments with more chances of illness and other health problems. Thus, in extreme environments, the need of efficient air conditioning becomes extremely important.

The air conditioning appliances have a fair amount of pollution effect as most of these systems use energy that is generated using fossil fuels [3]. The demand of electricity has an ever-increasing trend, as a result of which it has increased from 4661 MTOE in 1973 to 9384 MTOE in 2015 [4]. The availability of electricity as a source of energy has been strained due to ever increasing air conditioning demands. It has been reported that energy consumption for space conditioning will be increasing by up to 50% during next 15 years [5]. It is therefore the need of the time to evaluate alternate and renewable energy resources in all sectors, especially in air conditioning. Solar energy is one of the most efficient, clean, and affordable energy alternatives available today, and its use for space cooling and heating has proved to be feasible [6].

The utilization of renewable energy sources like solar energy is being given a serious consideration to meet the power requirements of the air-conditioning sector as energy demands drastic increase for air conditioning applications [7]. In addition, solar energy is both eco-friendly and energy efficient technology [8], which has motivated researchers toward development of hybrid air conditioning systems.

The air conditioning systems are classified into two main categories as shown in **Figure 1**.The first one is known as closed sorption technologies including absorption and adsorption systems, and second one is open sorption technologies including desiccant system. They are further classified as solid desiccant and liquid desiccant systems. However, these technologies are integrated with renewable energy sources especially solar energy source.

The energy saving potentials of absorption systems are more as compared with conventional systems for air conditioning and cooling applications [9]. These systems have main advantage of less moving parts [10]. To check the feasibility of solar-assisted absorption system under different climates was investigated by Baniyounes et al. [11], and results show that these systems have ability to save up to 80% when integrated with 50m^2 solar collector's area. Similarly, in another multi climate application study highlighted by Martínez et al. [12] of solar-assisted absorption system, it is shown that the system has ability to achieve 60–78% thermal comfort. In another study of two-stage solar absorption system, a maximum of 1.4 COP was reported [13].

Moreover, to improve the system performance, solar-assisted absorption system was coupled with fix speed and variable speed solar loop pump, and results showed that 11% increment was observed with variable speed pump [14]. The results of transient simulation-based parametric study of different configurations of solar-assisted absorption system show that reduced size system configuration gives 43% SF and 4.1 year payback period, which was found economically best among other configurations [15]. In another study, parabolic trough collector-assisted absorption system with a capacity of 16 kW was analyzed by simulations and experimentally. The results show that system achieves COP in the range of 0.65–1.29 with solar collector efficiency 26–35% and 82% PES when compared with conventional system [16]. Similarly, direct air cooled LiBr-H20 system integrated with solar collector was

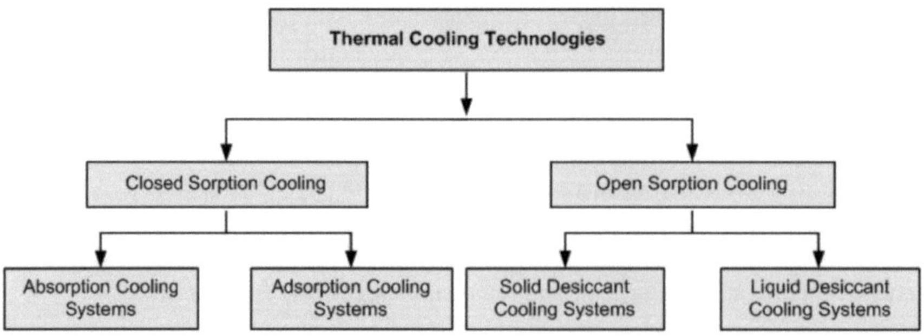

Figure 1.
Classification of thermal cooling technologies.

study experimentally for cooling season reported that 0.6 COP was achieved at 12.8° C temperature of chilled water [17].

The second type of closed sorption technique adsorption cooling systems is also evaluated by different researchers as solar-assisted adsorption cooling system was replaced by convention refrigeration system for the application of grain cooling and storage [18]. In another simulation study of solar-assisted adsorption system saves 23% primary energy as compared to conventional and achieves average COP in the range of 0.1–0.13 and provides 14-22oC chilled air temperature for domestic application [19]. Whereas the drawback of adsorption system was highlighted in [20] that these systems have complicated operating and maintenance mechanism with high cost and less efficient when used for cumulative loads [21, 22].

To avoid environmental hazards of absorption systems, desiccant systems are used as alternative for air conditioning purposes. Commercial conventional desiccant cooling systems are (1) liquid desiccant cooling system (LDCS) and (2) solid desiccant cooling system (SDCS). The liquid desiccant evaporative cooling system gives 68% of energy savings yearly compared to conventional system [23]. An experimental study show that average primary energy ratio was 1.6 and 30% of energy saving was achieved by liquid solar desiccant cooling system [24]. In another similar experimental study, results show that COP of the desiccant system increased about 54% over vapor compression system with reheat and achieved 33–60% energy savings [25]. In an economic comparison of proposed and conventional liquid desiccant system, results show that payback period of proposed system to return initial cost was 7 years and 8 months [26]. Significant energy savings were achieved in Hong Kong for three different commercial buildings where liquid desiccant system was deployed to handle latent and sensible loads [27].

However, performance of DCS can be improved by utilizing low grade renewable energy sources for regeneration purposes. Collector efficiency has been reported to increase further from 56% under hot and humid weather when desiccant system integrated with evacuated tube collectors was used [28]. PV panels have also been used for solar energy collection, which minimized the environmental pollution and maximized economic benefits [29].

Solar pond powered liquid desiccant evaporative cooling shows that indirect evaporator cooler was more effective than direct evaporative cooler [30]. However, the LDCS has disadvantages as crystallization risk and difficulty in design for small applications. Desiccant moves with supply air that is harmful for users. For large systems, cost of operating devices increased to handle large loads. To overcome these demerits, solar-assisted LDCS replaces by solar-assisted SDCS as SD cooling system has numerous advantages, for example, these systems are energy efficient, environment friendly with no contribution to ozone layer depletion, reduce electricity demands in hot and humid conditions and provide dry, clean, and comfortable environment, can handle latent and sensible loads separately, and cost effective as low grade energy can be used to remove moisture.

The SDCS has great potential to work efficiently in dry, humid, hot, and very hot climates, saves energy consumption, and provides clean environment. In humid climate, evaporative cooling has not been found efficient for greenhouses, poultries, vegetable, and fruits stores as compared to conventional vapor compression and vapor absorption systems [31]. Furthermore, studies show that solid desiccant cooling system provides CFCs free clean air conditioning [32–34]. Another feature of SDACS is that it can handle sensible and latent loads separately as compared to conventional systems [35, 36] and provides improved indoor air quality by controlling temperature and humidity. Desiccant systems have been reported to handle 51.7% humidity load. Conventional systems need more fossil fuel energy to control humidity and temperature, which pollute the environment [35, 37, 38], whereas desiccant system serves as an alternative to conventional systems for wet market applications, and results show that 1–13% less CO2 emissions can also be achieved by them [36]. In hot and humid climate, electric energy saving by desiccant system was found to be 24% [39], and 46.5% energy savings were achieved as compared to conventional systems [40]. It was predicted that desiccant system can efficiently use low grade renewable energy and increase COP as compared to conventional systems [35]. Furthermore, 50–120% increase has been reported in COP by utilizing solar energy, and reduced gas usage has also been achieved [41, 42]. Many experimental and simulation-based studies were carried out to make developments in standalone and hybrid desiccant air conditioning systems [43] as this technology development was started in 1979 by Shelpuk and Hooker [33], and its applications are expanding widely due to more efficient as compared with conventional systems [44].

2. Solar-assisted solid desiccant air conditioning

SASDAC system has four main components (1) desiccant dehumidifier, (2) sensible heat exchanger, (3) cooling unit, and (4) solar regeneration heat source. Main component of solid desiccant system basic working principle is elaborated below and pictorially presented in **Figure 2**. During process at stage (1–2) hot and humid air from outside enters in system and passed through desiccant wheel and becomes hot and dry as desiccant wheel absorbs moisture. This hot and dry air passes through heat recovery wheel (2–3) where heat exchange between return and primary air takes place. Then this air passes through humidifier at stage (3–5) moisture added to obtain desired cooling effect and enters in conditioned space. At stage (6–7), air returns from room and passed through humidifier where moisture added to reduce temperature. This moist air passes through heat recovery wheel at stage (7–8) and becomes hot. This hot air passes through heating coils at stage (8–10) and desiccant material regenerated by increasing the temperature using solar energy.

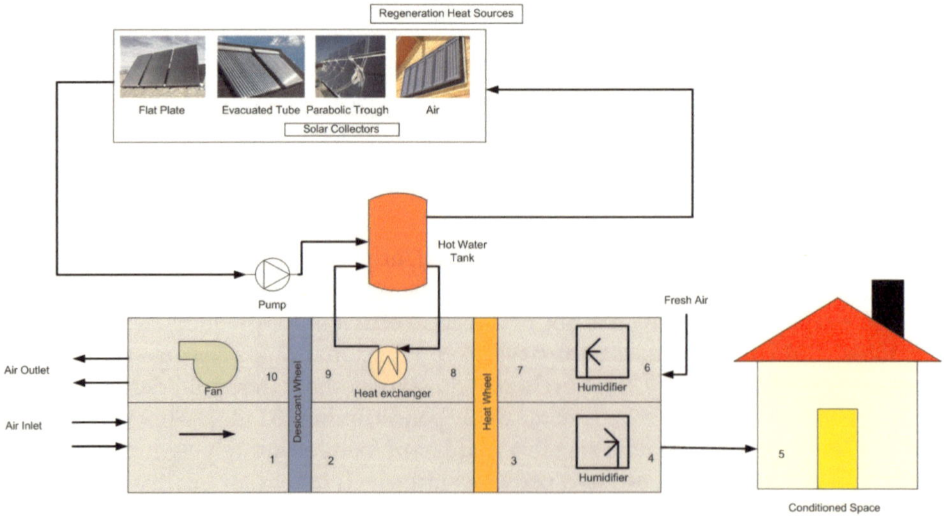

Figure 2.
Working principle of solar-assisted solid desiccant cooling system [20].

2.1 Classification of solar-assisted hybrid desiccant cooling system

The SAHSDCS is combined ability of air-conditioning system and cooling unit to remove latent and sensible loads separately by desiccant dehumidification process and cooling unit, respectively, while regeneration of solid desiccant is achieved by solar energy [45]. In other words, driving force for the process is water vapor pressure; moisture is transferred to the desiccant material from air when it is higher than on the desiccant surface, till an equilibrium is achieved. On the other side, desiccant material is regenerated by heating, and water vapor pressure increases on the surface of DW. When low vapor pressure air comes in contact, DW due to pressure gradient moisture transfers to the air, and desiccant material is regenerated.

The main classification of the hybrid solar-assisted solid desiccant cooling system is based on the cooling units used to reduce the temperature of dehumidified air and removes moisture to achieve comfort conditions. **Figure 3** presents a proposed classification for solar-assisted hybrid solid desiccant cooling system. Hybridization of SASDCS can be done with various conventional cooling technologies, which are DEC, VC, VA, and innovative modern evaporator cooler called Maisotsenko cycle (M-cycle).

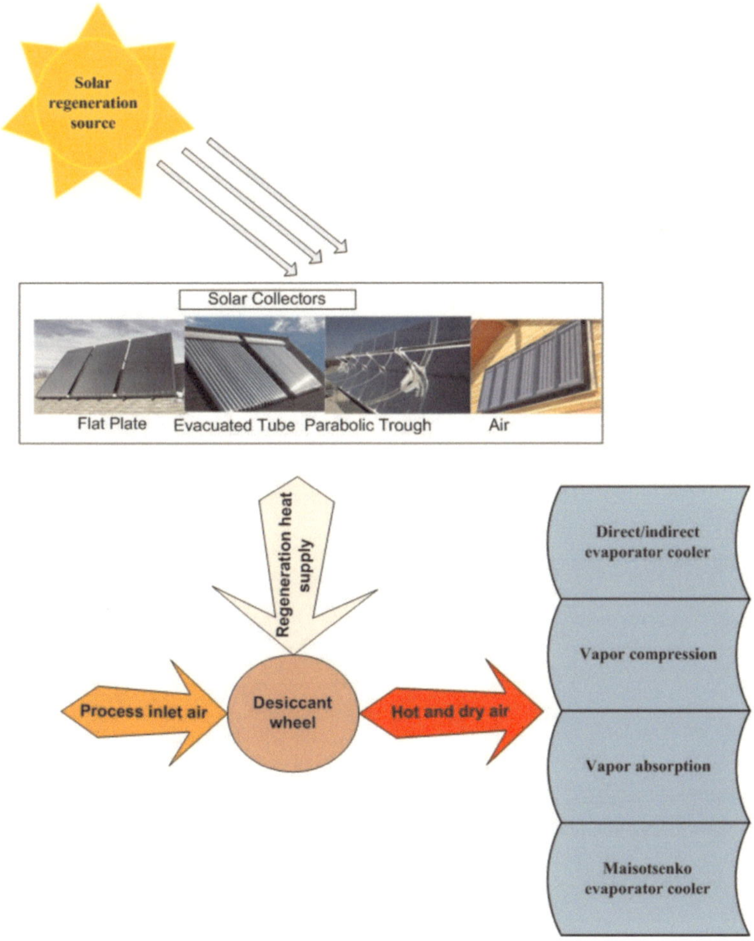

Figure 3.
Classification of solar-assisted hybrid desiccant cooling system.

2.2 Desiccant materials

Desiccant materials can be defined as materials that can adsorb water vapor from moist air and regenerated at low temperature [46]. Classification of desiccant materials is found in the literature as solid or liquid desiccant, natural or artificial desiccant, composite and polymer desiccant, bio or rock-based desiccant. **Figure 4** presents the classification of desiccant materials used in solid desiccant systems.

Silica gel is a granular or beaded form with amorphous microporous structure [47]. Large amount of water vapors can be adsorbed by desiccant material and can also be desorbed at low regeneration temperature. Similarly, composite desiccants are developed from synthetic zeolite and silica gel to achieve high dehumidification under different climatic conditions [43].

Studies have also shown that composite desiccants can give better results as compared to conventional silica gel, for example, [35]. Synthetic zeolite is suitable for different applications where dehumidification is required due to strong ability to adsorb moisture contents [48]. Water sorption analysis of clinoptilolite shows that less dehumidification capability is compared to silica gel and alumina [49]. Activated alumina has shown satisfactory results when used for desiccant dehumidification [50]. Furthermore, use of liquid desiccants, for example, lithium chloride, lithium bromide, and calcium chloride results in good COP of desiccant air conditioning because it regenerates at lower temperature [51]. Higher performance found at high humidity and low regeneration temperature [52].

As compared to silica gel, dry coconut performs better at low regeneration temperature [53]. Another naturally available porous adsorbent material is clay. The performance of this type of desiccant materials depends on their source and activation type. It was found that when bentonite clay was chemically treated with hygroscopic materials, their water vapor adsorption capacity increased by 20% [54].

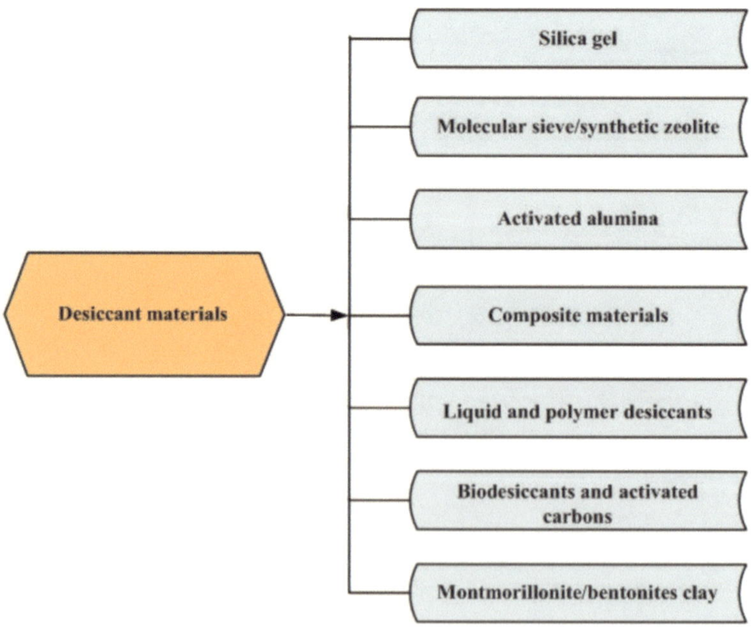

Figure 4.
Classification of desiccant materials.

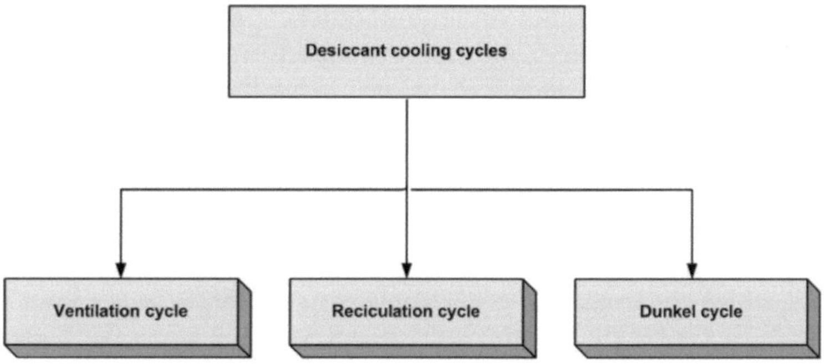

Figure 5.
Classification of desiccant cooling cycle.

2.3 Solid desiccant cooling cycles

Pennington [55] patented the earliest desiccant cooling mechanism in 1955. Since then many researchers have investigated the area. A desiccant can absorb water from its surrounding environment. The solid desiccant adsorbs moisture from air. Jain et al. [45] have classified the solid desiccant cooling cycles as shown in **Figure 5**.

Pingeton cycle is known as ventilation cycle in which air exhausted at the end of regeneration process and fresh air intake for further process. When building exhaust cannot be incorporated for coprocessing, a modified ventilation cycle also proposed but the drawback of this cycle is low cooling capacity and COP than standard cycle due to high temperature and humidity ratio. To increase the cooling capacity of the system, recirculation cycle was developed in which return air reused in process side and fresh air used for regeneration side but its COP not more than 0.8, the drawback of this cycle is lack of fresh air in conditioned space. Another cycle was developed by integrating an additional heat exchanger to take advantages of both ventilation and recirculation cycles named Dunkel cycle.

3. Hybridization of solar-assisted solid desiccant cooling system

This section presents recent research trends and literature review of SAHSDS. The major hybridization options for SADCS are already mentioned in Section 3.1.

Many research studies have shown that hybridization increases COP of SASDCS. An experimental investigation of SASDCS shows that COP of the system was increased due to solar energy utilization between 50 and 120% [41]. In another simulation study, the electrical COP of the system was found to be in the range of 1.22–4.07, and to regenerate desiccant, temperature range was 50–70°C, while at constant airflow rate, COP was found to be 3.2 [56]. Moisture control is an important aspect of the HVAC system. A two-stage air dehumidification system studied shows that this system has ability to remove moisture from incoming air by 8–10 g water per kg of dry air in tropical climate, and thermal COP of system was found to be 0.6 [57]. Similarly, in study of another two-stage SDACS COP was found 0.97 [58]. It was found that self-cooled solid desiccant coated heat exchanger system has higher thermal COP [59].

Use of solar energy reduced the 21% natural gas usage yearly, and experimental results showed that 35% of total cooling load was handled by solar energy [42]. Another simulation-based study reported that dehumidification decreased the latent load and provided humidity level for human comfort but increased the sensible load. It has also been observed that PV panels could easily meet the requirement of energy demand but they were unable to fulfill the air-conditioning demand [60]. For cooling and hot water production, it was reported that by using minimum backup electric energy, hybrid system performed better as SDCS reduced both the temperature and the moisture content of the incoming air using solar energy [61].

Bader et al. [62] presented their study for 17 cities in different regions of world and gave recommendations for the configurations and the design of solar desiccant system for different international regions. Impact of collectors on air conditioning system has also been studied. Evacuated tube collector was used to utilize 44% of solar energy, which achieved below 18% moisture content in 2 days [63]. Another study reported that solar air collector's efficiency was 50% when flat plate collector was used in Germany and Spain, whereas two-stage desiccant system provided 88% dehumidification efficiency in China [64].

System comparisons have been carried out along with financial analysis to assess the feasibility to show that SDACS performs more effectively than conventional systems with payback periods 4.7 years in Berlin and 7.2 years in Shanghai [65]. In experimental study, it was found that highest COP and exergy efficiency were achieved for Dunkel configuration in ventilation mode as 0.6 and 35%, respectively, while the Uckan and Dunkel configurations consumed 50% lower electrical energy [67].

3.1 Solar-assisted hybrid solid desiccant-based direct evaporator cooling system (SAHSD-EVC)

In SAHSD-EVC system, air passes through desiccant wheel where moisture is absorbed by desiccant material due to pressure difference, and temperature rises at the exit. This hot and dehumidified air then flows through heat recovery wheel and then DEC/IEC to cool the air at desired conditions for space. In regeneration side, return air flows through evaporator cooler, heat recovery wheel and then heating coil where temperature of air increases by using solar water heating system. This hot air passes through desiccant wheel and regenerates the desiccant material. A schematic diagram of such system is presented in **Figure 6**.

Literature reports various studies of these systems. Simulation results show that SAHDC-EVC for pre-cooling post-cooling of air achieved higher COP and payback period of about 14 years by economic assessment [68]. In other study, it was found that hybrid system provided comfort conditions in different climate zones and achieved highest and lowest COP values 1.03 and 0.15, respectively [69]. It has also been reported through simulation study that the cooling capacity of the system is increased by 40–60%, and energy consumption is reduced by 20–30% [70]. To achieve comfort conditions, SAHSD-EVC without thermal back up was analyzed for different cities of Australia, and it was found that ventilation cooling cycle-based desiccant system is not suitable for tropical climates [71]. SAHSD-EVC with active heat pump cooling and dehumidification can be achieved simultaneously by pre-heating regeneration air [72]. Full year performance with SAHSD-EVC was investigated under different climates, and primary energy savings were found up to 50% in south Europe and hot climatic conditions whereas in Frankfurt it was about 66% [73]. Furthermore, comparison between numerical and experimental results of SAHSD-EVC showed the latent load for 51.7% can be totally handled by the two-stage desiccant cooling unit [37]. Similarly, another SAHSD-EVC achieved a 0.7 COP with 22% of solar fraction during the cooling season, and COP can be increased by increasing collectors' area [74].

Seasonal analysis has predicted that 60% humidity load was efficiently handled by hybrid system and 70% of total cooling, and 40% heating load was handled by solar-assisted two-stage desiccant cooling system [75]. It has also been reported that air inlet velocity in regeneration side has strong effect on optimal rotational speed in case

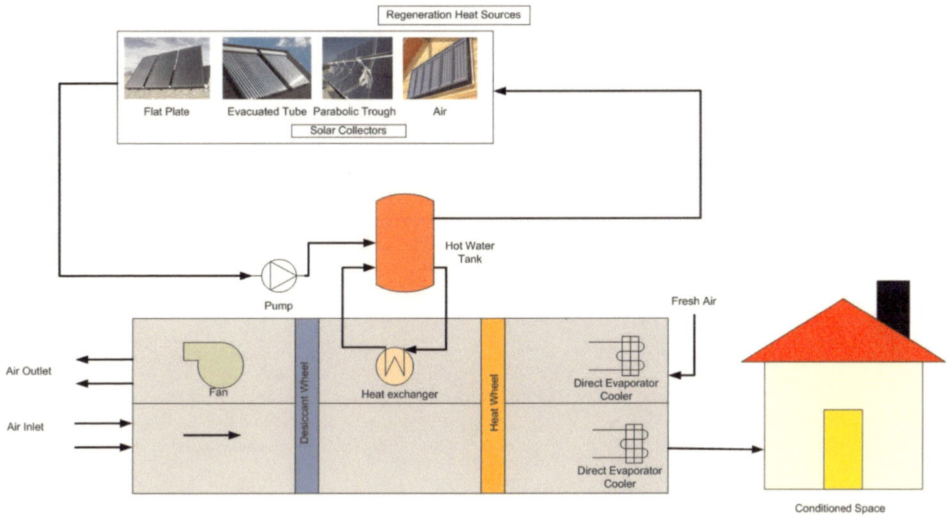

Figure 6.
Solar-assisted hybrid solid desiccant-based direct evaporator cooling system.

of one rotor six-stage solar desiccant cooling system [76]. Experimental investigation of SAHSD-EVC has revealed that thermal COP is strongly affected by optimal cycle time. System used 100% fresh air for mild conditions, and for high humidity, it was proposed to use primary return air with fresh to attain satisfactory supply air condition [77]. It was found that the energy performance of SAHSD-EVC system was more sensitive to outdoor humidity ratio as higher humidity ratio decreases the COP [78]. To investigate SAHSD-EVC by selecting optimum hot water and supply air conditions, system provides supply air 5.15 g/kg humidity ratio with supply air 28.3°C temperature and 1.78 COP [79].

3.2 Solar-assisted hybrid solid desiccant-based vapor compression cooling system (SAHSD-VC)

SAHSD-VC cooling system handles latent and sensible loads separately as desiccant wheel works to dehumidification of process air while vapor compression unit performs cooling operation as shown in **Figure 7**. In process side, ventilated or recirculated air first passes through desiccant wheel where moisture is absorbed due to pressure difference and dehumidifies the air. During this dehumidification process, temperature increases. This hot air passes through the heat recovery wheel where it is cooled and then passes through vapor compression unit to attain desired cooling and comfort conditions for selected space. In regeneration side sensibly, heated air from conditioned space passes through heat recovery wheel where it cools the air in process side, and temperature of the air rises at exit of heat wheel, but humidity remains constant. This hot air passes through heating coils of solar water heating system, which utilizes solar energy to elevate the temperature of water and transfers heat to regeneration air, and as result of it, desiccant material regenerated, so hot and humid air available at exit of desiccant dehumidifier.

In simulation-based study, it was found test control strategy for cooling season and compared with compression system that SAHSD-VC saves 40% energy in French climate [80]. Furthermore, another study results show that under Beijing, Shanghai, and Hong Kong, weather proposed system can remove 57, 69, and 55% moisture and reduce 32, 34, and 22% electric power. However, hybrid system is found feasible for humid, temperate, and extreme humid weather conditions.

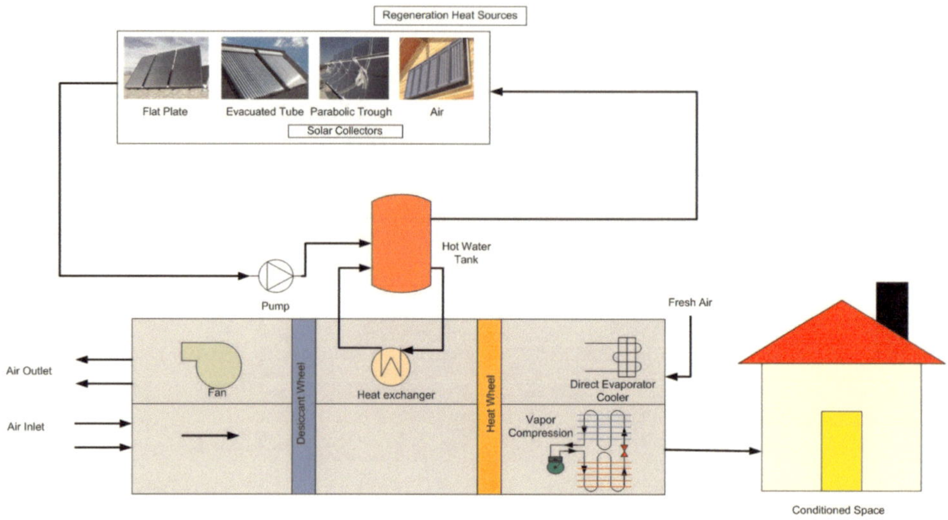

Figure 7.
Solar-assisted hybrid solid desiccant-based vapor compression cooling system.

In simulation-based study, it was found that SAHSD-VC operates under the condition with higher evaporation and condensation temperature to achieve COP of about 5.7 and adjustable MRC [81].

Another experimental study found that SAHSD-VC system performance increased as compared to VCS [82]. Similarly, in another study, SAHSD-VC is capable to handle high latent load and has energy saving potential than conventional system by 49.5% in the Chinese restaurant and 13.3% in the wet market [83]. In another study of two-hybrid cooling systems which were regenerated by solar and electric energy shows that solar SAHSD-VC saves more energy in humid climates than conventional vapor compression system [84]. It was reported in another study of SAHSD-VC that electric COP during summer operation was 2.4 and heat rejected by the chiller used for preheating airflow in regeneration side can reduce the collector area by about 30% [85]. Another experimental study conducted to examine the SAHSD-VC, 18% energy savings with 0.83 COP and 48% desiccant efficiency were achieved [86]. Similarly, experimental study shows that SAHSD-VC saves 46.5% energy than conventional system [39]. In experimental investigation of SAHSD-VC shows that process air humidity 61.7% reduces in hot and humid climates, and by varying the ambient conditions, results indicate that system performance is very sensitive to ambient conditions [87].

To predict the performance of rotary solid desiccant dehumidifier in SAHSD-VC using ANN shows that maximum percentage difference between the ANN predictions and the experimental values was found to be 7.27% for latent load handling and 3.22% for dehumidification effectiveness [88]. In another study, it was found that SAHSD-VC provides cold and dry supply air of 26°C, 8.9 g/kg and the corresponding COP reaches to 7.0 in summer, whereas in winter, supply air from the system is 26.6° C, 14.1 g/kg and the COP reaches up to 6.3 [89]. In another study, author reported that SAHSD-VC with solar panels having total collecting area of 102 m2 provides 77% of required regeneration heat to operate the system [90]. Similarly, SAHSD-VC using PV panels and PVT as power source, power consumption was 19.9 and 10.4% respectively. While in recirculation mode, 61.4 and 57.9% for ventilation and recirculation mode, respectively, less power as compared to reference system [91]. Furthermore, hybrid system was optimized by varying the temperature and humidity of the process air. Due to higher evaporation temperature, 75% share segment of the evaporator remains dry, therefore the consumption of electricity is reduced. The system required 37.5% lower energy as compared to standalone VCS [92]. In another SAHSD-VC study, capacity of VCS is reduced from 23 to 15 kW at the full demand, and the sensible capacity of the system is also improved from 0.47 to 0.73 with payback period is 5 years, and total savings for 20 years life cycle is 4295.19 USD [93]. In experimental comparison of VCS and SAHSD-VC by different operating parameters shows that at room temperature 26.7–10°C, the most suitable rotor speed is 40–50 rph, and moisture extraction ability of SAHSD-VC was improved by 17.6–27.1% as compared to the VCS [94].

3.3 Solar-assisted hybrid solid desiccant-based vapor absorption cooling system (SAHSD-VA)

SAHSD-VA cooling system as shown in **Figure 8** is designed to handle the latent load by desiccant and sensible cooling load by absorption, and the results show that proposed system feasible for high cooling demands with 36.5% lower energy consumption and reduces carbon emissions [95]. In an investigation of a SAHSD-VA shows that SAHSD-VA is environmental friendly and suitable for handling high latent loads. In comparison with other cooling technologies, SAHSD-VA with micro-generators reduces 34% emissions [96]. To improve the performance of solar-assisted absorption system by three integration strategies of components, they found that proposed strategies have less primary energy consumption and up to 50.6 and 25.5% year round energy savings than VCS and basic VA system, respectively [97]. In detail, a SAHSD-VA using six different configurations was investigated, and the results show that SAHSD-VA consumes 57.9% less power than SDCS [91].

3.4 Solar-assisted hybrid solid desiccant-based M-cycle cooling system (SAHSD-M)

SAHSD-M cycle cooling system has been schematically presented in **Figure 9**. The process side air flows through desiccant wheel where moisture is absorbed and its temperature increases. Hot and dry air then passes through heat exchanger and

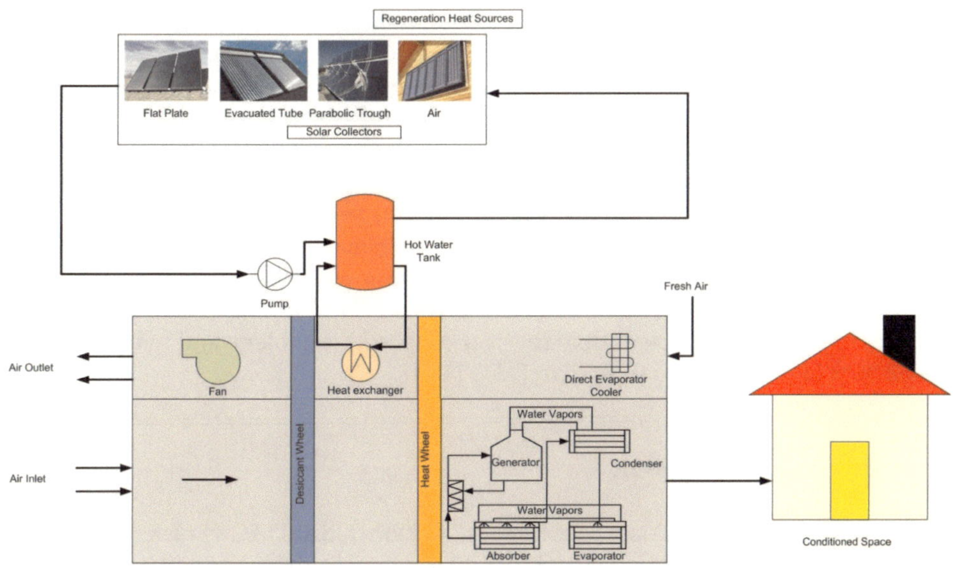

Figure 8.
Solar-assisted hybrid solid desiccant-based vapor absorption cooling system.

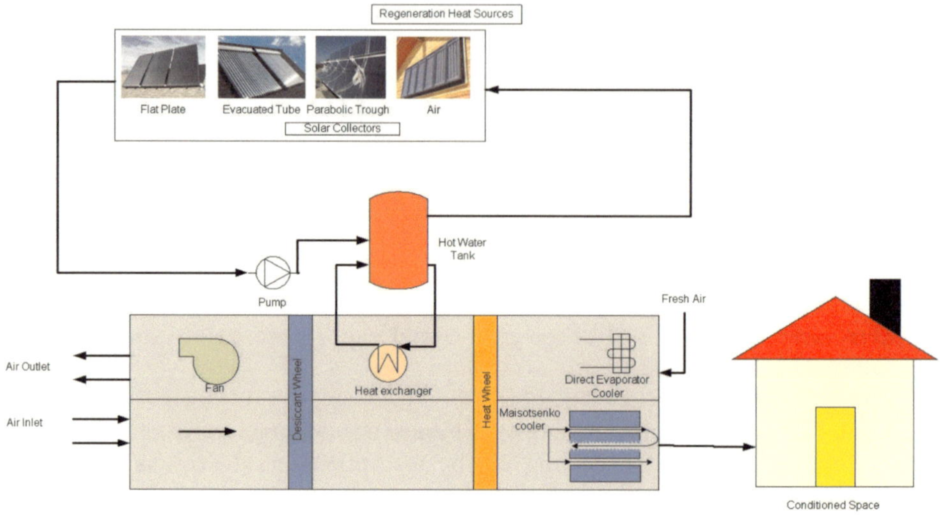

Figure 9.
Solar-assisted hybrid solid desiccant-based M-cycle cooling system.

M-cycle where air is divided in parts. Working air flows in wet channels, whereas product air travels through dry channels and gets moisturized, and sensible heat transfer takes place. As a result, this air becomes warm and saturated and discharges to atmosphere while remaining part of air moves in dry channels and cooled below the wet bulb temperature and delivered to conditioned space. In regeneration side, air passes through heat wheel and then through solar heating system and becomes hot and moves to desiccant wheel where desiccant material is regenerated.

The SAHSD-M is suitable for hot and dry climate and less suitable for hot and humid climate of Guangzhou and Shanghai [98]. To analyze proposed SAHSD-M, at low regeneration temperature 50–60°C, SAHSD-M provides comfort conditions for moderate climate [99]. Similarly in another numerical study, SAHSD-M with cross flow Maisotsenko cycle heat and mass exchanger was compared with a conventional system, and it was found that SAHSD-M system performance was in comfort zone in typical moderate climate conditions [100]. Furthermore, two-stage SAHSD-M for hot and humid climate and transient analysis show that system average COP was 0.46 [101]. Another study of SAHSD-M was designed to assess the solar energy utilization for two different configurations in humid climate. Average COP for two configurations is 0.2495 and 0.2713 and with solar shares 32.2 and 36.5%, respectively [102]. A numerical study of the different arrangements of the SAHSD-M under different inlet air conditions was carried out, and then based on results modified, the third configuration that provides thermal comfort regardless of the outdoor conditions [103]. Similarly, a hybrid system was compared with DAC under different operating parameters. It was found that Maisotsenko evaporative coolers are 16% more efficient than indirect evaporative coolers, and hybrid system has 62.96% higher value of COP than DAC [104].

4. Analysis and discussion

As noticeable from the data presented in Section 3, SASDAC systems are an important research area which is highly published, and efforts are still being made to attain good solutions to utilize freely available solar energy to develop systems which can perform efficiently in different climatic zones.

4.1 Performance-based studies on SAHSDCS

Table 1 presents performance-based studies conducted in past by different researchers to highlight different SAHSDCS in terms of COP, cooling capacity, energy savings, moisture removal, etc.

References	Research type	Climate	Desiccant wheel	System description	Findings
[71]	Experimental, simulation	Hot	Two stage	D + EV	Ventilation cooling cycle is not suitable for tropical climates
[77]	Experimental	Hot and humid	Single stage	D + EV	COP increases
[105]	Simulation	Humid	Single stage	D + EV	Energy saving high moisture removal
[106]	Experimental	Hot and humid	Single stage	D + EV	The COP was found 0.46 with a CC of 353.8 W
[78]	Experimental, simulation	Tropical climate	Single stage	D + EV	Comparative difference of experimental and simulation results varies from 0.2 to 3%, and the humidity ratio varies from 9 to 14%
[79]	Experimental	—	Single stage	D + EV	System supply air at 28.3°C, 5.15 g/kg with 1.78 COP
[73]	Simulation	Multiple climates	Single stage	D + EV	Save 50% primary energy
[74]	Simulation	Subtropical	Single stage	D + EV	Achieved 0.7 COP with 22% of solar fraction
[107]	Simulation	Multi climates	Single stage	D + EV	The maximum system COP is 7
[86]	Experimental, simulation	Hot and humid	Single stage	D + VC	18% energy savings with a COP of 0.83 and 48% efficiency
[108]	Experimental	Multi climate	Two stage	D + VC	35.7% of the CC provided by the SAHSD-VC
[84]	Numerical	Hot and humid, hot and dry	—	D + VC	SAHSD-VC saved more energy than VCS
[90]	Experimental	South European	—	D + VC	Innovative system is still very efficient as its PER is twice as high as the one of the considered reference systems
[100]	Numerical		Two stage	D + M	Higher temperature effectiveness than the traditional solution
[102]	Experimental	Humid	Two stage	D + M	COP for two configurations are 0.2495 and 0.2713, and solar shares are 32.2 and 36.5%, respectively
[103]	Simulation, modeling	Moderate climate	Single stage	D + M	Provide comfort conditions and desiccant wheel regenerated at low temperature

Table 1.
Performance-based studies on SAHSDCS.

Köppen climate classification	Average COP summer	Average COP winter
Csa (subtropical)	>2	≈0
Cfa (semiarid)	2.6	0.55
Aw (Tropical wet)	7	2
Cfb (oceanic climate)	>2	≈0

Table 2.
Performance comparison of SADCS for different climates [107].

Table 2 presents performance of SDEC system that was compared with conventional VAV system for office building for different climates. Solar collector area was taken 760m^2, 3 kg/s volume flow rate, and 3.5m^3 storage tank volume. A simulation model of the building is developed using Energy Plus software. Simulation results show that if economic factors are considered, the application of the SDEC technology would be more beneficial in Aw climate zone applications with an annual energy savings of 557 GJ and CO2 emission reduction of 121 tones. The maximum system COP is 7. For Cfb climate, the SDEC system is not as energy

References	Working fluid	Research type	Climate	System description	Findings
[80]	Silica gel	Experimental, simulation	Hot and humid	D + EV	Saves 40% energy for French climate.
[61]	Silica gel, titanium dioxide	Numerical, experimental	Multiple climates	D + EV	Titanium dioxide is more efficient than silica gel
[109]	Lithium chloride	Modeling, experimental		D + EV	A comparison of experimental and simulation results shows good compliance for wheel operation after adjusting relevant model parameters
[67]	—	Simulation	Hot and humid	D + EV	Dunckle cooling cycle has higher COP
[37]	—	Simulation	Hot and humid	D + EV	51.7% latent load totally handled by hybrid system, 49% solar energy used for heating
[83]	Silica gel	Experimental	Hot and humid	D + VC	Save energy consumption by 49.5% in the Chinese restaurant and 13.3% in the wet market
[39]	Silica gel	Numerical, experimental	Hot and humid	D + VC	20% energy consumption reduces at high humidity
[85]	Silica gel	Experimental	Humid	D + VC	Primary energy savings 50% achieved
[88]	Synthesized metal silicate	Simulation, experimental	Hot and humid	D + VC	Hybrid system saves primary energy
[110]	Silica gel	Experimental	Hot	D + AB	47.3% primary energy consumption lower than conventional

Table 3.
Comparison-based studies on SAHDAC.

efficient as the conventional VAV system. SDEC system is technically and environmentally more feasible for high cooling demand in hot and humid climates.

4.2 Comparison-based studies on SAHDAC

Literature survey shows that SAHDAC system performs efficiently as compared to conventional systems as listed in **Table 3** in different climatic conditions.

Table 4 presents a feasibility study of three different solar-assisted cooling technologies including SDEC system, SDCC system, and SAC system that was carried by [111]. These systems then compared to conventional VCS. Performance of each system was measured in terms of SF, COP, PBP, and annual energy savings. It was found that SDEC performs efficiently in hot and humid climate as it is most economical and environment friendly.

Different configurations of DEC based on operating cycle were investigated by Ali et al. [112] in different Köppen climate zones, and results show that performance of ventilated cycle is more suitable in BWh(arid) and Cfa (semiarid), while ventilated Dunkel cycle for Dfb (temperate), Cwa (dry summer), and Csa (subtropical) are weather conditions as shown in **Table 5**.

4.3 Economic and optimization-based studies

To evaluate the economic and optimal SAHSDCS, many researchers work in this area and find payback period of solar thermal source as well as cooling and dehumidification system, and also parametric analysis was performed to find optimal system for different climates and applications as shown in **Table 6**.

4.4 Effect of solar collector on SAHDCS

Table 7 presents summary of performance of solar collectors used in SAHSDC. It is based on the previous research work carried out in various climates in the world by researchers. The efficient utilization of solar energy for system performance is very encouraging to use solar energy.

Köppen climate classification	SF			COP			Annual energy savings (GJ)		
	SDEC	SDCC	SAC	SDEC	SDCC	SAC	SDEC	SDCC	SAC
Csa (subtropical)	0.68	0.45	0.6	2.9	1.9	2.9	196.88	34.14	211.22
Cfa (semiarid)	0.79	0.62	0.7	8.8	2.98	3.4	349.77	25.51	261.5
Cfb (oceanic climate)	0.55	0.4	0.43	2.1	1.8	1.9	141.52	11.75	158.03
Aw (Tropical wet)	0.81	0.6	0.68	25.5	6.2	3.6	855.88	384.34	277.64

Table 4.
Comparison of cooling technologies in different climates [111].

Configuration	Climate zones with Köppen climate classification				
	(Dfb)	(Cwa)	(Csa)	(BWh)	(Cfa)
Ventilation	0.19	0.76	0.65	2.46	3.03
Ventilated Dunkel	0.4	0.89	1.01	1.66	1.75

Table 5.
Operating cycle-based performance of DEC in different climate zones.

References	Research type	Climate	System description	Findings
[68]	Experimental, simulation	Hot and humid	D + EV	Payback period of solar collector 14 years and system 1 and 1.5 years, uncertainty in the COP was 11.76%
[113]	Numerical, experimental	Hot and humid	D + EV	4.86 years for the energy cost 0.45 LE/kW h
[101]	Experimental	Hot and humid	D + M	System average COP was found 0.46
[114]	Experimental	Hot and humid	D + EV	21–22°C temperature can be achieved with standalone optimized system
[75]	Numerical, experimental	Hot and humid	D + EV	60% of the humidity load can be handled by desiccant system and 40% of the heating load can be handled by collectors
[72]	Simulation	Hot and humid	D + EV	Hybrid system saves 45.5 MWh
[76]	Numerical	—	D + EV	Velocity of regeneration side air affects the moisture removal ability

Table 6.
Economic and optimization-based studies of SAHSDCS with findings.

Ref	Year	Collector type	Collector area	Outcomes
[68]	2009	FPC	$12m^2$	Payback period of solar collector 14 years
[80]	2008	FPC	$100\ m^2$	40% energy saving for French climate
[75]	2014	ETC	$15\ m^2$	Collectors contribute to handle 40% load
[61]	2012	FPC	$12,\ 14\ m^2$	Collector efficiency varies 50–70% for different locations
[106]	2016	ETC	$14\ m^2$	64.3°C attained by solar collectors for regeneration
[73]	2012	FPC	$285\ m^2$	Saves 60.5% primary energy
[37]	2013	ETC	$92.4\ m^2$	49% of total heating load handled by solar collectors
[74]	2012	FPC	$10\ m^2$	22% solar fraction during cooling season
[86]	2013	FPC	$10\ m^2$	Coefficient of performance of 0.83
[108]	2011	FPC	$90\ m^2$	Average efficiency of solar heating subsystem 0.32
[85]	2012	FPC	$22.5\ m^2$	Summer and winter collector efficiency 38 and 30%, respectively
[90]	2018	FPC	$102\ m^2$	Primary energy ratio improved
[110]	2010	ETC	$100\ m^2$	High solar thermal gain in cooling season
[66]	2016	ETC	$100\ m^2$	SF for Abu-Dhabi lower than Riyadh
[102]	2016	PV/T	$681,\ 656\ m^2$	Solar shares are 32.2 and 36.5% for proposed configurations

Table 7.
Performance of solar collectors used in SAHSDCS.

4.5 Applications of solar-assisted solid desiccant system

Fast technical developments in HVAC systems during last few years have produced severed environmental problems as these systems contribute to human comfort with harmful effects on environment through ozone depletion and global

Applications	References
Commercial	[42, 56, 58, 64, 65, 95–98, 101, 107, 108, 115, 116]
Residential, office, hospital buildings	[73, 89, 91, 94, 105, 117, 118]
Automobile, marine, and museum air conditioning	[119–123]
Storing food and fiber drying	[44, 63]
Hot water production	[115, 124]

Table 8.
Applications of SASDCS.

warming. So, some serious efforts put to develop ecofriendly and economic systems for different applications, and solar-assisted hybrid solid desiccant systems were found feasible where cooling and dehumidification required. **Table 8** shows the potential applications of SASDAC systems in different areas like commercial, domestic, and industry.

5. Conclusion

Performance of air conditioning systems can be enhanced by hybridization in terms of coefficient of performance, cooling capacity, and solar fraction as well as economically more feasible specially when integrated with renewable energy resources such as solar energy for regeneration purposes which cut down the peak electricity energy demand in hot and humid weather as compared to conventional systems.

As dehumidification in desiccant wheel results conversion of latent loads to sensible load and to remove this sensible load evaporator coolers are used to meet required cooling comfort conditions in hot and humid climates. When solar energy used as regeneration source of desiccant, it reduces the electricity cost, and these systems are environment friendly.

Hybridization of conventional vapor compression with solar-assisted solid desiccant results reduction in cost and improves the performance of system under various climatic conditions having high humidity and becomes environment friendly when freely available cheap solar energy uses to regenerate the desiccant wheel and auxiliary thermal energy requirement decreases.

Hybridization of solar-assisted solid desiccant with vapor absorption system results in reduction in source temperature as conventional vapor absorption system required high source temperature and system performance improved, and it became suitable for hot and humid climates.

Hybridization of solar-assisted solid desiccant system with Maisotsenko cooler results no moisture addition in process air, so more comfort conditions achieved easily as compared to simple evaporator cooler and solar-assisted solid desiccant-integrated Maisotsenko cooling systems are sensitive to environment, airflow rate, and rotational speed of desiccant wheel than humidity ratio change.

For right selection of solar-assisted hybrid cooling system in any climate, dry bulb temperature, relative humidity, and availability of solar energy are very important factors that should be considered.

Acknowledgements

Authors are thankful to their parent institutions for providing the support for the research.

Abbreviations

AHU	air handling unit
ANN	artificial neural network
CFC	chloro fluoro carbon
COP	coefficient of performance
D	desiccant
D + AB	desiccant absorption
D + EV	desiccant evaporative
D + M	desiccant Maisotsenko
D + VC	desiccant vapor compression
DAC	desiccant air conditioning
DEC	direct evaporator cooler
DINC	direct/indirect
ETC	evacuated tube collector
FPC	flat plate collector
GJ	giga joules
HD	hybrid desiccant
kW	kilo watt
m/s	meter/sec
MRR	moisture removal rate
MRC	moisture removal capacity
MWh	mega-watt hour
PBP	payback period
PES	primary energy saving
PV	photovoltaic
rph	revolution per hour
SAC	solar air conditioning
SAHSDCS	solar-assisted hybrid solid desiccant cooling system
SASDCS	solar-assisted solid desiccant cooling system
SCOP	system coefficient of performance
SDACS	solid desiccant air conditioning system
SDCC	solar desiccant compression cooling
SDEC	solar desiccant evaporative cooling
SF	solar fraction
USD	united states dollar
VAC	vapor absorption cooling
VAV	variable air volume
VCS	vapor compression system
W	watts

Author details

Rubeena Kousar[1], Muzaffar Ali[2], Nadeem Ahmed Sheikh[3], Faik Hamad[4*]
and Muhammad Kamal Amjad[5]

1 Department of Mechanical Engineering, University of Engineering and Technology Taxila, Pakistan

2 Department of Energy Engineering, Faculty of Engineering and Technology, International Islamic University, Islamabad, Pakistan

3 Department of Mechanical Engineering, Faculty of Engineering and Technology, International Islamic University, Islamabad, Pakistan

4 School of Science, Engineering, and Design, Teesside University, United Kingdom

5 School of Mechanical and Manufacturing Engineering, National University of Sciences and Technology, Islamabad, Pakistan

*Address all correspondence to: f.hamad@tees.ac.uk

References

[1] Vedavarz A, Kumar S, Hussain MI. The Handbook of Heating, Ventilation and Air Conditioning for Design and Implementation. New York: Industrial Press Inc.; 2007

[2] Gagge AP, Stolwijk J, Hardy J. Comfort and thermal sensations and associated physiological responses at various ambient temperatures. Environmental Research. 1967;**1**(1):1-20

[3] Enteria N, Akbarzadeh A. Solar Energy Sciences and Engineering Applications. London: CRC Press; 2013

[4] Birol F. Key World Energy Statistics. rue de la Federation, Paris, France: IEA Publications, International Energy Agency; 2017

[5] Pérez-Lombard L, Ortiz J, Pout C. A review on buildings energy consumption information. Energy and Buildings. 2008;**40**(3):394-398

[6] Raja I, Twidell J. Statistical analysis of measured global insolation data for Pakistan. Renewable Energy. 1994;**4**(2): 199-216

[7] Choudhury B et al. An overview of developments in adsorption refrigeration systems towards a sustainable way of cooling. Applied Energy. 2013;**104**:554-567

[8] Ghafoor A et al. Current status and overview of renewable energy potential in Pakistan for continuous energy sustainability. Renewable and Sustainable Energy Reviews. 2016;**60**:1332-1342

[9] Vazhappilly CV, Tharayil T, Nagarajan A. Modeling and experimental analysis of generator in vapour absorption refrigeration system. International Journal of Engineering Research and Applications. 2013;**3**(5):63-67

[10] Invernizzi C, Iora P. Heat recovery from a micro-gas turbine by vapour jet refrigeration systems. Applied Thermal Engineering. 2005;**25**(8–9):1233-1246

[11] Baniyounes AM, Rasul M, Khan MMK. Assessment of solar assisted air conditioning in Central Queensland's subtropical climate, Australia. Renewable Energy. 2013;**50**: 334-341

[12] Martínez PJ, Martínez JC, Martínez P. Performance comparison of solar autonomous and assisted absorption systems in Spain. International Journal of Refrigeration. 2016;**71**:85-93

[13] Cabrera F et al. Use of parabolic trough solar collectors for solar refrigeration and air-conditioning applications. Renewable and Sustainable Energy Reviews. 2013;**20**:103-118

[14] Shirazi A et al. Solar-assisted absorption air-conditioning systems in buildings: Control strategies and operational modes. Applied Thermal Engineering. 2016;**92**:246-260

[15] Shirazi A et al. Transient simulation and parametric study of solar-assisted heating and cooling absorption systems: An energetic, economic and environmental (3E) assessment. Renewable Energy. 2016;**86**:955-971

[16] Soussi M et al. Model performance assessment and experimental analysis of a solar assisted cooling system. Solar Energy. 2017;**143**:43-62

[17] González-Gil A et al. Experimental evaluation of a direct air-cooled lithium bromide–water absorption prototype for solar air conditioning. Applied Thermal Engineering. 2011;**31**(16):3358-3368

[18] Luo H et al. Experimental investigation of a solar adsorption chiller used for grain depot cooling.

Applied Thermal Engineering. 2006;**26** (11–12):1218-1225

[19] Angrisani G et al. Dynamic simulation of a solar heating and cooling system for an office building located in southern Italy. Applied Thermal Engineering. 2016;**103**:377-390

[20] Allouhi A et al. Solar driven cooling systems: An updated review. Renewable and Sustainable Energy Reviews. 2015; **44**:159-181

[21] Henning H-M. Solar assisted air conditioning of buildings–an overview. Applied Thermal Engineering. 2007; **27**(10):1734-1749

[22] Angelino G, Invernizzi C. Thermodynamic optimization of ejector actuated refrigerating cycles. International Journal of Refrigeration. 2008;**31**(3):453-463

[23] Kim M-H et al. Annual operating energy savings of liquid desiccant and evaporative-cooling-assisted 100% outdoor air system. Energy and Buildings. 2014;**76**:538-550

[24] Armanasco F et al. Performance analysis of a solar cooling plant based on a liquid desiccant evaporative cooler. International Journal of Refrigeration. 2015;**53**:163-176

[25] Bassuoni M. Experimental performance study of a proposed desiccant based air conditioning system. Journal of Advanced Research. 2014; **5**(1):87-95

[26] Keniar K, Ghali K, Ghaddar N. Study of solar regenerated membrane desiccant system to control humidity and decrease energy consumption in office spaces. Applied Energy. 2015;**138**: 121-132

[27] Qi R, Lu L, Yang H. Investigation on air-conditioning load profile and energy consumption of desiccant cooling system for commercial buildings in Hong Kong. Energy and Buildings. 2012; **49**:509-518

[28] Crofoot L, Harrison S. Performance evaluation of a liquid desiccant solar air conditioning system. Energy Procedia. 2012;**30**:542-550

[29] Buker MS, Mempouo B, Riffat SB. Experimental investigation of a building integrated photovoltaic/thermal roof collector combined with a liquid desiccant enhanced indirect evaporative cooling system. Energy Conversion and Management. 2015;**101**:239-254

[30] Elsarrag E et al. Solar pond powered liquid desiccant evaporative cooling. Renewable and Sustainable Energy Reviews. 2016;**58**:124-140

[31] Heidarinejad G et al. Feasibility of using various kinds of cooling systems in a multi-climates country. Energy and Buildings. 2008;**40**(10):1946-1953

[32] Mandegari MA, Pahlavanzadeh H. Introduction of a new definition for effectiveness of desiccant wheels. Energy. 2009;**34**(6):797-803

[33] Shelpuk B, Hooker D. Development programmes in solar desiccant cooling for residential buildings. International Journal of Refrigeration. 1979;**2**(5): 173-179

[34] Ge T et al. Performance investigation on a novel two-stage solar driven rotary desiccant cooling system using composite desiccant materials. Solar Energy. 2010;**84**(2):157-159

[35] Jeong J et al. Performance analysis of four-partition desiccant wheel and hybrid dehumidification air-conditioning system. International Journal of Refrigeration. 2010;**33**(3): 496-509

[36] Lee S, Lee W. Site verification and modeling of desiccant-based system as

an alternative to conventional air-conditioning systems for wet markets. Energy. 2013;**55**:1076-1083

[37] Li H et al. Simulation and parameter analysis of a two-stage desiccant cooing/heating system driven by solar air collectors. Energy Conversion and Management. 2013;**67**:309-317

[38] Panaras G, Mathioulakis E, Belessiotis V. Achievable working range for solid all-desiccant air-conditioning systems under specific space comfort requirements. Energy and Buildings. 2007;**39**(9):1055-1060

[39] Sheng Y, Zhang Y, Zhang G. Simulation and energy saving analysis of high temperature heat pump coupling to desiccant wheel air conditioning system. Energy. 2015;**83**:583-596

[40] Hirunlabh J et al. Feasibility study of desiccant air-conditioning system in Thailand. Building and Environment. 2007;**42**(2):572-577

[41] Hürdoğan E et al. Investigation of solar energy utilization in a novel desiccant based air conditioning system. Energy and Buildings. 2012;**55**:757-764

[42] Hands S et al. Performance analysis & energy benefits of a desiccant based solar assisted trigeneration system in a building. Renewable Energy. 2016;**85**:865-879

[43] La D et al. Technical development of rotary desiccant dehumidification and air conditioning: A review. Renewable and Sustainable Energy Reviews. 2010;**14**(1):130-147

[44] Ismail MZ, Angus D, Thorpe G. The performance of a solar-regenerated open-cycle desiccant bed grain cooling system. Solar Energy. 1991;**46**(2):63-70

[45] Jani DB, Mishra M, Sahoo PK. Solid desiccant air conditioning—A state of the art review. Renewable and Sustainable Energy Reviews. 2016;**60**:1451-1469. DOI: 10.1016/j.rser.2016.03.031

[46] Zheng X, Ge T, Wang R. Recent progress on desiccant materials for solid desiccant cooling systems. Energy. 2014;**74**:280-294

[47] Srivastava N, Eames I. A review of adsorbents and adsorbates in solid–vapour adsorption heat pump systems. Applied Thermal Engineering. 1998;**18**(9–10):707-714

[48] Rao Z, Wang S, Zhang Z. Energy saving latent heat storage and environmental friendly humidity-controlled materials for indoor climate. Renewable and Sustainable Energy Reviews. 2012;**16**(5):3136-3145

[49] White D, Bussey R. Water sorption properties of modified clinoptilolite. Separation and Purification Technology. 1997;**11**(2):137-141

[50] Hamed A et al. Theoretical and experimental investigation on the transient coupled heat and mass transfer in a radial flow desiccant packed bed. Energy Conversion and Management. 2013;**65**:262-271

[51] Koronaki I et al. Thermodynamic analysis of a counter flow adiabatic dehumidifier with different liquid desiccant materials. Applied Thermal Engineering. 2013;**50**(1):361-373

[52] Jia C et al. Use of compound desiccant to develop high performance desiccant cooling system. International Journal of Refrigeration. 2007;**30**(2):345-353

[53] Khedari J et al. Feasibility study of using agriculture waste as desiccant for air conditioning system. Renewable Energy. 2003;**28**(10):1617-1628

[54] Bulut G et al. Production of desiccants from Turkish bentonites.

Applied Clay Science. 2009;**46**(2):141-147

[55] Pennington NA. Humidity changer for air-conditioning, Google Patents; 1955

[56] Ruivo CR, Hernández FF, López JMC. Influence of the desiccant wheel effectiveness method approaches, with fix and variable effectiveness parameters, on the performance results of an airport air-conditioning system. Energy Conversion and Management. 2015;**94**:458-471

[57] Safizadeh MR et al. Two-stage air-dehumidification system for the tropics–experimental and theoretical analysis of a lab system. Energy Procedia. 2014;**48**:982-990

[58] Li H et al. Case study of a two-stage rotary desiccant cooling/heating system driven by evacuated glass tube solar air collectors. Energy and Buildings. 2012;**47**:107-112

[59] Wang H et al. Experimental investigation on solar powered self-cooled cooling system based on solid desiccant coated heat exchanger. Energy. 2016;**96**:176-186

[60] Milani D et al. Experimentally validated model for atmospheric water generation using a solar assisted desiccant dehumidification system. Energy and Buildings. 2014;**77**:236-246

[61] Enteria N et al. Performance of solar-desiccant cooling system with silica-gel (SiO 2) and titanium dioxide (TiO 2) desiccant wheel applied in east Asian climates. Solar Energy. 2012;**86**(5):1261-1279

[62] Bader T et al. Climate specific design and effectiveness of solar DEC-systems: A methodological zoning approach. Energy Procedia. 2014;**48**:778-789

[63] Misha S et al. Performance of a solar assisted solid desiccant dryer for kenaf core fiber drying under low solar radiation. Solar Energy. 2015;**112**:194-204

[64] Eicker U et al. Operational experiences with solar air collector driven desiccant cooling systems. Applied Energy. 2010;**87**(12):3735-3747

[65] Ge T et al. Performance comparison between a solar driven rotary desiccant cooling system and conventional vapor compression system (performance study of desiccant cooling). Applied Thermal Engineering. 2010;**30**(6):724-731

[66] Brumana G, Franchini G. Solar-powered air conditioning for buildings in hot climates: Desiccant evaporative cooling vs. absorption chiller-based systems. Energy Procedia. 2016;**101**:288-296

[67] Abbassi Y, Baniasadi E, Ahmadikia H. Comparative performance analysis of different solar desiccant dehumidification systems. Energy and Buildings. 2017;**150**:37-51

[68] Khalid A et al. Solar assisted, pre-cooled hybrid desiccant cooling system for Pakistan. Renewable Energy. 2009;**34**(1):151-157

[69] El-Agouz S, Kabeel A. Performance of desiccant air conditioning system with geothermal energy under different climatic conditions. Energy Conversion and Management. 2014;**88**:464-475

[70] Hwang W-B, Choi S, Lee D-Y. In-depth analysis of the performance of hybrid desiccant cooling system incorporated with an electric heat pump. Energy. 2017;**118**:324-332

[71] White S, Kohlenbach P, Bongs C. Indoor temperature variations resulting from solar desiccant cooling in a building without thermal backup. International Journal of Refrigeration. 2009;**32**(4):695-704

[72] Aprile M, Scoccia R, Motta M. Modelling and control optimization of a

solar desiccant and evaporative cooling system using an electrical heat pump. Energy Procedia. 2012;**30**:478-489

[73] Preisler A, Brychta M. High potential of full year operation with solar driven desiccant evaporative cooling systems. Energy Procedia. 2012; **30**:668-675

[74] Baniyounes AM et al. Analysis of solar desiccant cooling system for an institutional building in subtropical Queensland, Australia. Renewable and Sustainable Energy Reviews. 2012; **16**(8):6423-6431

[75] Zeng D et al. Numerical analysis and optimization of a solar hybrid one-rotor two-stage desiccant cooling and heating system. Applied Thermal Engineering. 2014;**73**(1):474-483

[76] Elzahzby AM et al. A mathematical model for predicting the performance of the solar energy assisted hybrid air conditioning system, with one-rotor six-stage rotary desiccant cooling system. Energy Conversion and Management. 2014;**77**:129-142

[77] Zhao Y et al. Experimental investigation on a desiccant dehumidification unit using fin-tube heat exchanger with silica gel coating. Applied Thermal Engineering. 2014; **63**(1):52-58

[78] Nie J et al. Theoretical modelling and experimental study of air thermal conditioning process of a heat pump assisted solid desiccant cooling system. Energy and Buildings. 2017;**153**:31-40

[79] Zhang J et al. Experimental investigation on solar powered desiccant coated heat exchanger humidification air conditioning system in winter. Energy. 2017;**137**:468-478

[80] Vitte T et al. Proposal for a new hybrid control strategy of a solar desiccant evaporative cooling air handling unit. Energy and Buildings. 2008;**40**(5):896-905

[81] Ge T, Dai Y, Wang R. Analysis on integrated low grade condensation heat powered desiccant coated vapor compression system. Applied Thermal Engineering. 2018;**138**:307-318

[82] Mandegari MA, Pahlavanzadeh H. Performance assessment of hybrid desiccant cooling system at various climates. Energy Efficiency. 2010;**3**(3): 177-187

[83] Fong K et al. Investigation on solar hybrid desiccant cooling system for commercial premises with high latent cooling load in subtropical Hong Kong. Applied Thermal Engineering. 2011; **31**(16):3393-3401

[84] Hong H, Guohui F, Hongwei W. Performance research of solar hybrid desiccant cooling systems. Procedia Environmental Sciences. 2012;**12**:57-64

[85] Beccali M, Finocchiaro P, Nocke B. Energy performance evaluation of a demo solar desiccant cooling system with heat recovery for the regeneration of the adsorption material. Renewable Energy. 2012;**44**:40-52

[86] Baniyounes AM, Rasul M, Khan MMK. Experimental assessment of a solar desiccant cooling system for an institutional building in subtropical Queensland, Australia. Energy and Buildings. 2013;**62**:78-86

[87] Jani D, Mishra M, Sahoo P. Experimental investigation on solid desiccant–vapor compression hybrid air-conditioning system in hot and humid weather. Applied Thermal Engineering. 2016;**104**:556-564

[88] Jani D, Mishra M, Sahoo P. Performance prediction of rotary solid desiccant dehumidifierin hybrid air-conditioning system using artificial neural network. Applied Thermal Engineering. 2016;**98**:1091-1103

[89] Hua L et al. Experimental investigation on a novel heat pump system based on desiccant coated heat exchangers. Energy. 2018;**142**:96-107

[90] Frein A et al. Field testing of a novel hybrid solar assisted desiccant evaporative cooling system coupled with a vapour compression heat pump. Applied Thermal Engineering. 2018;**130**: 830-846

[91] Fong K et al. Advancement of solar desiccant cooling system for building use in subtropical Hong Kong. Energy and Buildings. 2010;**42**(12):2386-2399

[92] Jia C et al. Analysis on a hybrid desiccant air-conditioning system. Applied Thermal Engineering. 2006;**26** (17–18):2393-2400

[93] Ghali K. Energy savings potential of a hybrid desiccant dehumidification air conditioning system in Beirut. Energy Conversion and Management. 2008; **49**(11):3387-3390

[94] Chung HJ et al. Performance characteristics of domestic hybrid dehumidifier combined with solid desiccant rotor and vapor compression system. Energy. 2017;**141**:66-75

[95] Fong K et al. Solar hybrid cooling system for high-tech offices in subtropical climate–radiant cooling by absorption refrigeration and desiccant dehumidification. Energy Conversion and Management. 2011;**52**(8–9): 2883-2894

[96] Angrisani G, Roselli C, Sasso M. Experimental assessment of the energy performance of a hybrid desiccant cooling system and comparison with other air-conditioning technologies. Applied Energy. 2015;**138**:533-545

[97] Fong K, Lee C. Performance advancement of solar air-conditioning through integrated system design for building. Energy. 2014;**73**:987-996

[98] Zhao X et al. Feasibility study of a novel dew point air conditioning system for China building application. Building and Environment. 2009;**44**(9):1990-1999

[99] Pandelidis D et al. Comparison of desiccant air conditioning systems with different indirect evaporative air coolers. Energy Conversion and Management. 2016;**117**:375-392

[100] Pandelidis D et al. Numerical analysis of a desiccant system with cross-flow Maisotsenko cycle heat and mass exchanger. Energy and Buildings. 2016;**123**:136-150

[101] Gadalla M, Saghafifar M. Performance assessment and transient optimization of air precooling in multi-stage solid desiccant air conditioning systems. Energy Conversion and Management. 2016;**119**:187-202

[102] Saghafifar M, Gadalla M. Performance assessment of integrated PV/T and solid desiccant air-conditioning systems for cooling buildings using Maisotsenko cooling cycle. Solar Energy. 2016;**127**:79-95

[103] Pandelidis D et al. Analysis of different applications of Maisotsenko cycle heat exchanger in the desiccant air conditioning systems. Energy and Buildings. 2017;**140**:154-170

[104] Shahzad MK et al. Experimental evaluation of a solid desiccant system integrated with cross flow Maisotsenko cycle evaporative cooler. Applied Thermal Engineering. 2018;**128**: 1476-1487

[105] O'Kelly M, Walter ME, Rowland JR. Simulated hygrothermal performance of a desiccant-assisted hybrid air/water conditioning system in a mixed humid climate under dynamic load. Energy and Buildings. 2015;**86**: 45-57

[106] Kumar A, Yadav A. Experimental investigation of solar-powered desiccant

cooling system by using composite desiccant "CaCl2/jute". Environment, Development and Sustainability. 2017; **19**(4):1279-1292

[107] Ma Y, Guan L. Performance analysis of solar desiccant-evaporative cooling for a commercial building under different Australian climates. Procedia Engineering. 2015;**121**:528-535

[108] La D et al. Case study and theoretical analysis of a solar driven two-stage rotary desiccant cooling system assisted by vapor compression air-conditioning. Solar Energy. 2011; **85**(11):2997-3009

[109] Wrobel J, Morgenstern P, Schmitz G. Modeling and experimental validation of the desiccant wheel in a hybrid desiccant air conditioning system. Applied Thermal Engineering. 2013;**51, 1**:1082-1091

[110] Fong K et al. Solar hybrid air-conditioning system for high temperature cooling in subtropical city. Renewable Energy. 2010;**35**(11):2439-2451

[111] Ma Y et al. Comparison of different solar-assisted air conditioning systems for Australian office buildings. Energies. 2017;**10**(10):1463

[112] Ali M et al. Performance investigation of solid desiccant evaporative cooling system configurations in different climatic zones. Energy Conversion and Management. 2015;**97**:323-339

[113] Elzahzby AM et al. Effect of inter-cooling on the performance and economics of a solar energy assisted hybrid air conditioning system with six stages one-rotor desiccant wheel. Energy Conversion and Management. 2014;**78**:882-896

[114] Finocchiaro P, Beccali M, Nocke B. Advanced solar assisted desiccant and evaporative cooling system equipped with wet heat exchangers. Solar Energy. 2012;**86**(1):608-618

[115] Enteria N et al. Development and construction of the novel solar thermal desiccant cooling system incorporating hot water production. Applied Energy. 2010;**87, 2**:478-486

[116] Czubinski FF, Oliveira RG, Pinho JT. Electricity reduction in building climatisation in hot and humid climate through the use of solar desiccant wheel. Energy Procedia. 2014; **57**:2572-2580

[117] Jani D, Mishra M, Sahoo P. Performance analysis of a solid desiccant assisted hybrid space cooling system using TRNSYS. Journal of Building Engineering. 2018;**19**:26-35

[118] Miyazaki T, Oda T, Ito M, Kawasaki N, Nikai I, editors. The possibility of the energy cost savings by the electricity driven desiccant system with a high performance evaporative cooler. In: International Symposium on Innovative Materials for Processes in Energy Systems. Singapore: Research Pub. Services; 2010

[119] Nagaya K et al. High energy efficiency desiccant assisted automobile air-conditioner and its temperature and humidity control system. Applied Thermal Engineering. 2006;**26**(14–15): 1545-1551

[120] Guojie Z et al. Development of a new marine rotary desiccant airconditioning system and its energy consumption analysis. Energy Procedia. 2012;**16**:1095-1101

[121] Zhu J, Chen W. Energy and exergy performance analysis of a marine rotary desiccant air-conditioning system based on orthogonal experiment. Energy. 2014;**77**:953-962

[122] Ascione F, Bellia L, Capozzoli A. A coupled numerical approach on

museum air conditioning: Energy and fluid-dynamic analysis. Applied Energy. 2013;**103**:416-427

[123] Ascione F et al. Energy saving strategies in air-conditioning for museums. Applied Thermal Engineering. 2009;**29**(4):676-686

[124] Enteria N et al. Construction and initial operation of the combined solar thermal and electric desiccant cooling system. Solar Energy. 2009;**83**(8): 1300-1311

Chapter 17

Thermal and Hydraulic Analysis of Transfer Medium Motion Regime in Flat Plate Solar Collector

Yedilkhan Amirgaliyev, Murat Kunelbayev, Kalizhanova Aliya, Ainur Kozbakova, Omirlan Auelbekov and Nazbek Katayev

Abstract

In the research herein, we have considered the thermal and hydraulic analysis of transfer media motion mode in the flat solar collector. We have substantiated the thermal and hydraulic parameters of the flat plate solar collector. Heat absorbing flat solar collector tubes hydraulic analysis has shown, that using the heat transfer standard size there might be located the pipeline, the length of which 2.5 times more than of the collector's body, sufficiently increasing at that insolation time on the transfer media.

Keywords: flat plate solar collector, heat, hydraulics, transfer media

1. Introduction

Flat plate solar collector is a type of the heat exchanger, in which the liquid absorbs the energy from the solid surface, exposed to the solar radiation. Dependent on the potentially reached temperature, those equipment types might be classified according to the following parameters: low temperature, average temperature and high temperature. In the work herein we concentrate at low temperature solar collectors, where the maximum temperature, reached by the working liquid, lower, than 100°C. Those types devices thermal characteristic is based on defining the collector's performance and heat losses for the environment. Growing interest in cutting the capital cost of the systems thereof is focused at raising the thermal efficiency at the expense of optimizing the usage of materials for collectors' construction [1].

The work [2, 3] considers industrial applying the solar heating systems which needs considerable amount of hot water. In the research there has been investigated the usage of the solar collector sets, located sequentially, and the system has been experimentally tested to check its adequacy for providing the demanded thermal loading. The work [4] has studied the energy efficiency of independent and centralized heating systems accounting the solar plants integration; specified the energy specific consumption, additional power consumption, which reduces the buildings energy performance. The work [4] has investigated the large-scale solar collector plant for using in hostels, pools, restaurants and production enterprises. The work has been focused at the assessing the performance, without accounting actual collectors' design and consumption capacity. The work [5] presents the integration of renewable power sources for a big meat company. There considered the renewable energy sources: solar power, biomass, some types of wastes and geothermal energy sources. Work [6] has studied the solar energy thermal integration in the process of fish canning applying the combined analysis and exergy analysis.

Pressure drop in water pipeline networks construction for energy distribution has been considered in [7], when there has been revealed, that the considerable energy saving might be reached using the pressure fall through the pressure management strategy in the network in the serial-parallel hydraulic circuit. The work [8] presents some solar water heating construction aspects, appropriate for big and regular demands to the hot water in hospitals and dormitories. There have been experimentally analyzed different schemes of the solar collectors switching ways, such as cascade, series, parallel and true parallel. Proceeding from the outcomes thereof the system with big number of solar collectors shows the maximum efficiency and economic feasibility. In the work [9] there has been conducted the economic assessment of industrial solar thermal plants in Greece. Also, we have considered the mathematical model of separate constructions and operation mode of thermosyphon circulation double circuit solar collector. Proceeding from the analysis results we have managed to optimize individual structural elements, as well, predicted the thermal regime and alternative solutions selection for designing the flat solar collectors and their operation regime selection [10]. In [11], the thermohydraulic performance parameter was calculated, which was used and presented to find a useful increase in thermal energy taking into account the equivalent thermal energy necessary for the production of working energy, as well as hydraulic losses as a result of expanded surfaces on the plate of the solar collector absorber were calculated. In [12], the influence of the shape of the absorber on the heat-hydraulic characteristics of the collector was studied for three general models with air flow. In [13], the experimental characteristics of the hydro-hydraulic parameters of compacted-bed solar air heaters were investigated. The parameter of thermohydraulic characteristics, called the "effective efficiency," was calculated, which was used to find a useful increase in thermal energy, taking into account the equivalent thermal energy. During the experimental work, it was observed that the thermo-hydraulic efficiency decreases with increasing values of the ratio of the layer depth to the size of the element and the porosity of the layer, but it increases with increasing mass air flow, reaches a maximum and subsequently decreases with a further increase in mass flow. In [14], optimization of thermohydraulic characteristics in tripartite artificially roughened solar air heaters was solved. The optimal thermo-hydraulic characteristics of such a solar air heater are both quantitatively and qualitatively better than that of single-sided rough solar air heaters. In [15], the thermohydraulic characteristics of a forced convection solar air heater using the surface of a fin absorber were investigated. The parameters of energy efficiency are experimentally calculated, such as absorbed thermal energy, thermal efficiency of a solar air heater, heat-hydraulic efficiency of an air heater. Verma and Prasad in [16] investigated the optimal thermo-hydraulic characteristics of solar air heaters, in which they determined the maximum heat transfer and the minimum pressure drop. Mittall and Varshney in [17] calculated the optimal thermo-hydraulic characteristics of a metal-mesh solar heater. In [18], an experimental study calculated the parameters of the thermal analysis of triangular glass covers with artificial triangular solar heaters.

The goal of the work hereby is to link the thermal project with hydraulic design to specify the location of the solar collectors which will correspond to both demanded thermal load and prescribed pressure drop.

2. Model of research

In the working hereby the researchers study the thermal and hydraulic analyses of transfer medium motion regime in the flat-plate solar collector.

To carry out the works the researchers analyzed a new flat solar collector, as well, experimentally investigated and substantiated the thermal and hydraulic transfer medium motion regime hydraulic parameters in the flat plate solar collector.

Figure 1 shows the flat plate solar collector's model. Concept and novelty is in the fact, that in distinction from the known design principle, the collector contains a transparent glazed unit 2 with double glass and with a reduced pressure, as well, a parametric frame 1. Wooden frame bottom 7 is made of 8 mm thickness plywood with an attached heat sealing film 5 with foil. In the gap between the glazed unit and frame bottom there is lied a flexible thin-walled stainless corrugated tube in the coil form tubes edges are attached to output and input protruding tubes 6.

Figure 1 demonstrates the flat plate solar collector mockup. Solar collector is the basic heat generating module of the solar plant. To achieve the set goal we have elaborated a principally new flat plate solar collector, based on which there will be constructed the standard series solar plants for water and buildings heating (**Figures 2–4** and **Table 1**).

Thermal mode in the solar collector elements is defined with the form and dimensions of its profile, thermal-physical properties of constituents parts and climatic conditions.

Heat exchanger's heat loading might be computed proceeding from the energy balance as [19]:

$$Q = mc\Delta T \tag{1}$$

where Q—heat loading, m—water flow, ΔT—water temperature raising, prescribed (T_{out}-T_{in}), where T_{out}—water output temperature, and T_{in}—inlet temperature. From the design equation and assuming, that the collector's absorbing surface is under the constant temperature (T), the demanded surface square equals to [20]:

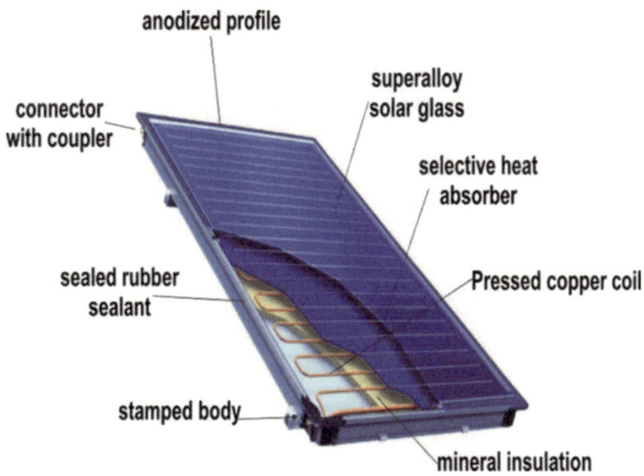

Figure 1.
Principal diagram of flat solar collector.

Figure 2.
Flat plate solar collector's mockup [12].

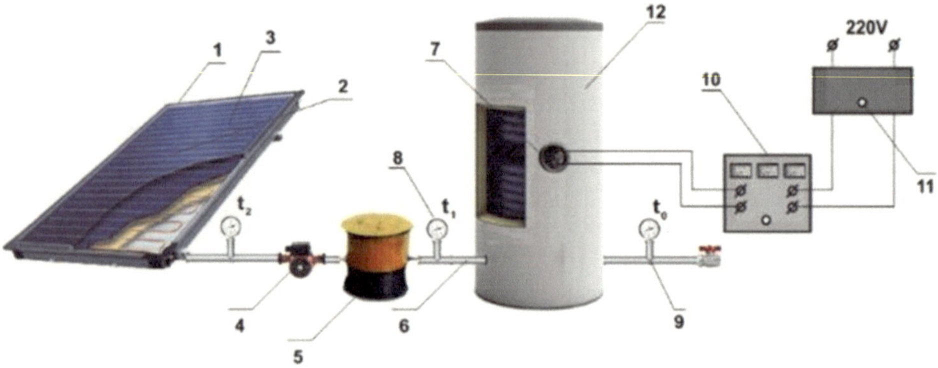

Figure 3.
Principal diagram of solar heat supply: (1) thermally insulated body; (2) translucent covering; (3) tank-absorber; (4) circulating pump; (5) thermal pump; (6) pipeline; (7) tubular energy heating; (8, 9) thermometers for measuring the water temperature at inlet and outlet from a tank absorber and environment; (10) set of electric meters K 501; (11) autotransformer; (12) tank-accumulator; and (13) controller.

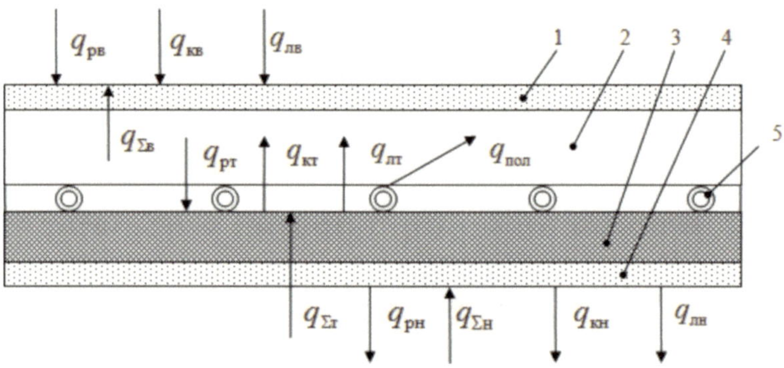

Figure 4.
Diagram of thermal flows, passing via collector elements surfaces: (1) upper cover, made of translucent material; (2) gap spacing; (3) absorber; (4) lower cover (heat insulator); and (5) copper tubular spiral [12].

$$A = \frac{Q}{h\Delta T_{Lm}} \tag{2}$$

where A—heat transfer surface square, h—heat transfer film factor, and ΔT_{Lm}—logarithmic mean temperature difference, accessible to the heat transfer. If the

Parameters	Value
Absorbing plate material	Copper
Absorber plate dimensions	2 m × 1 m
Plate thickness	0.4 mm
Glazing material	Hardened glass
Glazing dimensions	2 m × 1 m
Glazing thickness	4 mm
Insulation	Foam plex (foam, polyurethane)
Collector tilt angle	45°
Absorber's thermal conductivity	401 W/(m K)
Insulation thermal conductivity	0.04 W/(m K)
Transmittance-absorption factor	0.855
Sun apparent temperature	4350 K
Ambient temperature	303 K
Radiation intensity	1000 W/m^2

Table 1.
Specifications of flat plate solar collector.

collector (A) surface square is expressed as a member of the collector's geometry, then we obtain:

$$A = \pi d_i L_t N_t U_p \tag{3}$$

where d_i—tube inner diameter, N_t—tubes quantity per a collector, and U_p—collectors amount in parallel. Heat length (L_t) is an exchanger length, necessary to satisfy the demanded loading. Uniting the Eqs. (1)–(3) and rearranging, we obtain the thermal length, which equals to

$$L_t = \frac{mC_p(T_{out} - T_{in})}{\pi d_i N_t U_p h \Delta T_{Lm}} \tag{4}$$

The film heat transfer factor might be computed according to a formula [21]:

$$h = \frac{4200(1.35 + 0.02)v^2}{d_i^{0.2}} \tag{5}$$

It is important to note, that in the above equation di shall be used in (mm); T—water temperature, and v—speed, which can be computed as follows:

$$v = \frac{m}{\rho A_c} \tag{6}$$

If the term ρ—water density, and A_c—free flow square, which might be expressed as:

$$A_c = \frac{\pi d_i^2}{4} N_t U_p \tag{7}$$

Hydraulic length (L_h) is the heat exchanger length, necessary for conformity to the denoted pressure fall. Pressure fall at the heat exchanger core can be expressed as in [19]:

$$\Delta P = \frac{2f\rho L_h v^2}{d_i} \tag{8}$$

where *f*—friction factor, which for Reynolds numbers less than 2100 is specified as follows [19]:

$$f = \frac{16}{Re} \tag{9}$$

and for Reynolds numbers exceeding 2100, prescribed as in [19]:

$$f = \frac{0.054}{Re^{0.2}} \tag{10}$$

where Reynolds number is expressed as in [19]:

$$Re = \frac{d_i m}{\mu A_c} \tag{11}$$

Uniting and rearranging (6)–(8) we obtain:

$$L_h = \frac{\Delta P \rho d_i^{\,5} N_t^{\,2} U_p^{\,2} \pi^2}{32 f m^2} \tag{12}$$

In order to give the flat plate collector a standard dimension it is necessary to locate in its body an extended heat absorbing tube. If to add to the collector's heat absorbing tube a spiral form, then in the solar collector with a standard body dimension there might be placed the transfer media channel with a length, exceeding a body height several fold. Due to big amount of local resistances the hydraulic losses magnitude will be high. In that connection we forced to use the big capacity circulation pump which will increase financial expenditures. Therefore we shall minimize the hydraulic losses magnitude and at the same time maximize the collector's heat absorbing tubes length For that purpose there shall be used several heat absorbing tubes, connected between with rack-type tools and other fittings, as it is shown on the **Figure 1**. Such technical solution will broke down the overall flow of the transfer media in the collector into several smaller ones, which, in turn, will reduce amount of the pipeline local resistances, simultaneously having increased the time of the transfer media being under the solar radiation. Apart from that, the given solution will allow increase the transfer media speed, and accordingly, raise its flow, using a circulation pump with less capacity, cutting thereby the financial expenses. Speed increase and transfer media flow will give a possibility to upgrade the heat output factor from a tube to liquid, which, in turn, will increase the solar collector performance.

When water is heated and its density in the collector's circuit is decreased there appears surplus hydrostatic pressure *P*(Па)

$$\Delta P = g * [\rho_w(t_c(n_c) - \rho_w(t_b)] * F \tag{13}$$

$$F = \frac{d_b + d_c * \sin(S)}{2} + d_{c-b} \tag{14}$$

where *g*—gravity acceleration, m/s^2; *F*—vertical distance between the solar collector and tank-accumulator centers, m; d_b—tank-accumulator height, m; d_c—collector's length, m; d_{c-b}—distance between the tank-accumulator bottom and collector's upper part, m.

Such pressure is balanced with the head loss, caused with total collector circuit hydraulic resistance, consisting of the collector's copper tubular spiral resistances of input and output transfer media of pipelines and devices for their connecting with a collector and tank-accumulator. Upon calculating the pressure losses in the body the hydraulic circuit is broken down into linear and nonlinear parts.

Each element of a copper spiral consists of a linear part and two nonlinear parts (apart from the upper and lower coils, which contain one nonlinear element each).

Overall spiral elements quantity is computed according to a formula:

$$\mathrm{ne} = \frac{L - (d_c - 2 * dkk)}{dk - 2 * dkk} \tag{15}$$

where *dkk*—distance from the collector's edges to the linear part of the upper and lower spiral elements, m; *dkk*—distance from the right and left collector's sides to the copper spiral elements, m.

Due to the fact, that the spiral elements amount, in compliance with the collector's constructive properties shall be an integer and even number, the value *ne* is rounded off to the nearest even number.

Distance between the spiral elements is defined from the expression:

$$\mathrm{de} = \frac{dk - 2 * dkk}{ne} \tag{16}$$

Arc lengths of linear (*ll*, m) and nonlinear (*ln*, m) parts of the copper spiral element.

$$\mathrm{ln} = \pi * \frac{se}{2} \tag{17}$$

$$\mathrm{ll} = \frac{L}{de - 2 * ln} \tag{18}$$

Pressure drop in the spiral element linear part is described with an equation:

$$\Delta P = \frac{32\mu(t) * ll}{d^2} * v_{aver} \tag{19}$$

where μ—transfer media dynamic viscosity average along the spiral linear part length, Н·с/м^2; v_{aver}—heat transfer average speed.

Pressure loss in the nonlinear spiral element of the collector is computed from the expression:

$$\Delta P_{ln} = \xi * \frac{v_{aver}}{\rho} \tag{20}$$

where ξ—local resistance factor.

Collector's nonlinear spiral part local resistance represents the pipe rotation for 90° and it is computed according to the formula

$$\xi = 0.051 + 0.19\frac{d}{R} \tag{21}$$

where R—radius of the spiral copper tube rotation m, R = se/2.

For simulating hydraulic resistances fall in the copper tubular spiral at collector modules, obtained based on the equations.

- at the first area of the collector:

$$H_R = n * \frac{64 * (\mu(t_b(3) + \mu(t_c(1)) * ll}{\pi * d^4} + (2 * n - 1) * 2\xi * \frac{\rho(t_b(3) + \rho(t_c(1)}{\pi * d^4} \quad (22)$$

- at i-m collector's art (i = 2, ... , $n_{к}$ − 1):

$$H_c(t) = \mathrm{n} * \frac{64 * (\mu(t_c(t-1) + \mu(t_c(t))) * ll}{\pi * d^4} + 4n * \xi * \frac{\rho(t_c(t-1) + \rho(t_c(i0))}{\pi + d^4} \quad (23)$$

Heat amount $q_{к}$(Дж), incoming from a collector to a tank-accumulator we will define, using a formula

$$q_c = c_w(t_c) * \rho_w(t_c) * t_c * g_c \quad (24)$$

In flat plate solar collectors heat absorption area and working liquid conditions are not favorable enough, particularly, when the heat receiver (boiler) is placed horizontally. Actually, the liquid motion speed is low here and it is heated from top, and, as a result, convective currents mixing the liquid do not develop. At the boiler's sloping position under known conditions there might happen the natural convection improving the heat transfer [22].

Task solution of heat transfer between radiant heat absorbing surface and power liquid flat heaters upon applying the forced and natural convection (first type boundary conditions) presents sufficient difficulties. In case of applying the natural convection to the forced flow the velocity gradient on the wall will depend on the interrelated parameter part, defining both forced and free flow. In that case velocity gradient $A = \left(\frac{dW_A}{dy}\right)_{r0}$ is not known in advance (for circular tube) $A = \rho\overline{W}/d$, for flat one $A = 6\overline{W}/h$, (A) can be defined from the motion equation.

Let us use the motion equation in a simplified form, omitting inertial terms, but taking into account the upward force. Supposing the liquid physical properties being constant and presenting the density in the form of linear temperature, we will obtain [22]:

$$\frac{\partial^2 W_x}{\partial y^2} + \frac{g\beta(t_c - t_0)}{v} + \frac{1}{\mu}\frac{\partial P}{\partial x} = 0 \quad (25)$$

where β—volume coefficient; t_0—flow temperature far from the wall, admitted as equal to the inlet temperature, as within the thermal initial section the temperature in the flow core changes ineffectually.

The equation for temperature distribution is given by [11]:

$$\frac{\mathrm{t} - \mathrm{t_c}}{\mathrm{t_c} - \mathrm{t_0}} = \frac{\int_0^{\mathrm{n}} \mathrm{e}^{-\eta^3 \mathrm{d}\eta}}{\int_0^{\infty} \mathrm{e}^{-\eta^3 \mathrm{d}\eta}} \quad (26)$$

where t_c—wall temperature, $\eta = \left(\frac{A}{gdx}\right)_y^{1/3} y$—new independent variable.

Denominator (26) represents the gamma function, the values of which are tabulated [13]:

$$\int_0^\infty e^{-\eta^3} d\eta = \Gamma\left(\frac{3}{4}\right) = 0,8930 \tag{27}$$

Having substituted in (25) the value from (26) and upon integrating it we can define $A = \left(\frac{dW_x}{dy}\right)_{y=0}$. To carry out integration we approximate (26) and reduce the equation as follows:

$$\frac{t - t_c}{t_c - t_0} = 1 + b_1 e^{-\eta} + b_2 e^{-2\eta} + b_3 e^{-3\eta} \tag{28}$$

which meets boundary conditions and upon the constants appropriate selection conforms quite well with (25). Having inserted $t_c - t_0$ from (28) into (25), we find

$$\frac{\partial^2 W_x}{\partial y^2} \bullet \frac{g\beta(t_c - t_0)}{v}\left(1 + b_1 e^{-\eta} + b_2 e^{-2\eta} + b_3 e^{-3\eta}\right) - D \tag{29}$$

$$(\text{where})\ D = \frac{1}{\mu}\frac{\partial P}{\partial x} \approx const$$

Boundary conditions are in the following form: $x = 0, t = t_0$

$$W_x = 6W\left(\frac{Y}{h} - \frac{Y^2}{h^2}\right), W_x = 8W\left(\frac{Y}{d} - \frac{Y^2}{d^2}\right) \tag{30}$$

The first equation is for flat, the second—for circular tube. Having executed integration and used boundary conditions we obtain the expression for nondimensional velocity speed on the wall. For a flat tube

$$\varphi = \frac{Ah^2}{v} = 6\,Re + 0.43Gr\left(\frac{gx}{Pr\varphi h}\right)^{1/3} \tag{31}$$

$$Re = \frac{\overline{W}h}{v};\ Gr = \frac{g\beta(t_c - t_0)h^3}{v^2}$$

Unfortunately, the expression thereof contains non evident φ. Therefore in (31) it was substituted with an approximate relationship:

$$\varphi = 6\,Re + 0.92\frac{Gr^{3/4}}{Pr^{1/4}}\left(\frac{\alpha}{h}\right)^{1/4} \tag{32}$$

As you can see, the initial supposition about φ $\left(or\ \overline{A}\right)$ consistency is not justified. However, φ changes in length relatively ineffectually $\left(\varphi \sim x^{1/4}\right)$. Therefore, it will not be a great mistake if to use an average value in the section with the length:

$$\overline{\varphi} = \frac{\overline{A}h^2}{v} 6\,Re + 0.736\frac{Gr^{3/4}}{Pr^{1/4}}\left(\frac{e}{h}\right)^{1/4} \tag{33}$$

Let us define now local heat-transfer coefficient, assigning it to the difference between the wall temperature and liquid temperature at the inlet to the heating area. Such definition technique (α) is convenient hereby, as at small values $\frac{1}{Pe}\frac{x}{d}$ the liquid average mass temperature changes weakly in length

$$\alpha_0 = \frac{g_c}{t_c - t_0} = -\frac{\lambda}{t_c - t_0}\left(\frac{dt}{dy}\right)_{y=0} \tag{34}$$

Having used the correlation obtained previously, we find:

$$\left(\frac{dt}{dy}\right)_{y=0} = \left(\frac{A}{gax}\right)^{1/3}\left(\frac{dt}{d\eta}\right)_{\eta=0} = \left(\frac{A}{gax}\right)^{1/3}\left(\frac{t_c - t_0}{0.893}\right) \tag{35}$$

consequently α_0 from (32) and (33)

$$\alpha_0 = \frac{\lambda}{0.893}\left(\frac{A}{9ax}\right)^{1/3} \tag{36}$$

Having substituted the value A from (10) in (11) we find the expression of Nusselt criterion for flow in a flat tube [18]:

$$Nu = 1.467\sqrt[3]{Pe\frac{h}{l} + 0.123\left(GrPr\frac{h}{l}\right)^{3/4}} \tag{37}$$

where $Nu = \frac{\alpha d}{\lambda}$; $Pe = \frac{\overline{W}d}{d}$; $Gr = \frac{g\beta(t_c - t_0)d^3}{\upsilon^2}$

In case of flow in a circular tube the similar analysis brings to an expression [18]:

$$Nu = 1,615\sqrt{Pe\frac{h}{l} + 0,092\left(GrPr\frac{h}{l}\right)^{3/4}} \tag{38}$$

where $NU = \frac{\alpha d}{\lambda}$; $Pe = \frac{\overline{W}d}{d}$; $Gr = \frac{g\beta(t_c - t_0)d^3}{\upsilon^2}$

3. Results

For searching the optimal hydraulic regime in the pipeline system of the solar collector, presented on the **Figure 1**, it is necessary to carry out the hydraulic analysis. Its main point is in specifying the total pressure drop in collector tubes and defining optimal offset quantity and, accordingly, the length of heat absorbing tubes in the solar collector body with a square if 2 m^2. Pressure loss in the pipeline area (Пa)—linear and local resistances, are found according to the formula (24):

$$\Delta P = RL + z \tag{39}$$

where R—specific linear pressure losses per 1 m pipeline, Pa/m (depends on the pipeline diameter and water flow passing along those tubes). L—length of the area being computed, m; z—local pressure losses in the area, Pa. Defined according to the formula (25):

$$z = \sum \xi \frac{v^2}{2g} \tag{40}$$

where g—free fall acceleration, m/s^2; v—transfer media speed, m/s, $\sum\zeta$—local resistances factors sum.

Having assumed, that pressure losses in the areas A–G and A$'$–G$'$ is a constant magnitude, the hydraulic computation has been executed only for one standard coil and reduced to a **Table 2**. As a designed consumption value there has been selected the minimal magnitude of the circulation pump consumption with a minimal capacity UPS 25-40 (**Table 3**).

Figure 2. Hydraulic computation of heat absorbing tubes of a solar collector with several tubes.

As it is seen from the **Figures 1** and **5**, the values of total pressure losses will not allow using a low capacity pump, which will bring to increasing the heating system cost. Using bigger diameter tubes will allow cutting the total pressure losses, but it will considerably increase the solar collector cost. If to draw attention to the formulae (13) and (14), then we can see, that along with increasing the collector's insulated edge width, performance of the solar collector rectangular profile direct edge is reduced, and therefore reduced also the solar collector efficiency. Then it may be concluded, that using several collector tubes with less edge size allows upgrade the flat solar collector efficiency (**Figure 6**).

When using one pipe with the edge of a flat solar collector, the efficiency of the entire collector area will appear. That is, from **Figures 7** and **8** it can be seen that the water flow in the heat-absorbing single tube of a flat solar collector depends on the sum of the local resistance coefficients, which is an exponential function that increases to a critical maximum of the water velocity (**Figures 9–12**).

Using one tube as an edge of a flat plate solar collector is the efficiency of an overall collector square. That is, from the **Figures 7** and **8** it is seen, that the water consumption in one heat absorbing tube of a flat plate collector from the sum of the local resistance factors is an exponential function, which grows to the critical maximum water speed.

Figure 13 demonstrates the pressure loss in the linear part of the flat plate solar collector spiral element. It is known, that the pressure fall is an important factor for the thermosyphon systems performance. Accordingly, in the work herein a flat plate solar collector is being studied for predicting the pressure fall on it. Various dimensions collectors have been used as a control example, in which the model has been tested within 3–8% in terms of normalized mean square deviation.

Figure 14 demonstrates total amount of spiral elements of the flat solar collector copper spiral element. From the figure it is clear, that the more quantity of a flat plate solar collector copper spiral elements, the bigger the distance from the collector edges to the linear upper and lower spiral elements.

The study of convective heat transfer in flat solar collectors is considered, as can be seen from the analysis of heat transfer studies using round and flat pipes with matching forced and free convection, vertically or horizontally, with different liquids; flow direction (**Figure 15**).

Upon specifying the coefficient of heat transfer from the heating area to the heated water the thermal flow is defined according to the flow rate and temperature difference of the water being heated at outlet and inlet. Water temperature is specified at outlet and inlet and the surface temperature through thermocouples readings average [22].

Experiments have been carried out at the heat transfer surface's obliquing angles $\phi = 30°, 35°, 40°$.

At that, the number Re = 150–500, i.e., the experiments have been executed mainly at water flow laminar regime.

As a determining temperature there has been accepted the water temperature and as a characteristic dimension—the channel equivalent diameter $d = 4f/S$.

Accepted in the model dimension d secures its geometric similarity to experimental flat solar collectors.

Area	Water flow G, л/с	Area length L, m	Outer diameter D, mm	Water speed V, m/s	Specific linear pressure losses, Pa	Linear pressure losses RL, Pa	Local resistances factors sum $\sum \zeta$	Pressure losses on local resistances Z, Pa	Total pressure losses $\Delta P, Pa$	Local resistances factors sum ξ
A–A$^{/}$	0.08	6	7	0.08	345	1850	20	3020	4501.2	0.68

Table 2.
Hydraulic computation of heat absorbing tubes of a solar collector with several tubes.

Area	Water flow G, л/с	Area length L, m	Outer diameter D, mm	Water speed V, m/s	Specific linear pressure losses, Pa	Linear pressure losses RL, Pa	Factors sum in local resistances $\sum \zeta$	Pressure losses on local resistances Z, Pa	Total pressure losses $\Delta P, Pa$	Local resistances factors sum ξ
A–A$^{/}$	0.34	30	10	2.5	Tube diameter 10 mm	87,000	20.45	40215.00	125,460.00	0.68
A–A$^{/}$	0.34	28	20	1.5	Tube diameter 20 mm	10,421	16.45	7545.00	20,156.00	0.68

Table 3.
Hydraulic computation of heat absorbing tubes of a solar collector with one tube.

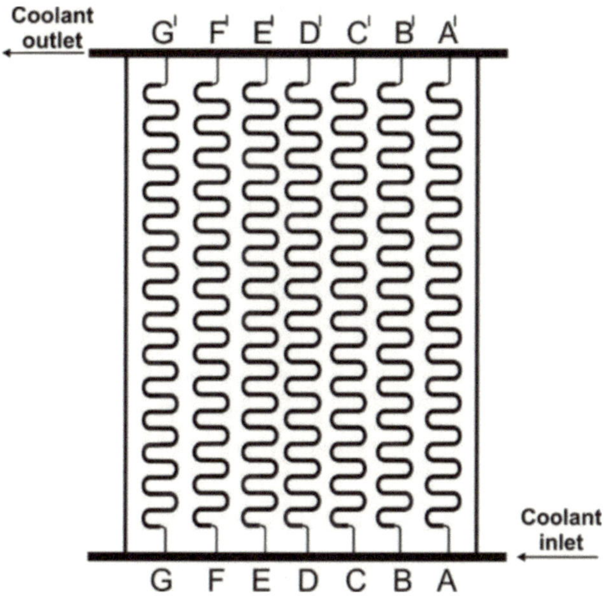

Figure 5.
Diagrammatic representation of flat plate solar collector with extended heat absorbing tubes.

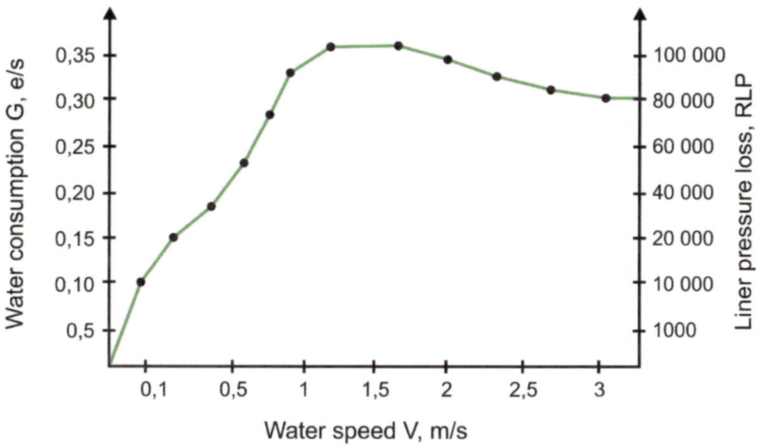

Figure 6.
Dependence of water consumption on the water flow and linear pressure drops of heat absorbing tubes of the flat solar collector with several tubes [12].

In the result, experimental data in the dependency form Nu = f(Pld/l) and $Nu/Pr^{0.44}$ = f(Re) is given in the **Figures 16** and **17** [22].

Figures 16 and **17** show number for the liquid plays an important role in the points spacing as regard to the curve. In all cases, the lines are with a slope 1/3, as in the research herein we have considered only laminar mode [22].

Obtained semiempirical formula allowing, in concept, processing and summarizing accumulated experimental data on the heat transfer upon convective heat transfer in the flat plate solar collectors as well gives a possibility to compare the results of theoretical investigations of heat transfer characteristics to experimental data.

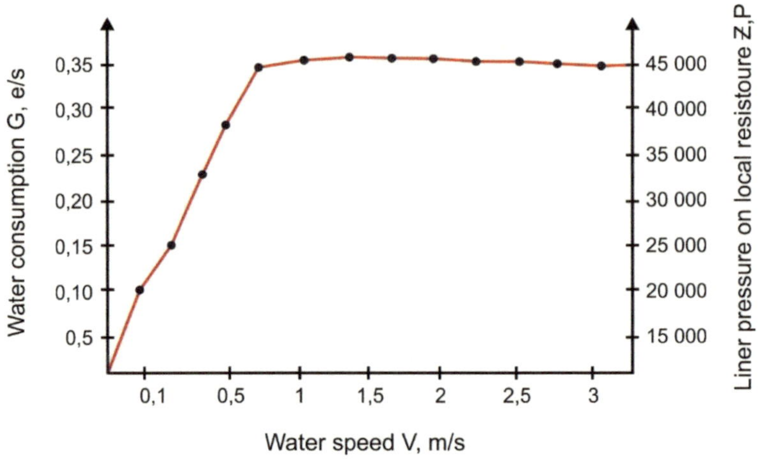

Figure 7.
Dependence of water consumption on the speed and losses on local resistances of heat absorbing tubes of flat solar collector with several tubes.

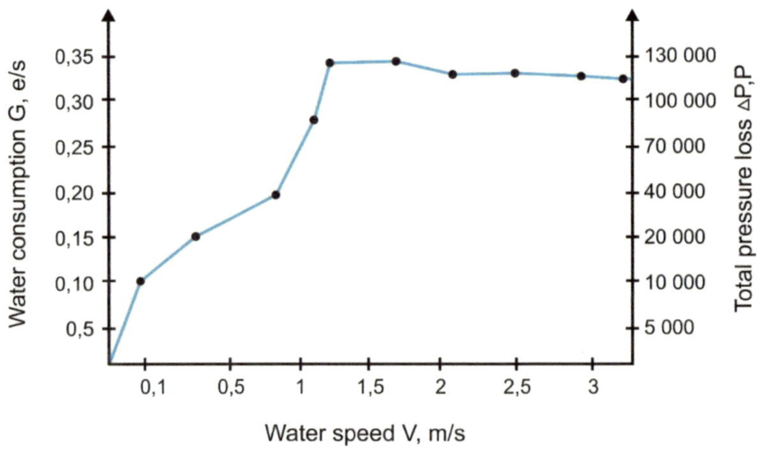

Figure 8.
Dependence of water consumption on the speed and total pressure losses of heat absorbing tubes of flat solar collector with several tubes.

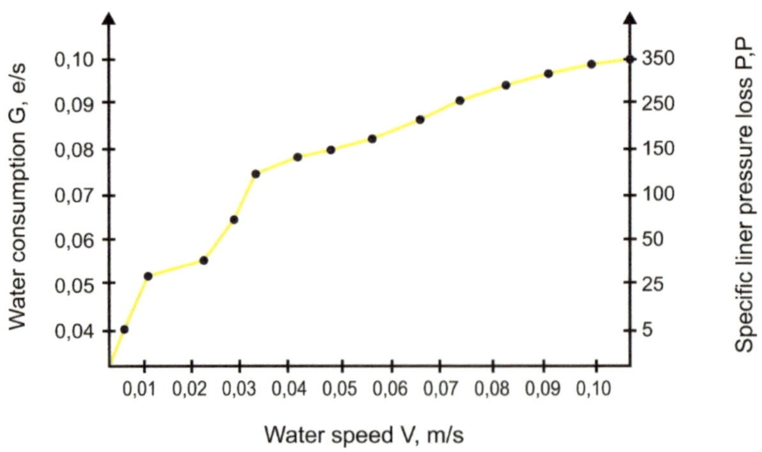

Figure 9.
Dependence of water consumption on the speed and specific linear pressure losses of heat absorbing tubes of flat solar collector with several tubes [21].

DOI: http://dx.doi.org/10.5772/intechopen.90451

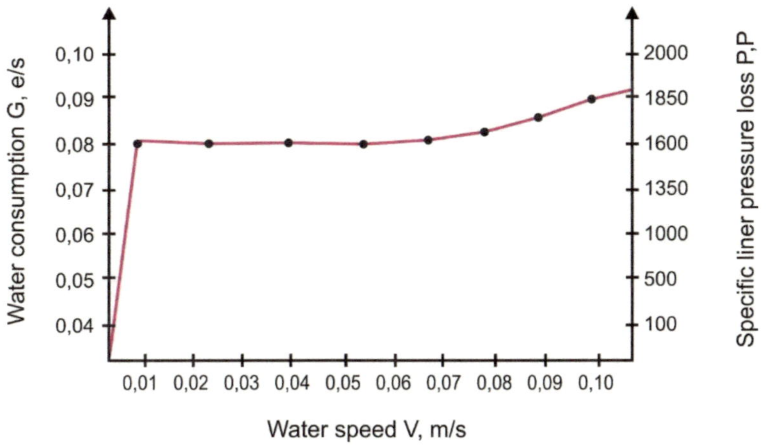

Figure 10.
Dependence of water consumption on the speed and specific linear pressure losses of heat absorbing tubes of flat solar collector with one tube.

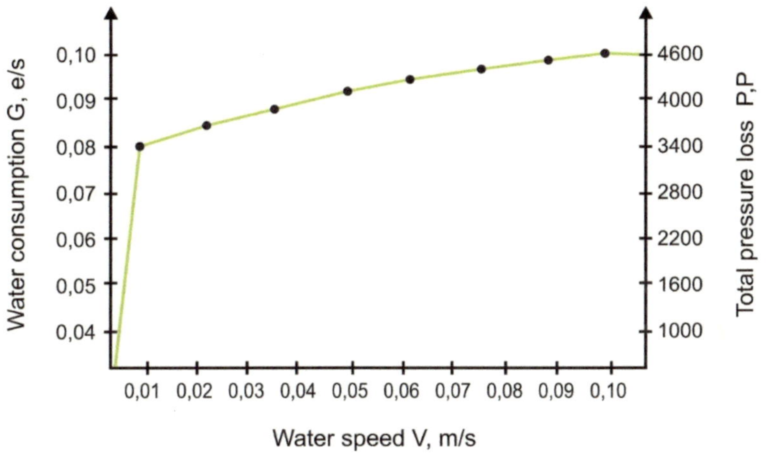

Figure 11.
Dependence of water consumption on the speed and total pressure losses of heat absorbing tubes of flat solar collector with one tube.

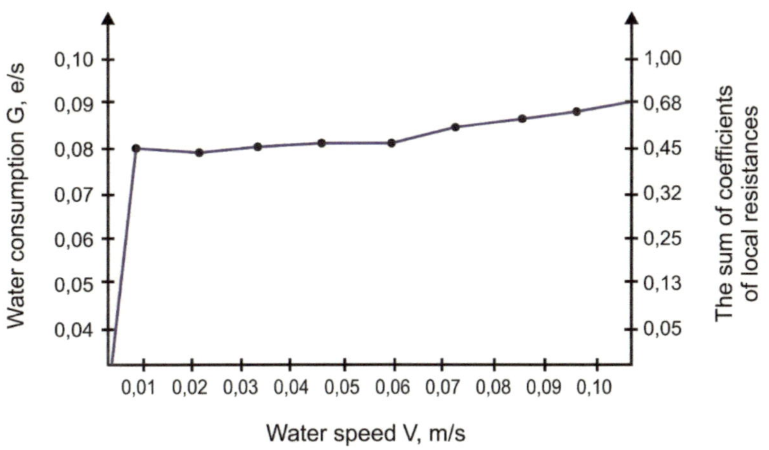

Figure 12.
Dependence of water consumption on the speed and local resistances factors of heat absorbing tubes of flat solar collector with one tube [21].

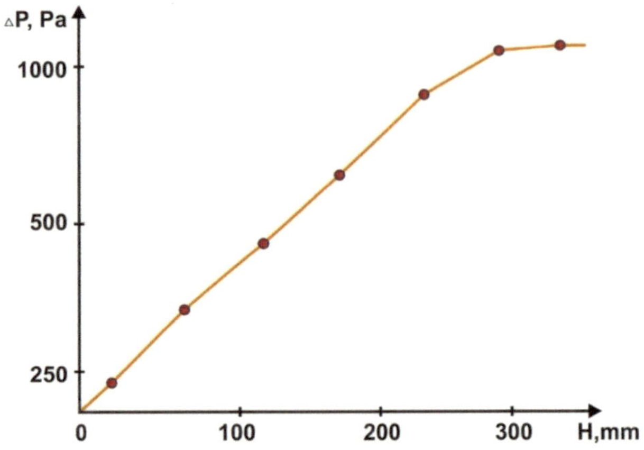

Figure 13.
Pressure loss in linear part of flat plate solar collector's spiral element linear part.

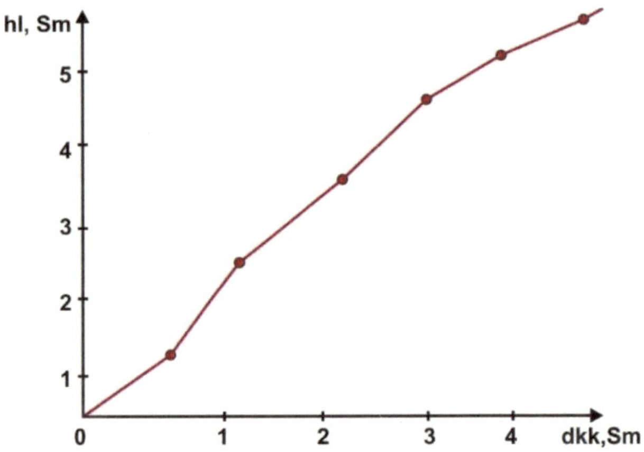

Figure 14.
Total amount of flat plate solar collector's copper spiral elements.

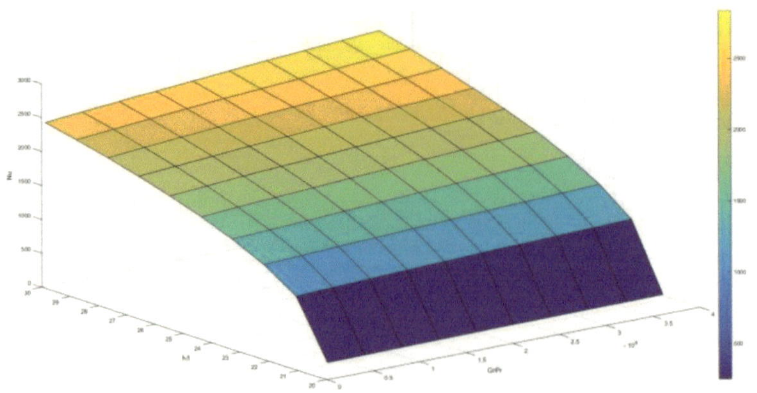

Figure 15.
Nusselt criterion for liquid flow in flat tube [21].

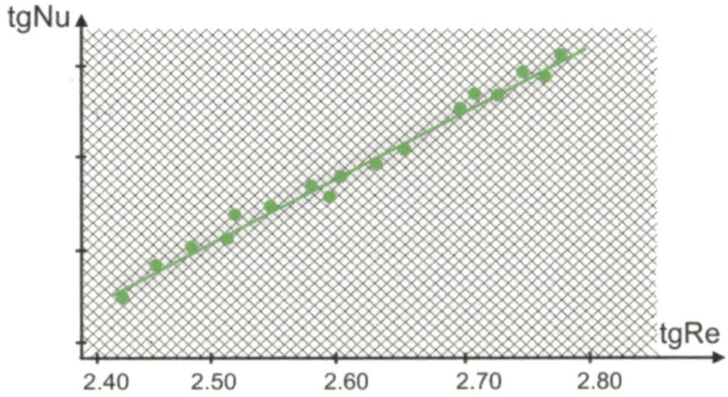

Figure 16.
Nusselt criterion dependence on Reynolds criterion for liquid flow in the flat solar collector [22].

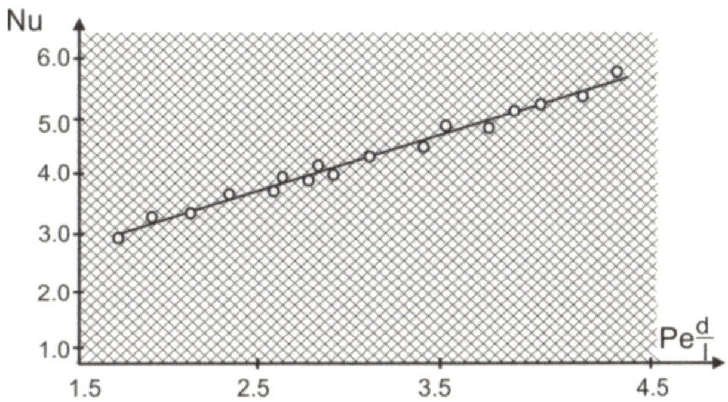

Figure 17.
Dependence of Nusselt-Prandtl criterion on Reynolds criterion for the liquid flow in the flat plate solar collector [22].

4. Conclusion

Flat plate solar collector heat absorbing tubes hydraulic analysis has shown, that using the standard size transfer media channels we can place a pipeline, the length of which 2.5-fold bigger than the collector's body length, and at that, the time of the transfer media being under solar radiation considerably increases. At that, the overall pressure losses in the collector's pipeline turned out to be relatively low, which allows increasing the transfer media speed with the aim to transfer from laminar flow to turbulent, which in its turn, raises the heat output factor from a tube to the liquid, and consequently, also, the flat plate solar collector's efficiency. Apart from that, the calculations have proved, that the optimal hydraulic regime, necessary for securing all above mentioned conditions in the pipeline system having been elaborated, can create even the least powerful circulating pump. As well, in the process of research it has been stated, that making use of several tubes will upgrade the performance of a flat plate solar collector.

Obtained semi empiric formulae, which allow, in principle, processing and generalizing accumulated experimental data on heat transfer at convective heat transfer in flat plate solar collectors, as well they give a possibility to compare the theoretical researches outcomes of various heat transfer characteristics to experimental data [22].

Author details

Yedilkhan Amirgaliyev[1,2], Murat Kunelbayev[1,2*], Kalizhanova Aliya[1,2], Ainur Kozbakova[1,2], Omirlan Auelbekov[1] and Nazbek Katayev[1]

1 Institute of Information and Computing Technologies CS MES RK, Kazakhstan

2 Al-Farabi Kazakh National University, Kazakhstan

*Address all correspondence to: murat7508@yandex.kz

References

[1] Carravetta A, Del Giudice G, Fecarotta O, Ramos HM. PAT design strategy for energy recovery in water distribution networks by electrical regulation. Energies. 2013;**6**:411-424

[2] Eisenmann W, Vajen K, Ackermann H. On the correlations between collector efficiency factor and material content of parallel flow flat-plate solar collectors. Solar Energy. 2004;**76**(4):381-387

[3] Garg HP. Design and performance of a large scale size solar water heater. Solar Energy. 1973;**14**:303-312

[4] Karagiorgas M, Botzios A, Tsoutsos T. Industrial solar thermal applications in Greece Economic evaluation, quality requirements and case studies. Renewable and Sustainable Energy Reviews. 2001;**5**:157-173

[5] Kiraly A, Pahor B, Kravanja Z. Integration of renewables for improving companies' energy supplies within regional supply networks. Chemical Engineering Transactions. 2012;**29**: 469-474

[6] Lin WM, Chang KC, Yi-Mei Liu YM, Chung KM. Field surveys of non-residential solar water heating systems in Taiwan. Energies. 2012;**5**:258-269

[7] Liu YM, Chung KM, Chang KC, Lee TS. Performance of thermosyphon solar water heaters in series. Energies. 2012;**5**:3266-3278

[8] Quijera JA, Garcia A, Labid J. Integration of solar thermal energy and heat pump in a fish canning process combining pinch and exergy analysis. Chemical Engineering Transactions. 2012;**29**:1207-1212

[9] Zago M, Casalegno A, Marchesi R, Rinaldi F. Efficiency analysis of independent and centralized heating, systems for residential buildings in Northern Italy. Energies. 2011;**4**: 2115-2131

[10] Amirgaliyev Y, Kunelbayev M, Kalizhanova A, Auelbekov O, Katayev N, Kozbakova A. Theoretical and mathematical analysis of double circuit solar station with thermo siphon circulation. Journal of Polytechnic-Politeknik Dergisi. 2019;**22**(2):485-493

[11] Priyam A, Chand P. Thermal and thermohydraulic performance of wavy finned absorber solar air heater. Solar Energy. 2016;**130**:250-259

[12] Hegazy AA. Thermohydraulic performance of air heating solar collectors with variable width, flat absorber plates. Energy Conversion and Management. 2000;**41**:1361-1378

[13] Ahmad A, Saini JS, Varma HK. Thermohydraulic performance of packed bed solar air heaters. Energy Conversion. 1996;**37**:205-214

[14] Prasad BN, Kumar A, Singh KDP. Optimization of thermo hydraulic performance in three sides artificially roughened solar air heaters. Solar Energy. 2015;**111**:313-319

[15] Balaji S, Saravanakumar K, Sakthivel M, Siva Kumar P. Thermo hydraulic performance of a forced convection solar air heater using pin-fin absorber surface. International Journal of Advanced Technology in Engineering and Science. 25 November 2015;**3**(11): 360-368

[16] Verma SK, Prasad BN. Investigation for the optimal thermo hydraulic performance of artificially roughened solar air heaters. Renewable Energy. 2000;**20**:19-36

[17] Mittall MK, Varshney L. Optimal thermo hydraulic performance of a wire

mesh packed solar air heater. Solar Energy. 2005;**80**:1112-1120

[18] Behura AK, Rout SK, Pandya H, Kumar A. Thermal analysis of three sides artificially roughened solar air heaters. Energy Procedia. 2017;**109**: 279-285

[19] Sourbron M, Ozalp N. Determination of heat transfer characteristics of solar thermal collectors as heat source for a residential heat pump. Journal of Solar Energy Engineering. 2016;**138**:1-8

[20] Ziemelis I, Kancevica L, Jesko Z, Putans H. Calculation of energy produced by solar collectors. Engineering for Rural Development. 2009:212-2018

[21] Chabane F, Moummi N, Benramache S. Experimental study of heat transfer and thermal performance with longitudinal fins of solar air heater. Journal of Advanced Research. 2014; **5**(2):183-192

[22] Amirgaliyev Y, Kunelbayev M, Kalizhanova A, Auelbekov O, Kataev N. Study of convective heat transfer in flat plate solar collectors. WSEAS Transactions on Systems and Control. 2019;**14**:129-137

Section 5

Wind Energy

Chapter 18

Offshore Renewable Energy

Giovanni Rinaldi

Abstract

Offshore renewable technologies hold the potential to satisfy a considerable amount of the global energy demand in the coming years. In this chapter, the main sources of renewable energy related to the oceans (waves, tides, and offshore winds) will be characterized and discussed, with reference to the challenges related to their use. Thus, the main devices capable of exploiting these resources will be presented. Their working principal, together with operational and technological requirements, will be described, highlighting strengths and weaknesses of each technology and providing examples of the past and current experiences. The elements of project management, as well as environmental impact and public perception, will be included. Finally, conclusions on the current viability of ocean energy devices will be drawn, together with guidelines for their future exploitation.

Keywords: offshore wind, wave, tidal, marine energy, reliability, availability, maintainability, economics

1. Introduction

While world population and global energy demand are rapidly increasing, the effects of carbon emissions and other contaminants are stimulating the quest for clean, carbon neutral, and renewable energy sources. Despite in the last few decades some technologies like photovoltaic and onshore wind have hugely progressed, these alone seem unlikely to satisfy all the electricity needs. Among the main reasons for this limitation are the unpredictability of the resource and the lack of suitable space in land. As a result, novel and reliable sources of energy are being proposed and investigated. Among these, offshore renewable devices, operating in the oceans and open seas, have been identified as suitable alternatives able to provide a substantial contribution to the energy mix. The theoretical resource from offshore technologies including offshore wind, wave, and tidal energy has been estimated between 260,000 and 330,000 TWh/year [1]. This represents a huge potential, which if properly exploited would be able to satisfy the electric energy demand of coastal locations and remote islands and more in general of all the countries having a direct access to sea. However, the oceans constitute a harsh environment, in which it is difficult to access and operate, with several limitations and augmented difficulties compared to onshore projects. For this reason, a number of technical, economical, logistical, and environmental challenges towards the successful exploitations of offshore resources exist.

In this context, this chapter aims at providing an overview of the current possibilities and experiences in the offshore renewable energy sector. In Section 2, the offshore resource in its various forms (wind, currents, and waves), as well as the most common parameters used to characterize it, will be presented and discussed.

In Section 3 the existing devices and technologies used to produce electricity from offshore resources, together with typical energy metrics, will be described. Among them, offshore wind represents the most advanced and successful offshore sector so far. This is mainly due to the experience with onshore wind projects, for what concerns the power production, and with oil and gas projects, for what concerns the foundation and platform of the device. Tidal energy, which uses a similar system to extract energy from the tides, is steadily advancing towards commercial availability. Despite several attempts and a handful of successful stories, nowadays wave energy still looks like the most challenging sector. Finally, other niche technologies that can be related to the use of offshore resources are ocean thermal energy conversion (OTEC) and salinity gradient (also known as "osmotic power").

In Section 4 other challenges related to the successful deployment of offshore renewable devices will be discussed, together with their implications on current academic investigation and industry efforts.

In Section 5, a summary on the current possibilities and challenges for the successful exploitation of offshore renewable energy devices will be provided, and a series of guidelines for future work considered.

2. Offshore resource

This section provides an overview of the offshore resources from which energy can be extracted. The physical phenomena involving the formation and propagation of winds, waves, and currents are hereinafter described with a view on energy production.

2.1 Wind

Wind is a direct effect of the thermal energy provided by solar radiation on earth. This generates gradients of temperature in the atmosphere, which in turn produce variations of pressure which move masses of air around the globe. Besides being affected by the Coriolis force, the intensity and direction of the propagation of these depend on several factors, such as the conformation of the territory, possible obstacles (e.g., buildings), and the roughness of the area on which the wind blows. Due to the lack of vegetation and other obstacles, the roughness over oceans and open seas is extremely low, allowing the wind to travel and develop undisturbed over long distances. This means that offshore winds are generally stronger but also steadier than onshore winds.

In terms of energy extraction, these features translate in an increased potential, as a result of the possibility to use bigger turbines and have less variability of the resource over time. In this regard, it must be remembered that the power extractable from the wind varies with the cube of the velocity, meaning that even small increases in the wind speed may generate large increases in the energy production. Besides, more stable winds generate less turbulence, with positive effects on the structural integrity of the devices and on the reliability of their components.

2.2 Currents

Ocean currents are driven and affected by primary and secondary forces. Primary forces are those that start the movements of masses of water and determine their velocity. These are mainly caused by the relative motions and gravitational effects of the sun, moon, and earth but also geothermal processes, such as variations in temperature and salinity, and tectonic movements (e.g., earthquakes).

Secondary forces are those which influence the direction and nature of the flow and are affected by the seabed composition, bathymetry, and gyres.

A distinction is often made between tidal currents, generated by the gravitational effects, and nontidal currents, associated to solar heating force mechanisms. Another important distinction is between deep-water circulation, mainly generated by temperature and salinity variations (and therefore also called "thermohaline"), and surface circulation, mainly generated by wind forces transmitting motion to layers of waters below through friction.

The power extraction phenomena are the same regulating wind energy. In this regard, it must be noticed that water current speeds are typically much lower than wind speeds but also that sea water is much denser that air. For instance, for the same area of flow, the energy contained in a current of 1 m/s is the same energy contained in a wind of approximately 10 m/s. As a result, tidal turbines are generally much smaller than a wind turbine of the same nominal capacity.

2.3 Waves

Waves are generated by the action of the wind blowing over the sea surface. The size and other characteristics of the waves are determined by a series of factors. Among these, the most important are the intensity and duration of the wind which generated them, as well as the distance over which the waves can propagate (called "fetch"). In this regard, waves generated at a long distance from the considered region are defined "swell," whereas waves generated by local winds are called "wind sea" or "wind-generated" waves. In general, swells are more energetic than local waves. Other factors which affect the propagation of waves are the water depth and the bathymetry.

Waves are generally measured in terms of their elevation, in meters, at a certain location as a function of time. This can be compared to the up and down movements of the sea surface experienced by a buoy. Other important parameters are the wave period (in seconds) and the wave length (in meters).

	T_e (s)																							
Significant wave height, H_{m0} (m)	<5.0	5.5	6.0	6.5	7.0	7.5	8.0	8.5	9.0	9.5	10.0	10.5	11.0	11.5	12.0	12.5	13.0	13.5	14.0	14.5	15.0	15.5	16.0	16.5
14.0	0.00	0.00	0.00	0.00	0.00	0.00	0.00	0.00	0.00	0.00	0.00	0.00	0.00	0.00	0.00	0.00	0.00	0.00	0.00	0.00	0.01	0.00	0.00	0.00
13.5	0.00	0.00	0.00	0.00	0.00	0.00	0.00	0.00	0.00	0.00	0.00	0.00	0.00	0.00	0.00	0.00	0.00	0.00	0.00	0.01	0.00	0.00	0.00	0.00
13.0	0.00	0.00	0.00	0.00	0.00	0.00	0.00	0.00	0.00	0.00	0.00	0.00	0.00	0.00	0.00	0.00	0.00	0.00	0.02	0.00	0.01	0.00	0.00	0.00
12.5	0.00	0.00	0.00	0.00	0.00	0.00	0.00	0.00	0.00	0.00	0.00	0.00	0.00	0.00	0.00	0.00	0.00	0.00	0.00	0.00	0.00	0.00	0.00	0.00
12.0	0.00	0.00	0.00	0.00	0.00	0.00	0.00	0.00	0.00	0.00	0.00	0.00	0.00	0.00	0.00	0.00	0.00	0.01	0.00	0.00	0.00	0.00	0.00	0.00
11.5	0.00	0.00	0.00	0.00	0.00	0.00	0.00	0.00	0.00	0.00	0.00	0.00	0.00	0.00	0.00	0.00	0.01	0.01	0.00	0.00	0.00	0.00	0.00	0.00
11.0	0.00	0.00	0.00	0.00	0.00	0.00	0.00	0.00	0.00	0.00	0.00	0.00	0.00	0.00	0.00	0.00	0.01	0.02	0.01	0.00	0.00	0.00	0.00	0.00
10.5	0.00	0.00	0.00	0.00	0.00	0.00	0.00	0.00	0.00	0.00	0.00	0.00	0.00	0.00	0.00	0.00	0.00	0.00	0.00	0.00	0.00	0.00	0.00	0.00
10.0	0.00	0.00	0.00	0.00	0.00	0.00	0.00	0.00	0.00	0.00	0.00	0.00	0.00	0.00	0.00	0.02	0.03	0.00	0.00	0.00	0.00	0.00	0.00	0.00
9.5	0.00	0.00	0.00	0.00	0.00	0.00	0.00	0.00	0.00	0.00	0.00	0.00	0.00	0.00	0.00	0.03	0.01	0.00	0.00	0.02	0.02	0.00	0.00	0.00
9.0	0.00	0.00	0.00	0.00	0.00	0.00	0.00	0.00	0.00	0.00	0.00	0.00	0.00	0.00	0.01	0.02	0.00	0.01	0.01	0.02	0.04	0.00	0.01	0.00
8.5	0.00	0.00	0.00	0.00	0.00	0.00	0.00	0.00	0.00	0.00	0.00	0.00	0.00	0.01	0.02	0.00	0.02	0.02	0.01	0.01	0.00	0.01	0.01	0.00
8.0	0.00	0.00	0.00	0.00	0.00	0.00	0.00	0.00	0.00	0.00	0.00	0.01	0.01	0.00	0.02	0.03	0.05	0.01	0.01	0.00	0.00	0.00	0.00	0.00
7.5	0.00	0.00	0.00	0.00	0.00	0.00	0.00	0.00	0.00	0.00	0.00	0.00	0.00	0.02	0.02	0.05	0.05	0.02	0.00	0.01	0.02	0.02	0.00	0.00
7.0	0.00	0.00	0.00	0.00	0.00	0.00	0.00	0.00	0.00	0.00	0.01	0.00	0.02	0.04	0.09	0.09	0.06	0.02	0.01	0.01	0.03	0.00	0.00	0.00
6.5	0.00	0.00	0.00	0.00	0.00	0.00	0.00	0.00	0.00	0.01	0.00	0.04	0.09	0.13	0.11	0.09	0.01	0.02	0.02	0.01	0.00	0.00	0.00	0.00
6.0	0.00	0.00	0.00	0.00	0.00	0.00	0.00	0.00	0.00	0.02	0.02	0.07	0.14	0.14	0.12	0.03	0.02	0.02	0.00	0.00	0.00	0.00	0.00	0.00
5.5	0.00	0.00	0.00	0.00	0.00	0.00	0.00	0.00	0.02	0.07	0.14	0.24	0.21	0.14	0.11	0.03	0.02	0.00	0.00	0.00	0.00	0.00	0.00	0.00
5.0	0.00	0.00	0.00	0.00	0.00	0.00	0.00	0.06	0.11	0.28	0.34	0.35	0.31	0.22	0.14	0.05	0.02	0.00	0.00	0.00	0.00	0.00	0.00	0.00
4.5	0.00	0.00	0.00	0.00	0.00	0.01	0.08	0.12	0.52	0.60	0.72	0.47	0.27	0.20	0.12	0.08	0.00	0.00	0.00	0.00	0.00	0.00	0.00	0.00
4.0	0.00	0.00	0.00	0.00	0.00	0.04	0.16	0.50	0.83	0.95	0.86	0.55	0.52	0.37	0.10	0.03	0.00	0.05	0.01	0.01	0.00	0.00	0.00	0.00
3.5	0.00	0.00	0.00	0.00	0.07	0.11	0.52	0.89	1.49	2.06	1.57	1.05	0.68	0.51	0.34	0.10	0.08	0.05	0.00	0.00	0.00	0.00	0.00	0.00
3.0	0.00	0.00	0.00	0.09	0.21	0.74	1.15	2.14	2.61	2.58	1.58	1.14	0.90	0.56	0.39	0.26	0.09	0.10	0.02	0.01	0.00	0.00	0.00	0.01
2.5	0.00	0.01	0.02	0.17	0.57	1.71	2.03	2.15	2.42	1.87	1.53	1.18	0.68	0.36	0.21	0.14	0.15	0.13	0.09	0.03	0.02	0.02	0.02	0.02
2.0	0.00	0.11	0.67	1.08	1.74	1.93	2.81	3.43	3.71	2.68	1.79	1.15	0.49	0.20	0.17	0.10	0.03	0.02	0.00	0.01	0.02	0.00	0.01	0.01
1.5	0.02	0.46	1.34	0.92	1.50	2.18	2.38	3.12	2.25	2.27	1.74	1.02	0.55	0.20	0.03	0.05	0.05	0.00	0.00	0.00	0.00	0.00	0.00	0.00
1.0	0.09	0.14	0.55	0.30	0.71	0.95	0.79	0.49	0.33	0.40	0.14	0.06	0.03	0.02	0.00	0.00	0.00	0.00	0.00	0.00	0.00	0.00	0.00	0.00
0.5	0.00	0.00	0.00	0.00	0.00	0.00	0.00	0.00	0.00	0.00	0.00	0.00	0.00	0.00	0.00	0.00	0.00	0.00	0.00	0.00	0.00	0.00	0.00	0.00

Figure 1.
Scatter diagram of the Atlantic Marine Energy Test Site (AMETS) in Belmullet, Ireland. Data recorded during 1 year. Wave occurrence is reported in terms of their significant wave height and energy period and as a percentage of the total sample recorded. Image retrieved from [2].

For simplicity, regular sinusoidal wave theory is often used to describe sea waves. However, real sea waves are characterized by high irregularity, and spectral analysis is needed for a thorough description. In this case, ocean waves can be represented with good approximation as a superposition of sinusoidal (regular) waves.

The amount of energy contained in waves is usually quantified through the energy flux across a vertical plane parallel to the wave crest. Most of this contribution is given by the upper layer of water, closer to the sea surface. The power or energy flux is proportional to the significant wave height *HS*, defined as the mean wave height (trough to crest) of the highest third of the waves, and the wave period *T*.

A common way of providing information about the wave climate at a certain location for marine energy applications is the "scatter diagram." This provides the joint occurrence of the combinations of HS and T over a certain period of time (e.g., 10–20 years). An example of scatter diagram is shown in **Figure 1**.

3. Offshore renewable energy technologies

Once the offshore resource is defined, the existing devices and technologies that can be used to convert it into electrical energy are presented. Thus, the three main types of offshore energy converters are described, and common elements of infrastructure for offshore renewables included. A brief description of less common technologies related to marine energy is provided at the end of the section.

3.1 Offshore wind energy

Offshore wind is without doubt the most established and mature among the offshore renewable sectors. It is the only offshore industry which has already reached a commercial stage, with offshore wind farms competing against other renewable energy technologies, even without any subsidized tariff in some countries. Europe is currently the biggest offshore wind market, with most of the turbines installed in the North Sea and countries like the UK, Denmark, Belgium, and Germany are among the top users. However, thanks to the quickly decreasing costs, offshore wind is rapidly spreading also in other countries (e.g., the USA and China). In addition, several offshore oil and gas companies are increasingly looking at diversifying their portfolio with offshore wind installations.

The reasons for this success are twofold. On one hand, as discussed in the previous section, stronger and steadier winds make the offshore wind resource extremely suitable to energy conversion, allowing for the use of devices with high nominal power (e.g., from 2 to 3 MW for the individual turbine, for projects of 10–20 years ago, to 15 MW, for present and future projects). On the other hand, the experience gained with the onshore wind sector allowed for the use of an already mature and established technology, i.e., the three-blade horizontal-axis wind turbine. As a result, the work towards making the technology economically viable could be focused mainly on the adaptation to the offshore environment, reducing the efforts and expenses in order to make the device technically viable.

The operational principle of offshore wind turbines is essentially the same as their onshore counterparts, with the wind acting on the blades of the turbine to make them rotate. Through a shaft, typically connected to a gearbox (except for "direct drive" machines), this kinetic energy is then converted into electricity by means of a generator. At this point, the electricity is transported onshore using subsea cables and then distributed to the grid. Sometimes offshore substations are used in order to collect the energy produced by several wind turbines before exporting it to the shore, in order to reduce losses by increasing the voltage (typically from 33 to 155 kV).

The trends for the future indicate a significant increase in offshore wind projects, with bigger turbines at increasing distances from the shore. However, this means also greater depths, posing new challenges for the installation, operation, and maintenance of the devices. As a result, floating platforms are being investigated as an alternative to the conventional fixed foundations used to install the wind turbine. Traditional bottom-fixed turbines are generally employed for depths up to 50 m, exploiting a series of structures (e.g., monopoles, tripods, caissons, and jackets) to fix the turbine to the seabed. Despite being technologically less mature, floating offshore wind installations are able to remove this limitation, permitting projects in water depths up to 600 m or in shallow waters with irregular seabed, by using different kinds of floating platforms. Some examples of offshore wind turbine foundations, for both bottom-fixed and floating devices, are shown in **Figure 2** [3].

3.2 Tidal energy

Tidal energy converters (TECs) are essentially rotatory machines, similar to wind turbines, but with the obvious difference that they operate underwater instead of open air. Despite a range of concepts and technologies exists, the most diffused design is also in this case the three-blade horizontal-axis turbine. Other concepts include two-blade turbines, vertical-axis turbines, and oscillating hydrofoils. For a comprehensive list of tidal energy concepts, see [4, 5]. Four prototypes for tidal energy conversion are shown in **Figure 3**.

TECs are generally anchored to the seabed by means of gravity-based foundations but can also be embedded in existing structures, e.g., dams (in which case they are called "tidal barrages"). In addition, TECs can be linked to floating structures, in order to take advantage of the same benefits available to floating offshore wind turbines, discussed in the previous section, as well as to allow for easier maintenance.

Their working principle is analogous to that of wind turbines, with blades able to pitch their blades depending on the flow speed, as well as the entire turbine able to

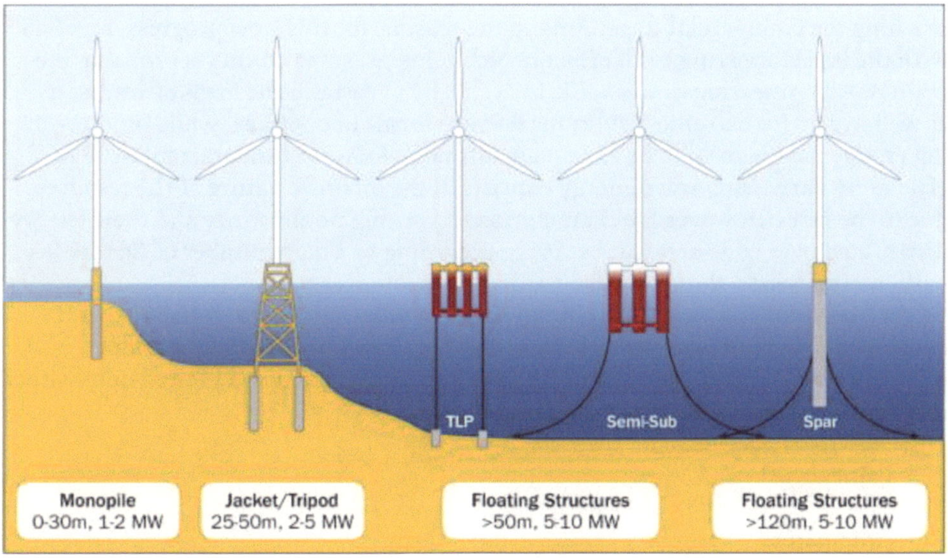

Figure 2.
Typical options for offshore wind foundations in function of water depth and rated capacity of the wind turbine. The first two designs (from the left) represent bottom-fixed solutions, whereas the other ones represent floating platforms. Source: EWEA (2013).

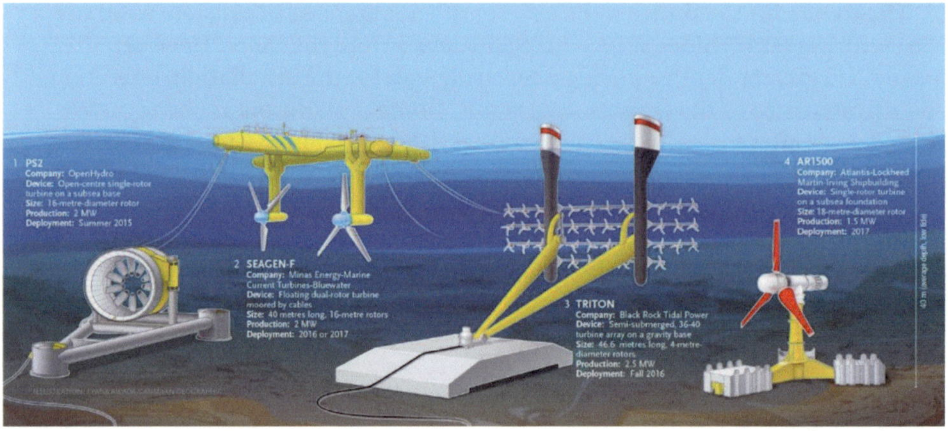

Figure 3.
Schematic drawing of different tidal energy concepts. Infographic retrieved from [6].

spin depending on the direction of the incoming flow. Also in this case, both designs with gearbox or direct drive exist, and the transmission of electricity to shore follows the same technical adjustments available to offshore wind turbines.

Due to the challenges posed by the marine environment (e.g., water tightness, corrosion, biofouling, etc.), the development of the tidal energy sector is being slower than that of offshore wind devices. However, several tidal power plants have been already operational for a number of decades (e.g., the La Rance project, in France, from 1966). More recently, four tidal turbines of 1.5 MW have been installed near the Pentland Firth (northern Scotland, UK) as part of the MeyGen project [7].

3.3 Wave energy

Wave energy can be certainly considered the most challenging of the offshore renewable technologies. Despite a high number of designs and concepts have been conceived to harness the huge potential of the waves, this sector is still far from reaching the commercial stage. Among the reasons for this slow progress, together with the issues applying to all offshore technologies, survivability is probably the main one. Engineering devices able to cyclically withstand the loads of waves, as well as survive extreme conditions during storms in open sea, while producing energy have proven to be extremely difficult. Another fundamental issue in efficiently harnessing wave energy consists in the intrinsic nature of the resource, due to the fact that waves are characterized by strong nonlinearity and their energy distributed over wide areas and varying according to a high number of factors (e.g., bathymetry, winds, depth, obstacles, distances, etc.).

As a result of this variability, hundreds of devices and prototypes have been developed to exploit wave energy. Nonetheless, depending on their working principle, wave energy converters (WECs) can be classified in a limited number of categories, of which the most popular are:

- Point absorbers
- Attenuators
- Oscillating water columns (OWCs)
- Oscillating surge converters

- Overtopping devices
- Submerged pressure differential devices

For a comprehensive review of wave energy concepts, the reader can refer to [8, 9]. A schematic classification of WECs is shown in **Figure 4**. Within the same category, both fixed (either to the seabed or to the shore) and floating devices exist. In addition, several types of power takeoff (PTO) systems exist to convert the kinetic energy of the waves into electricity (e.g., linear generators, hydraulic rams, air and water turbines, elastomers, etc.).

A brief description of the working principle of each of these technologies is hereinafter provided.

Point absorbers are characterized by being significantly smaller than the wavelength of the waves they operate in. These devices are generally similar to buoys, connected to the seabed, which exploit the upward and downward movements of the waves to produce energy. These can be directly connected to a linear generator, or pump water onshore where a conventional hydroelectric turbine is used.

Attenuators are devices composed of multiple floating bodies connected to one another, which flex in a relative motion as the waves pass by. The device is anchored at one hand, and it is designed to be always aligned to the direction of the coming waves. Hydraulic rams are actioned by the motion of the floating bodies, generating the pressure needed to activate a hydraulic motor which in turn is used to generate electricity.

Oscillating water columns are fixed or floating devices which are characterized by having an internal chamber. This can be partially filled with water as the waves approach the device. The remaining space of the chamber remains filled with air, which is compressed and sucked back according to the motions of the water column actioned by the wave. The air, in turn, drives a turbine which spins always in the same direction regardless of where the airflow comes from.

Oscillating surge converters are gravity-based devices sitting on the seabed, on which a float (or flap) is connected by means of a hinge and is free to move as the

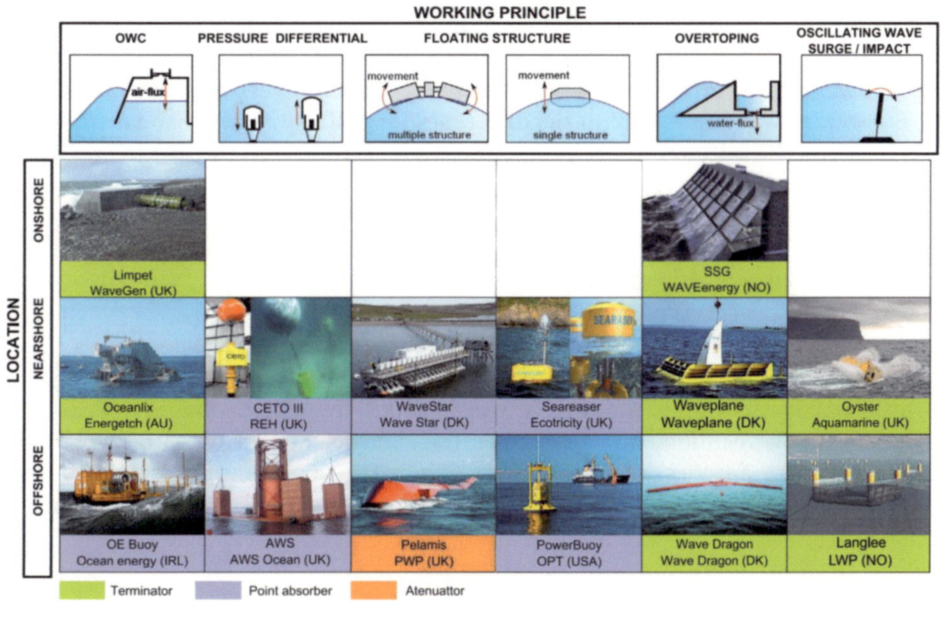

Figure 4.
Schematic illustration of different WECs categories, divided according to their working principle and working location (distance from shore). Image retrieved from [9].

waves push against it. The float is then connected to a pump, which brings pressurized water on the shore for the conversion of energy into electricity.

Overtopping devices are designed in such a way that waves can go over them, adding water in a reservoir above the sea level. Once in the reservoir, the water is then conveyed towards a conventional low-head hydroelectric turbine, which converts the potential energy of the accumulated water into electricity. Also for this kind of WECs, both onshore and offshore (floating) versions exist.

Submerged pressure differential devices are fixed on the seabed and exploit the pressure differential created by the passage of a wave over them. A flexible membrane (e.g., rubber or an elastomer) or an external displacer of the device is cyclically compressed by the waves. This functions as a pump which activates a fluid through a conventional hydroelectric system for energy conversion.

In order to estimate the energy that can potentially be produced by a WEC, the scatter diagram of the selected location is used in combination with the power matrix of the device. This expresses the amount of power that is generated by the WEC for a certain combination of HS and T. In most WECs, control algorithms are used in order to tune the response of the device to the frequency of the incoming waves. In order to maximize the energy extraction, the control system aims at maintaining the device in resonance, by making the natural frequency of the device as close as possible to that of the current sea state.

3.4 Other technologies

Other technologies which exploit the oceans in order to produce electricity are *OTEC* and *osmotic power* plants.

OTEC stands for "ocean thermal energy conversion," and it is a technology which exploits the temperature differential between the water on the surface and water at much higher depths (typically beyond 1500 m). A difference in temperature of at least 20°C is desired in order to activate a thermal cycle. As such, this technology is especially suitable for locations benefitting from a warm water surface all year round, e.g., the Maldives [10]. Through heat exchangers, a power turbine is then activated before condensing the fluid and restarting the cycle. The main advantage of this kind of technology is that the resource (the temperature gradient) is practically constant, allowing the power plant to reach a very high availability.

Osmotic power plants exploit the differential in salinity between the fresh water of a river and the salty water of the sea. Electricity is produced by using a series of membranes in a process of osmosis.

4. Offshore renewable device challenges

As discussed in the introduction and anticipated in the previous section, a number of challenges are preventing the offshore renewable sector from giving a significant contribution to the current energy mix. These limitations are mostly technical but also economical, environmental, and social which are currently being faced in order to support the ocean energy sector. The most common problematics to the deployment of offshore renewable devices will be discussed in this section.

4.1 Installation, operation, and maintenance

The oceans constitute a harsh and difficult environment, which requires a series of specific solutions in order to install and maintain the devices. These include offshore assets, like vessels, workboats, and helicopters, but also suitable onshore logistics, like maintenance harbors and electrical substations. Both onshore and offshore formed technicians and personnel are needed to take care of all the phases of an offshore renewable farm, from preliminary surveys to final decommissioning after the life cycle. Health and safety considerations and protocols need to be prepared and followed for each specific plant and operation.

A fundamental difference from onshore power plants is that marine renewables suffer from a higher degree of isolation. In other words, any operation is subject to strict meteorological conditions, which may significantly affect the start and duration of each activity. In fact, suitable conditions must be verified with respect to several variables, e.g., wind, waves, currents, and visibility. If any of the metocean parameters exceeds one of the pre-established thresholds, the operation has to be postponed or delayed until when all the parameters reach a suitable value.

Being huge machines, composed of a high number of different components, the installation of offshore renewable devices requires specialized vessels with lifting capabilities and able to operate far from shore. However, this provides the possibility to create a supply chain for those regions willing to invest in the infrastructures. Similarly, the operation and maintenance (O&M) of the devices is constituted by a series of preventive and corrective activities that require specific assets. Condition monitoring instrumentation is often integrated in order to monitor the status of the components and detect possible malfunctioning of the device. In this way, eventual repairs or replacements can be scheduled beforehand, reducing the risk for failures which might make the device unavailable for extended period of times.

4.2 Economics

The impact of the technical challenges described in the above section is firstly economic. Purchasing or chartering specialized vessels with specific capabilities has enormous repercussions on the finance of renewable projects. The costs of installing and operating a device offshore are also highly variable, depending on the location and the period of the year (prices generally increase in the summer, due to the major request as a consequence of more suitable operating conditions).

Before the installation, the device must be declared functional and technically ready to produce energy. With the exception of the offshore wind sector, which could benefit from analogous onshore machines, bringing tidal and wave energy machines to technical maturity will require a significant economic effort. Each device has to pass through a stage-gate process, during which the device is brought from the initial concept to the full-scale prototype, passing through various phases of numerical and experimental modelling. A technology readiness level (TRL) metric is often used to evaluate the stage of development of an offshore renewable device.

Apart from the device, a number of additional elements are needed to secure the device in the established location and allow for the transmission of electricity to the onshore grid. Moorings, anchors, foundations, electrical substations, and subsea cables are all items that must be specifically designed for each project, adding up to the final cost of the same. In most cases, station-keeping or transmission infrastructures play a pivotal role in the economic viability of an offshore project.

Lastly, the fact that most of the offshore renewable technologies are novel and innovative means that specific parts and components must be designed and manufactured ad hoc. This translates into high costs for each individual component, which cannot be purchased off the shelf and therefore for the final device. Once these technologies will be proven to be functional, series production will aid in reducing the price per component for most of these items. Besides, the novelty of the technology often affects the confidence of possible investors, posing barriers to the financial support of offshore renewables.

4.3 Environmental constraints

Several concerns have been raised with respect to the environmental impact of offshore renewable farms. These regard possible perturbations to the coastal dynamics (e.g., reduction of the wind and wave climate). Mitigation in this case could be provided through a proper siting and orientation of the offshore farm. However, both elements of the offshore and onshore infrastructure would modify the original landscape and could introduce dangers in the local ecosystem (e.g., spills and leakages).

Concerns exist also about marine organisms affected by the device, e.g., impacted by the blades of a TEC or trapped in the chambers of a WEC. Similarly, new elements and materials might provide the opportunity for invasive species to proliferate. Nonetheless, they might have beneficial effects on the local marine life, by creating new shelters for shellfish and crustacean.

For these reasons, every offshore renewable project is thoroughly assessed through an environmental impact assessment (EIA). This is constituted by a series of studies and surveys, sometimes lasting for years, which aim at identifying all the possible risks to the environment as a consequence of the offshore farm construction and operation. As a result, the viability of the project is assessed and the go-ahead given only in case of favorable conditions. Besides, mitigation measures are proposed in order to minimize the impact of all the identified risks.

Finally, life cycle analysis (LCA) studies are conducted in order to assess the environmental effects (including carbon emissions) related to all the stages of the project, from pre-development and consenting to final decommissioning.

4.4 Social impact

Another source of concern when an offshore renewable project is being developed is the social perception and its possible related consequences. The coastal communities living nearby the selected location might be reluctant to the construction of the offshore farm. Visual impact, noise, and possible effects on the local ecosystem are among the main reasons for this reluctance.

Besides, possible conflicts may arise from sharing the offshore location with existing activities in the same area. These regard competition with navigation (both commercial and touristic) and fishing activities.

Also in this case, suitable agreements and mitigation measures must be identified beforehand through communication and engagement with local public bodies and communities. However, if too complex, planning and permitting procedures may slow down the construction of an offshore farm and make possible investors desist.

5. Conclusions

In this chapter, an overview of the current possibilities and challenges in the exploitation of offshore renewable energy is presented. Oceans and marine areas provide a huge potential for electricity production. In this regard, an important differentiation between theoretical and economic potential must be taken into account. If the first represents the overall resource physically available, the latter represents the final potential of the resource after considering technical, social, environmental, and economic constraints. Thus, any assessment on the energy that can be converted from offshore winds, waves, and currents should take these limitations into account. Nonetheless, the marine environment can provide a reliable, continuous, and predictable source of energy; therefore any effort should be made in order to harness it.

As discussed, within the offshore industry, some sectors and technologies are at a more advanced stage of development than others. Therefore, the performance of some devices must be improved, while the gap between consolidated and novel technologies filled. This can be achieved in different ways, depending on the maturity of the technology. For established technologies like offshore wind turbines, efforts will be made towards improving the reliability and longevity of the devices, in order to maximize the availability and power production, increase the life cycle, and minimize operating costs. For novel technologies like wave energy converters, the focus will be on improving the power capture and conversion, while engineering devices are for surviving in the harshest conditions. For most devices, investigation and improvements of both kinds will be needed in order to obtain an effective and efficient technology.

In all cases, the main aim is the reduction of costs on all the phases of the project. This can be achieved by following the best practices and managing the project adequately. An effective project-management plan covering all the phases of the offshore farm, from pre-feasibility study to decommissioning or repowering, should be implemented. This process starts with studies assessing the feasibility of the project. It continues with the initial design, when not only functional but also easy to maintain (and possibly recycle) devices should be chosen. Thus, due to the risk and expenses related to the installation and maintenance of an offshore farm, a suitable holistic strategy for these phases should be chosen. The decision-making process in this case should be supported by tools specifically created for the offshore renewable sector. Assets management skills are thus required throughout the entire project life cycle. The final objective is to produce a reliable and cost-effective project, able to compete against both other renewable and nonrenewable developments.

Hence, a number of barriers preventing offshore renewables from significantly contributing to the global energy production remain. However, suitable mitigation measures exist and should be adopted. The support provided by governments and public bodies is and remains fundamental until the technologies are fully mature. This will provide the opportunity to meet sustainability objectives while creating new jobs and investment prospects.

Author details

Giovanni Rinaldi
Renewable Energy Group at the University of Exeter, Penryn Campus, Penryn, Cornwall, United Kingdom

*Address all correspondence to: g.rinaldi@exeter.ac.uk

References

[1] Kolliatsas C. Offshore Renewable Energy: Accelerating the Deployment of Offshore Wind, Tidal and Wave Technologies. Abingdon: Earthscan; 2012

[2] Sheng W, Li H, Murphy J. An improved method for energy and resource assessment of waves in finite water depths. Energies. 2017;**10**:1188. DOI: 10.3390/en10081188

[3] Arapogianni A, et al. Deep water—the next step for offshore wind energy. European Wind Energy Association (EWEA). 2013. Available from: http://www.ewea.org/fileadmin/files/library/publications/reports/Deep_Water.pdf

[4] Khan MJ, Bhuyan G, Iqbal MT, Quaicoe JE. Hydrokinetic energy conversion systems and assessment of horizontal and vertical axis turbines for river and tidal applications: A technology status review. Applied Energy. 2009;**86**:1823-1835. DOI: 10.1016/J.APENERGY.2009.02.017

[5] Güney MS, Kaygusuz K. Hydrokinetic energy conversion systems: A technology status review. Renewable and Sustainable Energy Reviews. 2010;**14**:2996-3004. DOI: 10.1016/J.RSER.2010.06.016

[6] Canadian Geographic. Tidal Turbine Prototypes. 2019. Available from: https://www.canadiangeographic.ca/article/nova-scotias-first-stream-tidal-turbine-starts-producing-power [Accessed: 10 July 2019]

[7] MeyGen Website. 2019. Available from: https://simecatlantis.com/projects/meygen/ [Accessed: 15 December 2019]

[8] Falcão AFDO. Wave energy utilization: A review of the technologies. Renewable and Sustainable Energy Reviews. 2010;**14**:899-918. DOI: 10.1016/j.rser.2009.11.003

[9] López I, Andreu J, Ceballos S, Martínez de Alegría I, Kortabarria I. Review of wave energy technologies and the necessary power-equipment. Renewable and Sustainable Energy Reviews. 2013;**27**:413-434. DOI: 10.1016/J.RSER.2013.07.009

[10] Rinaldi G, Crossley G, Mackay E, Ashton I, Campbell M, Wood T, et al. Assessment of extreme and metocean conditions in the Maldives for OTEC applications. International Journal of Energy Research. 2019;**43**:7316-7335. DOI: 10.1002/er.4762

Chapter 19

Wind Speed Analysis Using Signal Processing Technique

Omer Akgun and T. Cetin Akinci

Abstract

In this study, wind energy, the formation of this energy, and the necessary stages for the production of electrical energy will be discussed. Then, the countries' investments in wind energy will be mentioned. In the mathematical background, statistical methods and signal processing methods are used in the calculation of wind energy efficiency. In this chapter, a detailed analysis of the use of wind speed data with signal processing techniques will be made using the hourly wind speed data of Istanbul for the last 10 years. This data will be analyzed by the Fourier method. Afterward, analyses will be made with short-time Fourier transform (STFT) and bi-spectrum analysis method, and these results will be compared. The data obtained from the study can be considered as a framework for the wind farms to be constructed.

Keywords: wind speed, STFT, bi-spectrum analysis, data analysis

1. Introduction

The growth increase of the world population and the development of technology and industry cause the need for energy consumption. Energy consumption in the world increases by an average of 4–5% every year [1, 2]. Energy production from fossil fuels is less preferred due to reduced reserves and environmental sensitivity. Research has shown that humanity consumes a thousand years of fossil fuel formation in 1 day. Here, it shows that renewable energy sources are the most optimal solution in energy production [1–4].

Today, the most widely used renewable energy sources are hydraulic, wind, solar, geothermal, and biomass energy. Renewable energy sources are the ones that can renew themselves continuously, have no additional costs other than investment costs, and do not cause environmental problems. Among renewable energy sources, wind energy is the most popular today. Wind consists of solar radiation heating the ground surfaces differently [2, 4–7]. There are some advantages of using wind energy. Wind energy is a source of energy that does not create environmental pollution, is abundant in the atmosphere, and does not require very high technology for energy production. The conversion of wind energy to electrical energy is provided by using wind turbines [4–8]. The recent development of wind turbine technology increases the amount of energy produced and reduces initial costs. This form of energy production, which is also friendly to the environment, has been supported by many developed countries and turned into a state policy [1–3].

Determination of wind energy potential is the most important feasibility study before wind turbines are installed. There are many studies in the literature in different disciplines for determining wind potential. Wind power has reached a capacity of over 600GW in 2018 worldwide. According to 2018 statistics, China has a maximum capacity of 221 GW, using more than one-third of the world's capacity. America 96.4 GW, Germany 59.3 GW, India 35 GW, and Spain 23 GW capacity are ranked with [7–10].

However the advantages and disadvantages of wind energy vary according to sources [11]. The following are the benefits of wind energy: It is a clean source of energy and has no emissions. It is a local energy source, not dependent on external sources. It uses 1% of the investment area, and agriculture and animal husbandry activities can be done in these areas. It is a cheap source of energy. Idle areas can be used for turbine installation. It creates high employment. The drawbacks of wind energy are as follows: Image can create pollution. It may create noise pollution. It may disrupt radio and TV signals. In bird migration routes, it may harm birds [12–15].

As with most renewable energy sources data, wind speed is analyzed in non-stationary signals group. As can be expected, extreme winds during the typhoon and thunderstorms show non-stationary characteristics. The analysis of these data cannot be characterized by the use of static mathematical models. The frequency contents and amplitudes of these non-stationary signals may vary with time. In this study, signal-based analysis methods are used in the analysis of wind speed. Weibull distribution is known to be the most important statistics used in low wind speed analysis [16, 17]. Detailed data processing methods are needed to investigate the wind speed potential. In the analysis, the continuity of the wind speed is first investigated by statistical methods. The next stage is the efficiency and cost research. Fourier-based time-frequency methods with Weibull distribution provide a great convenience in wind speed analysis. The general purpose of this study is to analyze the wind speed data with spectral methods and to compare these analysis results.

2. Statistical and mathematical background for wind speed analysis

The wind speed is evaluated in a group of non-deterministic, nonlinear signals with no specific statistical and parametric values. Variation of wind speed according to time, day, and season requires statistical analysis to determine its potential. Anemometer is used to measure wind speed. Wind measurements in practice are minute, 10 minute, and hourly data. The data in the longer range is obtained by averaging these data. These data are divided into classes of wind speeds of 1 m/s each, and the energy in a given region is calculated by this frequency distribution. At this stage, Weibull distribution was found to be the most optimal approach for wind speed distribution [16].

2.1 Statistical analysis

The wind potential of a region is determined by a series of detailed statistical and mathematical methods to be performed in the region. Frequency of wind speed data and collection over a long period of time, data collection methods, etc. significantly affect the accuracy of the analysis result. Statistical parameters are used to analyze the data collected over time. It is possible for data to exhibit a sinusoidal property, while the mean and median values are expected to be close to zero, but it is possible that these values have moved away from zero. In the analyses, the most basic mean value μ and standard deviation, σ, can produce significant results in non-periodic signals [16–21]. For a particular set of signs {xi}, they are defined as in the following equations:

$$\mu = \frac{1}{N}\sum_{i=1}^{N} x_i \tag{1}$$

$$\sigma = \sqrt{\frac{1}{N}\sum\nolimits_{i=1}^{N}(x_i - \mu)^2} \tag{2}$$

Eqs. (1) and (2) above are *N* data numbers. Here, the standard deviation refers to the average distribution of variables in a data set. A small standard deviation value means that the values of the data group are distributed around the average. In practice, the data is expressed by the normal probability distribution (Gaussian) called the central limit theorem. Here, the two functions obtained from the Gaussian distribution, skewness (α) and kurtosis (β), are given in Eqs. (3) and (4):

$$\alpha = \frac{\left[\frac{1}{N}\sum\nolimits_{i=1}^{N}(x_i - \mu)^3\right]}{\sigma^3} \tag{3}$$

$$\beta = \frac{\left[\frac{1}{N}\sum\nolimits_{i=1}^{N}(x_i - \mu)^4\right]}{\sigma^4} \tag{4}$$

Frequency distribution of wind speed is determined by different statistical distributions such as Gamma, Rayleigh, and Weibull. Among these methods, Weibull distribution is the most preferred method. This method is well suited to wind distribution [21–25].

2.1.1 Weibull parameters method

The Weibull distribution was introduced by Waloddi Weibull in 1951 to estimate the life expectancy of machines. Continuous random variables are used to describe random events in which the variable can take any value in a given range. The Weibull distribution is continuous and at the same time flexible distribution and is applied in many random statistical applications.

In wind speed analysis, it can be used to identify wind speed changes in an area at an acceptable level of accuracy. Weibull distribution is determined by the shape and scale variable. The area expressed as a probability distribution is 1 in total. The Weibull distribution curve is skewed. As a result of the calculations, if the Weibull distribution parameter, *k* = 3 and 4, is obtained, the distribution is similar to the normal distribution. If parameter k takes the value of 1 as a result of calculations, the distribution is an exponential function. The Weibull distribution function is the most common method used to calculate wind speed and density in wind speed analyses [16–25]:

$$(v) = \frac{k}{c}\left(\frac{v}{c}\right)^{k-1} e^{\left(\frac{v}{c}\right)^k} \tag{5}$$

In Eq. (5) above, *v* is the wind speed, *f* is the frequency, *k* is the shape parameter, and *c* is the scale [22–25].

2.2 Time-frequency methods for wind speed analysis

The conventional Fourier transform only yields a spectral distribution over time and cannot represent the transient properties of a non-stationary process. In order to define the time-varying properties of non-stationary processes, methods such as short-term Fourier transform (STFT), Wigner-Ville distribution (WVD), and wavelet transform (WT) have been developed.

STFT uses a fixed-width window to capture the time-frequency distribution, so limited resolutions are captured in the analysis with STFT. In this sense, it may be possible to achieve better results with wavelet transform. Time-frequency methods are discussed in the following section [26].

2.3 Fourier transform and STFT

The Fourier transform (FT) method is used to extract information from the signal and to process the signals. This method is one of the most effective methods used for signal analysis. With FT, a signal can be represented as the sum of the different amplitudes, frequencies, and the fundamental components in the phase. Processing here provides a great convenience in the computer environment. The basic equations for Fourier transform are given in Eqs. (6) and (7) below [27, 28]:

$$f(x) = \frac{1}{\sqrt{2\pi}} \int_{-\infty}^{\infty} F(k) e^{ikx} dk \tag{6}$$

$$F(k) = \frac{1}{\sqrt{2\pi}} \int_{-\infty}^{\infty} F(x) e^{-ikx} dx \tag{7}$$

FT, short-time Fourier transform (STFT) and spectrogram were developed based on Eqs. (6) and (7). With this method, time-frequency localization can be obtained clearly. Here, an $x\ (t)$ signal of fixed window size and frequency resolution is used. The generalized equation for STFT is given below in Eqs. (8) and (9). In addition, the spectrogram equation is shown in Eq. (10) [27, 28]:

$$STFT\{x(t)\} \equiv X(\tau, f) = \int_{-\infty}^{\infty} x(t) g(t-\tau) e^{-j2\pi ft} dt \tag{8}$$

$$STFT\{x(n)\} \equiv X(m, f) = \sum_{n=-\infty}^{\infty} x(n) g(n-m) e^{-j\omega n} \tag{9}$$

$$\{x(t)\} \equiv |X(\tau, f)|^2 \tag{10}$$

2.4 Reassigned spectrogram

This analysis yields successful results for non-stationary signals. Contains a detailed analysis of spectral content of non-stationary signals. Within the limits of the analysis, it is possible to obtain better resolution using only more information from the power spectrum. The basis of the calculation is that the instantaneous frequency (IF) in each FFT frequency band is equal to the first derivative of the short-term FFT (STFT) phase at time T, and this is given in Eq. (11) [29, 30]:

$$\mathrm{IF} = \partial / \partial t \left(\arg \left(\mathrm{STFT}\left(\omega, T \right) \right) \right) \tag{11}$$

2.5 Welch analysis

Welch analysis is one of the most effective methods used for spectral estimation. The purpose of spectral estimation is to determine the frequency-dependent distribution of the power in a signal. The Welch method, a nonparametric method, is used as a more effective method. In this method, the time series is divided into sections that may overlap, and then the improved periodogram of each section is calculated and the average of the periodograms of these sections obtained is calculated. The average of these improved periodograms reduces the variance of all data over a single periodogram estimate. The Welch method estimates the power spectral density by averaging the improved periodograms. The first improved periodogram expression is given in Eq. (12):

$$\underset{xx}{\hat{P}}(i)(f) = \frac{T_S}{K.M}\left|\sum_{n=0}^{M-1} x_i(n)w(n).e^{-j2\pi fn}\right|^2 \tag{12}$$

where $f = fs$ is the normalized frequency variable, K is the normalized constant, and $w\ (n)$ is the windowing function. The expression of the constant K is given in Eq. (13). Power spectral density estimation is given in Eq. (14) [31–33]:

$$K = \frac{1}{M}\sum_{n=0}^{M-1} w^2(n) \tag{13}$$

$$\underset{xx}{\hat{P}}W(f) = \frac{1}{L}\sum_{i=0}^{L-1} \underset{xx}{\hat{P}}(i)(f) \tag{14}$$

2.6 Bi-spectral analysis

Phase relations between frequency components are not taken into account in signal analysis using quadratic statistics or power spectrum. Therefore, these methods do not provide information against phase information. In addition, second-order statistics and power spectrum cannot make statistical definitions of non-Gauss processes. High-level statistics and spectrum studies are conducted for the more precise definition of random processes and the processing of phase information.

In addition to suppressing Gaussian probability-distributed activity, the bi-spectrum reveals signals originating from the nonlinear process—bi-spectral analysis; the background is used to detect low-level but diagnostic important signs masked by FKG. The power spectrum of the random signals is defined by AFD in Eq. (15), and the third-order heap spectrum is shown in Eq. (16). The expression in this equation is called the bi-spectrum. In this case, when the sign is a stationary random process with real value, it can be expressed as in Eq. (17) [33–35]:

$$P_2^x(f) = AFD\left(C_2^x(m).e^{-j2\pi mf}\right. \tag{15}$$

$$B^x(f_1, f_2) = \sum_{m=-\infty}^{\infty}\sum_{n=-\infty}^{\infty} C_3^x(m,n).e^{-j2\pi(mf_1+nf_2)} \tag{16}$$

$$B(w_1, w_2) = X(w_1).X(w_2).X^*(w_1 + w_2) \tag{17}$$

3. Analysis and application

When applied as of 2014, Istanbul-Turkey wind data are used. Data are hourly wind speed from the State Meteorological Service. In addition, 2015 data were used for testing purposes for analysis. **Figure 1** shows the time-amplitude graph of the wind speed.

When the time-amplitude and time-energy graphs are evaluated together, the wind speed reaches the highest values of the whole year in the first months of the year (first 1500 samples) (**Figure 2**).

Figure 3 shows the frequency-amplitude graph of the wind speed. The graph shows an amplitude that continues up to approximately 0.015 Hz, followed by a noticeable peak at 0.045 Hz, decreasing frequency amplitude. This peak can be defined as the fundamental frequency of wind speed.

The power value generated by the annual wind speed in the Welch power analysis consists of a peak that reaches a maximum value of 340 units. Accordingly, it remains remarkably small on a hill of approximately 30 units (**Figure 4**).

When evaluated together with the time-frequency spectrogram (**Figure 5**) and the reassigned spectrogram (**Figure 6**), the red regions with high amplitude reaching up to 0.02 Hz in the first months of the year (first 1800 samples) are remarkable. The second high-amplitude region (between 6500 and 8000 data) corresponds to the last months of the year at frequencies of 0.01 Hz. In addition, the continuous distribution in the 0.045 Hz region, which spans almost the whole year, is noteworthy.

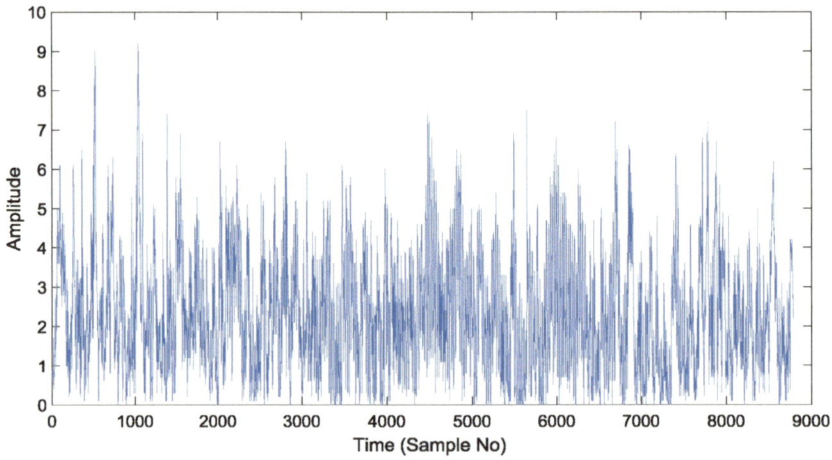

Figure 1.
Time-amplitude graph of wind speed.

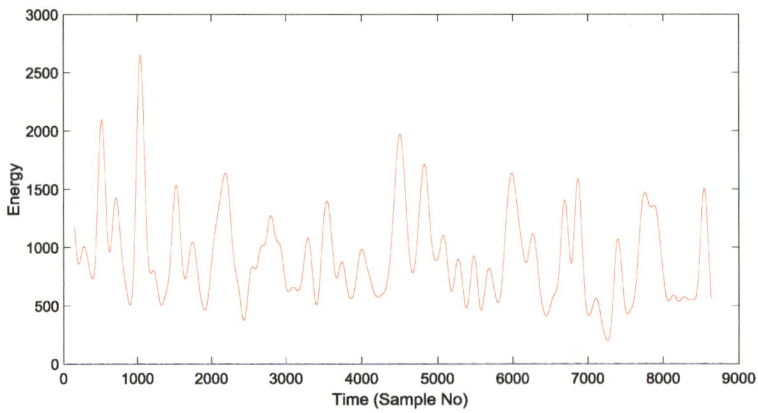

Figure 2.
Time-energy graph of wind speed.

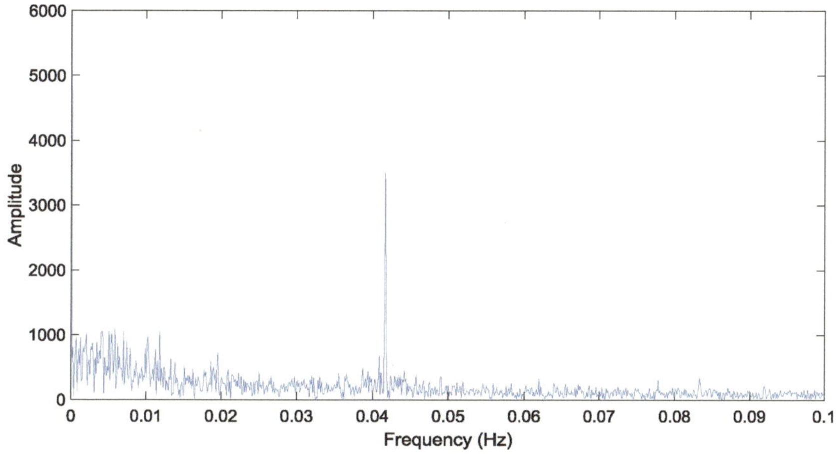

Figure 3.
Frequency-amplitude graph of wind speed.

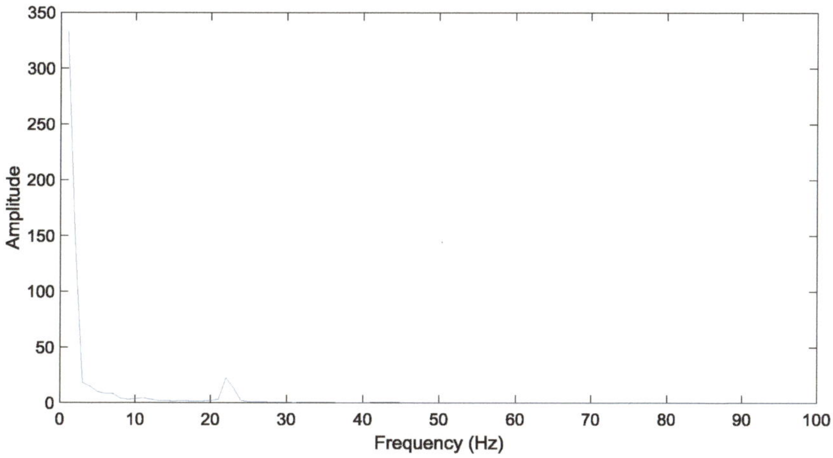

Figure 4.
Frequency-power graph of wind speed.

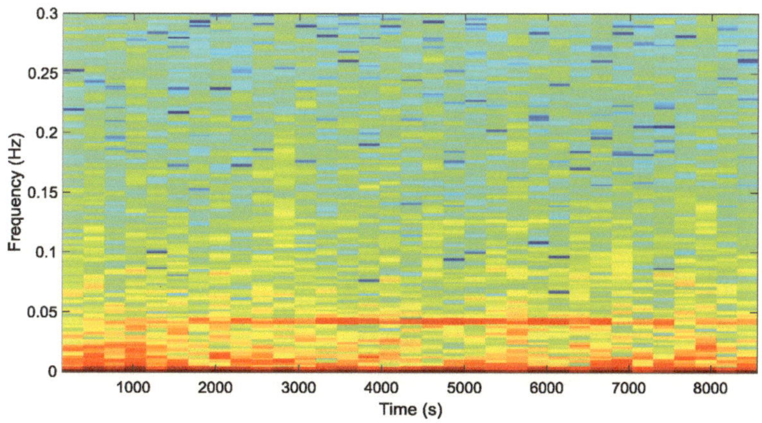

Figure 5.
Time-frequency spectrogram of wind speed.

The contour map of the bi-spectrum consists of two basic rings. The density ring consisting of high-amplitude peaks inside contains 0.1 Hz regions, and the outside small-amplitude ring covers 0.2 Hz regions (**Figure** 7).

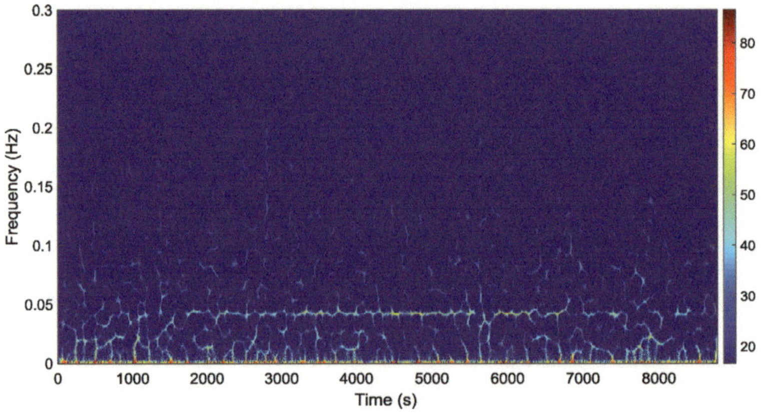

Figure 6.
Reassigned spectrogram of wind speed.

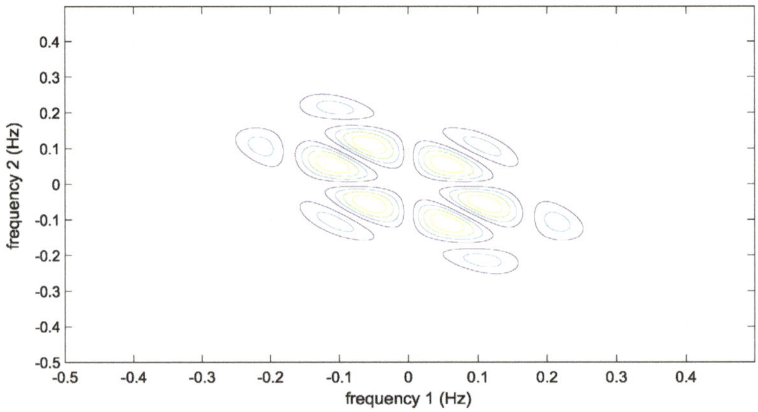

Figure 7.
Contour map of bi-spectrum estimation of wind.

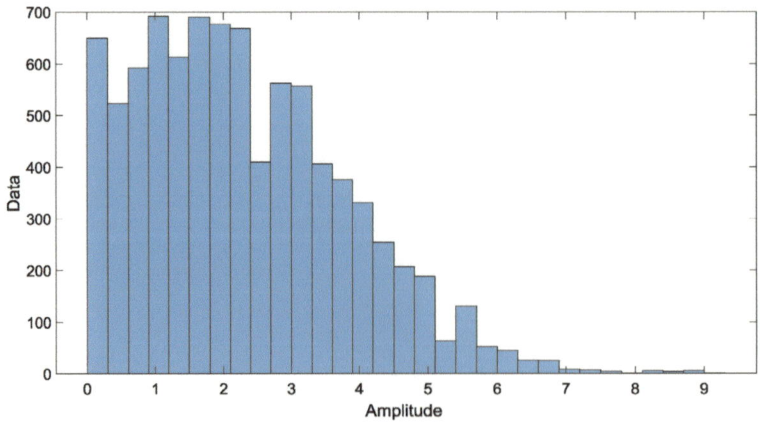

Figure 8.
Histogram of wind speed.

In **Figure 8**, wind velocities around 0–2.5 amplitude and 3 amplitude of wind velocity were observed at most. This can be seen with annual rh4mean = 2.3360 average value and rh4var = 2.2962 varying values.

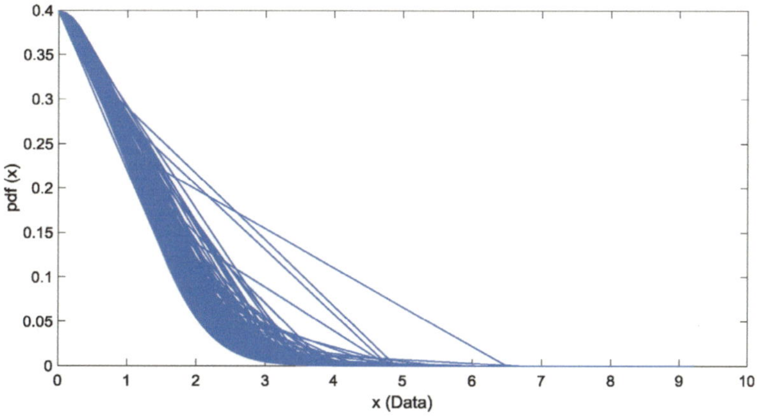

Figure 9.
Probability density function graph.

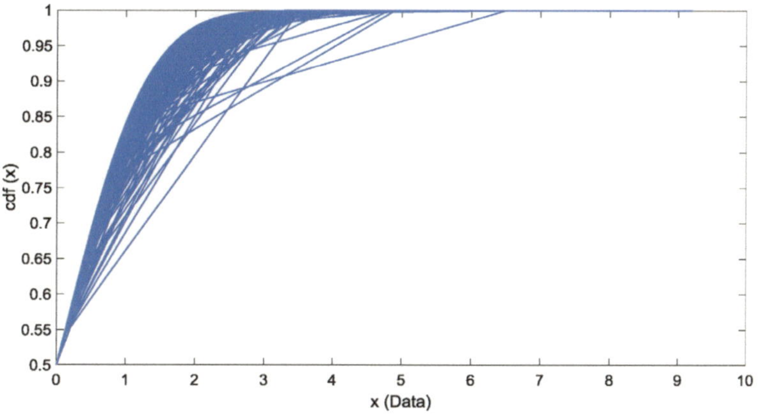

Figure 10.
Graph of probability distribution function of wind speed.

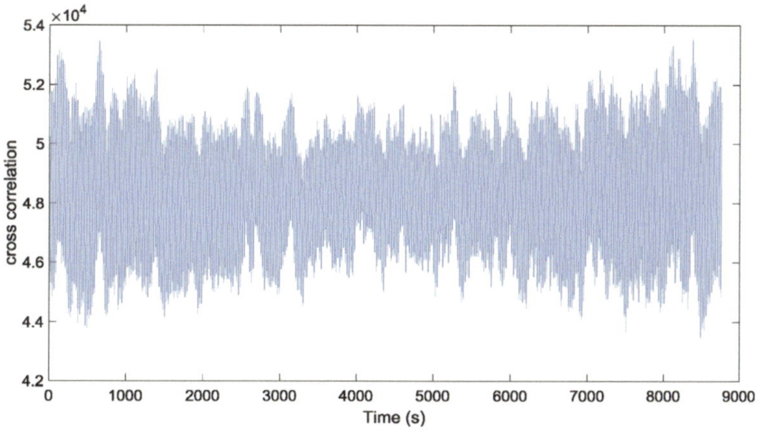

Figure 11.
Cross-relationship chart (2014–2015).

In **Figures 9** and **10**, when the probability distribution and density function are evaluated together, they are concentrated especially in 1–3 regions of the information signals.

When **Figure 11** is examined, it is seen that the cross-relationship is maximum in the first months and last months of the year (the season with the highest wind speeds).

4. Conclusions

In this study, Istanbul wind speed data were used. The data were obtained hourly from the national meteorological station. In the study, the seasonal variation of wind can be observed clearly from the time-amplitude graph of the wind speed. In addition, the wind speed data were analyzed using signal processing methods. These analyses are STFT, reassigned spectrogram, frequency power analysis, bispectral in-phase surface, and probability density function. In addition, cross-relationships and histogram between 2014 and 2015 were obtained by statistical analysis. In signal processing analyses, the distinction of seasonal transitions during the year can be determined by low- and high-frequency levels. Again in the autumn and winter seasons where the wind speed is higher, the frequency level at which the wind speed has reached can be clearly determined. The results of the statistical analysis of the histogram and cross correlation are quite satisfactory, and the relationship between the wind speed between 2014 and 2015 is quite similar, especially in autumn and winter seasons. Again, the dominant frequency was found to be around 0.04 Hz.

If this study evaluated as spectral analysis, it can be seen that the observed wind profile is a combination of some dominant modes defined by terms such as synoptic or large scale and mesoscale. With thıs assumptıon, these dominant scales are seen as the relative maximum ın power spectral analysis of wind speed. This assumption shows itself as the correct spectrum amplitudes ın this study. It shows consistency in terms of seasonal similarıty in the first months of the year and in the last month of the year, as seen from the cross-relation chart. The symmetrical distribution in **Figure** 7 also proves that the wind speed is seasonally similar. Similarly, **Figure 2** energy distribution also proves the seasonal wind speed distribution and the accuracy of the relationshıp between spectral analysis and energy. In this sense, thıs study shows that the analysis of spectral methods is parallel to statistical methods in the analysis of wind speed data.

Author details

Omer Akgun[1] and T. Cetin Akinci[2]*

1 Department of Computer Engineering, Marmara University, Istanbul, Turkey

2 Department of Electrical Engineering, Istanbul Technical University, Istanbul, Turkey

*Address all correspondence to: akincitc@itu.edu.tr

References

[1] Jianzhong X, Assenova A, Erokhin V. Renewable energy and sustainable development in a resource-abundant country: Challenges of wind power generation in Kazakhstan. Sustainability. 2018;**10**(3315):1-21

[2] Pryor SC, Barthelmie RJ. Climate change impacts on wind energy: A review. Renewable and Sustainable Energy Reviews. 2010;**14**:430-437

[3] Available from: https://www.irena.org/-/media/Files/IRENA/Agency/Publication/2019/Jun/IRENA_RE_Jobs_2019-report.pdf

[4] Varanasi J, Tripathi MM. A comparative study of wind power forecasting techniques — A review article. In: 3rd International Conference on Computing for Sustainable Global Development (INDIACom). New Delhi; 2016. pp. 3649-3655

[5] Nogay HS, Akinci TC. Application of decision tree methods for wind speed estimation. European Journal of Technique (EJT). 2019;**9**(1):74-83

[6] Gielen D, Boshell F, Saygin D, Bazilian MD, Wagner N, Gorini R. The role of renewable energy in the global energy transformation. Energy Strategy Reviews. 2019;**24**:38-50. DOI: 10.1016/j.esr.2019.01.006

[7] Usta I, Kantar YM. Analysis of some flexible families of distributions for estimation of wind speed distributions. Applied Energy. 2012;**89**(1):355-367

[8] Kantar YM, Usta I. Analysis of the uppertruncated Weibull distribution for wind speed. Energy Conversion and Management. 2015;**96**:81-88

[9] Zucatelli PJ, Sperandio Nascimento EG, Bandeira Santos AA, Gutiérrez Arce AM, Moreira DM. Study of the wind speed forecasting applying computational intelligence. DOI: 10.5772/intechopen.89758

[10] Available from: https://www.power-technology.com/features/wind-energy-by-country/

[11] Available from: http://www.tuba.gov.tr/files/yayinlar/raporlar/R%C3%BCzgar%20Enerjisi%20Teknolojileri%20Raporu.pdf

[12] Fauzan MF, Chelvan RK, Amirah A. Energy, economic and environmental impact of wind power in Malaysia. International Journal of Scientific Research and Management. 2017;**2**(7)

[13] Khahro SF et al. Techno-economical evaluation of wind energy potential and analysis of power generation from wind at Gharo. Sindh Pakistan, Renewable and Sustainable Energy Reviews. 2014; **35**:460-474. DOI: 10.1016/j.rser.2014.04.027

[14] Ghaith AF, Epplin FM, Frazier RS. Economics of household wind turbine grid-tied systems for five wind resource levels and alternative grid pricing rates. Renewable Energy. 2017; **109**:155-167

[15] Kaldellis JK et al. Environmental and social footprint of offshore wind energy. Comparison with onshore counterpart. Renewable Energy. 2016;**92**:543-556

[16] Guseinovienė E, Senulis A, Akinci TC, Seker S. Statistical and continuous wavelet analysis of wind speed data in Mardin-Turkey. In: Ninth International Conference on Ecological Vehicles and Renewable Energies (EVER), 25–27 March 2014. Monaco: Monte Carlo; 2014

[17] Sarkar A, Gugliani G, Deep S. Weibull model for wind speed data

analysis of different locations in India. DOI: 10.1007/s12205-017-0538-5

[18] Akinci TC. Wavelet analysis of wind speed and solar radiation data. In: 5th International Conference on Sustainable Development. Belgrade, Serbia; 2019. pp. 17-21

[19] Seker S, Akinci TC, Taskin S. Spectral and statistical analysis for Ferroresonance phenomenon in electric power systems. Electrical Engineering. 2012;**94**(2):117-124

[20] Idriss AI, Ahmed RA, Said RK, Omar AI, Barutçu B, Mohamed AA, et al. Suitability and evaluating wind speed probability distribution models in a hot climate: Djibouti case study. International Journal of Renewable Energy Research. 2019;**9**(3):1586-1596

[21] Dokur E, Kurban M. Wind speed potential analysis based on Weibull distribution. Balkan Journal of Electrical & Computer Engineering. 2015;**3**(4): 231-235

[22] Carrillo C, Cidrás J, Díaz-Dorado E, Montaño AFO. An approach to determine the Weibull parameters for wind energy analysis: The case of Galicia (Spain). Energies. 2014;**7**: 2676-2700

[23] Chang TP. Estimation of wind energy potential using different probability density functions. Applied Energy. 2011;**88**(5):1848-1856

[24] Chang TP. Performance comparison of six numerical methods in estimating Weibull parameters for wind energy application. Applied Energy. 2011;**88**(1): 272-282

[25] Costa Rocha PA, de Sousa RC, de Andrade CF, da Silva MEV. Comparison of seven numerical methods for determining Weibull parameters for wind energy generation in the northeast region of Brazil. Applied Energy. 2012; **89**(1):395-400

[26] Akinci TC. The defect detection in ceramic materials based on time-frequency analysis by using the method of impulse noise. Archives of Acoustics. 2011;**36**(1):77-85

[27] Vaseghi SV. Advanced Signal Processing and Digital Noise Reduction. John Wiley & Sons Inc; 1996. pp. 326-328

[28] Akinci TC. Time-frequency analysis of the current measurement by Hall effect sensors for electric arc welding machine. Mechanika;**85**(5):66-71

[29] Available from: https://people.ece.cornell.edu/land/PROJECTS/ReassignFFT/index.html

[30] Fulop SA. Speech Spectrum Analysis, Signals and Communication Technology. Springer-Verlag Berlin Heidelberg; 2011. pp. 127-165. DOI: 10.1007/978-3-642-17478-0_6

[31] Welch PD. The use of fast Fourier transform for the estimation of power spectra: A method based on time averaging over short, modified periodograms. IEEE Transactions on Audio and Electroacoustics. 1967;**AU-15** (2):70-73. DOI: 10.1109/TAU.1967.1161901

[32] Proakis JG, Manolakis DG. Digital signal processing: Principles. In: Algorithms and Applications (3 Ed.). Upper Saddle River, NJ: Prentice-Hall; 1996. pp. 910-913

[33] Akinci TC, Seker S, Akgun O, Dikun J, Erdemir G. Bispectrum and energy analysis of wind speed data. International Journal of Electrical Energy. 2017;**5**(1):87-91

[34] Alkan A, Kiymik MK. Comparison of AR and Welch methods in epileptic

seizure detection. Journal of Medical Systems. 2006;**30**(6):413-441. DOI: doi.org/10.1007/s10916-005-9001-0

[35] Nikias CL, Raghuveer MR. Bispectrum estimation: A digital signal processing framework. Proceedings of the IEEE. 1987:869-891

Chapter 20

Wind Turbine Integration to Tall Buildings

Ilker Karadag and Izzet Yuksek

Abstract

Having a far distance from the ground levels exposed to turbulent wind conditions, tall buildings have the potential of generating wind energy. However, there are many challenges to incorporating wind generation into urban areas. These include planning issues besides visual impacts. So, as to integration, there is a need for a combined approach that considers wind energy harvesting besides these issues. At this point, a multidisciplinary approach can fill the gap between the architectural design and the wind engineering processes. Based on this approach, this chapter presents design strategies from the literature to integrate wind energy to tall buildings using computational fluid dynamics (CFD) simulation. It is intended to guide further researches on wind energy and consequently to contribute to the environmental quality of urban areas and sustainable development of the cities.

Keywords: building-integrated wind turbines, building aerodynamics, wind energy, computational fluid dynamics (CFD), wind efficient design

1. Introduction

Tall buildings are common forms of urban settlements, and they involve too many functions with plenty of spaces on a low base area with the advantages of having too many floors. Besides, they have important impacts on local wind characteristics around the near-field urban area. Strong winds are usually accelerated at the roof level within the urban environment around tall buildings, due to particular aerodynamic configurations generally associated with this type of buildings. When it comes to a simple rectangular tall building, it is the flow separated from the windward side edge of the roof which led to shear layers with increasing wind velocities [1]. This flow may result in high wind velocities and turbulent wind conditions. For wind energy harvesting, higher wind speeds are mostly desired to a limit velocity, but higher turbulence levels are mostly not intended since the fluctuations in wind speed reduce the energy extracted from the wind turbine. Besides the turbulence, intensity increases the fatigue loads on the wind turbine. So, the assessment of wind flow around the building is very important for a thorough understanding of the flow characteristics.

Assessment of the wind flow around the buildings is conducted with both experimental and numerical methods. The experimental method ensures reliable information as to airflow in and around buildings; however, the available data is mostly limited due to the expensive experimental processes which require too many sensor locations for high-resolution data. Besides, the approach is not practical for an architect or engineer trying to determine the optimal location of the wind turbine. The parametric tools embodied in CFD codes make it very easy to try different geometrical forms or a principal form with different angle of attacks. However, the experimental method may extend the time consumed on optimization since it requires re-configuration of the experimental setup. In addition, wind tunnel testing has some limitations which CFD overcomes. As stated in [2], CFD does not have scaling problems and similarity constraints since simulations can be conducted at full scale.

The literature as to wind energy harvesting mostly constitutes of the planning and design issues of the wind farms; unfortunately, rare studies have been conducted on wind turbine integration to tall buildings located in dense urban areas. In this chapter, beyond describing wind energy systems of a number of tall buildings, a design methodology from architectural and engineering perspectives is proposed. Within this study, CFD assessment of the wind conditions around a number of tall buildings is given with the estimated amount of the energy generated with the wind turbine. First, the wind turbine integration strategy of these cases is described from the architectural design perspective. Then, brief information about these buildings and their surroundings are explained. Following that, the results of the numerical simulations are given and conclusions are derived.

2. Methodology for the assessment of wind energy potential in an urban area

Wind energy potential assessment mostly starts with climate data analysis. So, a thorough understanding of the local climate which is based on a detailed analysis of long-term (at least 30 years) meteorological data is required [3]. After this data has been achieved, a correlation between terrain roughness of the meteorological station site and the examined urban area is set. For this aim, aerodynamic information of the site and the building are investigated.

Mostly, meteorological stations are located on open areas which have aerodynamic roughness length of $y0 = 0.003$ m, where wind speed is measured at 10 m height and the mean wind speed values are taken on an hourly basis. To transfer this data to the site examined, a logarithmic wind speed profile can be used since it provides mean wind speed distribution with height taking aerodynamic roughness lengths into consideration. These data are used as the inlet boundary condition for CFD simulation to achieve wind speed velocities at different heights from the ground.

The result of the combination of weather data and the aerodynamic information is used in the CFD simulation as an input to assess wind velocities and turbulence conditions around the building accurately. At this point, it is required to use a convenient mathematical model. The turbulent flow within urban or industrial environments is in general modeled by the Navier-Stokes Equations [4]. In detail, as being an economical solution, steady-state 3D Reynolds-averaged Navier-Stokes (RANS) equations are used mostly. On the other hand, large eddy simulation (LES) undeniably has the potential to provide more accurate and more reliable results than simulations based on the Reynolds-averaged Navier-Stokes (RANS) approach. However, LES entails a higher simulation complexity and a much higher computational cost. The equations in these turbulence models are solved with commercial or open-source CFD codes. The most important point for the architect or engineer who will run the simulation is to comply with the CFD guidelines. The best practices largely advised in the literature are the Architectural Institute of Japan (AIJ) CFD guidelines [5] and the COST 732 Best Practice Guideline for CFD [4] simulation of flows in the urban environment. In both of these guidelines, the grid dependence of the solution is advised to be checked. It should be confirmed that the prediction result does not change significantly with different grid systems. In the best practice guidelines, it is indicated that at least three systematically and substantially refined grids should be used so that the ratio of cells for two consecutive grids should be at least 3.4 [4]. The value of 3.4 means finer grids with 1.5 times the grid number in three dimensions. As a practice, to improve the computational mesh in the necessary regions where solid geometry and fluid field contacts and separation of flow occur is advised.

Criteria for convergence are another important parameter of numerical simulations. For this purpose, as indicated in the Architectural Institute of Japan (AIJ) CFD guidelines, it is important to confirm that the solution does not change by monitoring the variables on specified points or by overlapping the contours among calculation results at different calculation steps [5].

A large computational model needs to be created, and different zones with respect to aerodynamic roughness lengths should be arranged. For instance, in the wind comfort study [6], the nearby area around the analyzed tall building is modeled explicitly (zone 2), and the far field containing an urban area with tall buildings is implicitly modeled (zone 3) using aerodynamic roughness length of $y0 = 2$ m (**Figure 1a**). While the whole domain area is 1400 m by 2600 m, the area with explicitly modeled buildings was 400 m by 400 m. An upstream domain extension of 5H and a downstream domain extension of 15H were left in the domain as advised in the guidelines [4]. The site of interest in the center of the computational domain is modeled in detail (zone 1). This zone includes smaller computational cells than the zone at the extension of the explicitly modeled region as illustrated in **Figure 1b**. A high-quality and high-resolution grid that consists of only cut cells is developed to gain fast convergence.

Once the meshing process was complete, boundary conditions were determined (inlet, outlet, etc.), the appropriate turbulence model and wind velocity profile were selected, and CFD simulations are run. Then the analysis results were examined in detail in a post-processing software, and the necessary conclusions are derived. At the last stage, an appropriate wind turbine model was chosen, and design proposals were developed to integrate the chosen wind turbine system to the case building. Then, the amount of energy that can be produced with the proposed system is calculated. In predicting power output from a wind turbine, a common approach in the literature is to take the power curve provided by the manufacturer

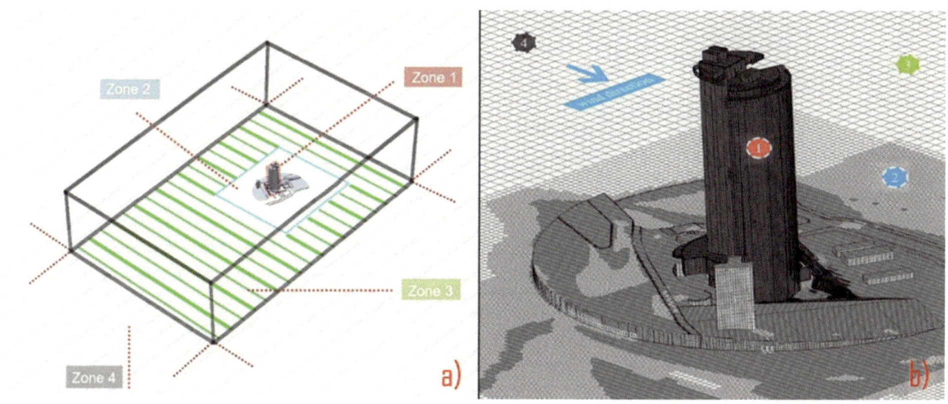

Figure 1.
Computational domain with zone management (a) and mesh refinement levels of different zones (b) [6].

and combine it with a wind speed frequency distribution function, such as a Weibull distribution or Rayleigh distribution, and integrating across the range of wind speeds in which generation takes place.

3. Building-integrated wind turbines

In most cases, the energy of the wind is gained through wind farms to supply economical clean power with a renewable energy source. However, the land is restricted in urban and suburban regions, and this is regarded to be a significant limitation on large-scale wind turbine installation. An alternative choice is to use building-integrated wind energy systems [7]. Wind energy installations near buildings have received far less attention [8–10]. The idea of the on-site generation of micro-wind energy is significant because the energy is being generated near where it is needed. In [8], the authors distinguish between three categories of possibilities for integration of wind energy generation systems into urban environments: (1) locating stand-alone wind turbines in urban locations, (2) retrofitting wind turbines onto existing buildings, and (3) full integration of wind turbines together with

Study	Data collection approach	Site model (concerned building)	Simulated flow	Key findings
Lu and Sun [11]	CFD	Lee Shau Kee building (97.4 × 38.7 × 69.9 m) (L, W, H). Kowloon, Hong Kong	RNG k-ε	Minimum height of 1.4 m above the roof is recommended for turbine installations. Building proximity effects can increase the wind power significantly (1.5–5.5 times) at 4 m above the building roof
Tabrizi et al. [12]	CFD field measurements from one location	Bunnings warehouse H ¼ 8.4 m. Port-Kennedy, Western Australia. The radius of the surrounding: 200 m	Steady RANS with k-ε. Computational domain (H): 0.2 km	A small HAWT is recommended to be installed **in the middle of the roof**
Al-Quraan et al. [13]	Wind tunnel experiments	EV Building (76 m high) Equiterre building (23 m high) Montreal, Quebec, Canada	Homogeneous terrain	Accuracy of wind tunnel results depends on the upstream terrain
Yang et al. [14]	Field measurements from two locations. CFD field measurements from 10 locations	The building has dimensions of 68.9 × 62.5 × 33.5 m (L, W, H). Central metropolitan Taipei, Taiwan	Nonhomogeneous terrain standard κ-ε. RNG κ-ε. Realizable κ-ε. Computational domain (L, W, H) 16 × 8 × 0.3 km	Recommend increasing the hub height and install microturbines on the windward side of the building. Curved roof edges increase the power density

Table 1.
Examples of recent studies for actual site locations [7].

architectural form. Categories 2 and 3 are often referred to as building-integrated wind turbines (BIWT). Within the scope of this chapter, category 1 is excluded since to position a stand-alone wind turbine in an urban area is not a form of the building-integrated wind energy system.

As to category 2, there are very few research studies on wind power utilization over existing buildings in urban areas. **Table 1** shows examples of recent actual building studies conducted by means of field measurements, wind tunnel experimentation, and CFD. Details of the considered models and test descriptions, as well as key findings, are provided in the table [7].

Category 3 includes aerodynamically designed building forms for full integration of wind turbines. As is well known, in urban environments the mean wind speed is lower than the one in rural areas. The wind speed in urban areas, however, is significantly high at specific locations near to tall buildings. It can be described as micro-generation to produce urban wind energy such as that generated by small-scale wind turbines installed on or around tall buildings. As previously stated, a main benefit of such systems is that the energy generated can be consumed straight at the assembly site, and the building's user obtains renewable additional energy. The use of wind power for distributed generation in tall buildings is becoming increasingly appealing. Since the theoretically produced energy is a function of the wind speed cube, a tiny rise in wind speed can lead to a significant difference in the generation of wind power. It is therefore of concern to correctly evaluate the wind resource in an urban area and try to improve it through different aerodynamic methods. This is opposite to the traditional strategy of wind engineering where the priority is on decreasing wind speeds and pressures in order to minimize wind-induced lateral loads to secure structural system.

When it comes to building-integrated wind turbines, the Bahrain World Trade Center, the Strata Tower in London, and the Pearl River Tower in Guangzhou, China, are very important examples and pioneers in this field. Therefore, their wind turbine integration strategies are given in the following subsections correspondingly to depict the interdisciplinary approach between architecture and wind engineering.

3.1 Bahrain world trade Center

The Bahrain World Trade Center is the focus of a master scheme to refurbish an existing hotel and shopping mall on a distinguished location overlooking the Arabian Gulf in Manama, Bahrain's central business district. The two 50-story sail-shaped office towers taper to 240 m high, supporting three horizontal-axis wind turbines with a diameter of 29 m (**Figure 2**). The elliptical plan forms and sail-like profiles function as airfoils, funneling the onshore breeze between them and producing a negative pressure behind them, thus accelerating the wind speed between the two towers. Vertically, tower form is also a consequence of the airflow dynamics. Their airfoil sections are decreasing in size as they taper upwards [15]. Combined with the increasing onshore breeze velocity at increasing heights, this impact produces a near-equal wind velocity regime on each of the three turbines. Understanding and using this phenomenon is one of the main variables that enabled the practical implementation of wind turbine systems into a design of a tall building. Wind tunnel experimentation verified how the towers' forms and spatial relationship reshape the airflow, forming a "S"-like flow whereby the center of the wind stream stays almost perpendicular to the turbine within a 45° wind azimuth, either side of the central axis. This improves the power generation capacity of the turbines while at the same time decreasing fatigue on the blades to acceptable levels during wind skewing across the blades [15].

Figure 2.
Wind turbines integrated in between the Bahrain world trade Center towers; elevation view (a) and a general view from the pedestrian level (b) [7].

The horizontal-axis wind turbines placed between the towers are normally pole-mounted and rotate to match the wind direction to maximize the energy output. Therefore, it is very hard to place such turbines to tall buildings located in a climate with variable direction wind conditions. Most architectural projects deploying building-integrated, horizontal-axis turbines, as in the case of the Bahrain World Trade Center, implement the concept of a fixed turbine. It seems more useful to deploy vertical-axis wind turbines since they advantage from the benefit of being fully omnidirectional. However, large-scale verified vertical-axis turbines were not accessible for building applications at the moment of design development for this building. The fixed horizontal turbine suffers the disadvantage of being able to function only with wind from a restricted azimuth range, if issues with blade deflections and stress are to be prevented due to excessive skew flow. The form of the towers was designed from the beginning of this project to catch the upcoming wind and channel it between the towers. Comprehensive wind tunnel testing latterly validated by CFD, illustrations of which are displayed in **Figure 3**, showed that the upcoming wind is in fact deflected by the towers in the form of a S-shaped streamline that runs through the space between the towers at an angle within the wind skew tolerance of the wind turbine. Engineering projections indicate that the turbine will be capable of operating between 270° and 360° of wind directions, but, to be on the safe side, turbine calculations and initial operating regimes are set on a more limited range between 285° and 345°. The turbine will automatically follow a "standstill" mode in all wind directions outside this range. It is no coincidence that the buildings are oriented towards the prevailing wind which is extremely dominant. The funneling of the towers amplifies the wind speed up to 30% at the turbine position. Together with the form of the towers (bigger impact at ground) and the velocity profile of the wind (lowest at ground), this amplification has the impact of balancing the energy output to the extent that the upper and lower turbines generate 109 and 93% compared to 100% for the turbine at the middle [15].

3.2 Strata tower, London

Strata Tower is a tall residential building in central London (**Figure 4**) which was the tallest one when it was constructed in 2005. The architects in the design team of the tower examined the effectiveness of several alternatives to achieve the most suitable lean, clean, and green option. As a residential tower, general energy

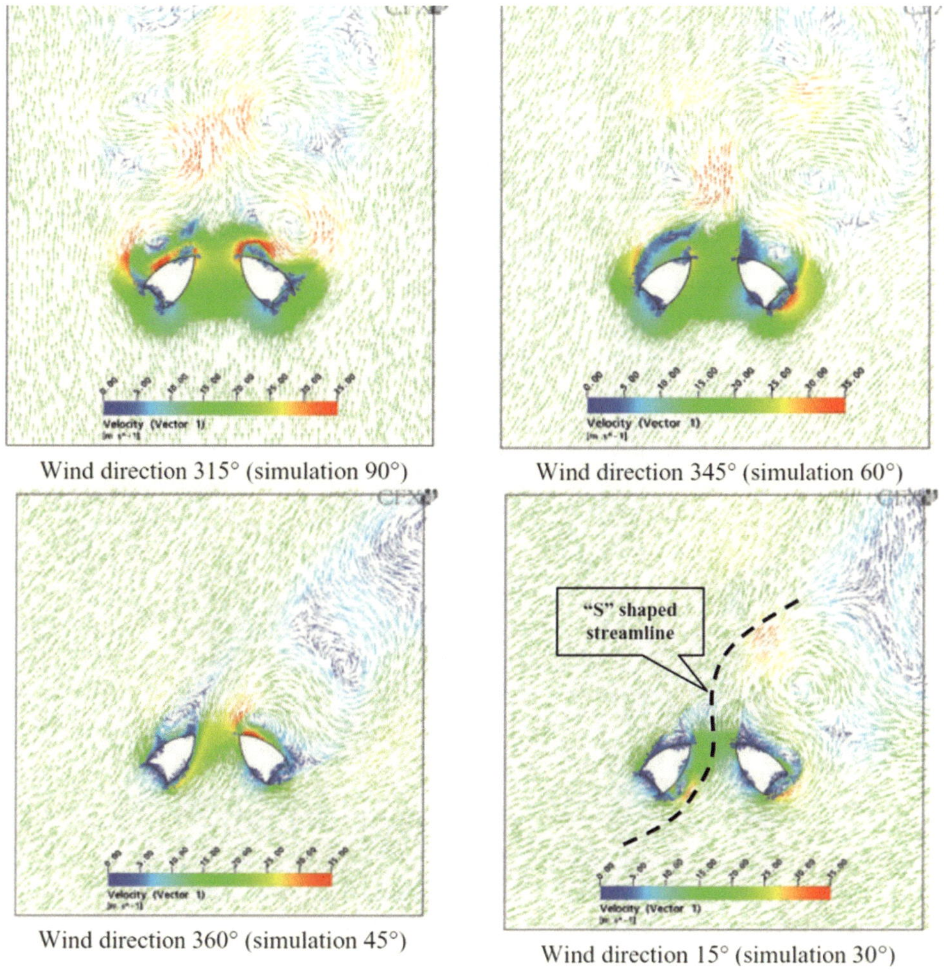

Figure 3.
CFD images by Ramboll, showing airflow patterns near towers, simulated at the level of the top turbine for different free, undisturbed wind incidence angles with respect to an "x" axis (i.e., horizontal line connecting towers).

consumption loads are significantly lower than similar retail or commercial office development, and thus the chance of achieving the required percentage of renewable energy on-site is comparably more possible [16].

The architects initially did not plan to design a building with wind turbines; however, they came through an extensive sequence of design alternatives, assessing each renewable option on their own merits and in the context of the site. For instance, ground source water solutions were thought, but the site's harsh limitations meant that the water pools would not be far enough from each other to remain feasible. Moreover, energy savings would be reduced by pumping through the entire building's height. Photovoltaic option was also thought, but the corresponding technology available at that time (2005) would have led to 80% of the southern facade being enclosed with photovoltaic (PV) cells, this too far compromising the amount of glazing needed to provide sufficient daylight into the spaces and views from the apartments. Commercial problems prevailing in 2005 would also have rendered this choice too costly, added to which photovoltaics have a life-span of about 15 years and must be kept carefully clean. This alternative would have had a major impact on service fees. Photovoltaic integration would also negatively affect the facade's cost per square meter. Similarly, biomass boiler option

Figure 4.
Strata SE1, London © will Pryce.

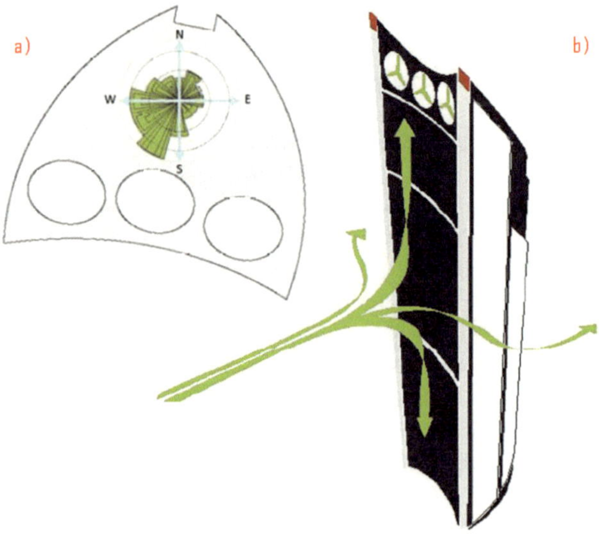

Figure 5.
Orientation study based on prevailing wind direction; site plan sketch with wind rose (a) and an illustration of airflow distribution around the tower (b) © BFLS.

was also considered, but the ongoing energy costs related with transporting and delivering the fuel, and the availability issues of such fuel, along with the need for a 150-m (492-foot) flue operating the entire building height, meant that this option was marked down. As such, a number of options inevitably lead to a wind energy-based solution [16].

The wind rose for London has a mainly south-westerly axis in summertime; therefore the curved facade was appropriately oriented to catch wind from this prevailing direction (**Figure 5**).

Once considering the introduction of wind turbines as a site-based renewable energy option, funneling and guiding the wind is a main design element that can considerably improve the wind turbines' prospective operational production. In fact, the design solution for Strata Tower is another iteration of that taken for the Bahrain World Trade Center also using the concepts of a "Venturi" to direct wind flow, but differently in between the two towers (**Figure 6**). There are also other significant distinctions—particularly the scale of the project, the reality that the three turbines are installed externally on linking bridges and the hot-humid site which required full air conditioning to help tackle any negative acoustic effects [16].

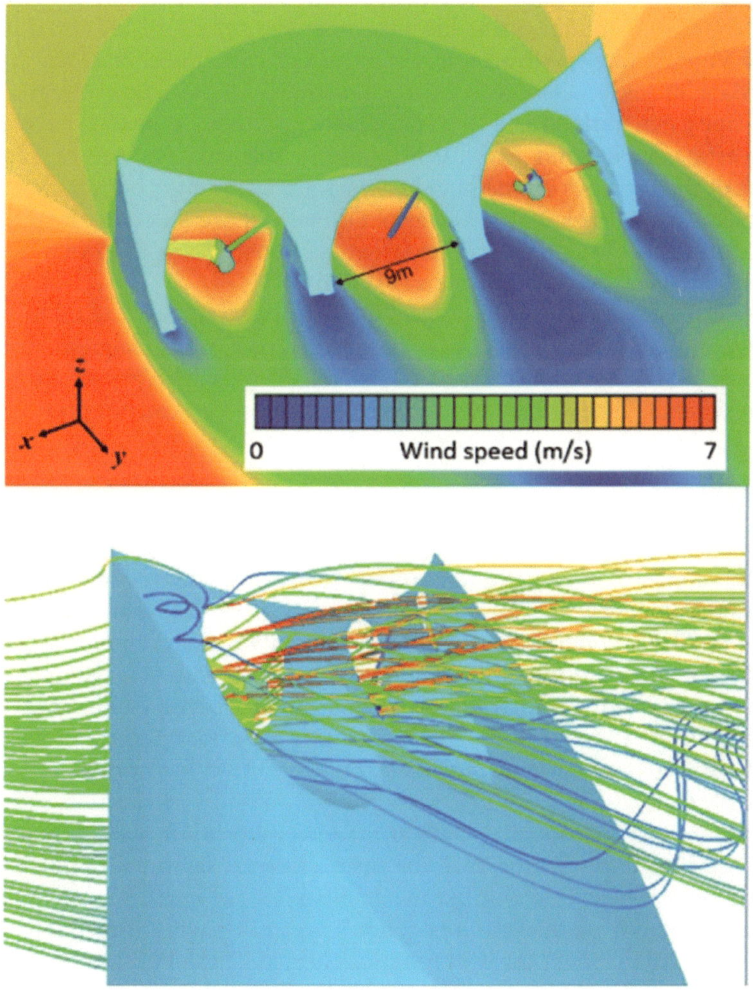

Figure 6.
Strata tower: Airflow patterns around the wind turbines [13].

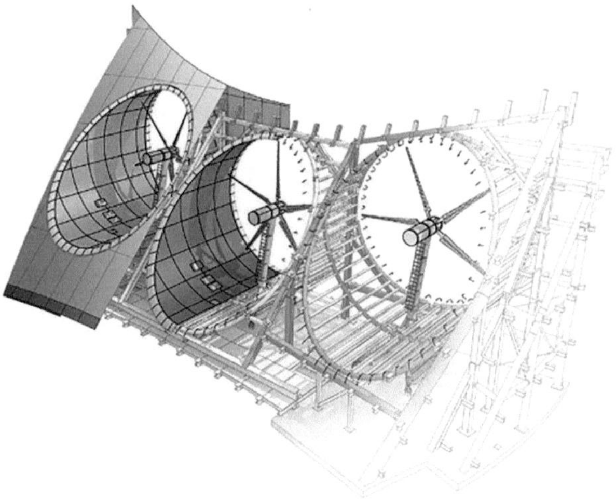

Figure 7.
Strata SE1's turbine deck © BFLS.

The turbines do produce noise, like any other typically roof-mounted piece of plant. The Venturi-like enclosures, however, effectively concentrate the noise away from the flats instantly below into two sound cones. All measures to regulate and minimize noise generation have been discussed, and these measures have actively improved efficiency in some cases. Within the Venturi-like enclosures, careful placement of the turbines on the plan has a significant impact both in general performance and in regulating noise output. Furthermore, in contrast to the more standard three-bladed turbine used on larger versions, opting for a five-bladed turbine provided further noise decreases (**Figure 7**) [16].

To sum up, the energy output needed for building-integrated wind turbine solution is completely case dependent. The overall production requirement for Strata Tower is 50 megawatt hours of electricity per year in the case of the design load requirement. This is about 8% of the complete energy consumption of the building.

3.3 Pearl River tower, Guangzhou, China

A typical instance of BIWTs and building-integrated photovoltaic (BIPV) in Asia is the Pearl River Tower in Guangzhou, China. It is a 309-m-high 71-story building completed in 2011 and designed to increase velocity through the nozzle effect by making flow openings in the building (**Figure 8**) [17]. Inside the building are mounted a total of four 8-m-high vertical-axis wind turbines, each at the inner opening on the left and right sides [18]. There are four openings in the Pearl River Tower adjacent to the two-level mechanical floors (Floor 24, from 104 to 111.3 m, and Floor 50, from 205 to 212.7 m), with a bell-mouthed form at both ends of each opening. Opening-1 and opening-2 are at the lower elevation, with opening-1 being placed on the west and opening-2 on the east. Likewise, opening-3 and opening-4 are located in pairs next to the upper mechanical floor (**Figure 9**). The Pearl River Tower's orientation is parallel to Guangzhou's prevailing wind direction, which seeks to maximize turbine power output. Furthermore, the Pearl River Tower is aerodynamically formed with a concave wall on the southern surface and a convex wall on the northern surface that enables better intensification and funneling of the wind through the openings (**Figure 9**).

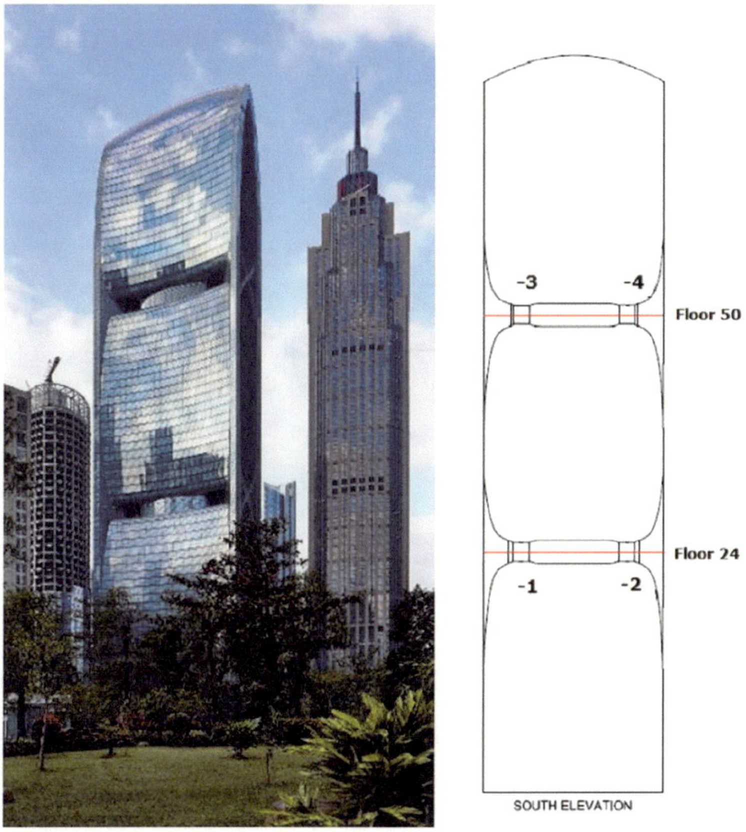

Figure 8.
Pearl River tower: Positions of the openings [18].

A helix-shaped vertical-axis wind turbine is installed on the floors inside each of the four openings (**Figure 10**). These turbines have an 8-kW rated power with 25 m/s rated speed. Such a turbine is designed to generate electricity when there is a

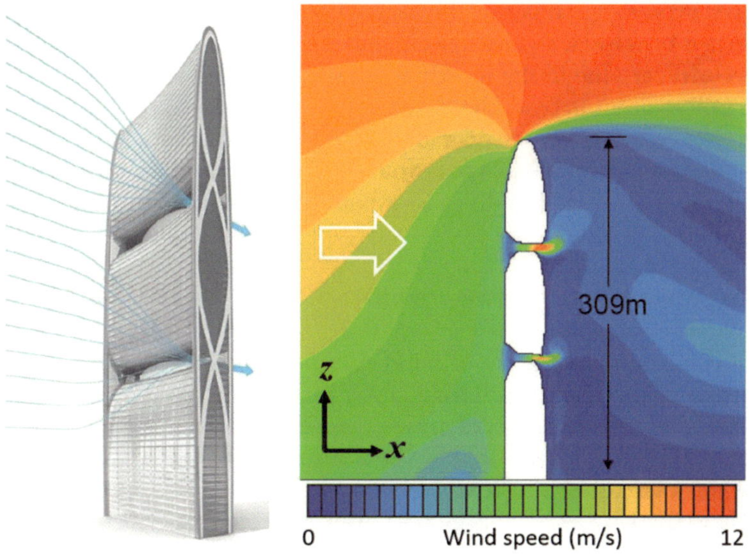

Figure 9.
Pearl River tower: Airflow patterns around the wind turbines [17].

a)
b)

Figure 10.
Pearl River tower; aerodynamic configuration on the facade (a) and the detailed view of the helix-shaped vertical-axis wind turbine (b) © SOM.

wind velocity range of 2.7–40 m/s. This turbine's swept area is about 10 square meters. It is worth noting that while vertical-axis wind turbines (VAWTs) may usually have lower energy efficiency than widely used horizontal-axis wind turbines (HAWTs), they appear to be particularly favored in urban wind energy development. Due to the lower tip velocity, the noise effect of VAWTs is relatively low, and VAWTs in urban settings can resist high turbulence. Most remarkably, VAWTs are omnidirectional, so wind and turbulence can be extracted more efficiently from all directions with reduced loss of effectiveness. This is particularly appealing for the installation of wind turbines in urban settings where wind direction is extremely variable owing to adjacent buildings and structures' interference effects [18].

It has been shown through the results of the study [18] that the design of Pearl River Tower makes a good use of the aerodynamic features of the super-tall building, which can result in a significant enhancement of wind speed in the four openings. However, due to the mostly weak wind condition in Guangzhou, the estimated power outputs from the four building-integrated wind turbines are relatively low. Nevertheless, it is expected that the integration of wind turbines into tall buildings will have a remarkable potential to contribute a significant part of the energy requirement of tall buildings in urban areas having high wind velocities.

4. Conclusions and further research

Urban wind energy is the use of wind energy technology in urban and suburban built environment applications. This chapter offers only some opinions on the progress made in the evaluation of wind resources in urban regions, the use of appropriate wind turbines to enhance the utilization of these resources, and the important role of architectural and urban aerodynamics in designing buildings accordingly to increase wind energy output. It is not meant to be comprehensive, but rather it is preferred from the point of view of architects in the context of buildings and cities to provide some perspectives on the abovementioned subjects. It is indicated that while important progress has been made in the field, there is a strong need for additional comprehensive research, especially in the area of urban aerodynamics and wind resource assessment, in order to optimize the generation and utilization of urban wind energy.

It is also concluded that wind conditions can change considerably depending on the building form. Therefore, CFD simulation is required to analyze the effect of the form of the building, accurately. Aerodynamically designing the building leads to improvement in power density and may potentially result in large changes in spatial variability. On the other hand, preliminary comprehensive studies as to the wind resource assessment are very important for the efficiency of a wind turbine system; otherwise, it would not be possible to benefit from this type of system, sufficiently. It is also derived that to accurately position the wind turbine, detailed analyses should be conducted on determining highly turbulent flow areas which can reduce the amount of energy yielded from the wind turbine.

Moreover, it should be noted that research on wind turbine integration to buildings has been mostly conducted by wind engineers. Wind engineering solutions are very crucial in developing building-integrated wind energy systems. The advanced analysis methods that combined with CFD simulations give important aerodynamic information about the building and the nearby field. However, when it comes to proposing aerodynamic form optimization based on the data produced, the solutions should not be free of architectural context, since a very basic change even affects the form of the building. At this point, cooperation with architects is significant when integrating wind energy systems to buildings, since an interdisciplinary approach may fill the gap between the wind engineering solutions and architectural design processes. In collective research, the effect of several factors may be assessed; consequently, this may widen the perspective of the contribution of wind engineering to the improvement of urban wind energy strategies.

The environmental benefits of using a sustainable energy system can be examined, further. For instance, it is very probable to see that using wind turbines leads to a significant decrease in carbon emissions. For this purpose, the elimination of carbon emissions which would be released for the equivalent energy yielded from fossil fuels can be calculated in further studies.

Author details

Ilker Karadag* and Izzet Yuksek
Manisa Celal Bayar University, Manisa, Turkey

*Address all correspondence to: karadagi@itu.edu.tr

References

[1] Stathopoulos T, Blocken B. Pedestrian wind environment around tall buildings. In: Tamura Y, Yoshie R, editors. Advanced Environmental Wind Engineering. Tokyo: Springer; 2016. pp. 101-127

[2] Blocken B, Janssen WD, van Hooff T. CFD simulation for pedestrian wind comfort and wind safety in urban areas: General decision framework and case study for the Eindhoven university campus. Environmental Modelling & Software. 2012;**30**(1):15-34. DOI: 10.1016/j.envsoft.2011.11.009

[3] Wu H, Kriksic F. Designing for pedestrian comfort in response to local climate. Journal of Wind Engineering and Industrial Aerodynamics. 2012; **104**(1):397-407. DOI: 10.1016/j.jweia.2012.02.027

[4] Franke J, Hellsten A, Schlünzen H, Carissimo B. The COST 732 best practice guideline for CFD simulation of flows in the urban environment - a summary. International Journal of Environment and Pollution. 2011;**44**(1–4):419-427. DOI: 10.1504/ijep.2011.038443

[5] Tominaga Y, Mochida A, Yoshie R, Kataoka H, Nozu T, Yoshikawa M, et al. AIJ guidelines for practical applications of CFD to pedestrian wind environment around buildings. Journal of Wind Engineering and Industrial Aerodynamics. 2008;**96**(10–11): 1749-1761. DOI: 10.1016/j.jweia.2008.02.058

[6] Serteser N, Karadag I. Design for improving pedestrian wind comfort: A case study on a courtyard around a tall building. Architectural Science Review. 2018;**61**(6):492-499. DOI: 10.1080/00038628.2018.1492899

[7] Stathopoulos T, Alrawashdeh H, Al-Quraan A, Blocken B, Dilimulati A, Paraschivoiu M, et al. Urban wind energy: Some views on potential and challenges. Journal of Wind Engineering and Industrial Aerodynamics. 2018;**179**: 146-157. DOI: 10.1016/j.jweia.2018.05.018

[8] Stankovic S, Campbell N, Harries A. Urban Wind Energy. London: Taylor & Francis; 2009. DOI: 10.4324/9781849770262

[9] Beller C. Urban Wind Energy - State of the Art 2009. Danmarks Tekniske Universitet, Risø Nationallaboratoriet for Bæredygtig Energi. Denmark. Forskningscenter Risoe. Risoe-R; No. 1668(EN); 2009. 43 p

[10] Sharpe T, Proven G. Crossflex: Concept and early development of a true building integrated wind turbine. Energy and Buildings. 2010;**42**(12): 2365-2375. DOI: 1016/j.enbuild.2010.07.032

[11] Lu L, Sun K. Wind power evaluation and utilization over a reference high-rise building in urban area. Energy Buildings. 2014;**68**:339-350

[12] Tabrizi AB, Whale J, Lyons T, Urmee T. Performance and safety of rooftop wind turbines: Use of CFD to gain insight into inflow conditions. Renewable Energy. 2014;**67**:242-251

[13] Al-Quraan A, Stathopoulos T, Pillay P. Comparison of wind tunnel and on-site measurements for urban wind energy estimation of potential yield. Journal of Wind Engineering and Industrial Aerodynamics. 2016;**158**:1-10

[14] Yang A, Su Y, Wen C, Juan Y, Wang W, Cheng C. Estimation of wind power generation in dense urban area. Applied Energy. 2016;**171**: 213-230

[15] Smith RF, Killa S. Bahrain world trade Center (BWTC): The first large-scale integration of wind turbines in a

building. The Structural Design of Tall and Special Buildings. 2007;**16**(4): 429-439. DOI: 10.1002/tal.416

[16] Bogle I. Integrating Wind Turbines in Tall Buildings. CTBUH Journal [Internet]. 2011 [cited: 01 September 2019]; 2011(IV):30-34. Available from: http://global.ctbuh.org/resources/papers/download/293-integrating-wind-turbines-in-tall-buildings.pdf

[17] Kim H-G, Jeon W-H, Kim D-H. Wind resource assessment for high-rise BIWT using RS-NWP-CFD. Remote Sensing. 2016;**8**(12):1019. DOI: 10.3390/rs8121019

[18] Li QS, Shu ZR, Chen FB. Performance assessment of tall building-integrated wind turbines for power generation. Applied Energy. 2016;**165**:777-788. DOI: 10.1016/j.apenergy.2015.12.114

Section 6

Energy Management

Chapter 21

Thermal Energy Storage for Solar Energy Utilization: Fundamentals and Applications

Kai Wang, Zhen Qin, Wei Tong and Chenzhen Ji

Abstract

Solar energy increases its popularity in many fields, from buildings, food productions to power plants and other industries, due to the clean and renewable properties. To eliminate its intermittence feature, thermal energy storage is vital for efficient and stable operation of solar energy utilization systems. It is an effective way of decoupling the energy demand and generation, while plays an important role on smoothing their fluctuations. In this chapter, various types of thermal energy storage technologies are summarized and compared, including the latest studies on the thermal energy storage materials and heat transfer enhancements. Then, the most up-to-date developments and applications of various thermal energy storage options in solar energy systems are summarized, with an emphasis on the material selections, system integrations, operational characteristics, performance assessments and technological comparisons. The emerging and future trends are finally outlined. This chapter will be a useful resource for relevant researchers, engineers, policy-makers, technology users, and engineering students in the field.

Keywords: thermal energy storage, solar energy utilization, sensible heat storage, latent heat storage, thermochemical heat storage, heat transfer

1. Introduction

Currently, energy, environment and sustainable development are the major concerns to the countries all over the world. The long term usage of fossil fuels (e.g., coal, oil and natural gas) has caused environment pollutions and greenhouse effects. The resulting global climate change and resource exhaustion can seriously threaten human survival and development. Renewable energy sources, including solar, wind, hydro, geothermal, etc. [1], have been increasingly attractive to many countries for their natural advantages on sustainability and cleanliness. Solar energy is considered as one of the most promising renewable energy sources, because of its abundance and easy-access to the most parts of the world. However, due to the intermittency in availability and constant variation of solar radiation, the output of a solar energy system is highly fluctuating if no or small inertia (thermal or electrical) is included. This poses strong needs for energy storage in solar energy systems to store energy at high solar irradiances for later uses whenever there is a demand while solar irradiance is not enough. With the energy storage systems associated to reduce the discrepancy between the demand and supply, solar energy systems can be more reliable and efficient in operation.

In order to develop efficient and economical energy storage systems, various energy storage technologies have been proposed [2], such as compressed air energy storage, pumped storage hydro-power, flywheel energy storage, thermal energy storage, electrochemical energy storage, hydrogen storage and so on. They are employed in different application fields depending on their specific characteristics on energy storage. Among all the storage methods, thermal energy storage (TES) is one of the most economical systems in practical applications, and it allows the storage of thermal energy by heating or cooling a storage medium to be used at a later time [3]. TES has been applied in a broad range of fields: district heating, domestic hot water, thermal comfort, concentrated solar plants (CSP), chemical industry, food industry, etc. [4–7]. It has also been proven that the usage of TES can improve the performance of whole system operation, save the fuel consumption, lower the investment and operation costs, increase the energy supply security and mitigate the pollution to the environment if it is appropriately designed. To obtain the above benefits for an energy system, it is critical to choose a suitable thermal energy storage approach. The selection of thermal energy storage depends on the type of energy source, required storage duration, operating condition, economic viability, etc. The most mature and widely used approach is sensible heat storage. Latent heat storage and thermochemical energy storage have also attracted extensive research and development efforts in the last two decades, as they can offer higher heat storage capacities and lower heat losses during the energy storage processes.

The aim of this chapter is to provide a state-of-the-art review of TES technologies and their applications in solar utilization systems. The principles of several energy storage methods and evaluation approaches of storage capacities are firstly described. Sensible heat storage technologies, including the solid and liquid storage methods, are briefly reviewed. Latent heat storage systems associated with phase change materials (PCMs) as well as thermochemical storage are also introduced and summarized. Further discussions on important criteria of energy storage technologies suitable for solar energy applications are also presented. Later, TES technologies applied in solar energy systems like solar power systems, solar heating/cooling systems are reviewed with the detailed analyses about the material selections and system integrations of TES and solar energy systems.

2. Thermal energy storage technologies

TES is one of the most practiced technologies to store energy in the form of heat to eliminate the gap between the energy supply and demand. As shown in **Figure 1**, there are three main thermal energy storage technologies [9]: sensible heat storage through a temperature change (sensible heat) of a material, latent heat storage through phase change (latent heat) of a material and thermochemical heat (chemical energy) by thermally inducing changes in materials' chemical states. As compared in **Table 1**, the choice of TES method depends on a variety of factors such as the storage capacity, cost, temperature range, duration requirement as well as the specific application.

2.1 Sensible heat storage

Solid sensible heat storage is an attractive option for thermal energy storage regarding the investment and maintenance costs. Sensible heat storage stores the thermal energy by varying the temperature of storage materials, without

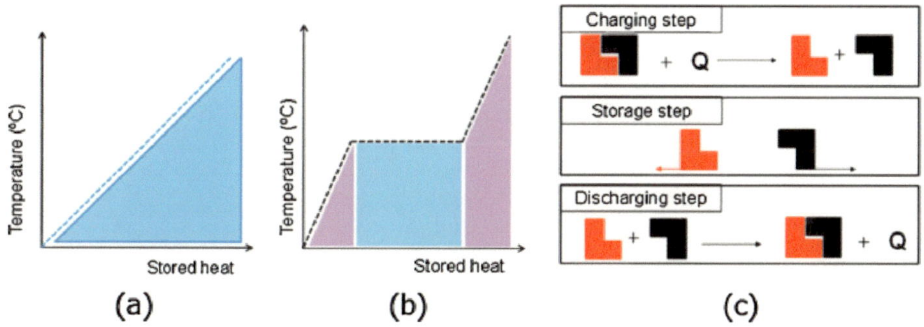

Figure 1.
Main approaches of thermal energy storage: (a) sensible heat, (b) latent heat, (c) thermo-chemical reactions [8].

TES technology	Capacity (kWh/t)	Cost (/kWh)	Storage period
Sensible	10–50	0.1–10	Days/months
Phase change materials	50–150	10–50	Hours/months
Chemical reactions	120–150	8–100	Months/seasons

Table 1.
Comparison of typical parameters of three TES technologies [10].

undergoing any form of phase change within the working temperature range. The amount of thermal energy stored or released is proportional to the density ρ, volume V, specific heat c_p, and temperature variation of the storage materials:

$$Q_{sen} = \int_{T_i}^{T_f} mc_p dT = \rho V c_p (T_f - T_i), \tag{1}$$

where Q_{sen} is the amount of sensible heat stored, dT is the temperature interval, T_i is the initial temperature and T_f the final temperature of storage medium during

Material	Type	Density (kg/m^3)	Thermal conductivity (W/m·K)	Heat capacity (kJ/kg·K)	Cost (€/m^3)
Rock	Solid	1500–2800	0.85–3.5	1	64–742
Concrete	Solid	2000	1.35	1	76
Sand and gravel	Solid	1700–2200	2	0.910–1.180	6–8
Ceramic tile	Solid	2000	1	0.8	1600–3500
Gypsum (coating)	Solid	1000	0.4	1	78
Ceramic brick	Solid	1800	0.73	0.92	36–64
Wood	Solid	450	0.12	1.6	404
Water	Liquid	990	0.63	4.19	1.6
Oil	Liquid	888	0.14	1.88	6560
Nitrite salts	Liquid	1825	0.57	1.5	2200
Carbonate salts	Liquid	2100	2	1.8	6050
Liquid sodium	Liquid	850	71	1.3	2000

Table 2.
Available sensible heat storage materials used in the thermal energy storage systems [11].

the storage process. Basically, specific heat c_p, density ρ and thermal conductivity k are the key thermal properties of sensible heat storage materials. According to the materials' phase state, sensible heat storage materials can be divided into two main categories: solid and liquid heat storage. **Table 2** lists the most common solid and liquid heat storage materials with their thermal properties.

2.1.1 Solid heat storage

Solid storage materials have been applied in many TES systems for their reliability, low cost, easy implementation and applicability in extensive practical cases [12, 13]. Different from liquid heat storage, there are no vapor pressure or leakage issues in solid heat storage. However, a fluid, usually air or oil, is needed to work as the heat transfer fluid (HTF) to transport the thermal energy that is to be stored into or released from the solid heat storage system. As listed in **Table 2**, the most frequently used solid heat storage materials include rock, concrete, brick, sand and so on [14–16].

- **Rock** is always loosely piled in a packed bed through which the HTF like air or oil can flow. Thermal energy is stored in the packed bed by forcing heated HTF flowing through the rocks and utilized again by recirculating the HTF through the heated rocks. Typically, the characteristic size of rock pieces varies from 1 to 5 cm. There is a large contact surface area available for heat transfer between HTF and rocks which is beneficial for the heat transfer. The amount and temperature level of energy stored in a packed bed storage system with rocks depend on the rock size and shape, packing density, HTF, etc. As a sensible energy storage option, rock has advantages like being non-toxic, non-flammable, cheap and easily available. This type of storage is operated very often for temperatures up to 100°C in conjunction with solar air heaters and thus convenient to be implemented in buildings. The heat storage with rocks can also be used for higher temperature applications, up to 1000°C. When rock is employed as thermal storage material, there are several drawbacks, including the poor thermal conductivity, high pressure drop under large flow rates of HTF.

- **Concrete** is a promising candidate as it has a low cost and is easy to obtain and process directly on site. Concrete is a construction material comprised of cementitious materials and/or calcium aluminate cement, coarse and fine aggregates, water and possibly chemical admixtures. Besides, it has relatively high specific heat and good mechanical properties. The heat exchanger between concrete and HTF is usually designed as the pipes embedded into the concrete block where HTF flows internally. As cracks may form after repeated cycles due to thermal expansion and contraction at high temperatures, research efforts have been devoted to developing appropriate concrete compositions, optimizing chemical–physical and durability performances at high temperatures. Long-term stability of concrete has been proven in oven experiments and through strength measurements up to 500°C. The main challenges to use the concrete as TES materials include: potential cracks, relatively low thermal conductivity, durability after long-term thermal cycling and high costs for heat exchangers to charge/discharge thermal energy.

- **Sand** grains are shown to be a promising low-cost candidate material that is suitable for concentrated solar power (CSP) applications with high-temperature thermal storage. The average size of sand grains is around 0.2–0.5 mm and they are commonly used in the form of packed beds for heat storage with air as HTF. It is possible to use desert sands directly as collected from the field of CSP, removing the need for third-party suppliers. Moreover, they can be used directly in solar receivers to collect solar thermal energy. After absorbing the heat of concentrated solar rays, due to the gravity forces, the sands can fall from the top of solar receiver tower and then they can be collected in an insulated storage tank below. Temperature of hot sands can go up to 700–1000°C which is appropriate for producing steam to drive a Rankine cycle [17].

2.1.2 Liquid heat storage

- **Water** is the most common liquid material for TES due to its high specific heat, none-toxicity, low-cost and easy-availability [18]. However, due to its high vapor pressure, water requires costly insulation and pressure withstanding containment for high temperature applications from 100 to 700°C (in the form of steam). Water in liquid phase is widely used for low temperature heat storage below 100°C in solar based applications, such as space heating and hot water supply [19]. Water in liquid state can also form thermal stratification or thermocline. Due to density difference caused by heating of liquid, the buoyancy force causes stratification of the water, forming a thermal gradient across the storage. Under such a condition, the hot fluid can be supplied to the upper part of a storage tank during charging, and the cold fluid can be extracted from the bottom part during discharging [20]. Thus, the efficiency of thermal energy store and release process can be improved. In some high temperature applications like CSP plants, water is stored in steam phase in high pressure tanks (steam accumulator) to work as TES systems. In additional, water can be also used in chilled water form or in ice form for cold energy storage, which is useful in refrigeration systems. The main drawbacks for using water as the TES material are its high vapor pressure and corrosiveness to the container above its boiling point.

- **Molten salt** is currently one of the most popular TES materials used in CSP plants. Compared to other liquid heat storage materials, molten salts have relative low costs, high energy storage densities, excellent thermal stabilities, low viscosities and non-flammabilities. Molten salts in liquid state can be operated at high temperatures of several hundred degree centigrade while its vapor pressure is much lower than that of water, so it is very suitable for high temperature CSP plants. The pure molten salt usually has a melting point above 200°C which hampers its further application at low temperatures. It is desirable to have a molten salt with a lower melting point so that it can remain the liquid state when storing the thermal energy. A new series of ternary salt mixtures have been proposed with ultra-low melting temperatures at 76°C, 78°C or 80°C, and they can prevent the solidification at low temperatures to enable the TES systems suitable for a wider applications. Molten salt also has several drawbacks that limit its application: low thermal conductivity, volume change during the melting and corrosivity to the container.

- **Thermal oil** is usually a kind of organic fluid and works as a HTF in many power and energy systems. When using as a thermal storage medium, thermal oil can remain in liquid phase at temperatures of 350–400°C with stable thermal properties, which is much higher than the liquid water. It means that thermal oil can store more thermal energy based on the wider temperature operation range. Compared to water, thermal oil also has a lower vapor pressure, which is beneficial for mechanical designs of relevant pipes and containers. Unlike molten salts, thermal oil does not freeze during the night in pipes so that it doesn't need any antifreeze system. However, the cost of thermal oil is usually higher than water and molten salts.

2.1.3 Pros and cons of sensible heat storage

Sensible heat storage materials are typically based on relatively low cost materials and thus extensively used, except the liquid metals. Due to the relatively good thermal stability, heat transfer performance and transport properties, sensible heat storage materials are the most used TES materials for high temperature applications. Compared to the latent heat storage, specific heat of sensible heat storage materials is 50–100 times smaller, leading to the requirement of large volumes or quantities in order to deliver the amount of energy storage necessary for high temperature thermal energy storage applications. The other main issue of sensible heat storage is that the temperature of the storage medium decreases during discharging process, so the HTF temperature also decreases with time.

2.2 Latent heat storage

Latent heat storage utilizing PCMs is an alternative TES technique compared to the sensible heat storage [11]. PCMs are substances which can absorb or release large amount of energy, i.e., so-called latent heat, when they experience phase transitions among solid, liquid and gas states. Although the highest latent heat of phase change is the liquid-gas phase change, it is hard to utilize this due to the enormous volume change associated with material evaporation. While another kind, 'solid-solid' latent heat storage material has its latent heat of transition one order of magnitude smaller than the solid-liquid PCMs [21], which is commonly applied for latent heat thermal energy storage. Solid-liquid PCMs should have a melting point near the required operation temperature range of the TES system, melt congruently with minimum subcooling [22], and also is desired to be chemically stable, cost competitive, non-toxic and non-corrosive. The amount of energy storage of the latent heat system with PCMs is calculated as:

$$Q_{lat} = m\left(c_{p,s}(T_m - T_i) + \alpha L_h + c_{p,l}\left(T_f - T_m\right)\right), \tag{2}$$

where Q_{lat} is the amount of heat stored, $c_{p,s}$ and $c_{p,l}$ are the specific heat of PCMs in solid and liquid state, L_h is the latent heat of fusion, α is the melting fraction, T_i and T_f are the initial and final temperatures of the storage materials, and T_m is the melting temperature. This section briefly introduces the classification of PCMs and the related heat transfer enhancement techniques.

2.2.1 Phase change materials

Solid-liquid PCMs are competitive alternatives to the sensible TES materials. Compared to sensible heat storage materials, PCMs can operate at the phase change temperature with small temperature variations between heat storage (charging) and heat releasing (discharging) as illustrated in **Figure 1(b)**, and **Figure 2** shows the classification of PCMs family for TES [23]. Different kinds of PCMs are introduced in the following subsections. **Table 3** presents the characteristics of several common PCMs.

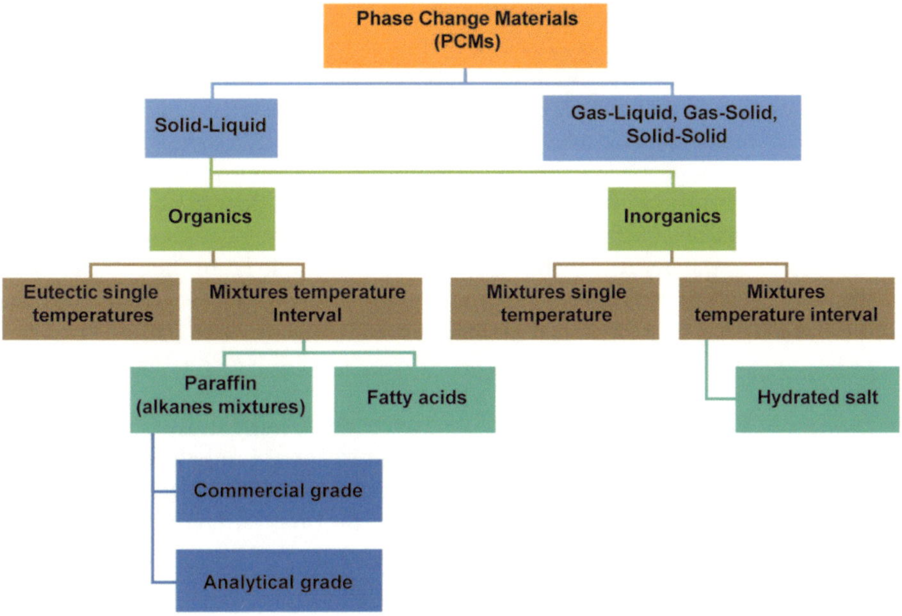

Figure 2.
Classification of latent heat materials with solid–liquid phase change behavior.

- **Organic PCMs** and their eutectic mixtures have been successfully implemented in many commercial applications, such as space heating in buildings, electronic devices, refrigeration and air-conditioning, solar air/water heating, textiles, automobiles, food and space industries [23]. Organic PCMs featured of congruent melting without phase separation usually have relative

Name	Melting point (°C)	Latent heat (kJ/kg)	Density (kg/m^3)	Thermal conductivity (W/m·K)	Heat capacity (kJ/kg·K)
(Organic)					
n-Octadecane	27.7	243.5	865/785	0.19/0.148	2.14/2.66
Paraffin wax	32	251	830	0.514/0.244	1.96/3.26
RT 55	55	172	880/770	0.2	2
RT 70 HC	69–71	260	880/770	0.2	2
(Inorganic)					
$CaCl_2{\cdot}6H_2O$	29.6	190.8	1562	N/A	N/A
$Ba(OH)_2{\cdot}8H_2O$	78	265–280	2070/1937	1.225/0.653	N/A
E117	117	169	1450	0.7	2.61
$LiNO_3$-$NaNO_3$	195	252	N/A	N/A	N/A
$NaNO_3$	306	172	2261	388.9	N/A
KNO_3	333	266	2110	N/A	0.5
KOH	380	150	2044	N/A	0.5

Table 3.
The thermal properties of organic and inorganic phase change thermal storage materials [24, 25].

low melting points. Commonly used organic PCMs are paraffin, fatty acids, esters, alcohols and glycols. Among them, paraffin wax is an excellent heat storage material and has been widely applied for low temperature heat storage applications [26]. It consists of straight n-alkanes chain (CH_3-(CH_2)-CH_3), featuring a high specific heat capacity (2.14–2.9 J/g·K), a low price (~1 USD/kg) with a moderate heat storage density (200 kJ/kg) and a narrow range of melting temperatures from −10 to 67°C [27], a small degree of subcooling, chemically stable and non-toxic properties. Due to the purity and specific composition, the organic PCMs show up a remarkable latent heat capacity in narrow temperature ranges. In addition, they are chemically inert and have an unlimited lifetime. However, their low thermal conductivities (0.1–0.35 W/m·K) limit their practical applications.

- **Inorganic PCMs** can be classified into two groups: salt/salt hydrates, and metals and their alloys. In general, inorganic PCMs not only have nearly doubled heat storage densities but also higher thermal conductivities, higher operating temperatures compared to the organic ones [28]. However, inorganic PCMs are corrosive to metals leading to a short service life of the system and a higher maintenance cost. The inorganic PCMs (salt/salt hydrates) can also suffer from phase segregation and supercooling, which would reversibly affect the energy storage capacity. For high temperature applications, however, metal and metallic alloys are potential PCM candidates as they don't suffer from these disadvantages. The inorganic salt means salt or its hydrates, which can be expressed as A_xB and $A_xB_y \cdot n(H_2O)$ respectively, where A_xB represents metal carbonate, sulfite, phosphate, nitrite, acetate or chloride and n represents number of water molecules. Although the inorganic PCMs show very promising and advantageous characteristics, these materials still face many problems to be commercial products for practical applications: (1) volume change at phase transition, (2) low thermal conductivity (nearly 1 W/m·K), (3) supercooling of salt hydrates, (4) corrosion with metal containers, (5) different melting temperatures of salt hydrates and (6) high cost of some specific salts.

- **Eutectic PCMs** are composites of two or more components, which usually do not interact with each other to form a new chemical compound but at certain ratios, inhabit the crystallization process of one another resulting in a system having a lower melting point than either of the components. The eutectic mixtures can be further classified into organic-organic, organic-inorganic and inorganic-inorganic PCMs. Eutectic PCMs generally melt and freeze congruently and leave no chances of separation of components. Molten salt is one of the best candidates for middle to high temperature applications in the range of 120–1000°C [28]. For solar energy utilization, normally middle-high temperature PCMs are applied and the "middle-high" temperature means the range of 100–300°C. The molten salts offer a favorable density around 1880 kg/m^3, a high specific heat around 1.5 kJ/kg·K, a very low chemical reactivity, a low vapor pressure and a low cost about 0.4–0.9 USD/kg [29]. A popular commercial molten salt used in the solar power generation as PCM is called "solar salt", which is a mixture of $NaNO_3$ and KNO_3 mixing at a weight ratio of 6:4 with a freezing point of 221°C. Despite its relatively high melting point, the low cost makes it widely utilized in CSP applications. Another similar molten salt product is named "HTEC", which is a ternary salt mixture system of $NaNO_3$, KNO_3 and $NaNO_2$, and has a freezing point of 141°C. Different salt combination brings the melting point down but the lack of combination of optimum thermal properties limits its further applications.

- **Composite PCMs** are the mixtures prepared by dispersing the high thermal conductive particles like carbon, graphite or metals into PCMs. One should note that the embedded thermal conductive materials should be compatible with the base PCMs. Although the nano-composite has less ability to store heat, it has higher ability to conduct heat. For example, the graphite based nano-composite has 12 times higher thermal conductivity than that of pure stearic acid [30]. Graphite can be applied as thermal promoters in various forms like graphite flakes (natural graphite), expanded natural graphite or the expanded graphite powder (50–500 nm). Expanded graphene is one of the most suitable PCM support materials due to its extraordinary thermal conductivity. The dispersion of expanded graphene to binary nitrate salts consisting of $NaNO_3$ and KNO_3 (6:4) by aqueous solution method adopting ultrasonic and the 2% integration enhanced the thermal conductivity to 4.9 W/m·K but reduced the latent heat by 11% [27]. It is also reported that the use of expanded graphene in molten salts can prevent the liquid leakage after the melting [31]. Different from expanded graphene, a highly conductive additive expanded natural graphite treated with sulfuric acid was introduced into the binary salt, KNO_3/$NaNO_3$ nitrate mixture and the additive establishes effective heat transfer matrix for more efficient heat transfer. The results showed that the thermal conductivity has been improved and the highest effective thermal conductivity is about 50.8 W/m·K, almost 110 times larger than the thermal conductivity of the salt powder. A slight decrease of latent heat was observed from the measurements with no obvious variation in the phase change temperature [32]. Another way to enhance the thermal conductivity is to add chloride as addictive into the nitrate salt composite by statical mixing method. It was found that an addition of 5% chlorides into KNO_3-$NaNO_3$-$NaNO_2$ composite increased the thermal conductivity, thermal stability with an higher operating temperature from 500 to 550°C. Lower freezing point was obtained and the loss of nitrite content was observed [33]. Those enhanced composite PCMs with enhanced thermal performance and stability can be used to create compact thermal energy storage systems when the space is limited. Not only different nanostructures but also different types of nanoparticles can be applied as the thermal conductivity promoters, such as, the carbon-based nanostructures, metals, metal oxides and silver nanowires. A review of the current experimental studies on variations in thermo-physical properties of PCMs due to the dispersion of nanoparticles is performed in the reference [30].

- **Microencapsulated PCMs (MEPCMs)** can be described as particles that contain core PCMs surrounded by a coating or a shell and have diameters in the scale of micrometers. The microencapsulated PCMs usually have required morphologies, uniform diameters, shell mechanical strengths, penetration abilities and thermal stabilities. Pouches, tubes, spheres, panels or other receptacles containing MEPCM can directly act as heat exchangers. They can be incorporated into the building materials for thermal energy storage. The shell can hold the liquid PCM inside and prevent changes in its composition. The encapsulation not only increases the contact surface area for heat transfer but also adds the mechanical stability with the rigid shell. Common encapsulation shell materials include urea-formaldehyde (UF) resin, melamine-formaldehyde (MF) resin and polyurethanes (PU). Specialized techniques to prepare the encapsulation with a polymer cover and a PCM core include coacervation, suspension polymerization, emulsion polymerization, polycondensation and polyaddition. The MEPCMs are widely applied into the building materials and are able to retain or improve the building structural performance, as well as the energy performance (**Figure 3**).

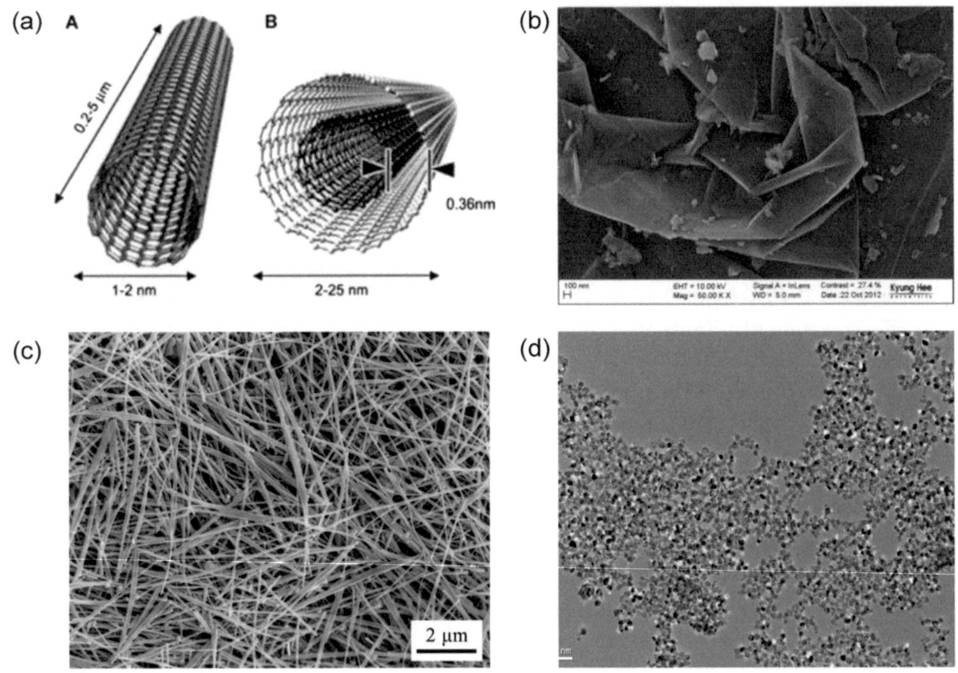

Figure 3.
Common applied thermal enhancement promoters: (a) carbon nanotubes/nanofibers [30], (b) graphite/graphene flakes [34], (c) metal nanoparticles (silver nanowires) [35], (d) metal oxide nanoparticles (γ-Al_2O_3) [36].

2.2.2 Heat transfer enhancement of PCMs

Although the usage of PCMs is attracting because of its high energy storage density, most PCMs have a low thermal conductivity (~0.5 W/m·K), which restrains the heat transfer rate during charging and discharging processes. Except the enhancement approaches by dispersing thermal enhancement promoters to form composite PCMs as introduced above, there are also other methods developed for heat transfer enhancement, such as adding metal fins to increase the heat transfer area or embedding metal fins (foam), using multiple PCMs, heat pipes, or manipulating the designs of the latent heat storage system to increase the natural convection heat transfer intensity, as shown in **Figure 4**.

- **Metal fins and extended surfaces** can effectively increase the contact area and enhance the heat conduction between the heat source and the PCMs. Commonly applied fin materials are steel (50.2 W/m·K), aluminum (205 W/m·K), copper (385 W/m·K) or graphite foil (140 W/m·K in x-y direction) [24]. Graphite foil is suitable for high temperature applications and has many advantages like high thermal conductivity, low density, good corrosion resistance against nitrate salts and nitrite at high temperatures. It occupies less volume for the same performance compared with steel fins. Aluminum is suitable for low temperature applications below 400°C and does not show corrosion effect when contacted with galvanized steel pipes. The introduction of fins into the PCM changes the PCM interior structure and hence affects the natural convection in the liquid phase of PCM during melting [37, 38].

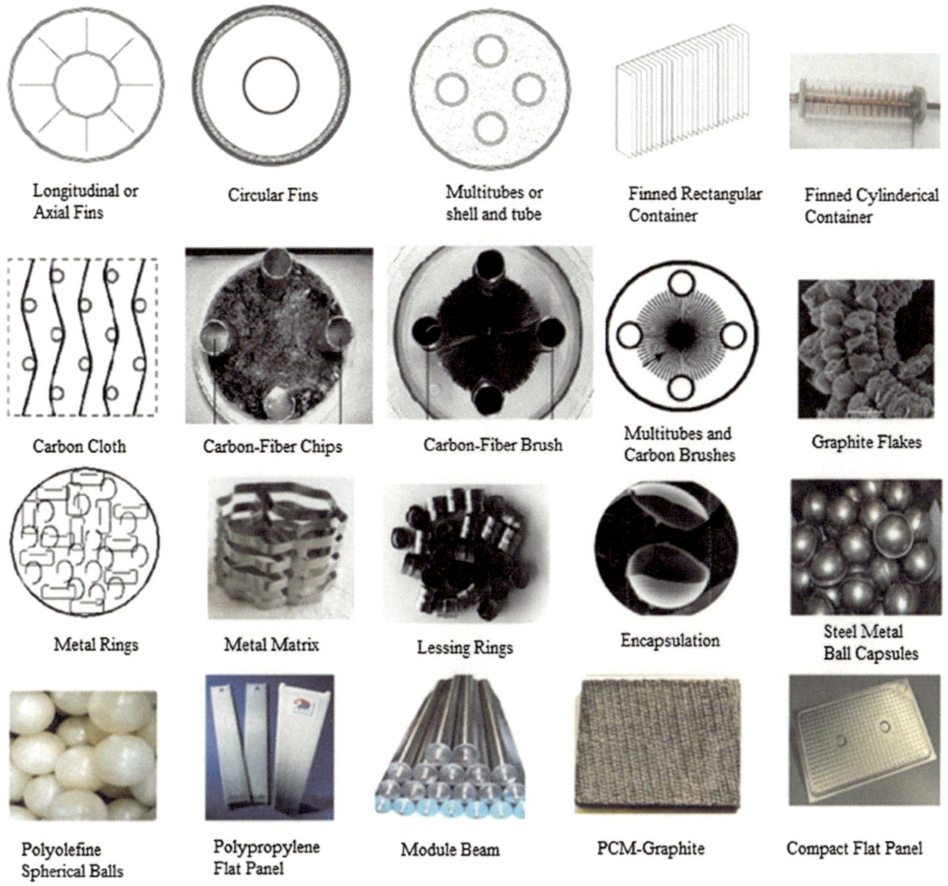

Figure 4.
Heat transfer enhancement techniques used in LHS systems [25].

- **Embedded porous matrices** made of copper, aluminum or a naturally porous material like graphite, metal foam can be embedded into interior PCM to enhance the heat transfer. The integration of metal foam helps the distribution of heat evenly into the PCM and is able to greatly enhance the PCM melting rate [24], even achieving as high as 41.2% enhancement on the melting process in a comparison with a pure PCM system [39]. One should note that the performance improvement with metal foam depends on the structure porosity and thermal conductivity.

- **Heat pipe** has a high thermal conductivity and it can be added to PCMs to increase the performance of the latent heat storage system. It serves as a thermal channel between the HTF and the PCMs. Within a heat pipe, there exist two regions, i.e., evaporation and condensation, and the working fluid like mercury evaporates and condensates for heat exchange. Metal fins can also be integrated with a heat pipe to further reduce the thermal resistance and a corresponding simulation showed that the addition of fins on a heat pipe leads to a threefold increase in the heat storage during the first 3 h of charging and a 79% increase in energy discharged within the first 3 h of releasing [40].

2.2.3 Pros and cons of latent heat storage

Latent heat storage with PCMs has a large latent heat of fusion so that it can store more amount of heat than sensible heat storage. This large difference gives PCMs the advantage of a high energy storage density, which reduces the volume of TES vessel and the outer wall surface area, and minimizes the heat loss. Compared to the sensible heat storage, the discharging process of the latent heat storage can maintain the temperature constant and make the contacting medium like HTF or the adjacent space temperature stable. However, the main drawback of the latent heat storage is the low thermal conductivity of PCMs. Moreover, the heat transfer design and media selection are more difficult, and the experience with low temperature PCMs like inorganic salts has shown that the performance of the materials can degrade after moderate number of freezing-melting cycles.

2.3 Thermochemical heat storage

Different from sensible heat storage systems and latent heat storage systems, a thermochemical heat storage system uses reversible endothermic/exothermic reactions to absorb, store and release energy between solar heat and chemical reactants. As depicted in **Figure 5**, during the charging process, heat is absorbed to dissociate material A, resulting in the products B and C. This reaction is endothermic and the amount of thermochemical heat stored is determined by the reaction enthalpy. It can be calculated by:

$$Q_{chem} = n_A \Delta H_r \tag{3}$$

where n_A is the mol number of A and H_r is the enthalpy of the endothermic reaction. During the discharging process, the products from the endothermic reaction, i.e., B and C, are mixed together at certain temperature and pressure conditions to reform the initial material A. This reaction is exothermic and the regenerated heat can be converted into electricity or used for other thermal powered systems.

To develop a thermochemical heat storage system, the first consideration is the selection of the reversible reaction and to characterize its operating temperature and pressure, the reaction rate, the reversibility and the kinetic properties, etc. For a high temperature energy storage, for instance, the endothermic reaction for the heat charging process should occur at temperatures below 1000°C to reduce the material restriction. The exothermic reaction, on the other hand, should be able to release heat at temperatures higher than 500°C [41]. Two parameters are usually used to assess the reactions from the thermodynamic perspective [42]. The first parameter is the turning temperature T^*. For a reversible reaction at a given

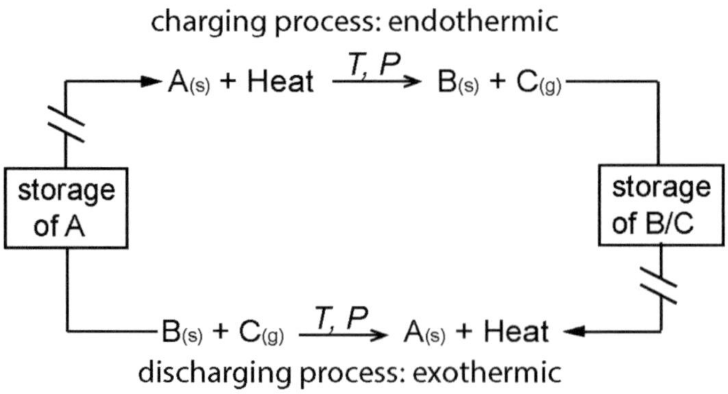

Figure 5.
Schematic of a thermochemical heat storage system.

pressure p, T^* is the temperature at which forward and reverse reactions are thermodynamically equilibrium. It can be approximated as

$$T^* = \frac{\Delta H_r(T^*, p)}{\Delta S_r(T^*, p)}, \quad (4)$$

where ΔH_r and ΔS_r are the reaction enthalpy and entropy at pressure p and temperature T^*. The other parameter is the ratio of the maximum recovery work to the actual recovery work, namely the maximum work recovery efficiency η_{max}. The value of η_{max} equals to 1 when the process is completely reversible. In the presence of irreversibility of a process, η_{max} is less than 1 due to the entropy generation and cannot be used for work. In order to achieve the maximum work, it is hence required that both endothermic and exothermic reactions should be completely reversible.

2.3.1 Reactions for thermochemical heat storage

A large number of reactions have been tested under the laboratory scale to determine their feasibility as thermochemical storage materials. These reactions can be classified according to their reaction family, namely carbonate system, hydroxide system, redox system, metallic hydride system, ammonia system and perovskite oxide system [43, 44].

- **Carbonate system** is operated by the decarbonation reaction of carbonate. The reaction can be expressed as

$$MCO_{3(s)} + \Delta H_r \rightleftharpoons MO_{(s)} + CO_{2(g)}. \quad (5)$$

In carbonate systems, chemical reactions are usually from inorganic oxide/carbon dioxide because of their high chemical reaction heat. These reactions generally occur at a temperature higher than 450°C. Many inorganic oxides can be used as thermal energy storage materials in carbonate systems [45]. Most attention is drawn on the calcium oxide (CaO)/carbon dioxide (CO_2) reaction for chemical reaction heat storage, especially for solar thermal energy storage [46]. Kyaw et al. [47, 48] carried out a series of investigations on the process of CaO/CO_2 reaction for chemical reaction heat storage. They listed the characteristics of this system such as high energy storage density, high working temperature, high reversibility, nontoxic process and low cost. More recently, Ortiz et al. [49] reported a study based on the CaO/CO_2 reaction for concentrated solar power plants. It is found that high overall efficiencies can be achieved by integrating the CaO/CO_2 reaction into concentrating solar power plants.

- **Hydroxide system** relies on the heat consumption and release when the reactant contacts with water. The universal reaction can be described as

$$M(OH)_{2(s)} + \Delta H_r \rightleftharpoons MO_{(s)} + H_2O_{(g)}. \quad (6)$$

This reaction usually occurs at moderate temperatures ranging from 250 to 450°C [43]. Using this reaction for thermal energy storage was first proposed by Ervin [50]. He successfully completed 290 reaction cycles with an average conversion rate of 95%. Criado et al. [51] analyzed the CaO/Ca$(OH)_2$ system on its reversibility and cyclicity. The author claimed that it is one of the best candidates for thermal energy storage. Brown et al. [52] discussed the feasibility of a thermochemical heat storage system using CaO/Ca$(OH)_2$ reaction from a technic-economic perspective.

Schaube et al. [53] experimentally explored this system on a fixed bed reactor and completed 25 cycles without reversibility loss.

- **Redox system** makes use of oxidation and decomposition reactions of oxide/peroxide for heat converting. The general reaction is

$$M_xO_{y(s)} + \Delta H_r \rightleftharpoons xM_{(s)} + y/2O_2. \tag{7}$$

These reactions occur at temperatures within 400–1000°C [43]. A couple of oxide/peroxide pairs include Co_3O_4/CoO, MnO_2/Mn_2O_3, CuO/Cu_2O, Fe_2O_3/FeO, Mn_3O_4/MnO and V_2O_5/VO_2 have been studies by Wong et al. [54]. Among the reactions suitable for thermochemical heat storage system, the Co_3O_4/CoO pair is considered as the most promising candidate due to its high reaction enthalpy, high reaction temperature and reversibility [43].

- **Metallic hydride system** is based on the reversible reaction between metals and hydrogen under certain conditions. For thermal energy storage process, the metallic hydride is thermally dehydrogenated and expressed as

$$MH_{y(s)} + \Delta H_r \rightleftharpoons M + n/2H_2. \tag{8}$$

The reversible metallic hydride reactions can be used in many industrial applications such as hydrogen storage [55], heat pumps [56], thermal energy storage [55], etc. The two metallic hydrides which have been intensively investigated for the solar thermal energy storage are magnesium hydride (MgH_2) and calcium hydride (CaH_2). Magnesium hydride reaction occurs at temperatures from 250 to 500°C and has a high energy storage density [47]. Attempts have also been extended by doping a third metal powder into Mg powder to form a Mg-based ternary hydride for thermal energy storage. It has been proved that the chemical reaction rate and cyclicity can be improved by mixing the metals such as Ni, Fe, Co, Na, etc. [57, 58].

- **Ammonia system** here refers to the dissociation and synthesis cycle of ammonia (NH_3) and the process has been known for hundred years. The reaction is

$$2NH_{3(g)} + \Delta H_r \rightleftharpoons N_{2(g)} + 3H_{2(g)}. \tag{9}$$

The reaction occurs at temperatures of 400–700°C. It was first considered for thermochemical heat storage by Carden [59] and Williams [42]. It is worth to note that catalysts should be applied for both forward and reverse reactions which increase the complexity of the system.

- **Perovskite oxide system** is a class of materials with similar crystal structure which can be generically expressed as ABO_3. The structure of perovskite is reviewed by Borowski [60]. Owing to the nature of its structure, perovskite oxides allow reversible accumulation and release of oxygen at high temperature with limited phase transition between different crystal structures [44, 61]. A general form of the reaction is given as

$$ABO_3 + \Delta H_r \rightleftharpoons ABO_{3-\delta} + \delta/2O_2. \tag{10}$$

The crystal structure facilitates the flexibility in replacement of the cations, i.e., A and B in the oxide. A variety of compounds have been synthesized and analyzed for thermochemical heat storage. Babiniec et al. [62] synthesized $La_xSr_{1-x}Co_yM_{1-y}O_{3-\delta}$ (M = Mn, Fe) materials for high temperature thermochemical energy storage owing to its good reaction reversibility. Albrecht et al. [63] evaluated the thermochemical energy storage system with the perovskite strontium-doped calcium manganite ($Ca_{0.9}Sr_{0.1}MnO_{3-\delta}$) in terms of specific storage and overall solar-to-electric efficiency. More recently, Gokon et al. [61] examined $La_xSr_{1-x}(Mn, Fe, Co)O_{3-\delta}$ and $Ba_ySr_{1-y}CoO_{3-\delta}$ redox powers in terms of reactivity and repeatability. They found that $Ba_{0.3}Sr_{0.7}CoO_{3-\delta}$ and $Ba_{0.7}Sr_{0.3}CoO_{3-\delta}$ powders were suitable thermochemical storage materials operating at above 600°C.

2.3.2 Pros and cons of thermochemical heat storage

Thermochemical heat storage system is unique and suitable for solar energy storage owing to its advantages: high volumetric storage density, low volume requirement, long energy preservation duration periods with limited heat loss, low storage temperature (ambient temperature) and unlimited transport distance. However, it also has some issues to be tackled, such as poor reactivity and reversibility of reactions, harsh reaction conditions, toxic and corrosive products from reactions, etc. Currently, most studies are conducted at laboratory scales. Large-scale tests are needed to verify the feasibility and the durability for long-term thermal energy storage. Furthermore, the criteria of material selections for thermochemical heat storage should be established.

3. Thermal energy storage for solar energy systems

Solar energy systems are found in many applications with a wide range of scales for different sectors, such as CSP, space heating for buildings, and hot water supply for houses. One major drawback of solar energy is its intermittence. To overcome this issue, one solution is to use backup energy sources or hybridisation of different energy sources, such as using gas as the backup solution for heating and electric grid for electricity. Another solution is to use a thermal energy storage system to store heat during sunshine periods and release it whenever demand is needed while solar irradiance is insufficient to cover the demand.

Thermal energy storage in solar energy systems usually has the following functions [5]:

- Mitigating short fluctuation of solar energy. Variations in solar irradiance can cause rapid fluctuations in energy generation (i.e., heat or power). These strong fluctuations influence not only the availability of energy, but also the stability of the power grid, heating network or gas-supplying systems. Integrating thermal energy storage units into the solar energy system can increase the thermal inertia, smooth the fluctuations and help the system continue supplying energy during short cloudy periods, thus increasing the system stability.

- Matching the energy generation and demand. Peak energy demand may not coincide with the peak solar insolation. Thermal energy storage can improve dispatchability of a solar energy system by storing heat during off-peak hours and discharging it during peak hours of demand. This helps match the energy generation and demand, and also increases the profitability as peak-hour electricity tariffs are higher. The duration of this energy-shifting mechanism can be within a day, a week, or seasons.

- Extending the energy delivery period. Solar energy is only available during some hours of the day and thermal energy storage can extend the energy delivery period to hours when no sun is available. If the thermal energy storage unit is large enough, the whole solar energy system, for either power generation or heating/cooling, can operate for 24 h.

In practical applications, solar energy is mainly used for delivering either electricity or heating/cooling [64]. Depending on applications, there are a wide range of technologies used for thermal energy storage. In CSP plants, thermal energy storage systems operate at relatively high temperatures as the thermal efficiency of power plants is proportional to the temperature. In solar heating/cooling systems, such as space heating in buildings, solar hot water supply, and solar absorption refrigeration systems, low-temperature thermal energy storage is often involved.

3.1 Thermal energy storage for solar power systems

One of the most important applications of solar energy is to generate electricity, which can be realized by either solar photovoltaic (PV) panels or solar-thermal driven power cycles [65]. To mitigate the intermittence of solar energy, PV systems usually use batteries to store energy in terms of electricity, while solar-thermal driven power cycles often store energy in terms of heat via thermal energy storage technologies. Comparisons between different energy storage technologies have shown that the thermal storage option is more cost-competitive than the battery option for large-scale energy storage [24, 66].

The most commonly known solar-thermal driven power technologies are CSP systems, in which high-temperature thermal energy storage is often involved [67, 68]. Another type is small-scale solar power systems driven by low-temperature solar heat [69], which is typically categorized as distributed solar power systems and where low-temperature thermal energy storage technologies are needed. Whatever the solar-thermal driven power cycle technologies are, the main processes in such systems include three steps: collection, storage and conversion, as shown in **Figure 6**. Solar energy is first collected via concentrated or non-concentrated solar collectors in terms of thermal energy, then transferred to and stored in thermal energy storage units through a heat transfer loop connecting the collectors and the storage units, and the thermal energy, either stored in the thermal energy storage units or collected directly from the solar collectors, is finally converted into electricity through heat-to-power conversion technologies, e.g., Rankine cycles.

3.1.1 Material selections

About half of the CSP plants in operation are integrated with a thermal energy storage system. There is an increase in the use of thermal energy storage for CSP

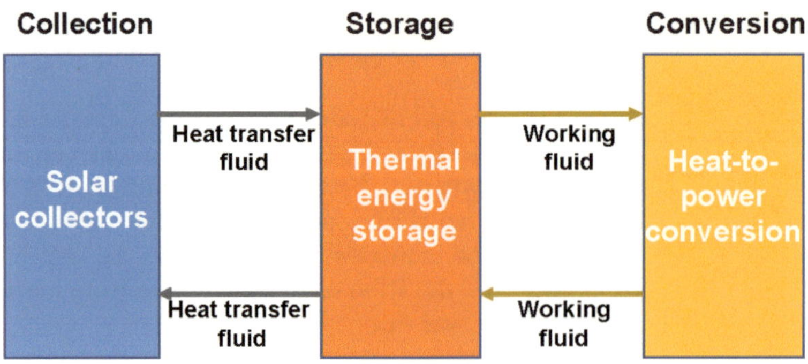

Figure 6.
The main processes of solar-thermal driven power technologies.

plants under construction and planned (>70%) [70]. The trends indicate the maturity and economic competitiveness of installing a thermal energy storage system in CSP plants. The most commonly used thermal energy storage materials in CSP applications include molten salt, water/steam, liquid sodium, thermal oil, concrete and rocks. Molten salt is the most widespread storage material in CSP commercial applications, due to its features of excellent thermal stability at high temperatures, low vapor pressure, low viscosity, non-flammability and non-toxicity [70]. **Table 4** lists the summaries of the world's operational CSP plants integrated with thermal energy storage. The molten salt fluids commonly used are nitrate mixtures with a weight composition of 60 wt% $NaNO_3$ and 40 wt% KNO_3, also called Solar Salt. The storage temperature of the molten salts in these commercial CSP plants are typically around 400°C and can go up to around 550°C. Steam is also used as the storage material for several CSP plants, such as Khi Solar One, Puerto Errado 2 and eLLO. The steam storage temperatures in these plants are normally around 270°C - 285°C. In Jemalong Solar Thermal Station in Australia, liquid sodium at 560°C is used as the storage material. Thermal oils have also been used in Dahan Power Plant in China and in many researches [67]. Apart from these fluid-type thermal energy storage materials, solid materials (concrete and rocks) are another option for thermal energy storage [71, 72]. Solid materials generally have a wide range of working temperatures (200–1200°C), with high thermal conductivities (from 1 W/m·K to 40 W/m·K) and relatively low costs (0.05–5 $/kg) [70]. Examples of such CSP plants using solid storage materials include Airlight Energy Ait-Baha Pilot Plant in Morocco (packed-bed rocks for parabolic trough CSP) and Jüich Solar Tower in Germany (ceramics for power tower CSP). In small-scale distributed solar power systems, such as solar-driven ORC systems [69, 73], low-temperature thermal energy storage materials can be used. For example, water, organic aliphatic compounds, inorganic hydrated-salt PCMs and thermal oils have been investigated for solar combined heat and power applications [74].

3.1.2 System integrations

Thermal energy storage materials are usually stored in tanks to form plant-scale storage subsystems for the CSP systems. Based on the motion state of storage materials during charging and discharging, thermal energy storage system can be categorized into active and passive systems. In active storage systems, the storage medium itself flows through a heat exchanger to absorb (charge) or release (discharge) heat by forced convection, while in passive storage systems, the storage medium stays stationary and is heated (charged) or cooled (discharged) by a HTF.

Active storage systems can be classified into direct and indirect systems. In active direct storage system, storage medium also has the function of transferring heat as a HTF. The use of molten salts or steam as a HTF or storage materials at the same time eliminates the need for an expensive heat exchanger. The cost of thermal energy storage system can be reduced and the system performance can be improved. One of the active direct systems is the two-tank direct storage system, which consists of a hot and a cold storage tank, as shown in **Figure 7(a)**. In such systems, molten salt or thermal oil collects heat from the solar field and the high-temperature fluid is directly stored in the hot storage tank for later uses, i.e., steam generation, in cloudy periods or nights. The cooled HTF (also the storage medium) after the steam generator is pumped into the cold storage tank. The two-tank direct storage concept is often used in tower CSP plants where molten salt is used as the storage medium and HTF [75], as solar tower can achieve higher temperatures, as shown in **Table 4**.

Another type of active direct storage system uses water/steam as the HTF and storage medium, as shown in **Figure 7(b)**. This system is also named as direct steam

CSP plant	Technology	Turbine capacity	Heat transfer fluid	Storage material	Storage capacity	Storage type	Inlet–outlet temperature
Xina Solar One (South Africa)	Parabolic Trough	100 MW	Thermal oil	Molten salt	5 h	Two-tank indirect	N/A–395°C
Termosol 2 (Spain)	Parabolic Trough	50 MW	Thermal oil	Molten salt	9 h	Two-tank indirect	293–393°C
Solana Generating Station	Parabolic Trough	250 MW	Therminol VP-1	Molten salt	6 h	Two-tank indirect	293–393°C
NOOR II (Morocco)	Parabolic Trough	200 MW	Thermal oil	Molten salt	7 h	Two-tank indirect	293–393°C
Manchasol-2 (Spain)	Parabolic Trough	50 MW	Diphenyl/diphenyl oxide	Molten salt	7.5 h	Two-tank indirect	293–393°C
Aste 1B (Spain)	Parabolic Trough	50 MW	Dowtherm A	Molten salt	8 h	Two-tank indirect	293–393°C
Bokpoort (South Africa)	Parabolic Trough	55 MW	Dowtherm A	Molten salt	9.3 h	Two-tank indirect	293–393°C
Kathu Solar Park (South Africa)	Parabolic Trough	100 MW	Thermal oil	Molten salt	4.5 h	Two-tank indirect	293–393°C
Ashalim (Israel)	Parabolic Trough	110 MW	N/A	Molten salt	4.5 h	Two-tank indirect	N/A
Archimede (Italy)	Parabolic Trough	5 MW	Molten salt	Molten salt	8 h	Two-tank direct	290–550°C
SUPCON Delingha 50 MW	Power Tower	50 MW	Molten salt	Molten salt	7 h	Two-tank direct	N/A
Tower (China)							
Shouhang Dunhuang 100 MW	Power Tower	100 MW	Molten salt	Molten salt	11 h	Two-tank direct	N/A
Phase II (China)							
Qinghai Gonghe 50 MW	Power Tower	50 MW	Molten salt	Molten salt	6 h	Two-tank direct	N/A
CSP Plant (China)							
Gemasolar Thermosolar Plant (Spain)	Power Tower	19.9 MW	Molten salt	Molten salt	15 h	Two-tank direct	290–565°C
Greenway CSP Mersin Tower	Power Tower	1.4 MW	Water	Molten salt	4 MWh	Single 3-pase tank	N/A
Plant (Turkey)							
Airlight Energy Ait-Baha	Parabolic Trough	3 MW	Air at ambient pressure	Rocks	5 h	Packed-bed	270–570°C

CSP plant	Technology	Turbine capacity	Heat transfer fluid	Storage material	Storage capacity	Storage type	Inlet–outlet temperature
Pilot Plant (Morocco)							
Dahan Power Plant (China)	Power Tower	1 MW	Water	Saturated steam and oil	1 h	N/A	104–400°C
Khi Solar One (South Africa)	Power Tower	50 MW	Water/steam	Steam	2 h	Steam accumulators	N/A
Puerto Errado 2 Thermosolar	Linear Fresnel	30 MW	Water	Steam	0.5 h	Single-tank thermocline	140–270°C
Power Plant (Spain)	Reflector						
eLLO Solar Thermal	Linear Fresnel	9 MW	Water	Steam	4 h	Steam drum	190–285°C
Project (France)	Reflector						
Jülich Solar Tower (Germany)	Power Tower	1.5 MW	Air	Ceramic	1.5 h	N/A	N/A
Jemalong Solar Thermal	Power Tower	1.1 MW	Liquid sodium	Liquid sodium	3 h	Two-tank direct	293–560°C
Station (Australia)							

Table 4.
In-operation CSP plants with thermal energy storage [https://solarpaces.nrel.gov/by-status/operational].

generation (DSG) plant. The only commercial thermal energy storage technology for DSG plants is using steam accumulators, where steam is stored at a high pressure in accumulator tanks [75, 76].

The dominant thermal energy storage technologies in parabolic trough CSP plants are the active two-tank indirect storage systems, as shown in **Figure** 7(c). The list of commercial parabolic trough CSP plants and their thermal storage types are given in **Table 4**. The two-tank indirect system uses different mediums for heat transfer and storage [77]. During charging phase, the storage medium in the cold storage tank is pumped through a heat exchanger, heated by the hot HTF flowing from the solar field, and then stored in the hot storage tank. During discharging phase, the flow direction of the storage material is reversed to release heat to the HTF to generate steam for power generation. In two-tank indirect systems, the storage medium is typically molten salt, and the HTF is often thermal oil.

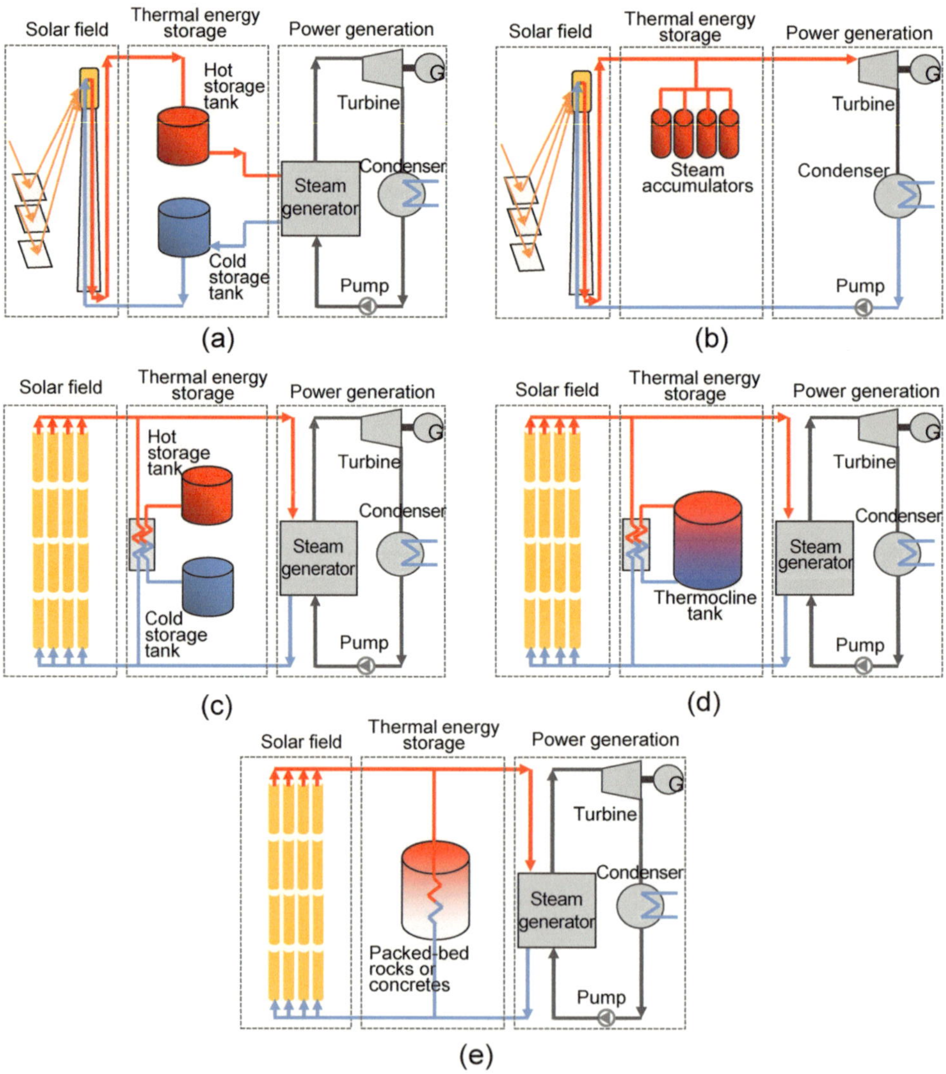

Figure 7.
Schematics of CSP systems integrated with different thermal energy storage technologies: (a) active two-tank direct storage system, (b) active direct steam storage system, (c) active two-tank indirect storage system, (d) active single-tank indirect storage system, (e) passive pack-bed rocks (or concrete, or castable ceramics) storage system. Note that although solar tower and parabolic trough collectors are illustrated here, other types of concentrated solar collectors (Linear Fresnel, dish) may also be used.

An alternative to two-tank indirect storage system is using a single tank (see **Figure 7(d)**) where hot fluid is stored at the top and the cold fluid is stored at the bottom by stratification. The zone between the hot and cold fluids is called the thermocline. A single tank system is about 35% cheaper than a two-tank system [11].

In passive storage system, solid materials are typically used as the thermal storage medium while a HTF passes through the storage medium to charge or discharge heat. Possible storage materials include concrete, castable ceramics, rocks, etc., while the HTF can be thermal oil or air. In the case of concrete, tubular heat exchangers are integrated with the concrete to enhance heat transfer rate. Rocks can be used in a packed bed after crushing to rough sizes of around 5 cm [70]. The hot fluid flows through the gaps of the packed rocks and heats the rocks during charging process, and the cold fluid flows through the gaps and absorbs heat from the hot rocks during discharging process. The typical system schematic is shown in **Figure 7(e)**.

3.2 Thermal energy storage for solar heating/cooling systems

Heating and cooling take a significant share of the total energy consumption in the world. For example, half of EU's primary energy is consumed for heating and cooling purposes. Currently, most of the heating and cooling demands are still met by fossil fuels, mainly natural gas. However, solar energy is starting to play an important role. In solar heating/cooling systems, solar energy is typically collected as low-temperature heat, for provisions of space heating, hot water, grain drying, etc. In hot seasons or hot regions, solar heat can be used to create cooling via thermally driven refrigerators, e.g., absorption chillers. Thermal energy storage is frequently involved in these solar heating/cooling systems, as it provides much more stable energy delivery capability and closes the gap between energy generation and demand.

3.2.1 Material selections

Water is regarded as a favorable storage medium for solar heating/cooling applications, due to its advantages of high specific heat, non-toxicity, low cost and easy availability. Therefore, water is the dominant material for space heating and hot water provisions. The capacity of the storage depends on the size of the water storage tanks, which could range from a few hundred liters to a few thousand cubic meters. In water tanks, thermal stratification can be formed due to the buoyancy effect, i.e., hotter water is lighter and gathered at the top while colder water drops at the bottom. Thermocline is then formed between the hot and cold water layers, which in-turn minimizes water mixing and heat losses. Apart from using water tanks, water is also used for large-scale seasonal thermal energy storage in underground aquifers where sand mixed with water is the storage medium [78]. This is a low-cost thermal storage option as it only uses natural materials and no tanks are involved. The problem of this aquifer thermal energy storage is the high heat losses, as the system cannot be insulated. It is also possible to use water tanks for seasonal thermal energy storage, although the tank size has to be large and it is often fully or partially buried in the ground. Borehole thermal energy storage is another competitive option for seasonal storage of heat or cold. It uses ground (soil) as the storage medium [78]. Multiple boreholes are drilled in the ground to a certain depth and U-shape pipes are inserted in those boreholes to charge or discharge thermal energy. PCMs have attractive extensive attentions in recent years as an effective thermal energy storage medium and have been integrated in various devices and applications. In some work, PCMs are directly integrated with the solar collectors for space heating or drying purposes [79, 80]. In other work, they can be embedded in concretes of walls for buildings [81], to store heat from daytime for night use. Rock is a low-cost thermal energy storage material which usually uses air as the HTF. Packed-bed rock thermal energy storage has been widely used for greenhouse heating [79].

3.2.2 System integrations

Water tanks are widely installed in solar heating system, including those for distributed and centralized heating for residential, industrial and commercial applications. **Figure 8** shows a typical solar heating system for the provisions of domestic space heating and hot water. Solar radiation is absorbed in terms of heat by the solar collectors, e.g. evacuated tube collectors, flat-plate collectors, or hybrid photovoltaic-thermal collectors [65]. A circulating HTF loop is used to collect heat from the solar collectors and store it in the water tank when the fluid temperature from the collectors is higher than the water temperature in the tank. The demands of the space heating and hot water are met by circulating or drawing hot water from the tank. Unlike the two-tank thermal energy storage systems, only one tank is typically involved in these applications and the water temperature thus fluctuates depending on the balance between energy storage and usage. Sizing of the water tank as well as the solar collectors is important and depends on various factors, including local weather conditions (solar irradiance, wind speed, air temperature, etc.) and demands (types, quantities and shapes). Annual simulations are often needed to fully assess the potential of a solar heating system [82]. Although water is the most popular storage material in such systems, PCM-based thermal energy storage has also been explored more recently [83]. The system integration is similar to water-based systems, but the storage temperature can be more stable, and its energy capacity can be larger for the same volume.

Solar energy has not only a daily period but also a seasonal period. For some period of a year, solar thermal production exceeds the demand for heating or cooling, while in other periods the production is less than the demand. Seasonal thermal energy storage would be a solution to store heat at the time that is not needed and use is for the time that is required. The concept of seasonal thermal energy storage is illustrated in **Figure 9(a)**. The implementation of seasonal thermal energy storage is shown in **Figure 9(b)**, with borehole thermal energy storage as an example. In summer period, excess heat from the solar thermal collectors, process heat, or other heat sources are transferred to the ground via borehole heat exchangers, while in winter the stored heat is released from the ground by a heat transfer loop for heating purposes. There are other types of seasonal storage systems, mainly including aquifer storages, cavern storages, pit storages, seasonal water storages, water-gravel storages,

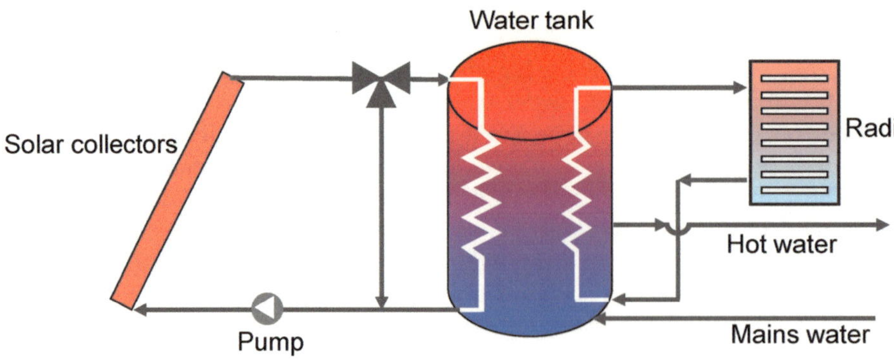

Figure 8.
Typical layout of a solar heating system for domestic space heating and hot water.

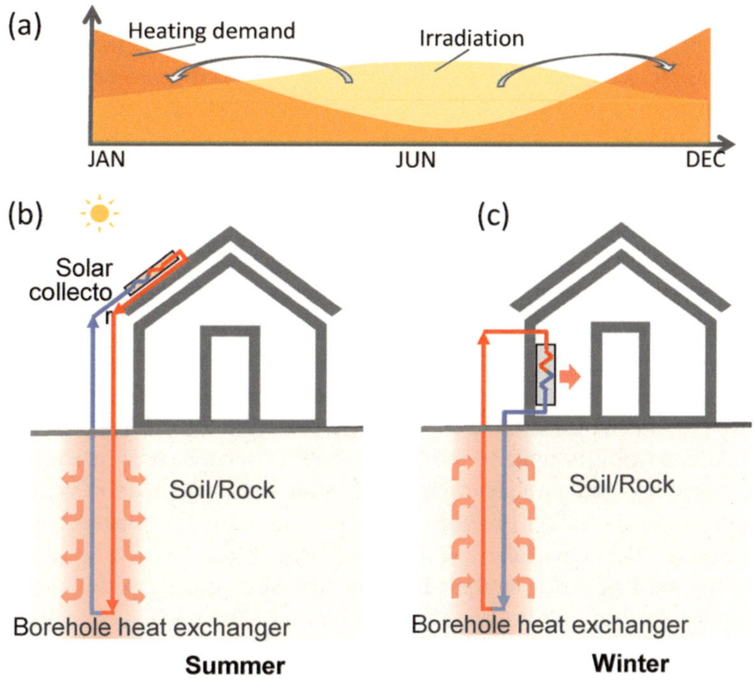

Figure 9.
Seasonal thermal energy storage based on borehole heat exchangers: (a) concept of seasonal thermal energy storage and (b) borehole thermal energy storage.

thermochemical storages, etc. [84]. The development of seasonal thermal energy storage is mainly motivated by the growing popularity of district heating [20]. Low energy and net-zero energy buildings are attracting increasing attention, through the incorporation of solar energy systems and thermal energy storage among others. Incorporating thermal energy storage into the building design, structure and facades have been explored to reach the above goals. These systems can be categorized into active and passive systems. In active systems, external mechanical driving systems are involved to transfer and distribute solar heat [85], such as an evacuated tube collector based space heating system, or a space cooling system based on solar-absorption refrigerators. In passive solar building systems, windows, walls and floors are designed to collect, store, reflect and distribute solar energy in the form of heat in the winter and reject heat in the summer. Thermal energy storage can be incorporated with these structures to enhance the energy efficiency. An example structure is given in **Figure 10(a)**, which is named as solar wall or Trombe wall [86]. The wall

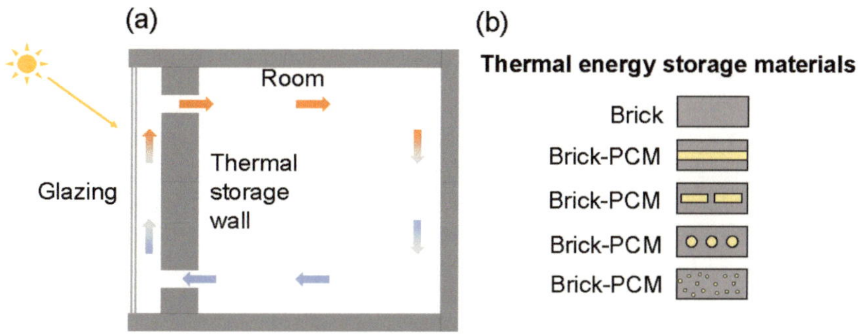

Figure 10.
Schematic of (a) a solar wall (Trombe wall), and (b) different thermal energy storage materials for solar wall.

with a high heat capacity is built on the winter sun side with a glass external layer and an air layer for thermal insulation. Sunlight passing through the glass is absorbed by the wall and stored for later uses at night. The stored heat then re-radiates in the far-infrared spectrum and trapped by the glass, and thus heats the inner side of the room. The materials used for storage could be bricks, stones, or bricks integrated with PCMs, as shown in **Figure 10(b)**.

4. Conclusions

This work reviewed some recent developments on the thermal energy storage technologies, where sensible, latent and thermochemical heat each offers a contribution to eliminate the mismatch between the energy supply and demand by different working principles. The thermal energy storage material categorization, long-term stability and compatibility with container materials, thermal performance analysis and thermal performance enhancement techniques were reviewed in detail. The sensible heat storage in solid or liquid is widely applied for thermal storage. Rock, sand and water are the typical storage mediums used in the solar energy systems from low to high temperatures due to their high specific heat, none-toxicity, low cost and easy availability. Molten salt in liquid is the most popular thermal energy storage material used in the medium and high temperature concentrated power plants for its excellent thermal stability, low cost and low viscosity. The main drawback of sensible heat storage is the temperature decrease during the discharging process. Latent heat storage with phase change materials is a popular energy storage technology today, as it brings higher storage density and nearly constant temperature. Several materials in organic, inorganic, eutectic and composite have been analyzed and identified. To overcome their low thermal conductivities, metal fins, embedded porous matrices, heat pipes and other heat transfer enhancement technologies have also been developed. The chemical storage technology is also a potential technology in terms of its high storage density and long energy preservation duration periods. Several reaction systems have been reviewed and discussed. However, it is much less developed than the sensible and latent heat ones for solar energy storage. Thermal energy storage provides a reliable technology for energy storage and security, but it also faces some barriers and needs to be further developed, such as material costs, thermal properties and stabilities, system integration and process parameters, especially for the latent and thermochemical storage systems.

Thermal energy storage for solar energy systems is then reviewed, focusing on the most common storage materials, components and their integrations with wider systems. In solar power systems, high-temperature thermal energy storage materials are widely used for concentrated solar power (CSP), including molten salt, water/steam, liquid sodium, thermal oil, concrete and rocks, etc. Molten salt remains as the dominant commercial storage option for CSP, while steam and concrete are also being demonstrated. Two-tank storage system is the main solution in CSP applications. In low-temperature solar power systems, water and low-temperature phase change materials (PCMs) are normally used and stored in tanks, which are integrated with low-temperature power cycles, such as organic Rankine cycles. These could be of interest to waste heat recovery and domestic applications. Compared to CSP systems, thermal energy storage in solar heating/cooling systems is mainly based on low-temperature materials, with water as the dominant storage material. Water tanks are widely used as a short-term storage option and typically coupled with solar thermal collectors for solar heating/cooling purposes. Long-term storage, i.e., seasonal storage, can be achieved by large water tanks or borehole thermal energy storage. The demand for seasonal thermal storage is mainly driven by district heating. Thermal energy storage can also be directly integrated into the building structures, e.g., walls, windows, and floors. In these systems, the storage medium is either the construction material itself or appropriately embedded/integrated with the construction materials. Although the development of thermal energy storage materials is very critical, further development of applications of more stable and efficient solar energy systems, system designs and integrations are equally important. Innovative technologies for thermal energy storage materials and solar energy systems are highly desirable.

Nomenclature

c_p	heat capacity of thermal storage material
H_r	enthalpy of the endothermic reaction
k	thermal conductivity of thermal storage material
L_h	latent heat of thermal storage material
m	mass of thermal storage material
Q_{chem}	amount of thermochemical heat stored
Q_{lat}	amount of latent heat stored
Q_{sen}	amount of sensible heat stored
S_r	entropy of the endothermic reaction
T_f	final temperature of thermal storage process
T_i	initial temperature of thermal storage process
T^*	temperature at thermodynamically equilibrium
V	volume of thermal storage material
α	melting fraction of PCM
ρ	density of thermal storage material

Abbreviations

CSP	concentrated solar plant
HTF	heat transfer fluid
MEPCM	microencapsulated PCM
PCM	phase change material
PU	polyurethanes
PV	photovoltaic
TES	thermal energy storage
UF	urea-formaldehyde

Author details

Kai Wang[1†], Zhen Qin[2†], Wei Tong[3†] and Chenzhen Ji[4*]

1 Institute of Refrigeration and Cryogenics, Zhejiang University, Hangzhou, China

2 Department of Mechanical and Industrial Engineering, University of Toronto, Toronto, Ontario, Canada

3 Department of Mechanical Engineering, University of Sherbrooke, Sherbrooke, Quebec, Canada

4 Dyson School of Design Engineering, Imperial College London, London, United Kingdom

*Address all correspondence to: c.ji@imperial.ac.uk

†These authors contributed equally.

References

[1] Aneke M, Wang M. Energy storage technologies and real life applications—A state of the art review. Applied Energy. 2016;**179**:350-377

[2] Amrouche SO, Rekioua D, Rekioua T, Bacha S. Overview of energy storage in renewable energy systems. International Journal of Hydrogen Energy. 2016;**41**(45):20914-20927

[3] Zhang H, Baeyens J, Caceres G, Degreve J, Lv Y. Thermal energy storage: Recent developments and practical aspects. Progress in Energy and Combustion Science. 2016;**53**:1-40

[4] Guelpa E, Verda V. Thermal energy storage in district heating and cooling systems: A review. Applied Energy. 2019;**252**:113474

[5] Kuravi S, Trahan J, Goswami DY, Rahman MM, Stefanakos EK. Thermal energy storage technologies and systems for concentrating solar power plants. Progress in Energy and Combustion Science. 2013;**39**(4):285-319

[6] Miró L, Gasia J, Cabeza LF. Thermal energy storage (TES) for industrial waste heat (IWH) recovery: A review. Applied Energy. 2016;**179**:284-301

[7] Lizana J, Chacartegui R, Barrios-Padura A, Valverde JM. Advances in thermal energy storage materials and their applications towards zero energy buildings: A critical review. Applied Energy. 2017;**203**:219-239

[8] de Gracia A, Cabeza LF. Phase change materials and thermal energy storage for buildings. Energy and Buildings. 2015;**103**:414-419

[9] Alva G, Lin Y, Fang G. An overview of thermal energy storage systems. Energy. 2018;**144**:341-378

[10] Sarbu I, Sebarchievici C. A comprehensive review of thermal energy storage. Sustainability. 2018;**10**(1):191

[11] Gil A, Medrano M, Martorell I, Lázaro A, Dolado P, Zalba B, et al. State of the art on high temperature thermal energy storage for power generation. Part 1—Concepts, materials and modellization. Renewable and Sustainable Energy Reviews. 2010;**14**(1):31-55

[12] Li G. Sensible heat thermal storage energy and exergy performance evaluations. Renewable and Sustainable Energy Reviews. 2016;**53**:897-923

[13] Lugolole R, Mawire A, Lentswe KA, Okello D, Nyeinga K. Thermal performance comparison of three sensible heat thermal energy storage systems during charging cycles. Sustainable Energy Technologies and Assessments. 2018;**30**:37-51

[14] Soprani S, Marongiu F, Christensen L, Alm O, Petersen KD, Ulrich T, et al. Design and testing of a horizontal rock bed for high temperature thermal energy storage. Applied Energy. 2019;**251**:113345

[15] John E, Hale M, Selvam P. Concrete as a thermal energy storage medium for thermocline solar energy storage systems. Solar Energy. 2013;**96**:194-204

[16] Diago M, Iniesta AC, Soum-Glaude A, Calvet N. Characterization of desert sand to be used as a high-temperature thermal energy storage medium in particle solar receiver technology. Applied Energy. 2018;**216**:402-413

[17] Mohan G, Venkataraman MB, Coventry J. Sensible energy storage options for concentrating solar power plants operating above 600°C.

Renewable and Sustainable Energy Reviews. 2019;**107**:319-337

[18] Furbo S. Using water for heat storage in thermal energy storage (TES) systems. Advances in Thermal Energy Storage Systems. 2015:31-47

[19] Simonetti M, Gentile VM. CFD optimization of large water storages for efficient cooling of high power intermittent thermal loads. Case Studies in Thermal Engineering. 2019;**14**: 100466

[20] Bott C, Dressel I, Bayer P. State-of-technology review of water-based closed seasonal thermal energy storage systems. Renewable and Sustainable Energy Reviews. 2019;**113**:109241

[21] Cárdenas B, León N. High temperature latent heat thermal energy storage: Phase change materials, design considerations and performance enhancement techniques. Renewable and Sustainable Energy Reviews. 2013; **27**:724-737

[22] Su W, Darkwa J, Kokogiannakis G. Review of solid-liquid phase change materials and their encapsulation technologies. Renewable and Sustainable Energy Reviews. 2015;**48**: 373-391

[23] Zalba B, Marın JM, Cabeza LF, Mehling H. Review on thermal energy storage with phase change: materials, heat transfer analysis and applications. Applied Thermal Engineering. 2003; **23**(3):251-283

[24] Alva G, Liu L, Huang X, Fang G. Thermal energy storage materials and systems for solar energy applications. Renewable and Sustainable Energy Reviews. 2017;**68**:693-706

[25] Khan Z, Khan Z, Ghafoor A. A review of performance enhancement of PCM based latent heat storage system within the context of materials, thermal stability and compatibility. Energy Conversion and Management. 2016;**115**: 132-158

[26] Ji C, Qin Z, Dubey S, Choo FH, Duan F. Three-dimensional transient numerical study on latent heat thermal storage for waste heat recovery from a low temperature gas flow. Applied Energy. 2017;**205**:1-12

[27] Farid MM, Khudhair AM, Razack SAK, Al-Hallaj S. A review on phase change energy storage: Materials and applications. Energy Conversion and Management. 2004;**45**:1597-1615

[28] Kenisarin MM. High-temperature phase change materials for thermal energy storage. Renewable and Sustainable Energy Reviews. 2010; **14**(3):955-970

[29] Herrmann U, Kelly B, Price H. Two-tank molten salt storage for parabolic trough solar power plants. Energy. 2004;**29**:883-893

[30] Kibria MA, Anisur MR, Mahfuz MH, Saidur R, Metselaar IHSC. A review on thermophysical properties of nanoparticle dispersed phase change materials. Energy Conversion and Management. 2015;**95**:69-89

[31] Xiao J, Huang J, Zhu P, Wang C, Li X. Preparation, characterization and thermal properties of binary nitrate salts/expanded graphite as composite phase change material. Thermochimica Acta. 2014;**587**:52-58

[32] Zhao YJ, Wang RZ, Wang LW, YuN. Development of highly conductive $KNO_3/NaNO_3$ composite for TES (thermal energy storage). Energy. 2014; **70**:272-277

[33] Peng Q, Ding J, Wei X, Yang J, Yang X. The preparation and properties of multi-component molten salts. Applied Energy. 2010;**87**(9):2812-2817

[34] Choi DH, Lee J, Hong H, Kang YT. Thermal conductivity and heat transfer performance enhancement of phase change materials (PCM) containing carbon additives for heat storage application. International Journal of Refrigeration. 2014;**42**:112-120

[35] Deng Y, Li J, Qian T, Guan W, Li Y, Yin X. Thermal conductivity enhancement of polyethylene glycol/expanded vermiculite shape-stabilized composite phase change materials with silver nanowire for thermal energy storage. Chemical Engineering Journal. 2016;**295**:427-435

[36] Li X, Zhou Y, Nian H, Zhang X, Ouyang D, Ren X, et al. Advanced nanocomposite phase change material based on calcium chloride hexahydrate with aluminium oxide nanoparticles for thermal energy storage. Energy & Fuels. 2017;**31**:6560-6567

[37] Ji C, Qin Z, Low Z, Dubey S, Choo FH, Duan F. Non-uniform heat transfer suppression to enhance PCM melting by angled fins. Applied Thermal Engineering. 2018;**129**:269-279

[38] Ji C, Qin Z, Dubey S, Choo FH, Duan F. Simulation on PCM melting enhancement with double-fin length arrangements in a rectangular enclosure induced by natural convection. International Journal of Heat and Mass Transfer. 2018;**127**:255-265

[39] Al-Jethelah M, Ebadi S, Venkateshwar K, Tasnim SH, Mahmud S, Dutta A. Charging nanoparticle enhanced bio-based PCM in open cell metallic foams: An experimental investigation. Applied Thermal Engineering. 2019;**148**:1029-1042

[40] Almsater S, Saman W, Bruno F. Performance enhancement of high temperature latent heat thermal storage systems using heat pipes with and without fins for concentrating solar thermal power plants. Renewable Energy. 2016;**89**:36-50

[41] Charvin P, Stephane A, Florent L, Gilles F. Analysis of solar chemical processes for hydrogen production from water splitting thermochemical cycles. Energy Conversion and Management. 2008;**49**:1547-1556

[42] Williams OM, Carden PO. Energy storage efficiency for the ammonia/hydrogen-nitrogen thermochemical energy transfer system. International Journal of Energy Research. 1979;**3**:29-40

[43] Pardo P, Deydier A, Anxionnaz-Minvielle Z, Rouge S, Cabassud M, Cognet P. A review on high temperature thermochemical heat energy storage. Renewable and Sustainable Energy Reviews. 2014;**32**:591-610

[44] Prieto C, Cooper P, Fernandez AI, Cabeza LF. Review of technology: Thermochemical energy storage for concentrated solar power plants. Renewable and Sustainable Energy Reviews. 2016;**60**:909-929

[45] Kato Y, Watanabe Y, Yoshizawa Y. Application of inorganic oxide/carbon dioxide reaction system to a chemical heat pump. In: Proceedings of the 31st Intersociety Energy Conversion Engineering Conference. Vol. 2. 1996. pp. 763-768

[46] Stanmore BR, Gilot P. Review—Calcination and carbonation of limestone during thermal cycling for CO_2 sequestration. Fuel Processing Technology. 2005;**86**:1707-1743

[47] Kyaw K, Matsuda H, Hasatani M. Applicability of carbonation/decarbonation reaction to high-temperature thermal energy storage and temperature upgrading. Journal of Chemical Engineering of Japan. 1996;**29**:119-125

[48] Kyaw K, Kubota M, Watanabe F, Matsuda H, Hasatani M. Study of carbonation of CaO for high temperature thermal energy storage. Journal of Chemical Engineering of Japan. 1998;**31**:281-284

[49] Ortiz C, Romano MC, Valverde JM, Binotti M, Chacartegui R. Process integration of Calcium-looping thermochemical energy storage system in concentrating solar power plants. Energy. 2018;**155**:535-551

[50] Ervin G. Solar heat storage using chemical reactions. Journal of Solid State Chemistry. 1977;**22**:51-61

[51] Criado JM, Macias M, Machin AM. Analysis of the system CaO-CO_2-H_2O for storage of solar thermal energy. Solar Energy. 1992;**49**:83-86

[52] Brown DR, Lamarche JL, Spanner GE. Chemical energy storage system for SEGS solar thermal plant. In: BATELLE. 1991

[53] Schaube F, Kohzer A, Schütz J, Wörner A, Müller-Steinhagen H. De- and rehydration of $Ca(OH)_2$ in a reactor with direct heat transfer for thermochemical heat storage. Part A: Experimental results. Chemical Engineering Research and Design. 2012;**91**:856-864

[54] Wong B, Brown L, Schaube F, Tamme R, Sattler C. Oxide based thermochemical heat storage. In: SolarPACES, Perpignan, France; 21–24 September. 2010

[55] Harries DN, Paskevicius M, Sheppard DA, Price TEC, Buckley CE. Concentrating solar thermal heat storage using metal hydrides. Proceedings of the IEEE. 2012;**100**: 539-549

[56] Yang FS, Wang GX, Zhang ZX, Rudolph V. Investigation on the influences of heat transfer enhancement measures in a thermally driven metal hydride heat pump. International Journal of Hydrogen Energy. 2010;**35**:9725-9735

[57] Sheppard DA, Paskevicius M, Buckley CE. Thermodynamics of hydrogen desorption from $NaMgH_3$ and its application as a solar heat storage medium. Chemistry of Materials. 2011; **23**:4298-4300

[58] Reiser A, Bogdanović B, Schlichte K. The application of Mg-based metal-hydrides as heat energy storage systems. International Journal of Hydrogen Energy. 2000;**25**:425-430

[59] Carden PO. Energy corradiation using the reversible ammonia reaction. Solar Energy. 1977;**19**:365-378

[60] Borowski M. Perovskite: Structure, Properties and Use. New York: Nova Science Publisher; 2011

[61] Gokon N, Yawata T, Bellan S, Kodama T, Cho HS. Thermochemical behaviour of perovskite oxides based on La_xSr_{1-x} (Mn, Fe, Co)$O_{3-\delta}$ and $Ba_ySr_{1-y}CoO_{3-\delta}$ redox system for thermochemical energy storage at high temperatures. Energy. 2019;**171**:971-980

[62] Babiniec SM, Coker EN, Miller JE, Ambrosini A. Investigation of $La_xSr_{1-x}Co_yM_{1-y}O_{3-\delta}$ (M = Mn, Fe) perovskite materials as thermochemical energy storage media. Solar Energy. 2015;**118**:451-459

[63] Albrecht KJ, Jackson GS, Braun RJ. Evaluating thermodynamic performance limits of thermochemical energy storage subsystem using reactive perovskite oxide particles for concentration solar power. Solar Energy. 2018;**167**:179-193

[64] Wang K, Markides CN. Solar hybrid PV-thermal combined cooling, heating and power systems. In: The 5th International Conference on Polygeneration (ICP 2019), Fukuoka, Japan, 15–17 May. 2019

[65] Wang K, Herrando M, Pantaleo AM, Markides CN. Technoeconomic assessments of hybrid photovoltaic-thermal vs. conventional solar energy systems: Case studies in heat and power provision to sports centres. Applied Energy. 2019;**254**:113657

[66] Chen HS, Cong TN, Yang W, Tan CQ, Li YL, Ding YL. Progress in electrical energy storage system: A critical review. Progress in Natural Science. 2009;**19**:291-312

[67] Song J, Simpson M, Wang K, Markides CN. Thermodynamic assessment of combined supercritical CO_2 (sCO_2) and organic Rankine cycle (ORC) systems for concentrated solar power. In: International Conference on Applied Energy 2019 (ICAE 2019), Västerås, Sweden, 12–15 August. 2019

[68] Pantaleo AM, Camporeale SM, Miliozzi A, Russo V, Shah N, Markides CN. Novel hybrid CSP-biomass CHP for flexible generation: Thermo-economic analysis and profitability assessment. Applied Energy. 2017;**204**:994-1006

[69] Freeman J, Hellgardt K, Markides CN. An assessment of solar-powered organic Rankine cycle systems for combined heating and power in UK domestic applications. Applied Energy. 2015;**138**:605-620

[70] Pelay U, Luo LA, Fan YL, Stitou D, Rood M. Thermal energy storage systems for concentrated solar power plants. Renewable and Sustainable Energy Reviews. 2017;**79**:82-100

[71] Al Kindi AA, Pantaleo AM, Wang K, Markides CN. Thermodynamic assessment of steam-accumulation thermal energy storage in concentrating solar power plants. In: International Conference on Applied Energy. 2019 (ICAE 2019), Västerås, Sweden, 12–15 August. 2019

[72] Bai F, Xu C. Performance analysis of a two-stage thermal energy storage system using concrete and steam accumulator. Applied Thermal Engineering. 2011;**31**:2764-2771

[73] Wang K, Pantaleo AM, Oyewunmi OA, Markides CN. Flexible PVT-ORC hybrid solar-biomass cogeneration systems: The case study of the University Sports Centre in Bari, Italy. In: The 5th International Seminar on ORC Power Systems, Athens, Greece, 9–11 September. 2019

[74] Freeman J, Guarracino I, Kalogirou SA, Markides CN. A small-scale solar organic Rankine cycle combined heat and power system with integrated thermal energy storage. Applied Thermal Engineering. 2017;**127**:1543-1554

[75] González-Roubaud E, Pérez-Osorio D, Prieto C. Review of commercial thermal energy storage in concentrated solar power plants: Steam vs. molten salts. Renewable and Sustainable Energy Reviews. 2017;**80**:133-148

[76] Islam MT, Huda N, Abdullah AB, Saidur R. A comprehensive review of state-of-the-art concentrating solar power (CSP) technologies: Current status and research trends. Renewable and Sustainable Energy Reviews. 2018; **91**:987-1018

[77] Fernández AG, Gomez-Vidal J, Oró E, Kruizenga A, Solé A, Cabeza LF. Mainstreaming commercial CSP systems: A technology review. Renewable Energy. 2019;**140**:152-176

[78] Rad FM, Fung AS. Solar community heating and cooling system with borehole thermal energy storage. Renewable and Sustainable Energy Reviews. 2016;**60**:1550-1561

[79] Alkilani MM, Sopian K, Alghoul MA, Sohif M, Ruslan MH. Review of solar air collectors with thermal storage units. Renewable and

Sustainable Energy Reviews. 2011;**15**(3): 1476-1490

[80] Khan MMA, Ibrahim NI, Mahbubul IM, Ali HM, Saidur R, Al-Sulaiman FA. Evaluation of solar collector designs with integrated latent heat thermal energy storage: A review. Solar Energy. 2018;**166**:334-350

[81] Berardi U, Gallardo AA. Properties of concretes enhanced with phase change materials for building applications. Energy and Buildings. 2019;**199**:402-414

[82] Herrando M, Pantaleo AM, Wang K, Markides CN. Solar combined cooling, heating and power systems based on hybrid PVT, PV or solar-thermal collectors for building applications. Renewable Energy. 2019;**143**:637-647

[83] Iten M, Liu SL, Shukla A. A review on the air-PCM-TES application for free cooling and heating in the buildings. Renewable and Sustainable Energy Reviews. 2016;**61**:175-186

[84] Dahash A, Ochs F, Janetti MB, Streicher W. Advances in seasonal thermal energy storage for solar district heating applications: A critical review on large-scale hot-water tank and pit thermal energy storage systems. Applied Energy. 2019;**239**:296-315

[85] Navarro L, de Gracia A, Colclough S, Browne M, McCormack SJ, Griffiths P, et al. Thermal energy storage in building integrated thermal systems: A review. Part 1. Active storage systems. Renewable Energy. 2016;**88**: 526-547

[86] Navarro L, de Gracia A, Niall D, Castell A, Browne M, McCormack SJ, et al. Thermal energy storage in building integrated thermal systems: A review. Part 2. Integration as passive system. Renewable Energy. 2016;**85**:1334-1356

Chapter 22

A Thermoelectric Energy Harvesting System

Khalid Yahya, Mohammed Salem, Nassim Iqteit and Sajjad Ahmad Khan

Abstract

Thermoelectric generators (TEGs) and their applications have gained momentum for their ability to use waste thermal energy. More contemporary technology must offer more exceptional energy-efficient applications at a lower cost. New technology must also have an ability to generate electric power through the conversion of wasted heat. The TEG has demonstrated its efficiency and how it can offer increased potential by adding an MPPT algorithm to increase the power flow while decreasing the cost of operation. The limitations can be offset by the use of lower cost manufacturing materials and automated systems in the TEG units. It is also important to note the cost per watt found in using a thermoelectric generator is estimated to be \$1/W for an installed device. To achieve this goal, the optimum operating point should be monitored by DC to DC converters. The DC to DC converters should also be driven through a generated pulse using an MPPT algorithm.

Keywords: DC to DC converter, maximum power point tracking (MPPT) algorithm, thermoelectric generator (TEG)

1. Introduction

By increasing the demand for electric energy and a greater societal sensitivity to environmental consequences, renewable energy sources attract attention. Examples of renewable sources of energy include geothermal, hydraulic, sun, and wind energy. Moreover, the production of renewable energy is quickly becoming one of the best ways for energy production (7.6% p.a.) and is currently an estimated two-thirds of the total global generation of power. This amount is expected to double by 2040 (as shown in **Figure 1**). This increase is keenly taken as a goal to cover a minimum of 30% out of the total consumed energy [1].

Due to the rapid artificial growth of the market, there is a huge rise in the amount of energy consumption and the necessity of alternative low-cost renewable energy becomes a must. Typically, producing electrical power energy necessitates both a chemical energy combustion process and usage of a generator. The cost and the efficiency of conversion systems have to be essentially taken into consideration in terms of investment and evaluation. Furthermore, the implementation of combined gas-steam cycles could increase the conversion efficiency, but this still collides with the limits imposed by the increasing installation costs. Also, increasing the overall plant efficiency without the conversion burden may be possible from the implementation of exhaust gases of the combustion process.

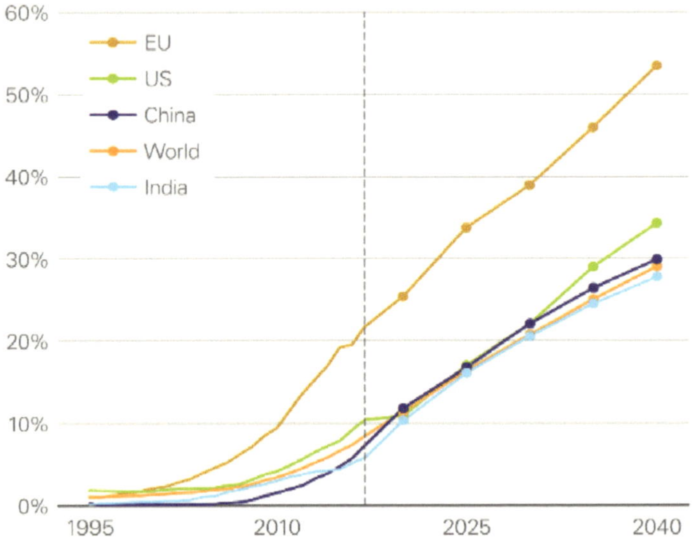

Figure 1.
Glimpses of energy future.

One of the noncostly renewable energy sources is the conversion of heat (a waste energy) where the thermal-electric energy can be automatically converted to electrical power without relying on the use of chemicals. This sort of energy is generated through a device called a thermoelectric generator (TEG).

TEG devices are cutting-edge technology, especially with recent applications such as cars exhaust, factories exhaust, and thermal panels each with a role in converting waste heat energy to applicable energy. Typically, the thermal energy production of the exhaust fumes utilized in the industries allows the insertion of a fumes/water heat exchanger to produce hot water for multi-industrial purposes. The application of such solutions go in contrast with the actual accessibility and then with investment sustainability.

Although marketing the heat for district heating is useful and attractive particularly in regard to economic values and the individual utilities' losses, but it is hard to be applied in undeveloped and urban areas where the classical energy sources appear strongly and occupy the people's priority and interest.

Another aspect has to be taken into account is the produced energy quality level; here, the heat is considered as a type of degraded energy and this thermodynamic point interprets the rational for less attraction for the market.

The sector of heat conversion attracts the huge interest of research, particularly the directly converting waste heat to electrical power. However, practical implementation is still being observed. The application of thermoelectric generators (TEG) in the field of large power generation introduces a unique step concerning the efficiency of the production system from fossil sources [2].

Feeding the power consumers through supplying a direct heat conversion is a favorable choice and attracts the global interest, especially in the realm of optimizing the production and enhancing the investment in new storage systems.

2. Thermoelectric generator (TEG)

TEG is a device for thermoelectric generation that uses semiconductors that provide an efficient system. This device is able to convert heat into electric power.

The construction of this device consists of N- and P-doped semiconductor pellets that are connected electrically through a series and further connected thermally parallel, to gain higher voltage. Also, mechanical substrate and electrical isolation are served by a ceramic layer. This device is able to operate in two different modes: an electrical power generating mode and in heating pumping mode.

In the power generation mode, the TEG can create voltage at different temperatures (ΔT). This is because the TEG is exposed on its sides according to the Seebeck effect. When the TEG is under a load, the terminals receive a current flow to the semiconductor couples while producing heat in Joules. The effect can be observed as heating while supplying more heat from the hot side to the cold side, as shown in the Peltier effect. The Peltier effect refers to the parasitic activity observed in generating power. The Peltier effect has shown the ability to increase the total thermal device conductivity.

Commercially, TEGs' come in a wide variety of sizes (a few millimeters to centimeters in size). Furthermore, the internal resistance and the current-voltage rating of the device are essentially affected by the cross-sectional area of the pellets. The wide pellets module can bear a limited number of pellets and provides a high output current, but it has small output voltage and internal resistance.

To meet the needed power level and to achieve a higher output voltage, various modules have to be connected either in series or parallel.

TEGs can be operated by utilizing different thermal energy sources. These devices possess extra advantages that other energy conversion methods lack such as light in weight, reliability, robust and have no mechanical moving parts, nonvibration, and muted operation.

In a steady-state condition, the equivalent circuit of TEG is presented as a voltage source connected to a resistance (internal resistance of TEG) in series as shown in **Figure 2**. Practically, the power generated by TEG is based on the temperature difference that imposes on its sides and the amount of the current which is drawn by the load. According to Thevenin's theorem, the maximum power can be harvested from the TEG device at any settled temperature difference, while the internal resistance of the source (TEG) should be matched with the load resistance. Characterizing the TEG and studying its performance under various temperature difference has a role in designing an efficient energy harvesting system. For the commercial TEG, the datasheet, in general, describes the operation performance under different conditions. On the other hand, it is hard to assign an optimal method of testing thermoelectric devices because of the datasheet record of the performance which is differing from the actual one.

2.1 Physical phenomena of thermos electrics

The three fundamental physical phenomena are related to a TEG's operation. The effect of Seebeck is the generated voltage when the change in temperature is sustained between two sides. The effect of Thomson is heating or the impact of cooling in the conductor that is homogeneous and it is observed at the time when the electric current is going through the electric current in the same gradient temperature direction. The effect of Joule is the heating effect that is observed in the conductor like the electric current that is passed by a conductor [3, 4].

2.1.1 Seebeck effect

It indicates that when any two dissimilar materials that are thermoelectric are connected in a single loop as shown in **Figure 3** with the help of two junctions that has the potential to maintain different temperature. In these, temperature has the

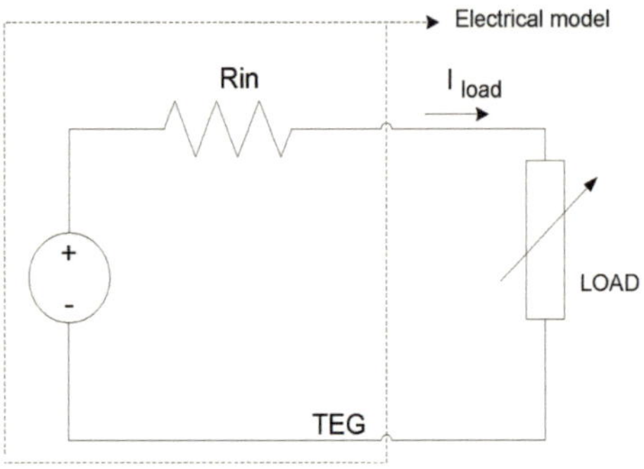

Figure 2.
Electrical model of a thermoelectric generator.

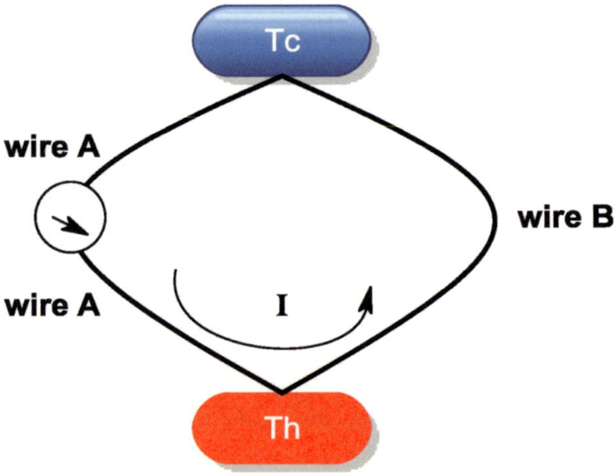

Figure 3.
Schematic of the Seebeck effect.

ability to induce the electrical current that is also known as electromotive forces is generated. These generated forces explain the voltage Seekbeck Voc and are recognized as Eq. (1) as shown below:

$$V_{oc} = \alpha_{ab}(T_H - T_C) \quad (1)$$

Here, α is the representation of the two coefficient that is not the similar conductor, and (TH-TC) represent the different temperatures between the stated junction. In the open circuit, the voltage is proportional to a difference in temperature and it depends on the material of conduction type with the fact that it is not the temperature function along with the distribution of conductor.

Usually, a semiconductor is used as a material that is thermoelectric. This is because such a material can show right properties that make them able to show the desired properties that also make them able to convert thermal energy in the form of electric and vice versa. Usually, impurities make the part of semiconductor, which is a phenomenon that is called doping that has the objective to increase the charge. Usually, doping is found in different two types. These are positive, p-type, and negative, n-type, as observed in **Figure 4**.

In the first type, the valance electron is removed, and in the second type of doping electrons, valence is made as the part of valence electrons. When there is a

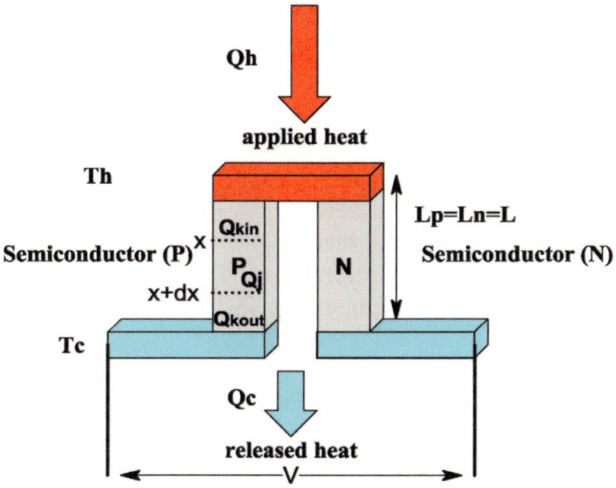

Figure 4.
Schematic of the thermocouple.

potential that semiconductor is related to two temperatures by which the flow of heat passes the pellets, then in the same direction, the charge flow carries, from the side that is the warm to the cold side. In the n-semiconductor, the tendency of electrons move to the cold side, while in the leg of p-semiconductor, the whole has the tendency to move the cold side.

This becomes the reason for the creation of gradient voltage between legs. It has not valued that the effect of Seebeck is the phenomena of contact by the connection of two nonsimilar material of thermoelectric.

2.1.2 Peltier effect

The effect of Peltier indicates that electrical current has the flows that are across from a junction between two materials that are not similar. That is why heat must continuously have subtracted or added to the junction for keeping the temperature constant. The heat rate absorbed at a junction because of the Peltier effect, which is proportional to the current flow, is presented as follows:

$$Q_{Pelter} = \pi_{ab} I \tag{2}$$

where π_{ab} (V) is the representation of relative coefficient of Peltier of two materials that are not similar, and it is a property that indicates the heating magnitude or cooling injunction that may occur in two material junctures. I (A) is the representation of electric current that uses thermocouple for flows. As the Seebeck effect, this effect is also the phenomena of contact that take place in the boundary of contact. It is essential that heating Peltier as reversible between electricity and heat that indicates heating may generate electricity that produces heat or cooling without the dissipation of energy.

2.1.3 Thomson effect

It states whenever in a wire, current flows with a gradient of temperature; then heat is absorbed across a wire that relay direction and material. The rate of heat absorbed along a length of wire is due to this effect. It can be explained as

$$Q_{Tho} = \tau I \tag{3}$$

where τ is the Thomson coefficient, while current is represented by I (A) and the temperature is by ΔT (K) which is the difference between the two ends' wire. The effect of Thomson heat can be altered and must not mix with Joule heating.

2.1.4 Joule heating

It is the effect that explains dissipated heat by the material that does not have resistance based on the availability of electric currents.

For high performance of TEGs devices, the materials should own a *significant Seebeck coefficients, high electrical conductivity, and low thermal conductivity.*

2.2 Structure of the TEG

The TEGs is constructed by various legs that are manufactured by n-type and p-type semiconductors that form thermocouples, and all these have the connection with the series as well as thermally electrically parallel. The legs of the semiconductor are linked with each other with the tabs copper and they are sandwiched

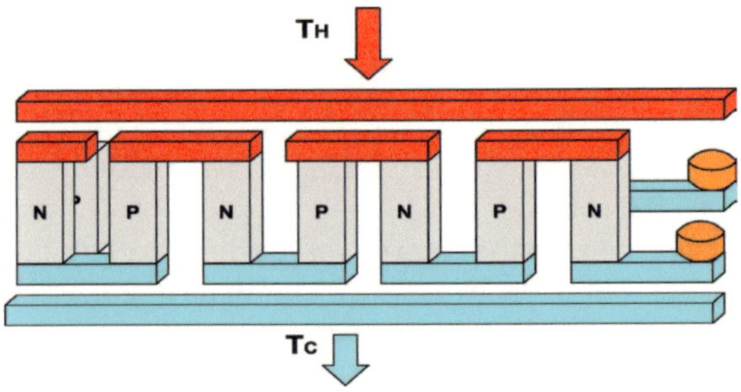

Figure 5.
TEG instructions.

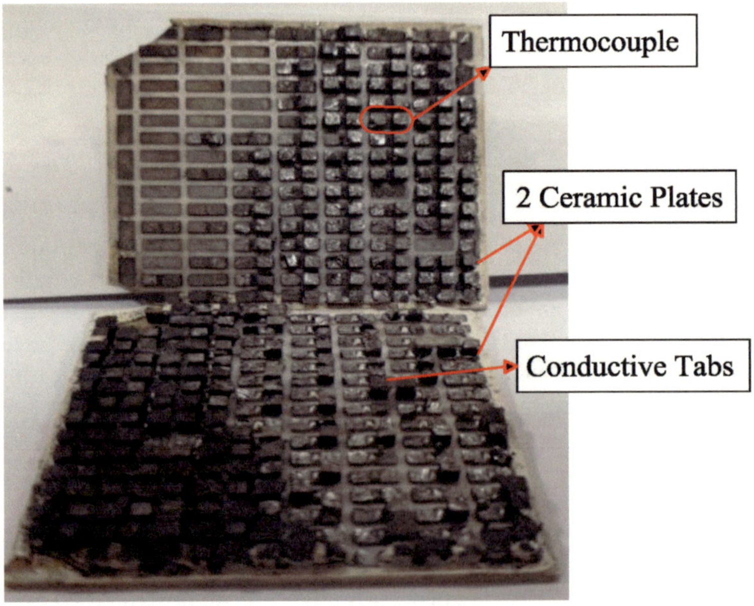

Figure 6.
Internal structure of commercial TEG.

between ceramic plates. These plates may be able to conduct heat while creating an electrical current. The diagram of schematic three-dimensional (3-D) thermoelectric generator of multi-element is shown in **Figure 5**. The waste of the heat from different sources like automobile exhaust engines, infrastructure-heating and industry and activities can be supplied to TEG's top ceramic.

In **Figure 5**, the flow of heat by the plate of ceramic as well as the conductive copper tabs before the time when it reaches on the top surface of n-type and p-type legs that are made of semiconductors that can be explained as TEG's hot side are explained. The heat flow by both legs that are semiconductor and pass by tabs that are copper-conductive and a bottom plate of ceramic, as shown in **Figure 6**. With the sink of heat, the ceramic at the bottom plate is vital due to lower temperature as compared to the top for the production of high temperature that moves toward the high output.

It explains the temperature that is applied to the bottom and top ceramic plates that are related to p- and n-types materials. The n-type and p-type materials are included in the process of designing by semiconductors to maintain the difference observed in the temperature of a hot and cold TEG. The distribution is pictorial along with the TEG legs at the difference that is conditional on temperature between cold and hot sides [5].

3. Energy harvesting system

There are several ways to increase the produced electrical energy by TEG devices, one of the ways is by boosting the produced voltage boost converters and also modeling different TEG arrays such as series, parallel, and series-parallel connections based on the required energy. A TEG device operates under various operating conditions due to the change of the temperature difference which is not easy to control it. The resistance found inside the device changes because of the difference of temperature resulting in a mismatch between the load and the TEG device. This mismatching will not allow the conditions for maximum power operation without the Maximum Power Point Tracking (MPPT) algorithm [6]. The MPPT algorithm would be applied to grab the Maximum Power Point (MPP). **Figure 7** shows the block diagram of the energy harvesting system.

MPPT algorithm improves the performance of the harvesting system and operates the system in the optimal operation point to generate the maximum power. MPPT is an algorithm which has reference variables (voltage, current) and according to variations of these variables, the algorithm will act to drive the converter through a generated pulse for tracking the maximum power point, i.e., changing the internal impedance to have matched load. Because of the linearity characteristic of the TEG devices, most of the researchers are shedding light on the MPPT algorithm. Based on an open-circuit voltage and a short-circuit current [7–9], the optimum value of the generated power is half of the short-circuit current (I_{sc})/open-circuit voltage (V_{oc}) of TEG.

A simple test on TEG's performance is normally made to provide accurate repeatable measurements and to get the TEG's electrical characteristic.

3.1 TEG characterization

A TEG device is found between the hot and cold blocks. The device contains high temperatures that are heated through the use of an electric stove. The cold block is kept cool by the use of a fan system. The TEG output is shown as a variable resistive load —-142T300 and can be identified by three separate gradients of temperature as ΔT: 80, 100, and 145°C. The test platform can be observed in **Figure 8**

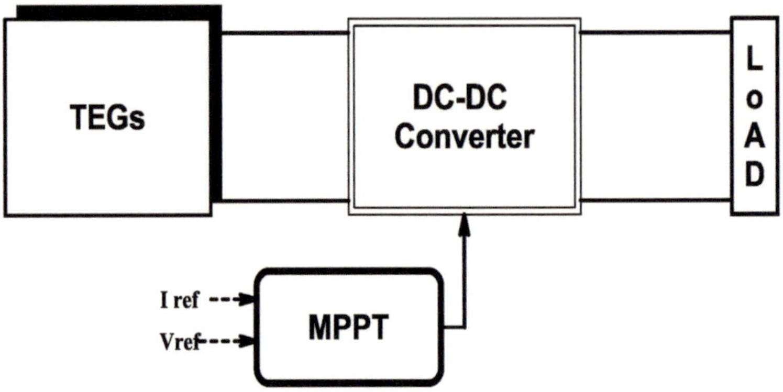

Figure 7.
The block diagram of the energy harvesting system.

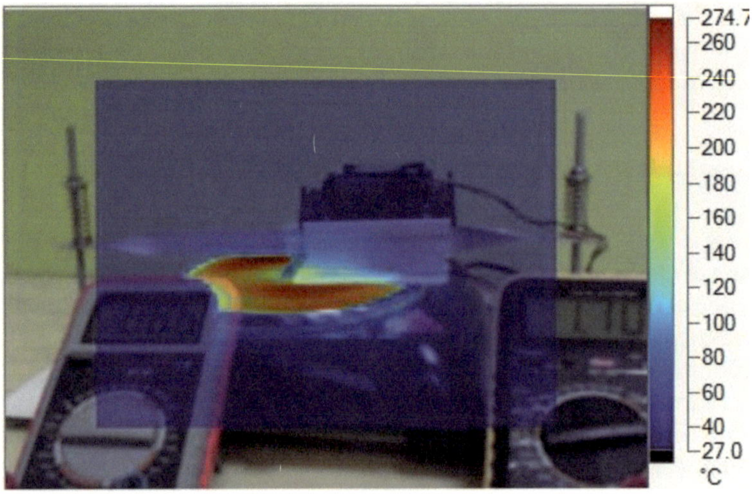

Figure 8.
The experimental test for TEG.

as it shows true measurements under different temperatures on the TEG's terminal while performing under a number of resistive loads.

In **Figure 9**, the power curve (P–I) for the device can be observed. In **Figure 10**, a straight line is shown to represent voltage versus current (V–I). The short-circuit current I_{SC} is the amount of the current when the resister of the load is equal to zero, i.e., TEG's terminals are connected together, while VOC (open-circuit voltage). Is where the voltage at the maximum and there is no load connected on the terminals. The MPP is at the medal of the curve on the peak point and that MPP can be tracked by knowing the V_{OC} or I_{SC} since the maximum power of TEG = $V_{oc}/2$ or $I_{SC}/2$ and is made when an equal amount of electrical load resistance of external circuits that are connected to the TEG equals that of the internal electrical resistance R_{int} of the TEG. The inverse slope (Rint) of the V–I line and the absolute value depends upon the temperature of the TEG while it is operating and is without a fixed value.

3.1.1 Series and parallel array configurations

T modules of thermoelectric are developed in different studies by connecting in parallel and/in series based on the required voltage and current.

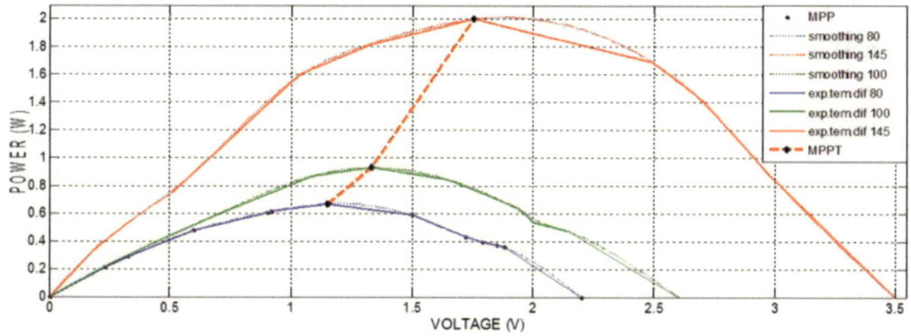

Figure 9.
P-I characterization of TEG mode (TEP1-142T300) for three different ΔT.

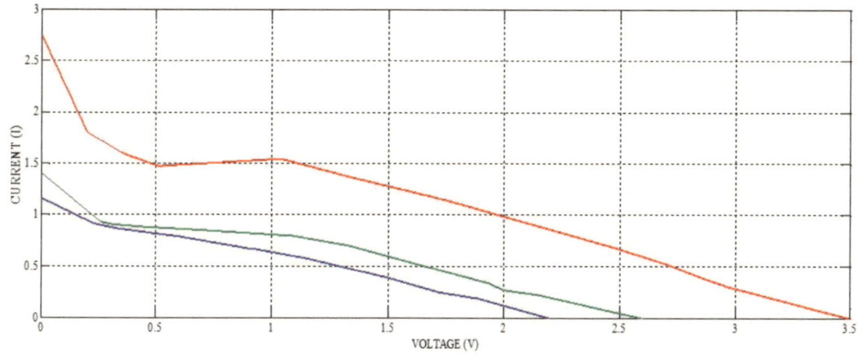

Figure 10.
V-I characterization of TEG mode (TEP1-142T300) for three different ΔT.

3.1.1.1 Series array configuration

In the series array configuration, TEGS that are three in number are represented in such a way that all are connected in series as shown in **Figure 11**. In this configuration style, voltage is kept in series as V_1, V_2, V_3,..., and V_n equal to the resistance; it is also placed in the series as R_1, R_2, R_3,..., and R_n.

In the ideal condition of the module, the performance and the outcome vary with the nonideal condition of the modules; here in ideal condition, the outcome will be equal to the ΔT. In the normal condition of the thermal balance, the equal voltage will be generated. However, on the other hand, in the normal and actual thermal condition, voltage and resistance will project different unequal values. For instance, MPP is considered as $nV_{oc}/2$.

In the nonideal thermal condition, the voltages will be in flow as Voc = V1 + V2 + V3. Eq. (4) elaborates further:

$$I_L = \left(\frac{V_{OC} - V_S}{R_1 + R_2 + R_3 + \ldots + R_n}\right) R_L \tag{4}$$

3.1.1.2 Parallel configuration

Three parallel configuration TEGs models are projected in **Figure 12**. To achieve the ideal conditions of operations TEG Models must be operate at same ΔT. The voltage and maximum power of the operations of these models are kept the same that is $I_1 = I_2 = I_3$, while TEG units will mismatch if the conditions are changed and rearranged in such a way that generates the condition of nonideal thermal conditions.

$$I_1 = \left(\frac{V_1 - V_L}{R_1}\right), I_2 = \left(\frac{V_2 - V_L}{R_2}\right), I_3 = \left(\frac{V_3 - V_L}{R_3}\right) \tag{5}$$

In the array terminals, the voltage is presented through the V_L.

3.2 MPPT converters

The design DC to DC converter connects between the TEG models and enables it to optimize the maximum power at all the time for voltage and current. Boost converters made up from the four elements are shown in **Figure 13**; it includes the inductor, MOSFET, diode, and capacitors. In DC to DC converters, it is attempted to control the output voltage to high or low values. This is accomplished through a control switch that feeds the load by a pulse voltage of different widths. The method is referred to as the pulse-width modulation (PWM). The PWM control method is one of the most commonly used applications. The method is also used in the case study of controlled DC to DC converter. To provide further explanations of this method, circuits observed in **Figure 13** are studied with the ideal switches and loads.

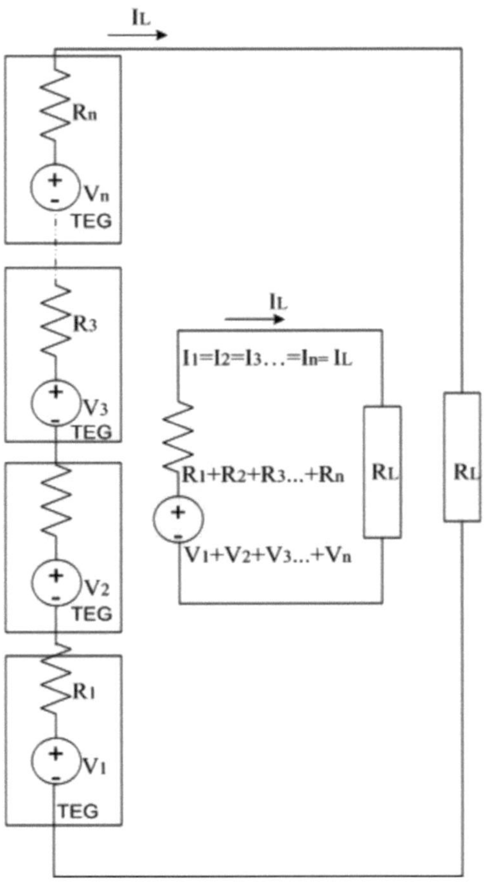

Figure 11.
TEG modules electrical schematic of the series array.

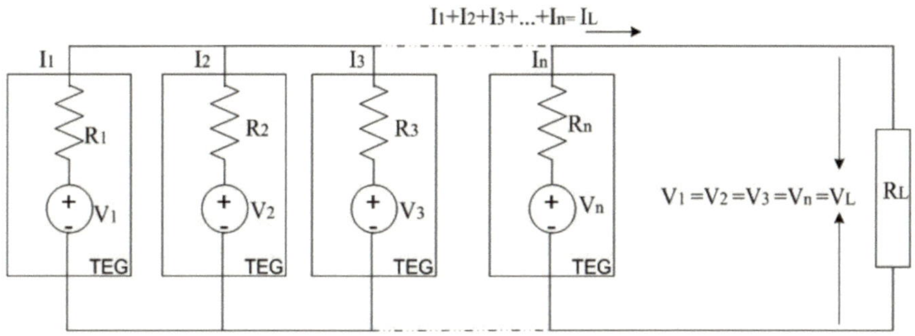

Figure 12.
Electrical schematic of parallel array of TEG modules.

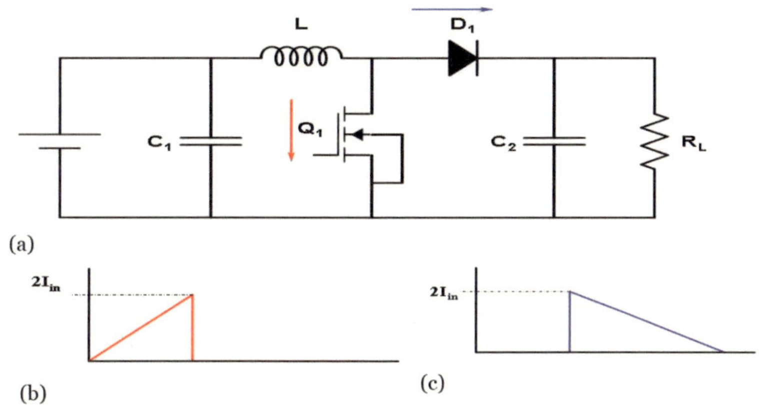

Figure 13.
(a) The schematic diagram for required components; (b) MOSFET currents; and (c) diode currents.

While it is switched "on," or the SW is closed, there is no conduction over the diode. The inductor pulls and gathers power from output of the TEG. The output of the converter as views at the load is gained through the energy inside the capacitor from previous cycles. In the "off" position, or when the SW is open, the diode is conducting energy. At this position, the inductor is load connected, and can show an output of power into the RC network. In the "off" setting, the power of the inductor is a combination of stored power and power supplied by the TEG. Part of the energy of the inductor is sent to the load and to charge the capacitor. As a result, the voltage of the output port may be observed as more than the input port.

The estimated resistance of the input of the converter is decided through the PWM signal duty cycle D and resistance of the load:

$$R_{in} \approx (1 - D)^2 R_L \tag{6}$$

3.3 Algorithm for maximum power point tracking (MPPT)

The main internal makeup of an MPPT's hardware is a DC to DC, switch-mode converter. It is more commonly found in DC supplies and DC motor drives for converting an unregulated DC input into controlled DC output, for a specified level of voltage. The MPPT converter is able to be used for a number of purposes. This includes the regulation of input voltage at PV MPP and can give a matching load during maximum power transfer.

A number of techniques have been offered for the MPPT, for the TEG in energy harvesting systems. There are a number of similarities in the methods that are found in PV systems and used in the TEG. One such algorithm is the Perturbation and Observation (P&O) MPPT technique. The P&O method is based upon two different power point observations of dynamic measurements and variables performed before and after a duty cycle of the converter that has been perturbed with both the short-circuit and the open-circuit-based algorithms. These are performed both before and after the duty cycle has been perturbed. The Incremental-Conductance (InC) algorithm has its basis in the principle in the MPP of TEG generators. There are less complicated maximum power point trackers found for thermoelectric generators [10]. A voltage-trend detection circuit is made [11], in which the gradient of the boost-converter output is drawn through the measurement of output voltage. This method allows a low-power application of the hill-climbing MPPT method. Circuit noise has been found to affect the MPPT efficiency in P&O and InC algorithms. It has also been discovered that it affects the accuracy in TEG output calculations or current gradient. As provided in [8, 9, 12, 13], increased perturbation steps will result in high steady-state oscillation around the MPP. This will reduce TEG power production. However, increased accuracy in the power or gradient measurements will also increase complexity and the consumption of power of the control unit in the MPPT. This characteristic of operation is important for low-power TEG processes for power production of the source of the TEG and power consumption of the control unit in the same order. Modern technology needs highly efficient devices with a low cost and the ability to convert waste heat into electric energy [14]. This deficiency can be covered by manufacturing low-cost materials and well-automated mechanisms for TEG units.

4. Case study: energy harvesting system

The performance of the thermoelectric campaign in different circuits configures with different temperatures. The measurement system can attach to the circuit in the experimental laboratory. This experiment is started to the test the actual environmental conditions and its effect on the real application, where the temperature is beyond the control.

As it can be seen in **Figure 14**, the TEG array ship is between a hot and cold block. In the experimenter, there are high temperature and high power provided by the direct current. The latter is air cooled by the attachment of the first device that is named in this experiment as TEG. TEG can attach to an electronic device or another desire component for the first circuit for one TEG device. A spring was used for load and weight over the TEG.

Moreover, a thermometer was used to sense the cold side (heat sink and fan unit) and the copper side as well (heat source) of the surface. The thermometer was used for obtaining the judgment of the temperature through the TEG device. The configuration of the thermoelectric array is connected to meet the required power with the different resistant load to get precise of different values according to the temperature range. The set consists in the determination of the ability of the system to operate in the MPP under different ΔTs. The need of graphing MPP under various conditions leads us to develop an MPPT algorithm.

4.1 MPPT technique development

The technique that has been developed in this case study to measure TEG's short-circuit current must be equipped with a capacitor in parallel, current sensor, switch,

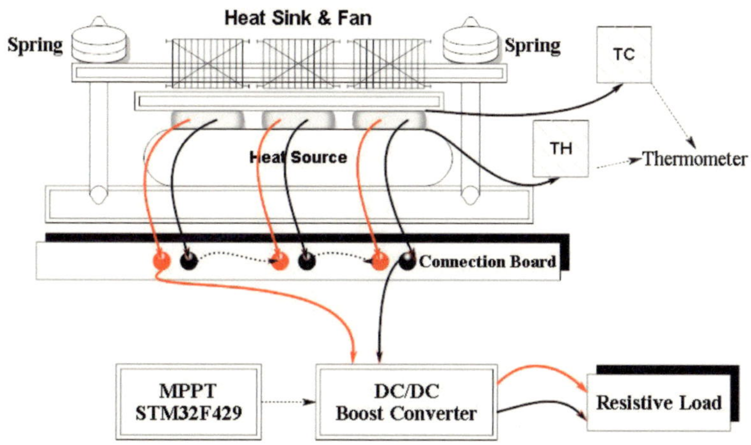

Figure 14.
The general schematic of the experiment.

and a diode in series provided with the output of TEG. This is equal to the time of input of the boost. The algorithm that was chosen for the case study was designed for smaller applications.

Short-Current Pulse (SCP) principal algorithms are dependent upon the relationship of optimum current I_m (current at maximum power point) and short-circuit current I_{SC}. This can be observed in section III as its fundamentals are seen in Eq. (7).

$$I_m = 0.5\, I_{sc} \tag{7}$$

The fundamentals of the Short-Current Pulse algorithm are shown in **Figure 15**; it performs a measurement while comparing the current and short-circuit current. If the current is more than half of the short-circuit current, the operating point of TEG is on the left side of the optimum power point. However, if the current is less, the operating point is on the right of the optimum power point. Based on this information, it can be observed that the MPP location can be confirmed and tracked [15, 16]. This operation can be done periodically in order to keep the MPP in a steady and dynamic state.

The power that is generated from the TEG module is considered to be low. There are a number of processes which should be kept to maximize output power. This makes the TEG array with n-TEG modules using a boost converter in the input voltage comparable to the circuit design. For this research, the MPPT system configuration was designed in the lab from the TEG array (a series of three connected TEG modules). These connections included a Short-Current Pulse circuit and a boost converter which was controlled by the designed MPPT linked to a resistive load. The schematic of the system can be seen in **Figure 16**.

To boost the converter when switched "on," the switch Q2 is closed without a current passing through the diodes D2. The stored energy from the output capacitor C2 will supply the load. When the circuit is open, the circuit position will cross the diode D2, while the output power of the TEG stores power through the input capacitor and inductor. While in the "off" position, the switch Q2 will not have a current passing through the switch and the diode D2. The stored energy is gathered during the "on" period of the inductor and the energy from the TEG will flow in the direction of the load. The observation made according to the previous steps is the output voltage that was found to be greater than the input.

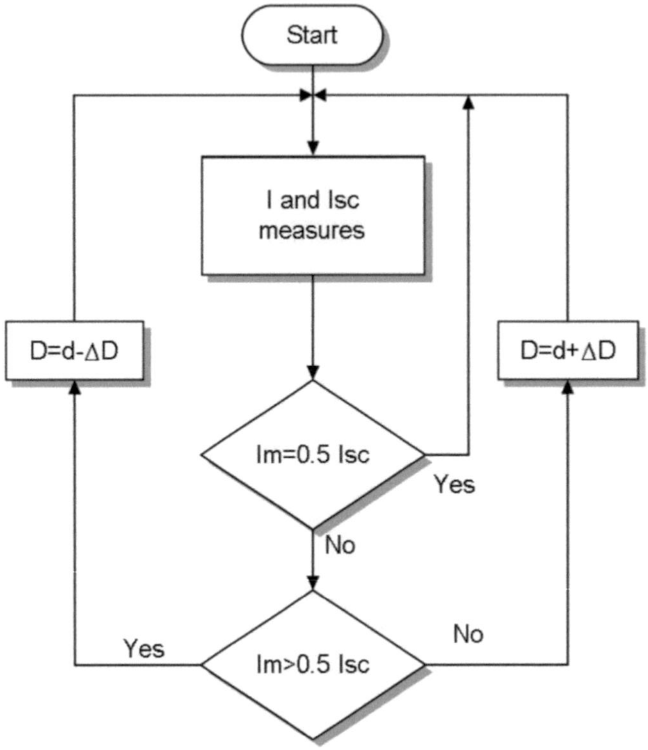

Figure 15.
Flow chart of SCP algorithm.

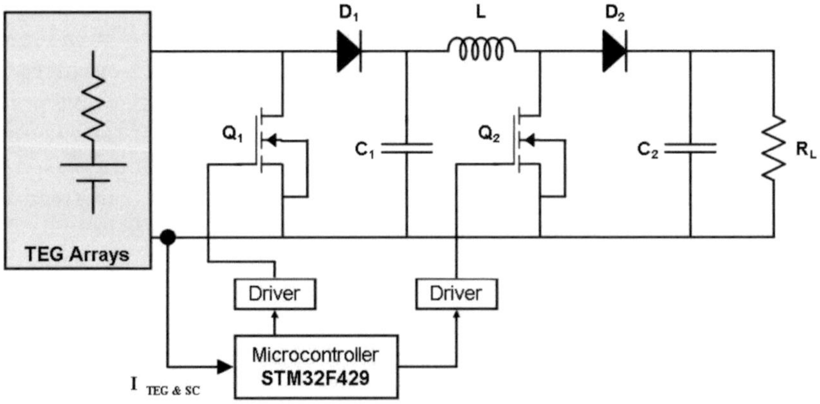

Figure 16.
The proposed energy harvesting system for TEG array.

4.2 Experimental results and discussion

The harvesting system hardware has the benefit of a lower cost but a high-performance element. This includes a high sensitivity INA250 used to sense the current; a STM32f429 discovery kit with an ARM® Cortex®-M4 core. The kit is also equipped with a monitor which provides a means to check the algorithm's accuracy of an operation. The measured parameters can also be tested to include the short-circuit current, output current of TEG, and duty and operating duty cycle of the boost converter at the MPP. In the experiment, in order to match the "on" voltage (microcontroller output voltage capability) of the Mosfet, two gate drivers

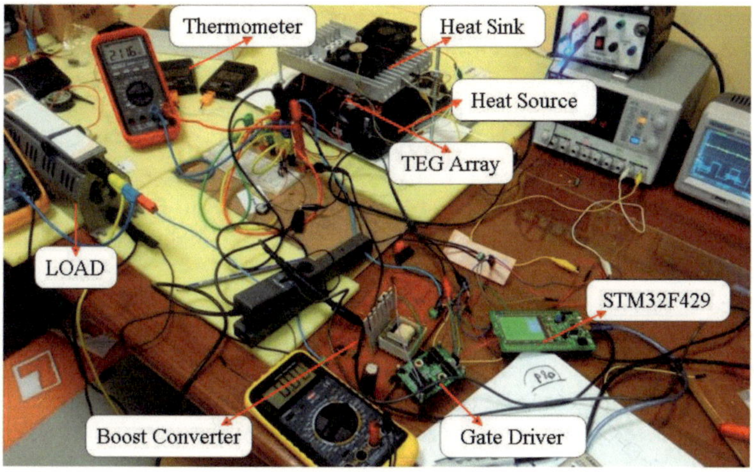

Figure 17.
Energy harvesting system prototype.

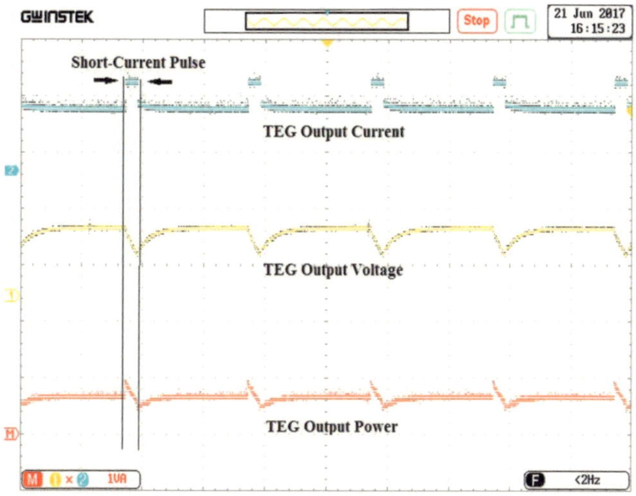

Figure 18.
The output power, current, and voltage of three series TEP1-142T300 with STM32F429 microcontroller.

were created to drive the Mosfet of SCP while boosting the converter circuit. The prototype of the system implementation can be observed in **Figure 17**.

As shown in **Figure 18**, the TEG takes a period of time in order to become stable when compared to response time of the boost converter. The SCP algorithm is discovered to have a high tracking performance and an efficiency of nearly 100% in measuring the short-circuit current at three times per second in order to make adjustments to the operating point while controlled by a generated pulse of a maximum power point tracking algorithm. It has been observed that the short-circuit current pulses can increase and decrease per second through the change set into the values of the code implemented in STM32f429.

5. Conclusion

Thermoelectric generators (TEGs) has opened the way for many new researchers by improving the coverage of renewable energy sources. There is no negative impact on the environment because TEGs have no pollution and are noiseless. It also has the advantages of low maintenance and the use of common waste heat energy. Temperature plays an important role in the TEG energy conversion process. In TEG performance, there are several other factors that have a significant effect, such as the effect of temperature difference ΔT between the two plates. To provide the desired voltage and/or current, the TEG device can be electrically connected in series and/or parallel form. TEGs can be applied at different temperature environments, different thermal forces, and different electrical loads. As a consequence of the operating conditions for each device in the TEG array, the changes of the temperature difference can affect on the system operation. For this reason, it is necessary to ensure that the energy conversion in the thermoelectric system is precisely monitored on a device basis and that maximum energy can be drawn from the system. For this operation, it is inevitable to use a power converter controlled by the Maximum Power Point Tracker (MPPT) algorithm between TEG and load.

Author details

Khalid Yahya[1*], Mohammed Salem[2], Nassim Iqteit[3] and Sajjad Ahmad Khan[4]

1 Faculty of Engineering and Architecture, Department of Mechatronic Engineering, Istanbul Gelisim University, Istanbul, Turkey

2 School of Electrical and Electronic Engineering, Universiti Sains Malaysia, Nibong Tebal, Malaysia

3 Department of Electrical Engineering, Palestine Polytechnic University, Hebron, Palestine

4 Faculty of Electrical and Electronics Engineering, Electronics and Communication Engineering Department, Istanbul Technological University, Istanbul, Turkey

*Address all correspondence to: khalid.omy@gmail.com

References

[1] Available from: https://www.bp.com/en/global/corporate/energy-economics/energy-outlook/demand-by-fuel/renewables.html [Accessed: 22 November 2019]

[2] LeBlanc S. Thermoelectric generators: Linking material properties and systems engineering for waste heat recovery applications. Sustainable Materials and Technologies. 2014;**1-2**:26-35. DOI: 10.1016/j.susmat.2014.11.002

[3] Rowe DM. CRC Handbook of Thermoelectrics. Boca Raton: CRC Press; 1995

[4] Siouane S, Jovanović S, Poure P. Fully electrical modeling of thermoelectric generators with contact thermal resistance under different operating conditions. Journal of Electronic Materials. 2016;**46**(1):40-50. DOI: 10.1007/s11664-016-4930-6

[5] Sales BC. Thermoelectric devices: Refrigeration and power generations with no moving parts. In: Encyclopedia of Materials: Science and Technology. Elsevier; 2001

[6] Yahya K, Zeki Bilgin M, Erfidan T. The effect of temperature variations over thermoelectric generator efficiency. In: International Engineering Research Symposium UMAS2017 Duzce; 2016. 2017. pp. 192-200

[7] Min G, Rowe D. Experimental evaluation of prototype thermoelectric domestic-refrigerators. Applied Energy. 2006;**83**(2):133-152. DOI: 10.1016/j.apenergy.2005.01.002

[8] Yahya K, Bilgin MZ, Erfidan T, Çakir B. Improving the performance of the MPPT for thermoelectric generator system by using Kalman filter. In: 2018 5th International Conference on Electrical and Electronic Engineering (ICEEE); Istanbul. 2018. pp. 129-132

[9] Im J, Wang S, Ryu S, Cho G. A 40 mV transformer-reuse self-startup boost converter with MPPT control for thermoelectric energy harvesting. IEEE Journal of Solid-State Circuits. 2012;**47**(12):3055-3067. DOI: 10.1109/jssc.2012.2225734

[10] Paraskevas A, Koutroulis E. A simple maximum power point tracker for thermoelectric generators. Energy Conversion and Management. 2016;**108**:355-365. DOI: 10.1016/j.enconman.2015.11.027

[11] Huang CJ, Chen WC, Ni CL, Chen K-H, Lu CC, Chu YH, et al. Thermoelectric energy harvesting with 1 mV low input voltage and 390nA quiescent current for 99.6% maximum power point tracking. In: 2012 Proceedings of the European Solid State Circuits Conference, ESSCIRC. European Solid-State Circuits Conference; 2012. pp. 105-108. DOI: 10.1109/ESSCIRC.2012.6341267

[12] Yahya K, Bilgin MZ, Erfidan T. Practical implementation of maximum power tracking based short-current pulse method for thermoelectric generators systems. Journal of Power Electronics. 2018;**18**(4):1201-1210

[13] Kwan T, Wu X. TEG maximum power point tracking using an adaptive duty cycle scaling algorithm. Energy Procedia. 2017;**105**:14-27. DOI: 10.1016/j.egypro.2017.03.274

[14] Alsharif MH, Yahya K, Geem ZW. Strategic market growth and policy recommendations for sustainable solar energy deployment in South Korea. Journal of Electrical Engineering and Technology. 2020;**15**:803-815. DOI: 10.1007/s42835-019-00331-6

[15] Al-Betar M, Alomari O, Abu-Romman S. A TRIZ-inspired bat algorithm for gene selection in cancer classification. Genomics. 2020;**112**(1):114-126

[16] Yahya K. Termoelektrik generatörler için değişken koşullar altında etkin maksimum güç noktası izleyicisinin geliştirilmesi [PhD]. Kocaeli University; 2018. Available from: http://dspace.kocaeli.edu.tr:8080/xmlui/handle/11493/1016

Chapter 23

Optimization Techniques of Islanded Hybrid Microgrid System

Sk. Shezan Arefin

Abstract

The utilization of energy is increasing day by day, it is an unavoidable truth. This large demand for energy cannot be satisfied by the conventional power sources alone. Sustainable power sources, for example, solar and wind turbine-based energy system are the most effective and both economically and environmentally feasible. The Hybrid renewable energy system (HRES) is a recent concept in the field of sustainable development which joins at least two renewable power sources like wind turbine, solar module and other inexhaustible sources such as ocean energy, fuel cell etc. This chapter starts with an audit by addressing the possible difficulties that can occur while a solar power plant and the wind farm will be integrated together to supply power to the main grid or in a islanded manner. This chapter gives an overview of the optimization techniques that can improve the integration of the hybrid systems to the grid as well as with the islanded load. This review also sheds significant light on the techniques to improve the equality of a grid integrated solar-wind hybrid energy system. It also includes the critical findings on the strategies that are necessary to build an efficient grid-connected and islanded solar-wind hybrid energy system.

Keywords: renewable energy, solar energy, wind energy, HRES, optimization, simulation

1. Introduction

Replace power is an essential factor for industrialization, urbanization and budgetary development of any nation. There are diverse sorts of conventional and non-conventional energy sources used to generate power. Solar and wind energy systems are a standout among the most prominent sources of energy [1]. The use of solar-based and wind energy system has turned out to be progressively well known because of particular and condition agreeable nature. The field of solar and the wind energy has experienced a pivotal advancement for late decades in this in all cases usage of independent to utility characteristic solar-wind frameworks. Sun-powered and wind vitality framework works much of the time in remain independent or network associated mode, yet the reasonability of these sources are less an immediate after-effect of the stochastic thought of sun-powered and wind assets [2]. The hybrid renewable power sources with framework coordination defeat this downside of being whimsical in nature. Half-breed sustainable power source framework (HRES) is a blend of unlimited and consistent vitality source, and it might besides consolidate no under two maintainable power sources that work in remain independent or system-related mode. The HRES that joins sun-based arranged and wind vitality sources works in two modes: concurrent and consecutive [3]. In synchronous mode, the sun-controlled and wind vitality frameworks produce vitality at the same time, while in a successive mode, they create a control on the other hand. The essential credits of HRES are to consolidate under two maintainable power age advancements to make fitting utilization of their working qualities and to get efficiencies higher than that could be gotten from a solitary power source. This paper shows the viewpoint of solar-wind half and half practical power source framework covering issues of achievability such as illustrating, controlling, change method, unwavering quality and power nature of the framework [4, 5]. The worldwide infiltration of sustainable power source in control systems is expanding quickly particularly for sunlight-based photovoltaic (PV) and wind systems. The sustainable power source meant around 19% of the total energy utilization worldwide in 2012 and kept on ascending amid the year 2013, according to the 2014 inexhaustible worldwide status report [6, 7]. The report highlighted that interestingly the PV establishment limit on solar and wind control is actually irregular and can make specialized difficulties to the network control supply particularly when the amount of solar and wind power integration increases or the grid is not sufficient to deal with fast changes in generation levels. Furthermore, if sun or wind is utilized to supply energy to a stand-alone system, energy storage system ends up with a guaranteed supply of power [8, 9]. The extent of the energy storage relies upon the irregularity level of the solar or wind energy systems. This paper gives a survey of difficulties and opens doors for combined PV and wind energy system. The paper audits the fundamental research works to identify the ideal measuring configuration, control hardware topologies and control strategies for both solar and wind energy systems [7, 10, 11]. The use of vitality has transformed into a fundamental matter of stresses in the latest decades because of speedy addition in vitality request. Furthermore, natural issues of customary vitality sources, for instance, ecological changes and an Earth-wide temperature boost, are perpetually compelling us for elective utilization of vitality assets. As indicated by the reports released by World Health Organization (WHO), brisk and twisting impacts of characteristic change prompt the death of 160,000 individuals for reliably and the rate is assessed to be increased by 2020 [12]. Environmental change causes disastrous incidents, for example, surges, dry spells and astounding changes in climate temperature. Also, there are a couple of sicknesses predominantly malaria, malnutrition, diarrhea and so forth getting to be plainly pandemic among the communities. One of the disasters was reported in 2003 which assaulted European nations and caused demising of 20 thousand individuals while remained $10 billion misfortunes in the plant part. Directly, conventional vitality sources constitute ideally around 80% of overall vitality usage [12]. Those fundamental compelling reason on substitute the energy sources might have been set off concurred for discovering nuclear noteworthiness in the mid-twentieth century, which might climb to ten on twenty times more than oil-based stock. In any case, there are a few confinements related to atomic energy generations. For instance, atomic fusion is the exposure of uranium and thorium minerals, which are viewed as petroleum derivatives also. Also, atomic plants are at present accessible just in vast scale power generations. Subsequently, for cooking, heating and other small-scale applications, sustainable power sources are the best solutions yet [13].

Reasonable power sources like daylight-based, wind, biomass, hydropower and tidal vitality are promising CO_2-free decisions. Despite the general recognition with central purposes of practical power source utilization, this wellspring of vitality contributed to just around 1.5% of world vitality as of 2006. The example is assessed to rise to 1.8% of each by 2030 [14, 15].

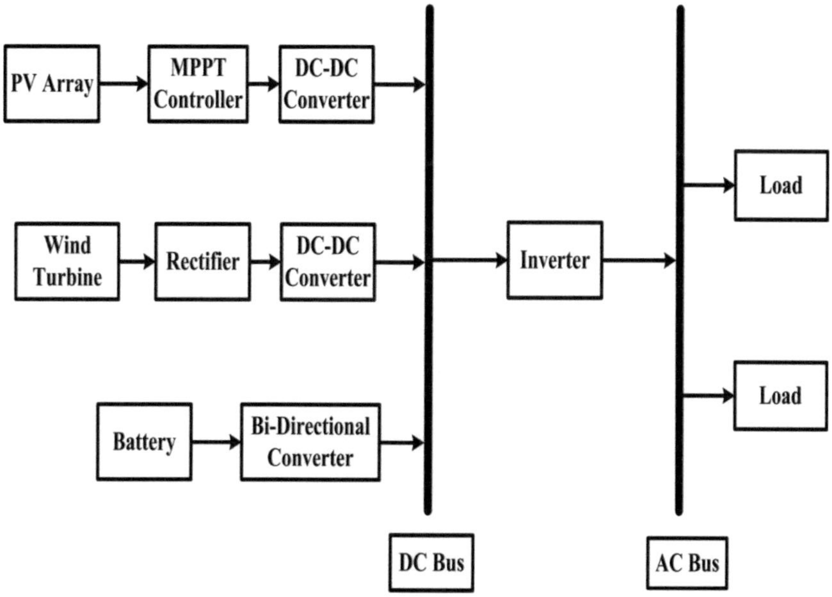

Figure 1.
A complete block diagram of a hybrid PV-wind renewable energy system [16, 17].

Figure 1 shows the complete block diagram of a hybrid PV-wind renewable energy system. The basic diagram gives the general idea of the architecture of an HRES.

2. Evolutionary of optimization methods used in HRES

The developmental calculation is one of the subclasses of counterfeit consciousness technique to decrease the streamlining trouble. The use of various algorithms and optimization methods are involved with the evolutionary technique in simulating the HRES.

2.1 Optimization with genetic algorithm

Genetic algorithm is an indiscriminate chase and enhancement framework guided by the profound quality of the common hereditary framework. Genetic algorithm is being widely used for optimization of various characterizations. Applications of renewable energy technologies with GA are gaining popularity day by day. Solar and wind energy, as well as the geothermal energy technology, is gaining more popularities in terms of simulation and optimization techniques. GA is a versatile heuristic hunt calculation in view of the transformative thoughts of normal choice and hereditary qualities. Liu et al. clarified an elitist methodology of ideal measuring of independent mixture of PV-wind control frameworks utilizing hereditary calculation with loss of energy supply likelihood as a limitation and limits the aggregate capital of the whole framework [18].

Zhao et al. used a hereditary calculation for PSO to see the most ideal capacity model of a daylight-based breeze blend with inexhaustible essentialness structure with quick overall combining [9]. Zhou et al. utilized GA with the plan to also supply fluctuating load that was spotted in the region from Xuzhou, China [19]. From the demonstration it can be said that the framework might convey vitality for a standalone establishment for a worthy cost. Fadale et al. Produced ideal measuring of mixture of PV-wind battery framework utilizing fuzzy logic with GA, which chooses the ideal number about PV panels, WT and battery units, and further, GA is utilized to assume perfect control design from guaranteeing an off-network blend of inexhaustible essentialness system [20]. Turku et al. created true energy planning from claiming an off-grid HRES utilized to warming plus lighting previously, a prototype private house. In this chapter, double-coded GA may be utilized as opposed to arranged basic straight customizing on minimizing that operational unit cost of HRES [21].

2.2 Optimization with particle swarm optimization

Particle swarm optimization (PSO) is a populace built static streamlining act enthused toward social conduct for winged animal flocking or fish schooling, the place winged animal flocking or fish educating is the aggregate movement of a vast number for self-impelled substances. Zhao et al. offered enhancing the PSO algorithm for the ideal limit plan from claiming a free wind-PV mixture force supply framework. The technique of PSO completely depended on various types of data set based on the practically implemented parameters. Various researchers are using PSO as the perfect prediction tool for various purposes. An integration of the hybrid energy framework ability ideal configuration may be an ordinary non-linear incorporated basic streamlining issue. An algorithm is anticipated and also tried to look into the framework found in an island. The drawbacks and strength of the algorithm can talk about shortages gotten turns out its possibility and effectiveness [22]. Dehgan et al. created the majority positive position measuring of hydrogen built wind/PV plant permitting for dependability indices by applying a molecule swarm streamlining. Wang et al. altered the PSO calculation by creating a multicultural plan of the coordinated force area framework. Affectability contemplates may be likewise conveyed out to analyze those effects from claiming different framework parameters on the general configuration execution [23]. Measuring of solar-wind renewable vitality framework will be done by Sanchez et al. [24]. Also, an evolutionary calculation strategy known as PSO may be utilized with those expenses of framework as a goal capacity. PSO calculation is utilized by Ardakani et al. for ideal measuring of the system's part [25]. Thus as an eventual outcome, the perfect number of PV modules, wind turbine and battery close to the inverter capacity is obtained. Bansal et al. made streamlining of blend PV/wind/batteries imperativeness structure using multi-target atom swarm streamlining (MOPSO) [26]. An objective work to cosset of the mixture framework may be framed that incorporates starting costs; yearly working costs furthermore support expenses. Because of that many-sided nature for mixture renewable vitality framework with nonlinear essential analytics planning, MOPSO is used to take care of the issue. Those enhanced PSO might stay away from that alternative of a neighborhood base trap. Keyrouz et al. formed a bound together MPPT to controlling a mixture of wind-solar and power module framework [27]. A following calculation method has been constructed Bayesian lion's share of the information joined with swarm quickness is used concerning delineation a streamlining, looking calculation by Amer et al. For limiting the hardship from guaranteeing imperativeness with a satisfactory choice of the era taking secured close by contemplating those incidents the center of planning additionally ask for sides to diminish the cost [28]. The conclusion reveals that PSO execution may be quicker, and it is also skilled at providing a streamlining design that recovers around 10% of the aggregate expense of the incorporated framework. Borhanazad et al. created streamlining for microgrid framework utilizing MOPSO and also discovered those are the best setup for the mixture framework utilizing energy oversaw economy algorithm [29].

2.3 Optimization with fuzzy logic

Nema et al. have shown a perfect money-related operation from asserting sharp network by fluffy pushed quantum developmental technique. Adhikari et al. Elucidated examination, design moreover control of an independent composed non-regular essentialness change system for the perspective of those fluffy basis control strategy toward detecting those dc voltage besides current yield about sun situated and the revised yield voltage about enduring request short of breath quick present (PMBLDC) generator driven by a breeze turbine. Chakraborty et al. created insight activity speculation operation of a sharp matrix using soft moved quantum developmental framework [30].

2.4 Optimization with artificial neural network

Neural networking system will be an interconnectedness get together of re-enacted neurons that occupy a logical model or computational model for information changing in light of an association with approximate estimation. Fidalgo et al. connected an artificial neural networking (ANN)-based way to deal with applying preventive control techniques to an extensive hybrid energy system. ANNs are a fundamental piece, which is superior to standard measurable strategies in the progressive security design population and additionally evaluates that degree of security [31]. Martin et al. suggested a neural system control technique for multi-energy regular dc transport mixture energy supply eventually Tom's perusing examining that uniqueness for sun oriented energy, wind vitality. Levenberg-Marquardt algorithm that is interfaced with the neural system may be utilized and the energy component is acquainted in the preparation. During et al. created a mixture model to an hourly numeral for PV-wind renewable vitality framework that can also be utilized with the computational brain-power of PSO for registering distinctive definitions of the cost slip [32].

2.5 Optimization with game theory in HRES

For the 1838 diversions seemed too commercial concerns writing furthermore toward that time cornet also how produced An model which may be In light of oligopoly estimating and production, yet the model needs the detriment of the confined player with methodologies that just included amount or cost choice. Afterwards, in 1944, John Von Neumann and Oskar Morgenstern demonstrated the fundamental hypothesis of amusements which came out with the investigation for all the perplexing arrangements and methodologies [33].

Ogino et al. introduced non-cooperative amusement methodology on an electric generating framework, which will be identified with government-funded cooperation and asset supply. A principle drawback for this framework may be that it does not provide for whatever control component of the whole framework [34].

On the other hand Saad et al. introduced a dimension for a coalitional amusement hypothesis for agreeable micro-grid conveyance networks of sun-based panels, wind turbines and so forth throughout this way, observing and stock arrangement of all instrumentation may be enhanced [35]. Baeyens et al. furthermore Bitar clarified those wind vitality amassed for those assistance of a coalitional amusement methodology.

Many researchers brought different thinking about the degree on which a gathering for a wind constrains creators could abuse those truthful benefits of amassed and aggregate peril might be figured by coalitional entertainment approach. Mei and Wang made the approach for facilitated control structures. Beguilement

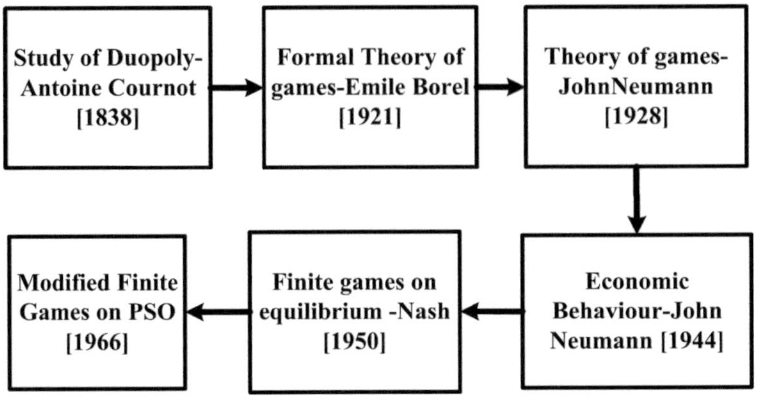

Figure 2.
Revolution of game theory for hybrid energy system [39].

Authors	Optimization techniques	Parameters optimized	Highlights
Tao Ma et al. [41]	Genetic algorithm	PV-pump turbine	The genetic algorithm (GA), along with Pareto optimality concept, is used for the system techno-economic optimization: to maximize power supply reliability and minimize system lifecycle cost simultaneously
Ogunjuyigbe et al. [42]		PV-wind-diesel	
Alireza Askarzadeh et al. [43] and Akbar Maleki et al. [44]	Particle swarm optimization	PV-wind-battery	In order to find the optimal values of the variables, particle swarm optimization (PSO) and some of its variants are proposed. Due to the non-linearity and non-convexity of the sizing problem, PSO, which is an efficient population-based heuristic technique, can be a good candidate
Caballero et al. [45]	Loss of power supply probability (LPSP)	Hybrid PV-wind	The proposed method allows the possibility to supply excess power generated by the HES to the utility grid at a fixed sales price or through a net metering scheme. The system and design method is proposed to represent a viable alternative for grid-only power supply in rural/remote communities
Kerim Karabacak et al. [46]	Artificial neural network	Wind-PV	Artificial neural network applications of PV, WECS and hybrid renewable energy systems, which consist of PV and WECS, are presented. Usage of neural network structures in such types of systems has been motivated
Kefayat et al. [47]	Artificial bee colony	Wind energy	An efficient point estimate method (PEM) is employed to solve the optimization problem in a stochastic environment by using artificial bee colony
Abdolvahhab Fetanat et al. [48]	Ant colony algorithm	PV-wind	Ant colony optimization for continuous domains application to reservoir (ACOR)-based integer programming is employed for size optimization in a hybrid photovoltaic (PV)-wind energy system. ACOR is a direct extension of ant colony optimization (ACO)

Authors	Optimization techniques	Parameters optimized	Highlights
Chedid et al. [49]	Linear programming optimization	Solar-wind	Unlike the traditional 2D simulation, a novel modeling of a trade-off surface in 3D space is presented where the knee set is determined using the minimum distance approach. Robust and inferior plans are segregated based on their frequent occurrence in the conditional decision set of each future and hedging analysis to reduce risk is performed in order to assign alternative options in case risky futures occur

Table 1.
Various types of optimization techniques applied by the researcher for hybrid solar-wind energy system [40].

speculation is displayed in this paper to demonstrate the modeling of hybrid energy system, which includes wind turbines, PV boards, etc. with the storage batteries. For both islanded and grid connected system can be agreeable amusement hypothesis models need aid fabricated by bringing wind turbines, PV panels furthermore stockpiling batteries Concerning illustration players and their life cycle salary similarly as payoffs [36].

Alaska et al. recommended an amusement built stochastic modifying on produce ideal offering methodologies should amplify the aggregate benefits about wind furthermore routine energy makers have done both the vitality advertise along with a reciprocal save market, the place the save value may be settled the middle of wind also customary energy makers Toward utilizing diversion hypothesis [37]. Vikas et al. created and examined diversion hypothesis based on cornet's model for sun-based wind HRES and also gives some key choices that will figure out the best reaction starting with sun oriented furthermore wind vitality framework. An amusement methodology is utilized to investigate the sun-based wind mixture framework. Different strategies, for example, diversion hypothesis logic, Nash equilibrium, and non-cooperative diversion hypothesis, can also be compared [38]. **Figure 2** demonstrates the revolution of the game theory algorithm for the hybrid energy system.

Table 1 describes various types of optimization techniques applied by the researcher for a hybrid solar-wind energy system.

3. Software tools used in modeling and simulation of hybrid systems

3.1 HYBRID 2

The HYBRID 2 software tool may be a Recreation device around that means will furnish a versant model for those specialized foul furthermore monetary examination of renewable mixture vitality framework. The gadget may have been developed in NREL, Canada, a long time ago in 1993. This altering model uses both those time courses of actions and besides a quantifiable strategy ought to survey the operation of the inexhaustible energy system. This permits the model to focus on long haul execution at the same time still taking into account the impact for fleeting variability for sun-based and wind information. The modification of the system components, control furthermore dispatch alternative could make modeled with clients specified occasion when steps. HYBRID 2 comprises various sorts for vitality dispatch methodologies scrutinized by grain (1995). HYBRID 2 will be an extensively approved model. However, the specialized foul precision of the model is high, yet the model will be unable to streamline the vitality framework. The HYBRID 2 code utilizes an easy-to-understand graphical client interface (GUI) and also a glossary of terms ordinarily connected with mixture energy frameworks. HYBRID 2 will be additionally bundled with a library of supplies that should aid the client over planning mixture of energy frameworks. The combination of all equipments and components is economically accessible also by utilizing the manufacturer's determinations. Besides this, the library incorporates example energy frameworks with the activities that will be followed as a standard applicable framework by the clinets. Two levels from claiming yield are provided: a rundown judgment and a nitty-gritty time step toward occasion when step depiction about energy streams. A graphical repercussions interface (GRI) does not consider not a difficult and in-depth survey of the point by point reproduction outcomes [50].

3.2 PVSYST

PVSYST 4.35 (2009) made by Geneva College in Switzerland is an item package for the examination measuring, re-building, and data examination of complete PV structures. It gives the exact PV module requirement and battery size, etc. The product offers an extensive database of PV segments, meteorological destinations, a specialist framework, and a 3D apparatus for close shading nutty-gritty investigations. This product is situated for planners, designers, and analysts and holds exceptionally supportive apparatuses for training. It incorporates a broad logical help, which clarifies in detail the techniques and the models utilized. The device plays out the database meteorological and segments administration. It gives additionally a wide decision of general sun-based instruments (sun-oriented geometry, meteorological on tilted planes, and so forth.), and an intense mean of bringing in genuine information measured on existing PV frameworks for close correlations with mimicked esteems [51].

3.3 INSEL

Incorporated reproduction condition and a graphical performing dialect (INSEL) is a redoing made by College of Oldenburg, Germany, in which age models can be delivered utilizing existing pieces in the reasonable supervisor HP VEE with a few mouse clicks (Quick and Holder, 1988) [52]. The increase of structures like on-compose PV generators with MPP tracker and inverter, for example, winds up being essentially a representation work out. This thing fortifies the coordinator with database for PV modules, inverters, warm gatherers, and meteorological parameters [53]. Inside and out, INSEL offers a programming interface for the improvement of the square library. The major favored viewpoint of this model is the adaptability in the affecting system to model and arrangement showed up diversely in connection to spread contraptions with settled associations. An issue is that INSEL does not perform structure change; regardless, it finishes or even replaces the exploratory research office for a feasible power source framework since parts can be interconnected like constant [54].

3.4 SOLSIM

Reproduction and advancement demonstrate for sustainable power source systems (SOLSIM) (Schaffrin, 1998) are made at Fachhochschule Konstanz, Germany [55]. SOLSIM is a multiplication gadget that draws in clients to set-up, dissect, and streamline off-cross area and arrange related hybrid sun-based significance structures. It has requested particular models for PV, wind turbine, diesel generator, and battery sections and moreover for biogas and biomass showing [56]. SOLSIM programming pack contains diverse devices: the basic augmentation program called SOLSIM; the unit to overhaul the tilting reason for PV module called SolOpti; the unit to figure life cycle cost called SolCal; and the unit to repeat wind generators called SolWind. This program is also inadequate to locate the ideal size of hybrid vitality framework for any region on the techno-economical-related ground [57].

3.5 WATSUN-PV

WATSUN-PV 6.0 (Tiba and Barbosa, 2002) made by College of Waterloo, Canada, is a program proposed for hourly redirection of different PV frameworks: self-governing battery go down, PV/diesel half-breed, utility cross-segment-related structure and PV water pumping structure re-foundations [58]. The autonomous battery modules go down and PV/diesel cross-breed structure re-authorization modules are especially whole; on the other hand, the module that courses of action with PV water pumping systems just allows the examination of set-ups using DC electric motors, which is not a plan a great part of the time used nowadays. The showing structures for daylight-based radiation, PV course of action, and the battery are extremely pointed by point and revived [59]. The model utilized for DC engines is a reasonable relationship between the voltage and current given by the show and the torque and spruce speed of the engine. WATSUN-PV 6.0 has a library containing data on PV modules, batteries, inverters, and diesel and fuel generators. The database rejects data on engines or pumps [60].

3.6 PV-DESIGN PRO

The PV-DESIGN PRO augmentation program (arranging and displaying PV structure: a guide for installers, fashioners and artists, 2005) consolidates three assortments for impersonating self-ruling framework, cross-section-related framework, and PV pump framework [61]. For autonomous structures, a spare generator and a breeze generator can be consolidated into the PV system, and a shading examination can be finished. The system can be progressed by contrasting the individual parameters. Bitty lumpy calculations are performed for working data and traits twists. The module and air database are to a great degree broad. This program is endorsed for the PV systems that have battery accumulating. Diversion is finished on hourly introduction. The use of PV design expert is that its database starting at now consolidates most information required for PV structure diagram [62].

3.7 RAPSIM

RAPSIM (Pryor et al., 1999) or remote range control supply test system is a PC showing program made at the Murdoch College Vitality Exploration Foundation, Australia. It is proposed to re-establish elective power supply choices, including PV, wind turbine, battery, and diesel structure [63, 64]. The customer picks a system and working method from a few pre-portrayed options and streamlining is searched for by changing part sizes and by attempting distinctive things with the control factors that settle on-off cycles of the diesel generator. Battery developing effect isn't considered in this model [65].

3.8 RETScreen

RETScreen is created and kept up by the Administration of Canada through Normal Assets Canada's Canmet Vitality inquiry about focus in 1996. RETScreen programming is competent to ascertain the vitality productivity, sustainable power source and hazard for different sorts of sustainable power source, vitality proficient innovations and furthermore break down the cost capacity of the planning framework and cross-over framework plausibility (RETScreen, 2009). RETScreen working depends on Microsoft exceed expectations programming instrument. The fundamental qualities of this product are to limit the ozone-depleting substance discharge, life cycle cost and vitality era [66].

3.9 PHOTO

The PC code PHOTO (Manninen et al., 1990) made at the Helsinki University of Technology in Finland imitates the execution of feasible power source structure, including PV-wind cross-over vitality framework plot. A move down diesel generator can in the same manner be fused into the structure plan. The dynamic technique influenced utilizations to revise framework part models addressing piece cooperation and hardships in wiring and diodes [67]. The PV group can work in the craziest power mode with trade subsystems. Various control procedures can likewise be considered. Single sub-structure models can be checked against veritable estimations. The model can be used to reproduce diverse system outlines unequivocally and evaluate structure execution, for instance, essentialness streams and influence hardships in PV show, wind generator, fortification generator, wiring, diodes, and most extraordinary impact point GPS reference point, inverter, and battery. A cost investigation should be possible by PHOTO. This code has an office to make a stochastic atmosphere period database in the circumstances where hourly data are not available. The re-authorization occurs to differentiate well and consider execution of a PV test plant [68].

3.10 SOMES

The PC show SOMES (recreation and advancement demonstrate for sustainable power source frameworks) made at the University of Utrecht Netherlands (RETScreen, 2009) can duplicate the execution of economic power source structures [69]. The vitality framework can incorporate monetary power sources (PV bundles, wind turbines), diesel generator, a cross-section structure, battery putting away, and several sorts of converters. An examination of the outcomes gives particular and money related execution of the framework and the reliability of vitality supply. The diversion is done on hourly explanation behind the re-establishment time of, for example, one year. Hourly ordinary power conveyed by sun fuel and wind structure is settled. Hourly results are accumulated for the re-establishment period. The gathered regards are used to evaluate particular and productive execution of the system. The model contains a streamlining routine to search for the system with most lessened power cost, given the customer's desired steady quality level [70].

3.11 HOMER

HOMER is a PC display that rearranges the assignment of assessing plan choices for both off-lattice and network-associated control frameworks for remote, independent, and circulated era (DG) applications [71, 72]. HOMER was produced by the National Renewable Energy Laboratory (NREL, USA). HOMER's enhancement and affectability investigation calculations enable us to assess the monetary and specialized achievability of a substantial number of innovation alternatives and represent the variety in innovation expenses and energy asset accessibility [73]. HOMER models both customary and renewable energy advances, for example, PV, wind turbine, keep running of-stream hydro-power, diesel or biogas generator, power device, utility framework, battery bank, small scale turbine, and hydrogen stockpiling. HOMER performs re-enactment for the majority of the conceivable framework arrangements to decide if a set-up is achievable [74]. At that point, HOMER gauges the cost of introducing and working expense of the framework, and presentations a rundown of designs arranged by their life cycle cost. This device offers an intense user interface and exact estimating with a detailed investigation of the framework. **Figures 3–5** demonstrate the optimization model for solar-wind hybrid energy system, PV-diesel hybrid energy system and wind-diesel hybrid energy system by HOMER subsequently [75–80]. Shezan et al. demonstrated PV-wind, PV-diesel and wind-diesel hybrid energy systems by using HOMER for techno-economic assessment issues [46, 58, 71, 81–87].

3.12 RAPSYS

RAPSYS (adjustment 1.3) was created in the University of New South Wales, Australia, in the year 1987. This item can emulate a broad assortment of unlimited structure sections that may be joined into a half and half option vitality technique. The item can be used just by the people who are authorities in remote domain control supply structure [91]. RAPSYS does not update the degree of portions. The customer is required to pre-describe the structure set-up. The multiplication recommends the turn ON and OFF timings of diesel generator. RAPSYS does not

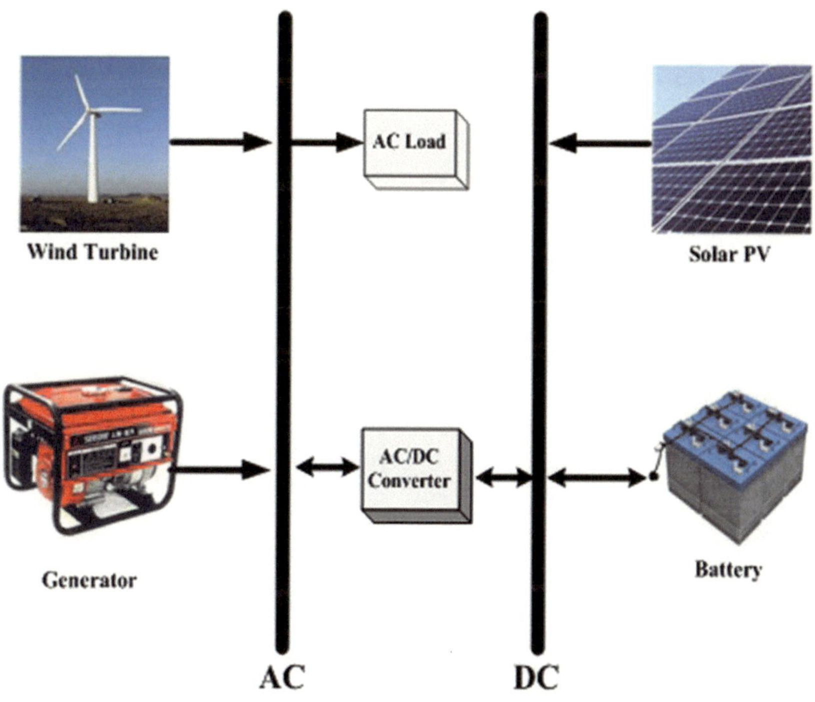

Figure 3.
Optimization model of the solar-wind hybrid energy system by HOMER [88].

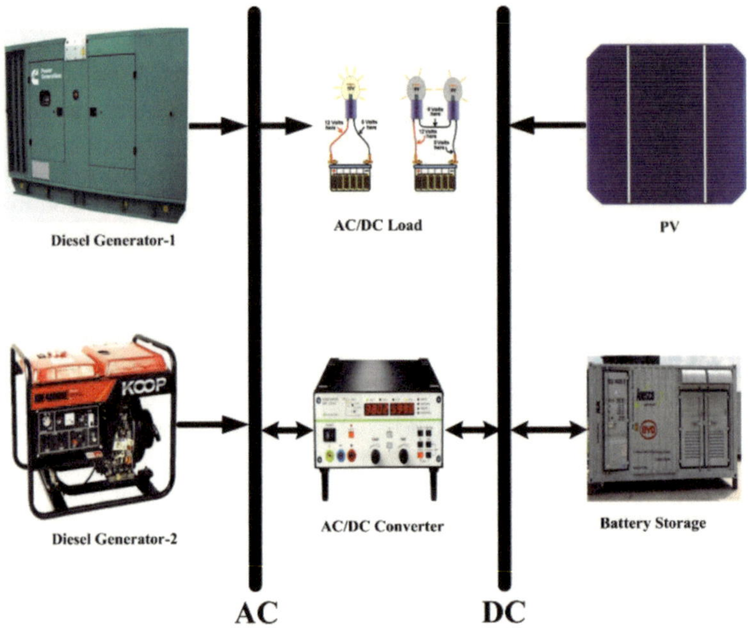

Figure 4.
Optimization model of PV-diesel hybrid energy system by HOMER [89].

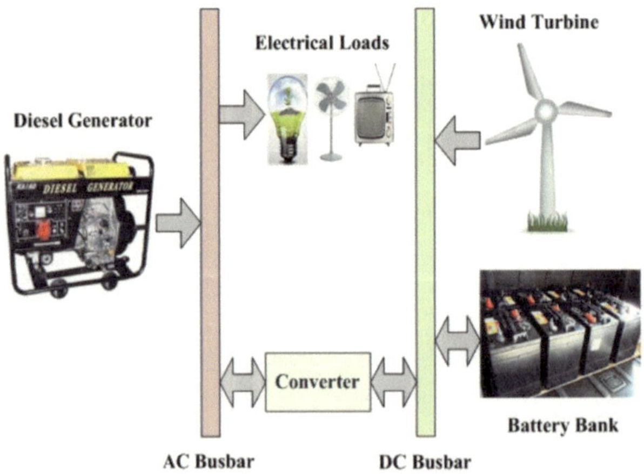

Figure 5.
Optimization model of the wind-diesel hybrid energy system by HOMER [90].

determine the life cycle COE structure; in any case, it is skilled to give positive information about the working cost of the system [92].

3.13 ARES

A refined entertainment program for assessing and streamlining of self-overseeing cross-breed imperativeness systems (ARES) made at University of Cardiff, the UK chooses if a structure meets the pined for reliability level while meeting the wander spending design in perspective of customer demonstrated cost data (Morgan et al., 1995; Morgan et al., 1997). This program, not in the least likely the predominant piece of different cross-breed re-order program, predicts the battery state of voltage (SOV) rather than its state of charge (SOC). The nonappearance of data concerning charge qualities and temperature impacts is considerably more glaring. It would be extremely valuable if information saving money with such parameters were to be made accessible. The battery creating and its impacts on structure execution have not been tended to as an important piece of this program. The precision and unflinching nature of the generation coming to fruition depend generally on the exactness of the drawing in parameters [93, 94].

3.14 PVF-CHART

The PC program PVF-CHART (Klein and Beckman, 1993; arranging and introducing PV framework: a guide for installers, modelers and fashioners, 2005) made by F-plot PC programs is sensible for wanting the entire arrangement as conventional execution of PV utility interface structure, battery putting away framework, and structure without interface or battery stockpiling. It is a sweeping PV system examination and design program. The program offers month-to-month ordinary execution examinations to each hour of the day [95].

In **Table 2**, a brief idea of the invention of each optimization software for HRES has been given. From a brief discussion of each and every software tool used for renewable energy system modeling and optimization, we have come to know the common things of the optimization and sizing methodology. Among all the potential software, HOMER has been used most extensively and continuously because HOMER has already brought all the renewable energy resources together with each technical part of each renewable energy resource [96].

Software	Invented country	Invented year
SOLSIM	Germany	1987
RAPSYS	Australia	1987
SOMES	Netherlands	1987
PHOTO	Finland	1990
PVSYS	Switzerland	1992
HYBRID 2	Canada	1993
PVF-Chart	USA	1993
HOMER	USA	1993
ARES	UK	1995
INSEL	Germany	1996
RETScreen	Canada	1996
RAPSIM	Australia	1997
WATSUN-PV	Canada	2002
PV-DESIGN-PRO	USA	2005
REopt	USA	2007
CREST	USA	2009–2010
PVSYST	Switzerland	2009

Table 2.
Software tools have been used for conducting the optimization in HRES [97].

4. Mathematical modeling

4.1 Equations for diesel generator

HOMER assumes the fuel curve of a generator is a straight line with a y-intercept. The following equation has been used for the generator's fuel consumption:

$$F = F_O Y_{gen} + F_1 P_{gen} \tag{1}$$

where F_o is the fuel curve intercept co-efficient, F_1 is the fuel curve slope, Y_{gen} is the rated capacity of the generator (kW) and P_{gen} is the electrical output of the generator (kW).

To calculate the generator's fixed cost of energy, the following equation can be used:

$$C_{gen,fixed} = C_{om,gen} + \frac{C_{rep,gen}}{R_{gen}} + F_O Y_{gen} C_{fuel,eff} \tag{2}$$

where $C_{om,gen}$ is the operation and maintenance cost per hour, $C_{rep,gen}$ is the replacement cost, R_{gen} is the generator lifetime in hours, F_O is the fuel curve intercept co-efficient, Y_{gen} is the capacity of the generator (kW) and $C_{fuel,eff}$ is the effective price of fuel per quantity of fuel.

The marginal cost of energy can be calculated by using the following equation for the generator:

$$C_{gen,mar} = F_1 C_{fuel,eff} \tag{3}$$

where F_1 is the fuel curve slope in quantity of fuel per hour Kilowatt-hour and $C_{fuel,eff}$ is the effective price of fuel.

4.2 Equations for cost of energy

To calculate the optimum cost of energy for a hybrid system in HOMER, the following equation has been used:

$$COE = \frac{C_{ann,tot}}{E_{prim} + E_{def} + E_{grid,sales}} \tag{4}$$

where $C_{ann,tot}$ is the total annualized cost, E_{prim} is the total amounts of primary load, E_{def} is the total amounts of deferrable load and $E_{grid,sales}$ is the amount of energy sold to the grid per year.

4.3 Equations for net present cost

To calculate the total net present cost, the following equation has been used:

$$C_{NPC} = \frac{C_{ann,tot}}{CRF(i, R_{proj})} \tag{5}$$

where $C_{ann,tot}$ is the total annualized cost, i is the annual real interest rate, R_{proj} is the project lifetime and $CRF(.)$ is the capital recovery factor.

4.4 Operating cost

$$C_{operating} = C_{ann,tot} - C_{ann,cap} \tag{6}$$

where $C_{ann,tot}$ is the total annualized cost [$/yr], $C_{ann,cap}$ is the total annualized capital cost [$/yr] and $C_{ann,cap}$ is $C_{ann,cap} \times CRF\ (.)$

To calculate the CO_2 emissions from the hybrid energy system, the following supporting equations have been introduced:

$$tCO_2 = 3.667 \times m_f \times HV_f \times CEF_f \times X_c \tag{7}$$

where tCO_2 is the amount of CO_2 emissions, m_f is the fuel quantity (litter), HV_f is the fuel heating value (MJ/L), CEF_f is the carbon emission factor (ton carbon/TJ) and X_c is the oxidized carbon fraction.

Another factor must be considered that in 3.667g of CO_2 contains 1g of carbon.

4.5 Cost function

To obtain the optimal values for the size and number of each power generating module, the following optimization problem needs to be solved:

$$\min_{a,b,c,d,e,f \in N^0} \begin{pmatrix} w_1(a \cdot LCOE_{PV,5} + b \cdot LCOE_{PV,18} + c \cdot LCOE_{PV,30} + d \cdot LCOE_{WT} + e \cdot LCOE_{DG} + f \cdot LCOE_{BT}) \\ +w_2(a \cdot NPC_{PV,5} + b \cdot NPC_{PV,18} + c \cdot NPC_{PV,30} + d \cdot NPC_{WT} + e \cdot NPC_{DG} + f \cdot NPC_{BT}) \\ +w_3(e \cdot GHG_{DG}) \end{pmatrix} \tag{8}$$

subject to:

- The power generated constraint: The power generated from each source $P_{gen(i)}$ must be less than or equal to the maximum capacity of the source:

$$P_{gen(i)} \le P_{gen,\,max\,(i)} \tag{9}$$

- Power balance constraint: The total power generation of the energy sources must cover the total load demand (P_{demand}), the total power losses (P_{losses}) and storage power ($P_{storage}$) being used.

$$\sum_i P_{gen(i)} \ge P_{demand} + P_{losses} + P_{storage} \tag{10}$$

In Eq. (8), $a, b, c, d, e, f \in \mathbb{N}^0$ refers to the numbers (integer including 0) of 5 kW PV module, 18 kW PV module, 30 kW PV module, wind turbine, diesel generator and battery unit, respectively, whereas, w_1, w_2 and w_3 are used to weigh the importance of the individual criterion.

4.6 Solar energy

The electrical energy generation as an output of a photovoltaic system can be estimated by a worldwide formulated equation as follows [32]:

$$E = A \times r \times H \times PR \tag{11}$$

where E is the energy (kWh), A is the total solar panel area (m^2), r is the solar panel yield (%), H is the annual average solar radiation on tilted panels (shadings not included) and PR is the performance ratio, the coefficient for losses.

The annual average solar radiation data can be collected from the meteorological department.

The maximum power from a solar panel can be calculated using the following equation:

$$P_{mp} = \eta_{PV} G_{\beta} A \tag{12}$$

where A is the surface area of the PV module, P_{mp} is the maximum power from a solar panel, η_{PV} is the efficiency of the silicon-based PV cell and G_{β} is the global horizontal solar irradiation.

Figure 6 represents the basic design of a PV-based HRES where charge controller, general control strategy, storage management system and load demand are also included.

4.7 Wind energy

A rotor consolidating of at least two cutting edges mechanically joined to an electrical generator can create wind's motor vitality that can be caught by the breeze turbines. From the condition, it can be discovered that the mechanical power caught from twist speed by a breeze turbine can be figured as shown by [34]:

$$P_m = 0.5 \rho A C_p v^3 \tag{13}$$

where P_m is the maximum power from wind turbines, ρ is the air density, A is the swept area, C_p is the power coefficient and v is the average wind speed.

The most astounding estimation of the power coefficient has been favored as 0.59 hypothetically. It is dependent on two variables: the tip speed ratio (TSR) and the pitch angle. The pitch angle alludes to the angle in which the turbine blades are aligned with respect to its longitudinal axis. The linear speed of the rotor to the wind speed has been addressed by TSR.

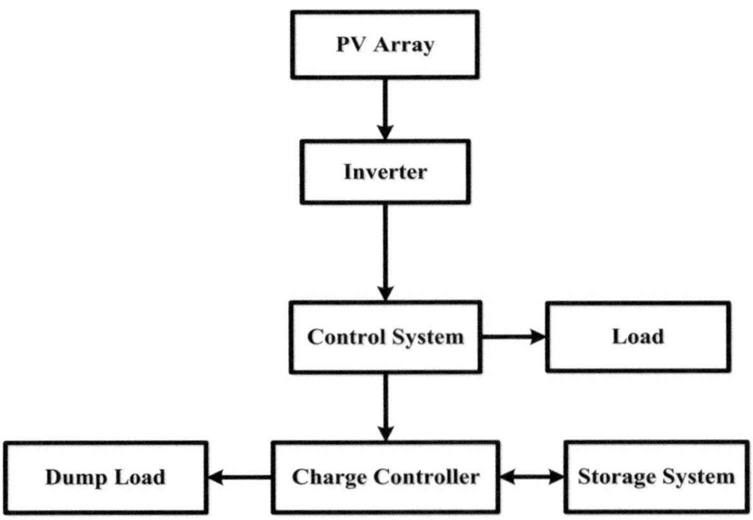

Figure 6.
Design of a PV hybrid energy system [33].

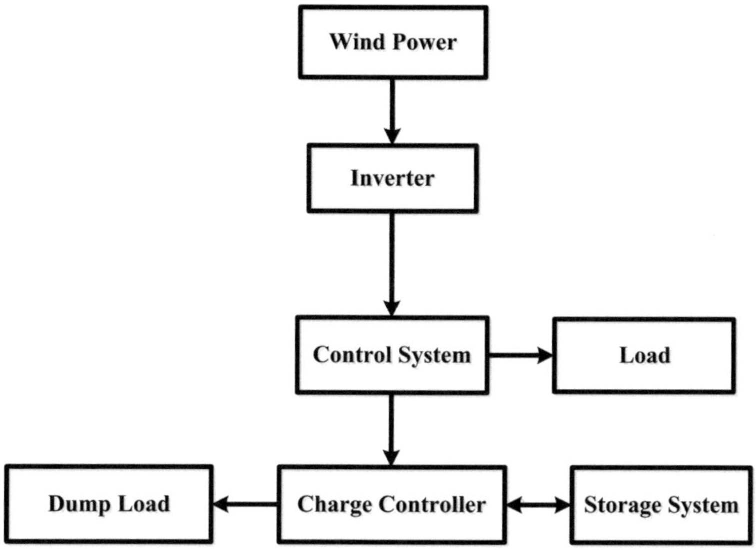

Figure 7.
Architectural design of a hybrid wind-based energy system [35].

$$TSR = \lambda = \frac{\omega R}{v} \quad (14)$$

For the practical planning of wind turbines, two distinct conditions have been actualized, for example, the scope of 0.4–0.5 for the quick breeze turbines and the scope of 0.2–0.4 for generally slower wind turbines.

Figure 7 represents the basic design of a wind-based HRES where charge controller, general control strategy, storage management system and load demand are also included.

5. Conclusion

The application, demand and popularity of solar-wind hybrid renewable energy system are on the rise for the last few decades due to the advancement in human civilization. Due to the change of new advancements in the field of solar-wind HRES, some technical and in addition unwavering quality issues have emerged. From the case study, the identical results have been found as the levelized COE and NPC. Moreover, the greenhouse gas (GHG) emissions also reduced noticeably in comparison to the conventional power plants. These issues will be repaid by some future research in the individual field.

6. New research ideas for the future generation solar-wind HRES

Some necessary steps need to be taken for to figure out the correct area also environmental condition, the site with site information may be needed, which is was troublesome to get for remote area. Subsequently, it is necessary to create an accurate streamlining method; furthermore, the geological product should figure out the possibility of claiming sun-based radiation and also wind speed.

There need to aid distinctive sorts from claiming to measure systems constantly utilized, for example, iterative method, counterfeit consciousness method, at these strategies do not speak to exact dynamic execution of sun oriented furthermore wind vitality framework. Henceforth, it will be important to create a unit measuring system that dodges multifaceted nature over the planning of the framework and clarifies superbly recurrence reaction of the Iframework done by changing execution criteria.

It will be essential to create an incorporated multilevel controlling method that evades that possibility intricacy about correspondence framework and also huge calculation load that will be subjected with a solitary side of the point disappointment.

Still, the voltage fluctuation problem due to the inconsistent average wind speed and solar radiation has been solved with a practical and feasible approach. With a proper control strategy, an alternate energy storage system can be introduced to solve this problem.

Conflict of interest

The authors declare no conflict of interest.

Author details

Sk. Shezan Arefin
Department of Electrical and Biomedical Engineering, School of Engineering, RMIT University, Melbourne, Australia

*Address all correspondence to: shezan.ict@gmail.com; shezan.arefin@rmit.edu.au

References

[1] Sun S, Liu F, Xue S, Zeng M, Zeng F. Review on wind power development in China: Current situation and improvement strategies to realize future development. Renewable and Sustainable Energy Reviews. 2015;**45**:589-599

[2] Suganthi L, Iniyan S, Samuel AA. Applications of fuzzy logic in renewable energy systems—A review. Renewable and Sustainable Energy Reviews. 2015;**48**:585-607

[3] Hu X, Johannesson L, Murgovski N, Egardt B. Longevity-conscious dimensioning and power management of the hybrid energy storage system in a fuel cell hybrid electric bus. Applied Energy. 2015;**137**:913-924

[4] Fathima AH, Palanisamy K. Optimization in microgrids with hybrid energy systems—A review. Renewable and Sustainable Energy Reviews. 2015;**45**:431-446

[5] Sinha S, Chandel S. Review of recent trends in optimization techniques for solar photovoltaic–wind based hybrid energy systems. Renewable and Sustainable Energy Reviews. 2015;**50**:755-769

[6] Bhandari B, Lee K-T, Lee G-Y, Cho Y-M, Ahn S-H. Optimization of hybrid renewable energy power systems: A review. International Journal of Precision Engineering and Manufacturing-green Technology. 2015;**2**:99-112

[7] Mahesh A, Sandhu KS. Hybrid wind/photovoltaic energy system developments: Critical review and findings. Renewable and Sustainable Energy Reviews. 2015;**52**:1135-1147

[8] Arul P, Ramachandaramurthy VK, Rajkumar R. Control strategies for a hybrid renewable energy system: A review. Renewable and Sustainable Energy Reviews. 2015;**42**:597-608

[9] Zhao H, Wu Q, Hu S, Xu H, Rasmussen CN. Review of energy storage system for wind power integration support. Applied Energy. 2015;**137**:545-553

[10] Wang X, Palazoglu A, El-Farra NH. Operational optimization and demand response of hybrid renewable energy systems. Applied Energy. 2015;**143**:324-335

[11] Kalinci Y, Hepbasli A, Dincer I. Techno-economic analysis of a stand-alone hybrid renewable energy system with hydrogen production and storage options. International Journal of Hydrogen Energy. 2015;**40**:7652-7664

[12] Ma T, Yang H, Lu L, Peng J. Technical feasibility study on a standalone hybrid solar-wind system with pumped hydro storage for a remote island in Hong Kong. Renewable Energy. 2014;**69**:7-15

[13] Alotto P, Guarnieri M, Moro F. Redox flow batteries for the storage of renewable energy: A review. Renewable and Sustainable Energy Reviews. 2014;**29**:325-335

[14] Kim D, Resasco J, Yu Y, Asiri AM, Yang P. Synergistic geometric and electronic effects for electrochemical reduction of carbon dioxide using gold-copper bimetallic nanoparticles. Nature communications. 2014;**5**:4948

[15] Wang C, Chen J, Zou J. Decomposition of energy-related CO_2 emission in China: 1957–2000. Energy. 2005;**30**:73-83

[16] Hosenuzzaman M, Rahim N, Selvaraj J, Hasanuzzaman M, Malek A, Nahar A. Global prospects, progress,

policies, and environmental impact of solar photovoltaic power generation. Renewable and Sustainable Energy Reviews. 2015;**41**:284-297

[17] Mi Z-F, Pan S-Y, Yu H, Wei Y-M. Potential impacts of industrial structure on energy consumption and CO_2 emission: A case study of Beijing. Journal of Cleaner Production. 2015;**103**: 455-462

[18] Liu Z, Geng Y, Lindner S, Zhao H, Fujita T, Guan D. Embodied energy use in China's industrial sectors. Energy Policy. 2012;**49**:751-758

[19] Chen S, Chen B. Urban energy consumption: Different insights from energy flow analysis, input-output analysis and ecological network analysis. Applied Energy. 2015;**138**:99-107

[20] Liu Z, Guan D, Wei W, Davis SJ, Ciais P, Bai J, et al. Reduced carbon emission estimates from fossil fuel combustion and cement production in China. Nature. 2015;**524**:335-338

[21] Ellabban O, Abu-Rub H, Blaabjerg F. Renewable energy resources: Current status, future prospects and their enabling technology. Renewable and Sustainable Energy Reviews. 2014;**39**:748-764

[22] Hordy N, Rabilloud D, Meunier J-L, Coulombe S. High temperature and long-term stability of carbon nanotube nanofluids for direct absorption solar thermal collectors. Solar Energy. 2014; **105**:82-90

[23] Bertani R. Geothermal power generation in the world 2010–2014 update report. Geothermics. 2016;**60**: 31-43

[24] Nematollahi O, Hoghooghi H, Rasti M, Sedaghat A. Energy demands and renewable energy resources in the Middle East. Renewable and Sustainable Energy Reviews. 2016;**54**:1172-1181

[25] Kannan N, Vakeesan D. Solar energy for future world: A review. Renewable and Sustainable Energy Reviews. 2016;**62**:1092-1105

[26] Maity SK. Opportunities, recent trends and challenges of integrated biorefinery: Part I. Renewable and Sustainable Energy Reviews. 2015;**43**: 1427-1445

[27] Nejat P, Jomehzadeh F, Taheri MM, Gohari M, Majid MZA. A global review of energy consumption, CO_2 emissions and policy in the residential sector (with an overview of the top ten CO_2 emitting countries). Renewable and sustainable energy reviews. 2015;**43**:843-862

[28] Zhang Y-J, Da Y-B. The decomposition of energy-related carbon emission and its decoupling with economic growth in China. Renewable and Sustainable Energy Reviews. 2015; **41**:1255-1266

[29] Sumathi S, Kumar LA, Surekha P. Solar PV and Wind Energy Conversion Systems: An Introduction to Theory, Modeling with MATLAB/SIMULINK, and the Role of Soft Computing Techniques. Springer; 2015

[30] Bigdeli N. Optimal management of hybrid PV/fuel cell/battery power system: A comparison of optimal hybrid approaches. Renewable and Sustainable Energy Reviews. 2015;**42**:377-393

[31] Ghafoor A, Munir A. Design and economics analysis of an off-grid PV system for household electrification. Renewable and Sustainable Energy Reviews. 2015;**42**:496-502

[32] Lin Q, Armin A, Nagiri RCR, Burn PL, Meredith P. Electro-optics of perovskite solar cells. Nature Photonics. 2015;**9**:106-112

[33] Rezzouk H, Mellit A. Feasibility study and sensitivity analysis of a stand-alone photovoltaic–diesel–battery

hybrid energy system in the north of Algeria. Renewable and Sustainable Energy Reviews. 2015;**43**:1134-1150

[34] Yuan X, Chen C, Yuan Y, Huang Y, Tan Q. Short-term wind power prediction based on LSSVM–GSA model. Energy Conversion and Management. 2015;**101**:393-401

[35] Ata R. Artificial neural networks applications in wind energy systems: A review. Renewable and Sustainable Energy Reviews. 2015;**49**:534-562

[36] Lu Y, Wang S, Shan K. Design optimization and optimal control of grid-connected and standalone nearly/net zero energy buildings. Applied Energy. 2015;**155**:463-477

[37] Asrari A, Ghasemi A, Javidi MH. Economic evaluation of hybrid renewable energy systems for rural electrification in Iran—A case study. Renewable and Sustainable Energy Reviews. 2012;**16**:3123-3130

[38] Varma RK, Rahman SA, Vanderheide T. New control of PV solar farm as STATCOM (PV-STATCOM) for increasing grid power transmission limits during night and day. IEEE Transactions on Power Delivery. 2015;**30**:755-763

[39] Clarke DP, Al-Abdeli YM, Kothapalli G. Multi-objective optimisation of renewable hybrid energy systems with desalination. Energy. 2015;**88**:457-468

[40] Kolhe ML, Ranaweera KIU, Gunawardana AS. Techno-economic sizing of off-grid hybrid renewable energy system for rural electrification in Sri Lanka. Sustainable Energy Technologies and Assessments. 2015;**11**:53-64

[41] Nugent D, Sovacool BK. Assessing the lifecycle greenhouse gas emissions from solar PV and wind energy: A critical meta-survey. Energy Policy. 2014;**65**:229-244

[42] ElNozahy M, Salama M. Technical impacts of grid-connected photovoltaic systems on electrical networks—A review. Journal of Renewable and Sustainable Energy. 2013;**5**:032702

[43] Nayar C, Lawrance W, Phillips S. Solar/wind/diesel hybrid energy systems for remote areas. In: Energy Conversion Engineering Conference, 1989 IECEC-89, Proceedings of the 24th Intersociety. IEEE; 1989. pp. 2029-2034

[44] Khare V, Nema S, Baredar P. Optimization of hydrogen based hybrid renewable energy system using HOMER, BB-BC and GAMBIT. International Journal of Hydrogen Energy. 2016;**41**:16743-16751

[45] Saidi A. Solar-wind hybrid renewable energy systems: Evolutionary technique. Electrotehnica, Electronica, Automatica. 2016;**64**:24

[46] Sinha S, Chandel S. Review of software tools for hybrid renewable energy systems. Renewable and Sustainable Energy Reviews. 2014;**32**:192-205

[47] Atici KB, Simsek AB, Ulucan A, Tosun MU. A GIS-based multiple criteria decision analysis approach for wind power plant site selection. Utilities Policy. 2015;**37**:86-96

[48] Abdin Z, Alim M, Saidur R, Islam M, Rashmi W, Mekhilef S, et al. Solar energy harvesting with the application of nanotechnology. Renewable and Sustainable Energy Reviews. 2013;**26**:837-852

[49] Bajpai P, Dash V. Hybrid renewable energy systems for power generation in stand-alone applications: A review. Renewable and Sustainable Energy Reviews. 2012;**16**:2926-2939

[50] Bronicki LY. Geothermal power conversion technology. Encyclopedia of Sustainability Science and Technology. 2015:1-123

[51] Giraud F, Salameh ZM. Steady-state performance of a grid-connected rooftop hybrid wind-photovoltaic power system with battery storage. IEEE Transactions on Energy Conversion. 2001;**16**:1-7

[52] Chauhan A, Saini R. A review on integrated renewable energy system based power generation for stand-alone applications: Configurations, storage options, sizing methodologies and control. Renewable and Sustainable Energy Reviews. 2014;**38**:99-120

[53] Alsayed M, Cacciato M, Scarcella G, Scelba G. Multicriteria optimal sizing of photovoltaic-wind turbine grid connected systems. IEEE Transactions on Energy Conversion. 2013;**28**:370-379

[54] Sovacool BK, Hirsh RF. Beyond batteries: An examination of the benefits and barriers to plug-in hybrid electric vehicles (PHEVs) and a vehicle-to-grid (V2G) transition. Energy Policy. 2009;**37**:1095-1103

[55] Mohammed Y, Mustafa M, Bashir N. Hybrid renewable energy systems for off-grid electric power: Review of substantial issues. Renewable and Sustainable Energy Reviews. 2014;**35**: 527-539

[56] Ardizzon G, Cavazzini G, Pavesi G. A new generation of small hydro and pumped-hydro power plants: Advances and future challenges. Renewable and Sustainable Energy Reviews. 2014;**31**: 746-761

[57] Murphy PM, Twaha S, Murphy IS. Analysis of the cost of reliable electricity: A new method for analyzing grid connected solar, diesel and hybrid distributed electricity systems considering an unreliable electric grid, with examples in Uganda. Energy. 2014; **66**:523-534

[58] Maheri A. Multi-objective design optimisation of standalone hybrid wind-PV-diesel systems under uncertainties. Renewable Energy. 2014;**66**:650-661

[59] Zakeri B, Syri S. Electrical energy storage systems: A comparative life cycle cost analysis. Renewable and Sustainable Energy Reviews. 2015;**42**: 569-596

[60] Bhandari B, Lee K-T, Lee CS, Song C-K, Maskey RK, Ahn S-H. A novel off-grid hybrid power system comprised of solar photovoltaic, wind, and hydro energy sources. Applied Energy. 2014; **133**:236-242

[61] Erdinc O, Uzunoglu M. Optimum design of hybrid renewable energy systems: Overview of different approaches. Renewable and Sustainable Energy Reviews. 2012;**16**:1412-1425

[62] Bayod-Rujula AA, Haro-Larrodé ME, Martinez-Gracia A. Sizing criteria of hybrid photovoltaic–wind systems with battery storage and self-consumption considering interaction with the grid. Solar Energy. 2013;**98**: 582-591

[63] Maleki A, Pourfayaz F. Optimal sizing of autonomous hybrid photovoltaic/wind/battery power system with LPSP technology by using evolutionary algorithms. Solar Energy. 2015;**115**:471-483

[64] Hassan A, Saadawi M, Kandil M, Saeed M. Modified particle swarm optimisation technique for optimal design of small renewable energy system supplying a specific load at Mansoura University. IET Renewable Power Generation. 2015;**9**:474-483

[65] Mokheimer EM, Al-Sharafi A, Habib MA, Alzaharnah I. A new study for hybrid PV/wind off-grid power

generation systems with the comparison of results from homer. International Journal of Green Energy. 2015;**12**: 526-542

[66] Shi Z, Wang R, Zhang T. Multi-objective optimal design of hybrid renewable energy systems using preference-inspired coevolutionary approach. Solar Energy. 2015;**118**:96-106

[67] Ma T, Yang H, Lu L. A feasibility study of a stand-alone hybrid solar–wind–battery system for a remote island. Applied Energy. 2014;**121**: 149-158

[68] Sharafi M, ELMekkawy TY. Multi-objective optimal design of hybrid renewable energy systems using PSO-simulation based approach. Renewable Energy. 2014;**68**:67-79

[69] Zhou W, Lou C, Li Z, Lu L, Yang H. Current status of research on optimum sizing of stand-alone hybrid solar–wind power generation systems. Applied Energy. 2010;**87**:380-389

[70] Abbes D, Martinez A, Champenois G. Life cycle cost, embodied energy and loss of power supply probability for the optimal design of hybrid power systems. Mathematics and Computers in Simulation. 2014;**98**:46-62

[71] Belmili H, Haddadi M, Bacha S, Almi MF, Bendib B. Sizing stand-alone photovoltaic–wind hybrid system: Techno-economic analysis and optimization. Renewable and Sustainable Energy Reviews. 2014;**30**: 821-832

[72] Hosseinalizadeh R, Shakouri GH, Amalnick Mohsen S, Taghipour P. Economic sizing of a hybrid (PV–WT–FC) renewable energy system (HRES) for stand-alone usages by an optimization-simulation model: Case study of Iran. Renewable and Sustainable Energy Reviews. 2016;**54**: 139-150

[73] Ramli MAM, Hiendro A, Al-Turki YA. Techno-economic energy analysis of wind/solar hybrid system: Case study for western coastal area of Saudi Arabia. Renewable Energy. 2016;**91**:374-385

[74] Blechinger P, Cader C, Bertheau P, Huyskens H, Seguin R, Breyer C. Global analysis of the techno-economic potential of renewable energy hybrid systems on small islands. Energy Policy. 2016;**98**:674-687

[75] Jung J, Villaran M. Optimal planning and design of hybrid renewable energy systems for microgrids. Renewable and Sustainable Energy Reviews. 2017;**75**: 180-191

[76] Bahramara S, Moghaddam MP, Haghifam MR. Optimal planning of hybrid renewable energy systems using HOMER: A review. Renewable and Sustainable Energy Reviews. 2016;**62**: 609-620

[77] Olatomiwa L, Mekhilef S, Ismail MS, Moghavvemi M. Energy management strategies in hybrid renewable energy systems: A review. Renewable and Sustainable Energy Reviews. 2016;**62**: 821-835

[78] Maleki A, Pourfayaz F, Rosen MA. A novel framework for optimal design of hybrid renewable energy-based autonomous energy systems: A case study for Namin, Iran. Energy. 2016;**98**: 168-180

[79] Karellas S, Braimakis K. Energy–exergy analysis and economic investigation of a cogeneration and trigeneration ORC–VCC hybrid system utilizing biomass fuel and solar power. Energy Conversion and Management. 2016;**107**:103-113

[80] Herrando M, Markides CN. Hybrid PV and solar-thermal systems for

domestic heat and power provision in the UK: Techno-economic considerations. Applied Energy. 2016; **161**:512-532

[81] Ciez RE, Whitacre JF. Comparative techno-economic analysis of hybrid micro-grid systems utilizing different battery types. Energy Conversion and Management. 2016;**112**:435-444

[82] Romero Rodríguez L, Salmerón Lissén JM, Sánchez Ramos J, Rodríguez Jara EÁ, Álvarez DS. Analysis of the economic feasibility and reduction of a building's energy consumption and emissions when integrating hybrid solar thermal/PV/micro-CHP systems. Applied Energy. 2016;**165**:828-838

[83] Chen J, Rabiti C. Synthetic wind speed scenarios generation for probabilistic analysis of hybrid energy systems. Energy. 2017;**120**:507-517

[84] Siddaiah R, Saini RP. A review on planning, configurations, modeling and optimization techniques of hybrid renewable energy systems for off grid applications. Renewable and Sustainable Energy Reviews. 2016;**58**:376-396

[85] Shezan SA, Hossain A, Ishrak A. A complete off-grid PV-diesel-battery hybrid energy system with feasibility analysis, system modeling and optimization. In: International Conference on Mechanical, Industrial and Materials Engineering 2015 (ICMIME2015). 2015. pp. 1-7

[86] Shezan SA, Das N. Optimized hybrid wind-diesel energy system with feasibility analysis. Technology and Economics of Smart Grids and Sustainable Energy. 2017;**2**:1-9

[87] Mahmudul H, Rahman MM, Shezan SKA, Metselaar HSC, Mekhilef S, Sohel R, et al. Temperature regulation of photovoltaic module using phase change material: A numerical analysis and experimental investigation. International Journal of Photoenergy. 2016;**2016**:1-8

[88] Shezan SA, Julai S, Kibria M, Ullah K, Saidur R, Chong W, et al. Performance analysis of an off-grid wind-PV (photovoltaic)-diesel-battery hybrid energy system feasible for remote areas. Journal of Cleaner Production. 2016;**125**:121-132

[89] Shezan SA, Saidur R, Ullah KR, Hossain A, Chong WT, Julai S. Feasibility analysis of a hybrid off-grid wind-DG-battery energy system for the eco-tourism remote areas. Clean Technologies and Environmental Policy. 2015;**17**:2417-2430

[90] Shezan S, Saidur R, Hossain A, Chong W, Kibria M. Performance analysis of solar-wind-diesel-battery hybrid energy system for KLIA Sepang station of Malaysia. In: IOP Conference Series: Materials Science and Engineering. IOP Publishing; 2015. p. 012074

[91] Shezan SA, Salahuddin A, Farzana M, Hossain A. Techno-economic analysis of a hybrid PV-wind-diesel energy system for sustainable development at coastal areas in Bangladesh. In: Development in the in Renewable Energy Technology (ICDRET) 2016 4th International Conference on the. IEEE; 2016. pp. 1-6

[92] Shezan S, Farzana M, Hossain A, Ishrak A. Techno-economic and feasibility analysis of a micro-grid wind-dg-battery hybrid energy system for remote and decentralized areas. International Journal of Advances in Engineering & Technology (IJAET). 2015;**8**:874-888

[93] Shezan S, Khan NH, Anowar MT, Delwar MH, Islam MD, Reduanul MH, et al. Fuzzy logic implementation with MATLAB for solar-wind-battery-diesel hybrid energy system. Imperial Journal

of Interdisciplinary Research (IJIR). 2016;**2**:574-583

[94] Rashid S, Rana S, Shezan S, Karim SAB, Anower S. Optimized design of a hybrid PV-wind-diesel energy system for sustainable development at coastal areas in Bangladesh. Environmental Progress & Sustainable Energy. 2016;**36**:297-304

[95] Shezan S, Das N, Mahmudul H. Techno-economic analysis of a smart-grid hybrid renewable energy system for Brisbane of Australia. Energy Procedia. 2017;**110**:340-345

[96] Palash MM, Shezan SA, Kusum F, Arnat R, Shaila S. Feasibility and techno-economical analysis of an off-grid solar-wind biomass hybrid energy system. Imperial Journal of Interdisciplinary Research. 2016;**2**: 914-920

[97] Shezan S, Delwar MH, Anowar MT, Islam MDI, Kabir MA, Reduanul MH, et al. Technological analysis of a solar-wind-battery-diesel hybrid energy system with fuzzy logic controller. International Journal of Research in Engineering and Science (IJRES). 2016; **4**:46-57

Chapter 24

Performances Analysis of a Micro-Grid Connected Multi-Renewable Energy Sources System Associated with Hydrogen Storage

Salah Tamalouzt, Nabil Benyahia,
Abdelmounaim Tounzi and Amar Bousbaine

Abstract

This work highlights the modeling and simulation of a micro-grid connected renewable energy system. It comprises of wind turbine (WT) based on doubly fed induction generator (DFIG), photovoltaic generator (PV), fuel cell (FC) generator, a Hydrogen tank, a water electrolyzer used for long-term storage, and a battery bank energy storage system (BBESS) utilized for short-term storage. In this paper, a global control strategy and an energy management strategy are proposed for the overall system. This strategy consists in charging the BBESS and producing hydrogen from the water electrolyzer in case of power excess provided from WT-DFIG and photovoltaic generators. Therefore, the FC and the BBESS will be used as a backup generator to supply the demand required power, when the WT-DFIGs and the PV energy are deficient. The effectiveness of this contribution is verified through computer simulations under Matlab/Simulink, where very satisfactory results are obtained.

Keywords: AC and DC micro-grid applications, multi-renewable energy sources, energy management system, hybrid energy storage system, hydrogen production and storage, doubly fed induction generator, direct active and reactive powers control, local reactive energy compensator

1. Introduction

A growing interest in renewable energy resources (RESs) has been observed these last years. The alternative energy sources (AESs) are non-polluting, free in their availability and continuous. These advantages make this type of energy attractive in remote locations and for microgrid applications [1, 2]. To experience the best benefits of these sources, many energy sources can be used for hybrid energy generators systems, including wind turbine (WT), Hydro-electric power (HEP), photovoltaic panels (PV), Fuel Cells (FC) and micro-turbines [1–38]. Nevertheless, the major RESs used and reported in the literature are WT and PV [1–33].

Indeed, a high RESs penetration can be created reliability, power quality and stability problems in the main electrical grid. We must think of more effective solutions to remedy these problems. For these, the microgrid (MG) is one of the solutions. In addition, for efficient and economical utilization of these RESs, some forms of backup systems are almost universally required. In the case these RESs are based on WTs or PVs, then Energy Storage Systems (ESSs) are necessary to further mitigate the effects of the seasonal and daily randomness fluctuations in the climatic (solar radiation, wind speed, temperature) and geographical conditions as well as the profiles of the required power. In addition, to maintain the power and energy balance as well as to improve the output power quality [2–6].

To ensure proper operation of the ESSs, they must have a high-power density in order to face fast power variations, and at the same time, they must have a high energy density to give autonomy to the MG. For these reasons, it is necessary to associate more than one storage technology creating a Hybrid Energy Storage Systems (HESSs). Knowing that the energy storage system as battery banks (BBs) is very important for the efficient and economical utilization of the MG system [25]. However, the lifetime of these systems decreases according to their operating modes, particularly according charge and discharge cycles [5, 6, 37]. To improve the energy supply reliability of WTs and PVs hybrid system, a fuel cell (FC) can be combined with an electrolyzer system, thus constituting a third energy source, to ensure the need for any additional, backups or supply power system [26].

Thus, the fuel cell (FC) and electrolyzer technology ideally fulfill the need for any backups the supply power system, and hydrogen is a suitable option and it can be used as a fuel to get a reliable power for almost every application that used fossil fuels [3, 38]. The coupling of the differents energy sources, using a combination of the WT, PV array, FC with electrolyzer containing a BBESS is one of the most efficient procedures to produce very high-quality energy [2, 3, 38]. This configuration brings us to speak of an electrical MG, which is associated with distributed generation and decentralized management, directed at developing small electrical grids that are normally connected to the general electricity grid. The associated energy storage devices make possible to balance out the power exchange between the MG supply and demand, and also between the MG and the grid [3].

Several studies and various configurations of an hybrid micro-grid applications-based WTs and PVs generations have been proposed in the litterature [1–33]. However, a little attention has been given for the integration of a DFIG on the MG system. Indeed, the DFIG driven by wind turbine (WT-DFIG) brings more benefits to the wind energy conversion system such as direct access on the AC grid, the reactive power compensation unit is not needed, the ability of opertating at three modes (sub, super and synchronous mode including the over-speed mode), the capability of producing electric power more than the rated power and a flexible active and reactive power control. Furthermore, the size of the AC/AC frequency converter with its DC-bus capacitor, the rotor side controller, the converter costs and power losses are reduced [2–6].

The power converter with its control and protection are the key of the DFIG advantages for electrical grid integration. Knowing that the variable speed constant frequency operation of DFIG can be realized by adjusting the amplitude, frequency, phase and phase sequence. If an appropriate control strategy is adopted, active and reactive powers of the DFIG can be regulated independently, which makes this machine as an active energy production system and a local reactive energy compensator. The AC excitation performance has a direct influence on the WT-DFIG operation. In the aim to develop efficient and economical wind energy conversion systems, different algorithm control applied to WT-DFIG have been proposed in the literature [2–23]. Conventional control strategies are generally based on two most families, which are; the direct control techniques namely the direct torque control or the direct power control (DPC) [2–16], and indirect control strategies based on field-oriented control (FOC) [17–23]. In this case, we have the FOC Linear Vector Control (FOC-LVC) methods [17, 18] and the FOC non-linear Vector Control (FOC-NLVC) [19–23], which are have been widely applied to WT-DFIG in differents operating quadrants and modes in many research works. The variable-structure control technique based on sliding mode control (SMC), is the most widely used nonlinear control technique applied to wind turbine control, due to these advantages over other nonlinear techniques [19–21].

Indeed, the advantage of the proposed micro-grid topology is the easier accessibility to AC and DC grid. However, the AC/DC and DC/DC converters required for supplying the DC loads are not needed. So, the main objectives of this contribution are: achieving an optimal and efficient control of the coordination between the different sources; WT-DFIG, PV and FC, BBESS and electrolyzer, ensuring better power quality for customers (for DC or AC grid) and regulating the output voltage amplitude and frequency (AC grid voltage). Moreover, to ensure continuity of service (provide the output energy needed) and a local reactive power compensation, while operating maximum use of benefits of the proposed topology, even in extreme operation conditions. In addition, the micro-grid based on the multi-source topology can increase the flexibility of grid connection power management and vice versa. Indeed, assessing the usefulness of the proposed system depends on the applied system-control effectiveness, in particular the active and reactive power control of the DFIG, and the applied management algorithm in very severe situations (change in power demand, the random variations and sudden conditions of meteorological, ensure continuity of service, ...).

This work focuses on the development of simulation model to design and analyze the micro-grid system overall performances. A combination of the WT-DFIG, PV array, BBESS, and FC-electrolyzer is used. The numerical simulation model is used not only to analyzing hybrid system performances, but also for designing this last one, to meet for all load demands for any available meteorological condition. For this, the simulation of applied management algorithm of the whole system in very severe situations, such as a change in power demand random variations and sudden meteorological conditions, is presented and the results are discussed.

The paper is summarized as follows: In Section 2, a brief description of the proposed micro-grid is presented. In Section 3, a DRPC principle applied to the WT-DFIG, and the reactive power local compensation system sought is explained. Unit sizing of wind generation, PV, electrolyzer, and FC are given in this section. In addition, the control strategy is developed to regulate the voltage and hence coordination among sources, storages (battery bank and electrolyzer) is explained. The detailed simulation results of the different situations studied with their comments are presented in Section 4. Finally, the main paper conclusions are given.

2. Description of the proposed micro-grid

The proposed hybrid energy system, shown in **Figure 1**, consists of a Doubly Fed Induction Generator (DFIG) based variable speed wind energy conversion, PV array, battery bank, fuel cell, and electrolyzer. The primary sources, wind speed turbine and PV, are connected to the common DC-grid and equipped with optimum powers algorithms extracting. For the PV generator, the maximum power point tracking (MPPT) is applied, whereas, in the WT source case, it depending on the operating zone of the WT, MPPT technique and pitch angle control are considered to optimize the wind energy efficiency. BBESS is used as a storage device and is

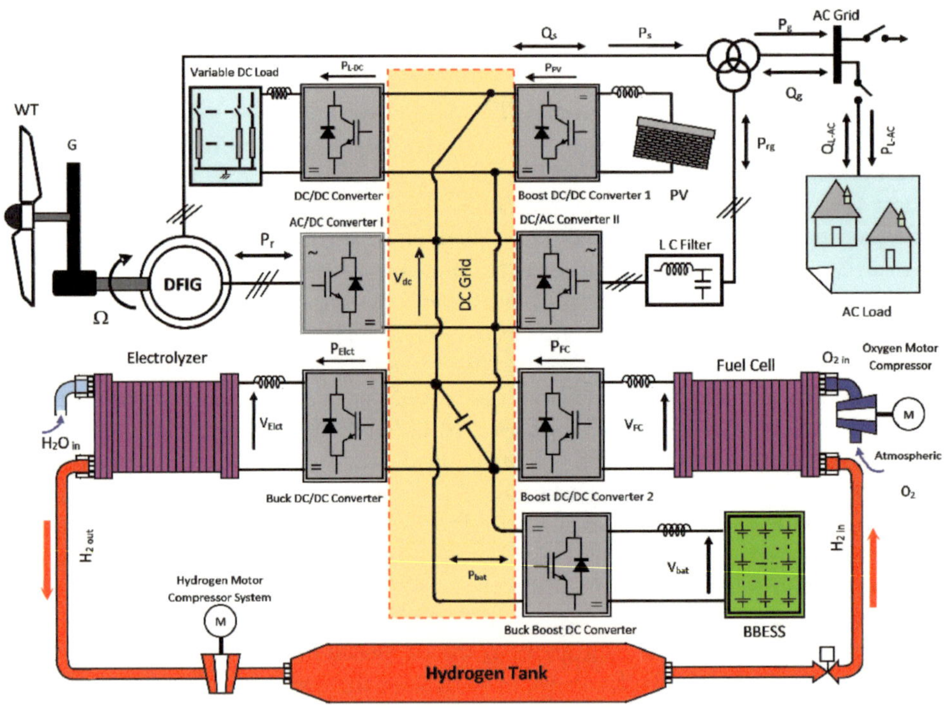

Figure 1.
Global control scheme combining RESs (WT-DFIG with PV) with HESS associated with hydrogen energy production (HEP).

connected to DC-grid through DC/DC buck-boost converter (BBDC). Wind and solar powers depend on weather conditions and during night hours solar power is zero. Therefore, under the situation of long-term no-wind and solar or low-wind and solar condition, the BBESS alone cannot cater to the load demand. Hence, fuel cell (FC) is integrated to make the system more sustainable.

In case of high-power generation from wind and solar for a long time and the BBESS hits its upper limit of charge-storage, the electrolyzer comes into effect and consumes the surplus power, while generating hydrogen. This hydrogen produced from the electrolyzer is stored in the tanks, where can be used as input to supply the FC.

3. Modeling and control of the proposed micro-grid

3.1 Modeling and control of the wind energy conversion system

3.1.1 Wind energy conversion system modeling

The wind turbine mathematical model consists of the modeling of its aerodynamic torque T_t which is given by the following relation [3, 10]:

$$T_t = \frac{1}{2} C_p \frac{\rho \pi R^2}{\Omega_t} \nu^3 \tag{1}$$

where, ν, ρ and R are respectively the wind speed, the air density and the rotor radius. Besides, Ω_t and C_p are the turbine speed and the power coefficient. The tip-speed ration (λ) is given by the Eq. (3) [3, 10]:

$$\lambda = \frac{\Omega_t \cdot R}{\nu} \tag{2}$$

Figure 2 shows the wind turbine characteristic, where, in the region (I) (MPPT control region), the tip speed ratio is maintained at its optimal value $\lambda_{opt} = 9$ and the power coefficient is always maintained at its maximum value ($C_{pmax} = 0.49$). In this case, the reference electromagnetic torque applied as being an input to the proposed DFIG control in the control is given by:

$$T_{em\ ref} = \frac{1}{2} C_{p-max} \frac{\rho \pi R^5}{G^3 \lambda_{opt}^3} \nu^3 \tag{3}$$

In the regions (II) and (III), called Pitch angle control regions, the generated active power is maintained constant. The power coefficient (C_p) for wind turbine model is given as a function of tip-speed ratio (λ) and the blade pitch angle (β). In this work, the following expression is used [2, 3, 10]:

$$\begin{aligned} C_p(\lambda, \beta) = 0.5176 \cdot [116 \cdot (1/\lambda_i) - 0.4 \cdot \beta - 5] \\ \exp(-21/\lambda_i) + 0.0068 \cdot \lambda_i \end{aligned} \tag{4}$$

with: $\frac{1}{\lambda_i} = \frac{1}{\lambda + 0.08 \cdot \beta} - \frac{0.035}{1+\beta^3}$.

3.1.2 Control of wind turbine generator (WT-DFIG)

When the DFIG is connected to an existing AC grid, as shown in **Figure 1**, this connection must be established in the following done in three steps. The first step is the stator voltages regulation with the AC grid voltages as reference synchronization. The second step is the stator connection to this grid. As the voltages of the two devices are synchronized, the connection can be effectively established. Once this connection is achieved, the third step is the regulation of the transit of the power between the DFIG and the AC grid.

In this work, the WT-DFIG control is based on the control of the exchanged powers between the AC-grid and the WT-DFIG. To provide independent control of the active power P_g and reactive power Q_g exchange between the AC-grid and WT-DFIG, by means of rotor voltage regulation, it is necessary to define the dq

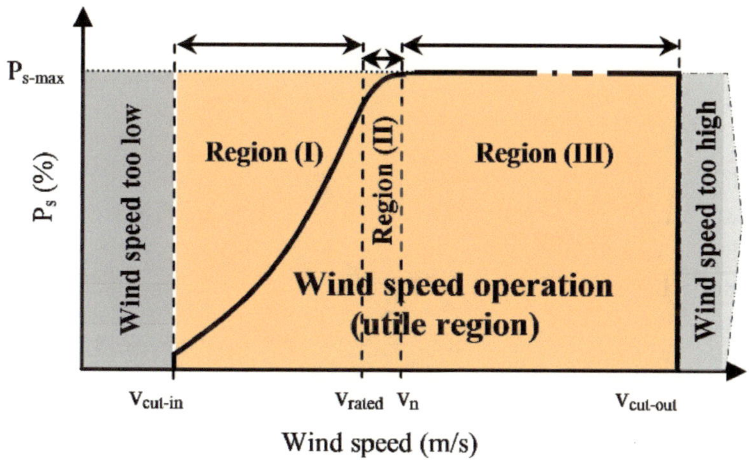

Figure 2.
Operation regions of wind turbine.

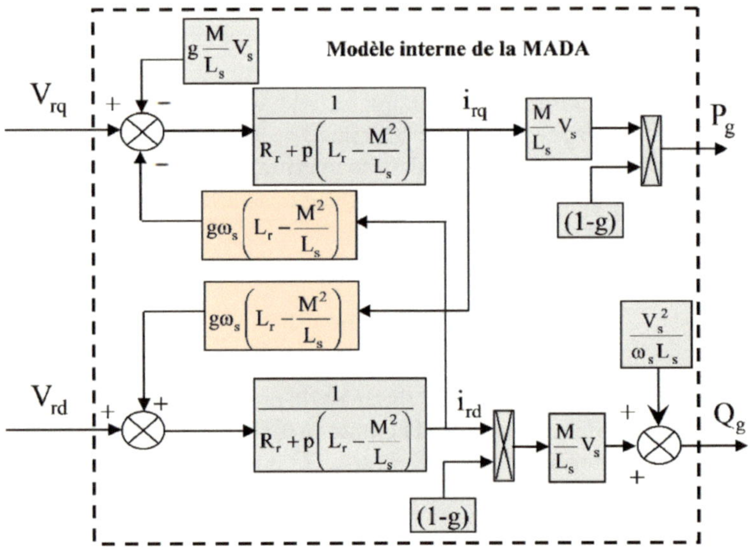

Figure 3.
Simplified scheme model of DFIG.

components of the rotor voltages in the stator-flux oriented reference frame and show that P_g and Q_g can be represented as functions of the individual voltage components. Subsequently, the P_g and Q_g control can be used to determine the reference rotor voltages.

The AC grid active and reactive powers are directly controlled, assuming that the optimal mechanical power developed by the turbine is known, as shown in **Figure 2**. The reference AC grid active and reactive powers, denoted by (P_g^*, Q_g^*) are used as the reference value for the WT-DFIG power control loop. In the inner voltage control loop, the method of stator flux oriented vector control, combining with the Variable structure control (VSC) possesses based on the nonlinear SMC is applied. They are used to establish a reference frame, that allows the d and q axis components of the rotor voltage to be controlled independently. Adjustment of the q-axis component of the rotor voltage, V_{qr}, controls the developed generator torque or the AC grid-side active power of the WT-DFIG (P_g). Regulating the d-axis component, V_{dr}, controls directly the grid-side reactive power flow (Q_g), as illustrated in **Figure 3**.

The two-phase model of the Doubly Fed Induction Machine (DFIM) expressed in the rotor frame is used. The classical electrical equations are written in this frame, as follows [3]:

$$\begin{cases} \frac{di_{sd}}{dt} = \frac{1}{L_s}\left[V_{sd} - M\frac{di_{rd}}{dt} - R_s i_{sd} + \omega_s L_s i_{sq} + \omega_s M i_{rq}\right] \\ \frac{di_{sq}}{dt} = \frac{1}{L_s}\left[V_{sq} - M\frac{di_{rq}}{dt} - R_s i_{sq} - \omega_s L_s i_{sd} - \omega_s M i_{rd}\right] \\ \frac{di_{rd}}{dt} = \frac{1}{L_r}\left[V_{rd} - M\frac{di_{sd}}{dt} - R_r i_{rd} - (\omega_s - \omega) L_r i_{rq} - (\omega_s - \omega) M i_{sq}\right] \\ \frac{di_{rq}}{dt} = \frac{1}{L_r}\left[V_{rq} - M\frac{di_{sq}}{dt} - R_r i_{rq} + (\omega_s - \omega) L_r i_{rd} + (\omega_s - \omega) M i_{sd}\right] \end{cases} \tag{5}$$

with: $\begin{cases} L_s = l_s + M \\ L_r = l_r + M \end{cases}$.

where R_s, R_r, l_s and l_r are the phase resistances and leakage inductances respectively. The M represents the magnetizing inductance, Ω is the mechanical speed and p the pair pole number. Besides, v_{sd}, v_{sq}, i_{sd} and i_{sq} are the d-q stator voltage and currents, respectively. While the v_{rd}, v_{rq}, i_{rd} and i_{rq} are the rotor voltage and currents, along the d-q axis, respectively.

The vector control strategy, using the Park transformation, allows to decoupling the control of active and reactive powers [3, 19]. By setting the stator flux vector aligned with d-axis, we have:

$$\begin{cases} \psi_{ds} = \psi_s = L_s i_{ds} + M i_{qs} \\ \qquad \text{and} \\ \psi_{qs} = L_s i_{qs} + M i_{ds} = 0 \end{cases} \tag{6}$$

The electrical active and reactive powers at the rotor, delivered by the stator as well as those provided for the AC grid are defined as:

$$\begin{cases} P_s = v_{ds} i_{ds} + v_{qs} i_{qs} \\ Q_s = v_{qs} i_{ds} - v_{ds} i_{qs} \end{cases} \tag{7}$$

$$\begin{cases} P_r = v_{dr} i_{dr} + v_{qr} i_{qr} \\ Q_r = v_{qr} i_{dr} - v_{dr} i_{qr} \end{cases} \tag{8}$$

$$\begin{cases} P_g = P_s + P_r \\ Q_g = Q_s + Q_r \end{cases} \tag{9}$$

The electromagnetic torque is defined as follows:

$$T_{em} = p\left(\psi_{ds} i_{qs} - \psi_{qs} i_{ds}\right) \tag{10}$$

$$T_{em} = p\frac{L_s}{M}\left(\psi_{ds} i_{qs}\right) \tag{11}$$

The electromagnetic torque and then the active power are controlled through the q-axis rotor current and d-axis rotor current control the reactive power. Neglecting the phase stator resistance R_s (that's the case of medium power machines used in wind energy conversion systems). In order to calculate Park transformation angles for the stator and rotor variables, the stator pulsation and the mechanical speed must be sensed. By choosing this reference frame, stator voltages and fluxes can be rewritten as follows:

$$\begin{cases} v_{ds} = 0 \\ \text{and} \\ v_{qs} = v_s = \omega_s \psi_{ds} \end{cases} \tag{12}$$

With these relations, as given in Eq. (12), the rotor fluxes, the stator active and reactive power and voltages can be written versus rotor currents as:

$$\begin{cases} \psi_{dr} = \sigma_r\, i_{dr} + \dfrac{M}{\omega_s L_s} v_s \\ \psi_{qr} = \sigma_r\, i_{qr} \end{cases} \tag{13}$$

where, $\sigma_r = L_r\left(1 - \frac{M^2}{L_r L_s}\right)$ and $s = \frac{\omega_r}{\omega_s}$;

The AC grid active power reference to be injected into the AC-grid is determined by multiplying the value of the efficiency (η) of the DFIG by the optimal turbine mechanical power. The latter is sought by the optimization algorithm: The MPPT technique and pitch angle control are considered to optimize the wind energy efficiency. In addition, The generator mechanical speed remains at its optimum.

$$P_{g\_ref} = \eta\left(-P_{mec\_opt}\right) \tag{14}$$

$$\Omega_{ref} = \Omega_{opt} \tag{15}$$

The AC-grid reactive power reference is determined according to the reactive energy that it is desired to compensate locally.

The DFIG efficiency η is estimated at 96%, knowing that the machine mechanical losses are neglected and static power converters are considered ideal [3, 18].

We first recall the expressions of powers, active and reactive, exchanged with the AC-grid, as well as the rotor currents in the Park coordinate system with stator flux orientation, as given in Eqs. (18) and (20).

$$\begin{cases} P_s = -v_s \dfrac{M}{L_s} i_{qr} \\ Q_s = \dfrac{v_s \psi_{ds}}{L_s} - \dfrac{v_s M}{L_s} i_{dr} \end{cases} \tag{16}$$

$$\begin{cases} v_{dr} = R_r i_{dr} + \sigma_r \dfrac{d}{dt} i_{dr} \text{-} s\omega_s \sigma_r i_{qr} \\ v_{qr} = R_r i_{qr} + \sigma_r \dfrac{d}{dt} i_{qr} \text{-} s\omega_s \sigma_r i_{dr} + s\omega_s \dfrac{L_m}{\omega_s L_s} v_s \end{cases} \tag{17}$$

For relatively weak slip values, Eq. (9) can be simplified by Eq. (18):

$$\begin{cases} P_g = (s-1)\, v_s \dfrac{M}{L_s} i_{qr} \\ Q_g = \dfrac{v_s^2}{\omega_s L_s} + (s-1)\, v_s \dfrac{M}{L_s} i_{dr} \end{cases} \tag{18}$$

$$\begin{cases} \dfrac{di_{rd}}{dt} = \dfrac{1}{\sigma_r}\left(V_{rd} - R_r i_{rd} + \sigma_r \cdot (\omega_s - \omega) \cdot i_{rq} - \dfrac{M}{L_s}\dfrac{d\Phi_s}{dt}\right) \\ \dfrac{di_{rq}}{dt} = \dfrac{1}{\sigma_r}\left(V_{rq} - R_r i_{rq} - \sigma_r \cdot (\omega_s - \omega) \cdot i_{rd} - (\omega_s - \omega) \cdot \dfrac{M}{L_s}\Phi_s\right) \end{cases} \tag{19}$$

$$\Rightarrow \begin{cases} \dfrac{di_{rd}}{dt} = \dfrac{1}{\sigma_r}\left(V_{rd} - R_r i_{rd} + \sigma_r \cdot g \cdot \omega_s \cdot i_{rq}\right) \\ \dfrac{di_{rq}}{dt} = \dfrac{1}{\sigma_r}\left(V_{rq} - R_r i_{rq} - \sigma_r \cdot g \cdot \omega_s \cdot i_{rd} - g \cdot \dfrac{M}{L_s} V_s\right) \end{cases} \tag{20}$$

3.1.2.1 AC-grid active power control

The grid active power surface is given as follow:

$$S(P) = P_{g\_ref} - P_g \tag{21}$$

The derivative of the grid active power surface:

$$\dot{S}(P) = \dot{P}_{g\_ref} - \dot{P}_g \tag{22}$$

$$\Rightarrow \dot{S}(P) = \dot{P}_{g\_ref} - \left((g-1) \cdot V_s \cdot \frac{M}{L_s}\right) \cdot \dot{i}_{rq} \tag{23}$$

We replace the currents of Eq. (20) in Eq. (23), gives us:

$$\dot{S}(P) = \dot{P}_{g\_ref} - (g-1) \cdot V_s \cdot \frac{M}{L_s} \cdot \frac{1}{\sigma_r} \left(V_{rq} - R_r \cdot i_{rq} - \sigma_r \cdot g \cdot \omega_s \cdot i_{rd} - g \cdot \frac{M}{L_s} \cdot V_s \right) \quad (24)$$

We have;

$$V_{rq} = V_{rq(eq)} + V_{rq(n)} \quad (25)$$

Then;

$$\begin{aligned} \dot{S}(P) = \dot{P}_{g\_ref} - (g-1) \cdot V_s \cdot \frac{M}{L_s} \cdot \frac{1}{\sigma_r} \Bigg(V_{rq(eq)} + \\ V_{rq(n)} - R_r \cdot i_{rq} - \sigma_r \cdot g \cdot \omega_s \cdot i_{rd} - g \cdot \frac{M}{L_s} \cdot V_s \Bigg) \end{aligned} \quad (26)$$

During the slip mode and steady-state, we have:

$$S(P) = 0; \dot{S}(P) = 0; V_{rq(n)} = 0 \quad (27)$$

$$\Rightarrow \begin{cases} V_{rq(eq)} = -\left(\frac{L_s}{M} \cdot \frac{\sigma_1}{(1-g) \cdot V_s} \right) \dot{P}_{g\_ref} + R_r \cdot i_{rq} + \sigma_r \cdot g \cdot \omega_s \cdot i_{rd} + g \cdot \frac{M}{L_s} \cdot V_s \\ V_{rq(n)} = K_{rq} \cdot Sat(S(P)) \\ K_{rq} > 0 \end{cases} \quad (28)$$

The convergence condition is $S(P) \cdot \dot{S}(P) < 0$, is ensured if:

1. $S(P) > 0; \dot{S}(P) < 0$, then:

$$K_{rq} > \left(\left(\frac{L_s}{M} \cdot \frac{\sigma_r}{(g-1) \cdot V_s} \right) \dot{P}_{g\_ref} + R_r \cdot i_{rq} + \sigma_r \cdot g \cdot \omega_s \cdot i_{rd} + g \cdot \frac{M \cdot V_s}{L_s} \right) \quad (29)$$

2. Or $S(P) < 0; \dot{S}(P) > 0$, then:

$$K_{rq} > -\left(\left(\frac{L_s}{M} \cdot \frac{\sigma_r}{(g-1) \cdot V_s} \right) \dot{P}_{g\_ref} + R_r \cdot i_{rq} + \sigma_r \cdot g \cdot \omega_s \cdot i_{rd} + g \cdot \frac{M \cdot V_s}{L_s} \right) \quad (30)$$

From Eqs. (29) and (30), we deduce the K_{rq}:

$$K_{rq} > \left| -\left(\left(\frac{L_s}{M} \cdot \frac{\sigma_r}{(g-1) \cdot V_s} \right) \dot{P}_{g\_ref} + R_r \cdot i_{rq} + \sigma_r \cdot g \cdot \omega_s \cdot i_{rd} + g \cdot \frac{M \cdot V_s}{L_s} \right) \right| \quad (31)$$

3.1.2.2 AC-grid réactive power control

The grid réactive power surface is given as follow:

$$S(Q) = Q_{g\_ref} - Q_g \quad (32)$$

The derivative of the grid active power surface:

$$\dot{S}(Q) = \dot{Q}_{g\_ref} - \dot{Q}_g \quad (33)$$

$$\Rightarrow \dot{S}(Q) = \dot{Q}_{g\_ref} - \left((g-1)\cdot V_s \cdot \frac{M}{L_s}\right)\cdot \dot{i}_{rd} \tag{34}$$

We replace the currents of Eq. (20) in Eq. (34), gives us:

$$\dot{S}(Q) = \dot{Q}_{g\_ref} - (g-1)\cdot V_s \cdot \frac{M}{L_s}\cdot\frac{1}{\sigma_r}\left(V_{rd} - R_r i_{rd} + \sigma_r \cdot g \cdot \omega_s \cdot i_{rq}\right) \tag{35}$$

where;

$$\begin{aligned}\dot{S}(Q) = \dot{Q}_{g\_ref} - (g-1)\cdot V_s \cdot \frac{M}{L_s}\cdot\frac{1}{\sigma_r}\left(V_{rd(eq)} + \right.\\ \left. V_{rd(n)} - R_r i_{rd} + \sigma_r \cdot g \cdot \omega_s \cdot i_{rq}\right)\end{aligned} \tag{36}$$

During the slip mode and steady-state, we have:

$$S(Q) = 0; \dot{S}(Q) = 0; V_{rd(n)} = 0 \tag{37}$$

$$\Rightarrow \begin{cases} V_{rd(eq)} = -\left(\frac{L_s}{M}\cdot\frac{\sigma_r}{(1-g)\cdot V_s}\right)\dot{Q}_{g\_ref} + R_r \cdot i_{rd} - \sigma_r \cdot g \cdot \omega_s \cdot i_{rq} \\ V_{rd(n)} = K_{rd}\cdot \mathrm{Sat}(S(Q)) \\ K_{rd} > 0 \end{cases} \tag{38}$$

The convergence condition is $S(Q)\cdot\dot{S}(Q) < 0$, is ensured if:

1. $S(Q) > 0; \dot{S}(Q) < 0$, then:

$$K_{rd} > \left(\left(\frac{L_s}{M}\cdot\frac{\sigma_r}{(g-1)\cdot V_s}\right)\dot{Q}_{g\_ref} + R_r \cdot i_{rd} - \sigma_r \cdot g \cdot \omega_s \cdot i_{rq}\right) \tag{39}$$

2. Or $S(Q) < 0; \dot{S}(Q) > 0$, then:

$$K_{rd} > -\left(\left(\frac{L_s}{M}\cdot\frac{\sigma_r}{(g-1)\cdot V_s}\right)\dot{Q}_{g\_ref} + R_r \cdot i_{rd} - \sigma_r \cdot g \cdot \omega_s \cdot i_{rq}\right) \tag{40}$$

From Eqs. (39) and (40), we deduce the K_{rd}:

$$K_{rd} > \left|-\left(\left(\frac{L_s}{M}\cdot\frac{\sigma_r}{(g-1)\cdot V_s}\right)\dot{Q}_{g\_ref} + R_r \cdot i_{rd} - \sigma_r \cdot g \cdot \omega_s \cdot i_{rq}\right)\right| \tag{41}$$

The schematic diagram of the AC grid active and reactive powers sliding mode control structure, denominated DRPC, is illustrated in **Figure 4**.

3.2 Photovoltaic system modeling and control

The PV cell mathematical model description is represented in [3, 31, 33], where the typical PV module power characteristic is given in [3, 33]. In order to extract the optimum power of the PV array, it is necessary to use an algorithm to operate the

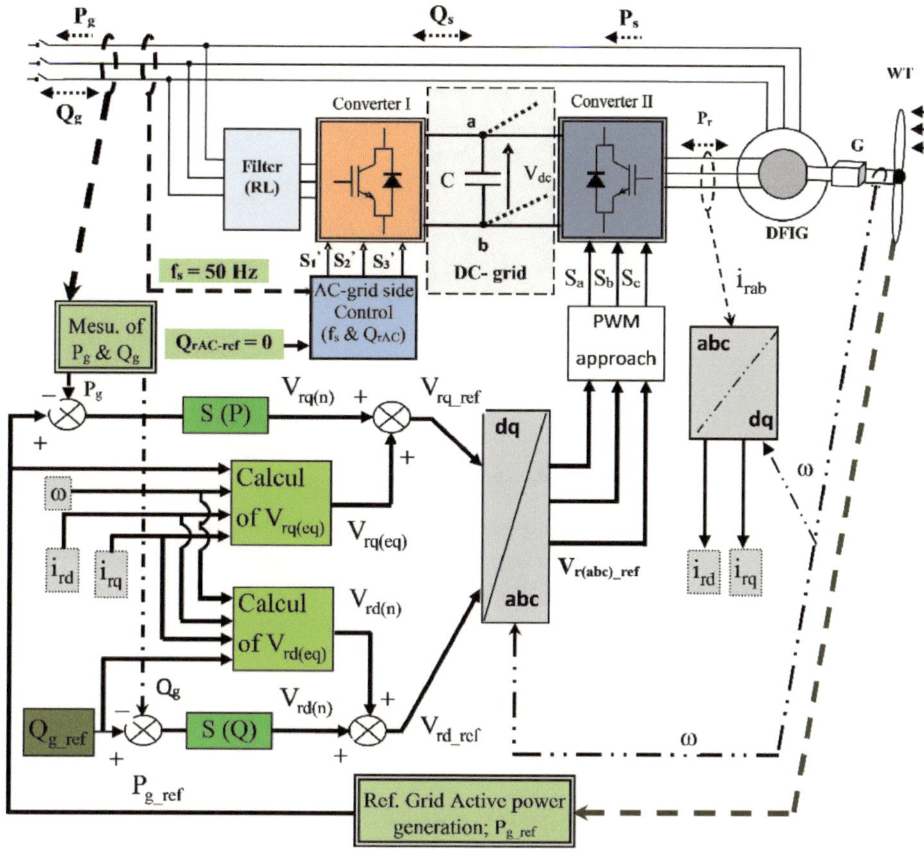

Figure 4.
Schematic diagram of the AC grid active and reactive powers, exchanged with the proposed micro-grid, SMC structure.

PV generator at its maximum power point, called MPPT technique. The MPPT device is a high-frequency boost DC/DC converter (BDCC) inserted between the PV array and the DC-grid, and it takes the DC input from the PV array, convert it to a different DC voltage and current to exactly match the PV array to the DC-grid, as illustrated in **Figure 5**.

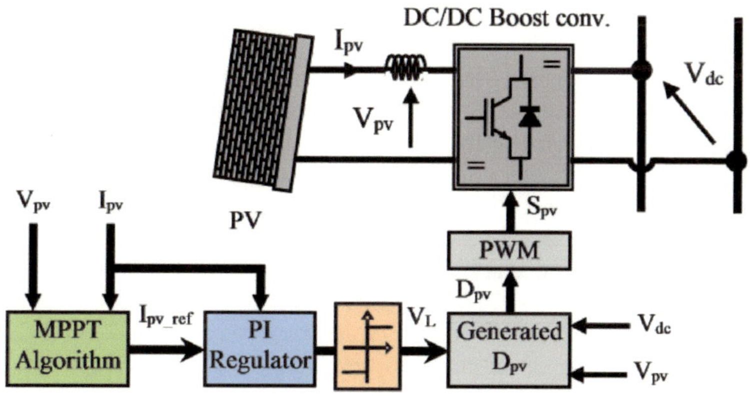

Figure 5.
MPPT algorithm control scheme of PV array.

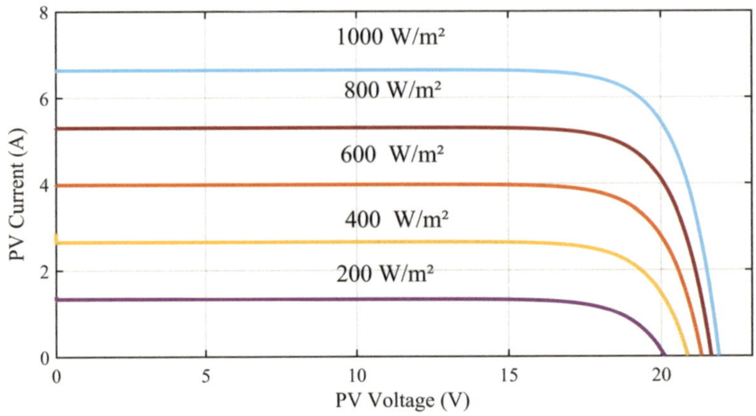

Figure 6.
Typical PV module unit I-V characteristic showing the effects of irradiance.

Figure 6 shows that the PV output power is directly proportional to the irradiance, as in [3, 33]. As such, a smaller irradiance will result in reduced power output from the PV module. However, it is also observed that only the output current is affected by the irradiance. This is expected because the generated current is proportional to the flux of photons.

The number of the PV cell interconnected for the proposed micro-grid is given according to V_{mpp} they result 04 arrays are connected in parallel, each array consists of 27 modules connected in series, each of them delivers 108 W peaks.

3.3 Fuel cell (FC) system modeling and control

The chemical energy of a fuel is directly converting to electrical energy by the PEM FC. The output voltage of a single cell can be defined by Eq. (42), as given in [3, 24, 25]:

$$E_{cell} = E_{Nernst} - E_{Act} - E_{Con} - E_{Ohm} \tag{42}$$

where E_{Con} is the concentration voltage drop, E_{Ohm} is the Ohmic voltage drop, E_{Nernst} is the reversible voltage and E_{Act} is the activation voltage drop. The terminal voltage of PEM fuel cell stacks is defined by Eq. (43), as developed in [2, 3].

$$V_{FC} = N_s E_{Cell} \tag{43}$$

Figure 7 shows voltage and power versus current polarization curves of a PEM FC. It is shown that the activation voltage drop dominates at low current, the Ohmic drop voltage dominates at mid-range current, and the concentration drop voltage dominates at high current. The voltage deviates further as the current is increased, illustrating the effects of drop-voltages. In concentration voltage drop region, the FC output power occurs near the FC rated power.

The FC proper operation is achieved by controlling those auxiliaries, in particular, the air supply system [2–4]. The air consumption system is about 20% of the power supplied by the FC, which reduces the effective capacity of the FC. This is why; the air system control is an important challenge for the development of the FC on micro-grid applications. Indeed, maintaining an adequate level of the oxygen partial pressure in the cathode during rapid variations of FC current in rapid operation mode is important to avoid degradation of the membrane and a decrease in the system efficiency. The excess oxygen ratio control is depicted in **Figure 8**.

Figure 9 shows the FC controller through which the boost converter switch is controlled. Controllers associated with the BBESS and FC are developed in such a fashion that when there is a sudden load power change, the BBESS provides the power instantaneously. In order to make the proper coordination between BBESS

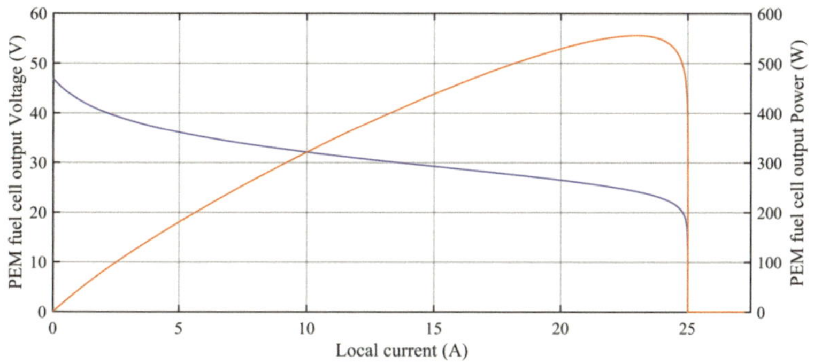

Figure 7.
Power and voltage versus current polarization curves of PEM FC [3].

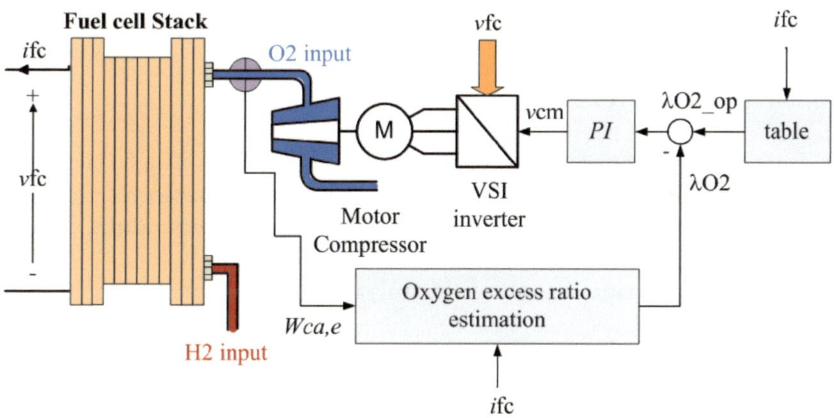

Figure 8.
Excess oxygen ratio control.

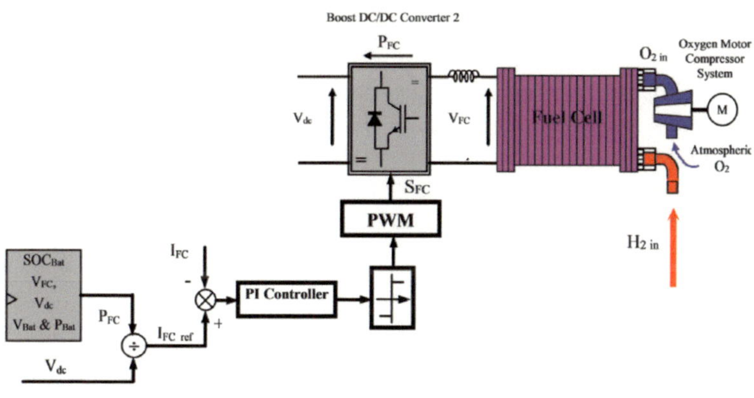

Figure 9.
Boost converter controller associated with fuel cell.

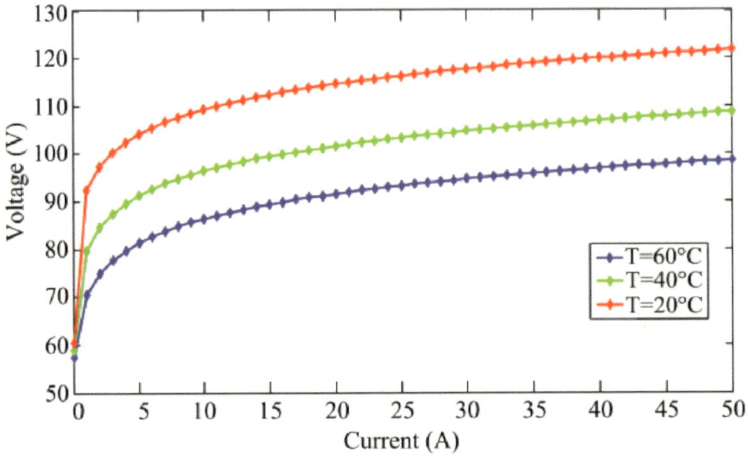

Figure 10.
Typical characteristic of an electrolyzer under different temperatures [3].

and FC, the boost converter controller associated with FC is developed based on assuming BBESS current is ideally zero. Moreover, when SOC of the battery reaches its minimum at 20%, the fuel cell should be generated the required power.

In this work, the PEM fuel cell is composed of 2 × 6 stacks with a power rating of 10 kW.

3.4 Modeling and control of electrolezer system

The water electrolyzer description is presented in [3, 5, 6]. The U(I) electrolyzer model characteristics used in this study at different cell temperatures are shown in **Figure 10**. At a given current, the higher of the operating temperature, the lower is the terminal voltage needed.

The empirical U(I) equation of an electrolyzer cell can be expressed as [3]:

$$U_{cell} = U_{rev} + (r_1 + r_2 T) I_{elect}/A + k \tag{44}$$

$$k = k_{elec} \ln\left(\left(k_{T_1} + k_{T_2}/T + k_{T_3}/T^2\right) I_{elect}/A + 1\right) \tag{45}$$

where U_{cell} is the cell terminal voltage; U_{rev} is the reversible cell voltage; r_1 and r_2 are the parameters for the Ohmic resistances inside the electrolyzer; k_{elec}, k_{T1}, k_{T2}, and k_{T3} are the parameters for the overvoltage; A is the area of the cell electrode; I_{elect} is the electrolyzer current and T is the electrolyzer cell temperature.

The excess energy produced is first pushed into the battery until it reaches its upper limit of charge carrying capacity and then the excess power is fed to the electrolyzer and is regulated via the buck DC/DC converter, as illustrated in **Figure 11**. The decision about switching on the control action is carried out by comparing the upper limit of the battery state of charge (SOC) and the present status of SOC. When the SOC reaches its maximum limit at 80%, the controller will increase the duty cycle as a function of overvoltage in the DC-grid voltage.

3.5 Battery bank (BB) system modeling

In this work, we have implemented a standard battery model, as given in [2–6]. The lead-acid battery is modeled by putting in series an electromotive force corresponding to the open-circuit voltage when it is charged E_{bat}, a capacity

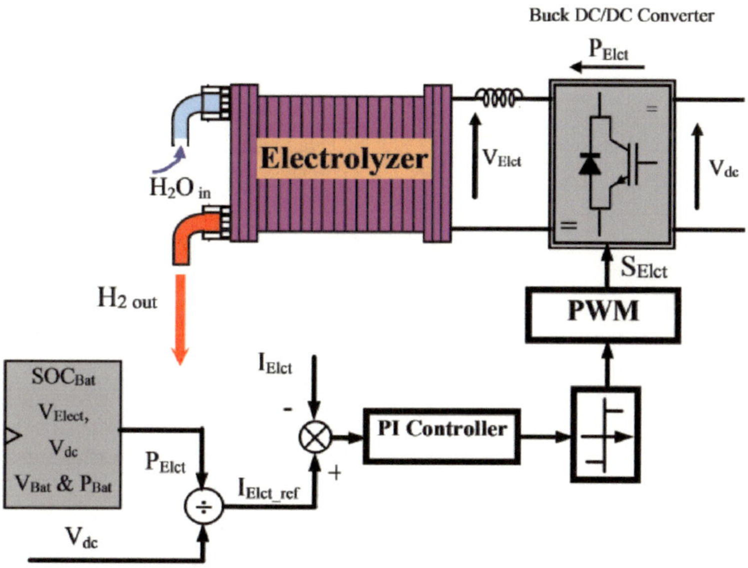

Figure 11.
Buck converter controller associated with an electrolyzer.

indicating the internal capacity of a battery (C_{bat}) and an internal resistance R_{bat}, as illustrated in **Figure 12**.

The battery terminal voltage is given by Eq. (46), as follow:

$$V_{bat} = E_{bat} - R_{bat} \cdot I_{bat} \tag{46}$$

$$E_{bat} = E_0 - V_{cbat} \tag{47}$$

$$V_{cbat} = K_{bat} \cdot \frac{Q}{Q - K_{bat} \cdot \int i_{bat}\, dt} + A_{bat} \cdot \exp\left(-B_{bat} \cdot \int i_{bat}\, dt\right) \tag{48}$$

where E_0 is the no-load battery voltage (V), K_{bat} is the polarization voltage (V), Q is the battery capacity (Ah), A_{bat} is the exponential zone amplitude (V), B_{bat} is the exponential zone time constant inverse $(Ah)^{-1}$, V_{Bat} is the battery voltage (V), R_{bat} is the battery internal resistance (Ω), I_{Bat} is the battery current (A), and $\int i_{bat}\, dt$ is the charge supplied and drawn by the battery (Ah).

The battery state of charge (SOC) is an important parameter to be controlled. It is the amount of electricity stored during the charge; the Q_d is the ampere-hours stored in the battery during a time (t) with a charging current I_{bat} and C_{bat}.

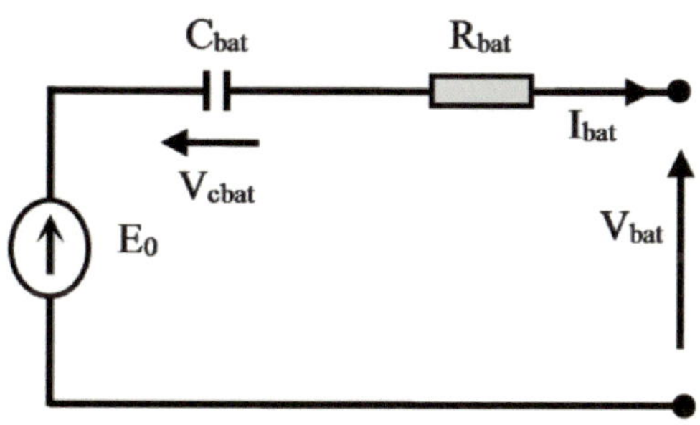

Figure 12.
Electrical model of lead-acid battery.

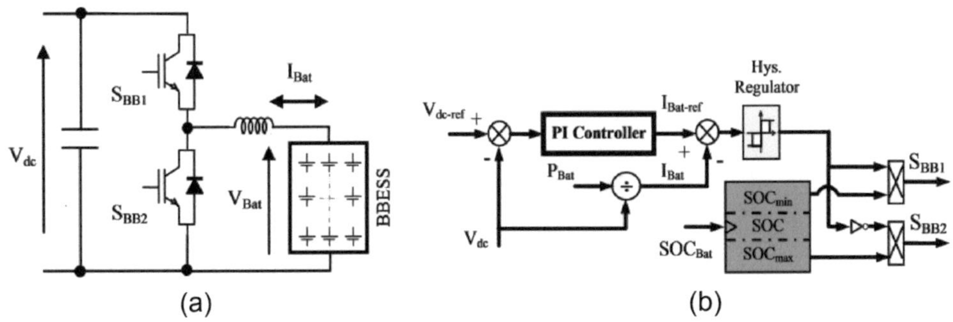

Figure 13.
Schematic diagram of BBDC converter controller associated with BBESS. (a) Power system conversion, and (b) Control system.

The supervisory system must know the battery state of charge to make a decision according to its status and the load power demand.

In case of modern lead acid batteries, charge acceptance is very high; typically, the SOC range is between a low limit around 20% and upper limit around 80% [2–6], where the SOC is given by Eq. (49).

$$SOC(\%) = 100 \cdot \left(1 - \frac{Q_d}{C_{batt}}\right) = 100 \cdot \left(1 - \frac{I_{batt} \cdot t}{C_{batt}}\right) \quad (49)$$

The primary objective of the Buck-Boost DC converter (BBDC) control is to maintain constant DC-grid voltage, as a reference value, and to discharge/charge power from/into BBESS according to the required power. The schematic diagram of the BBESS Buck-Boost DC converter and its control are presented in **Figure 13**. The BBESS voltage can be kept lower as compared to the DC-grid voltage reference ($V_{dc\text{-}ref}$) using BBDC. Hence, several batteries are required to be connected in serial. In the proposed micro-grid, the V_{bat} is kept around 420 V. Therefore, it is required that 36 cells of batteries are connected in series.

3.6 Energy management system

The BBESS can act either as a power supply or as a dump load, it's should discharge/charge within specified limits when there is deficit/surplus of hybrid power, due to the weather conditions or perturbation in power requirement. Knowing that, due to BBESS dynamics, it cannot feed the power instantaneously and hence unable to stabilize the DC-grid voltage during transient state. In order to overcome this constraint, a FC is incorporated using a Boost converter (connected between FC and DC-grid), as given in **Figure 9**, in such a manner that the FC provides the power instantaneously. Consequently, this FC should increase the BBESS power which is decreasing. The flow chart of the above-discussed control coordination among various sources is depicted in **Figure 14**.

The DC-grid control voltage is assured by the Buck-Boost DC converter associated with BBESS, as illustrated in **Figure 13**. The BBESS reference power is given by the power flow available in the DC-grid while respecting the flowchart given in **Figure 14**, imposed by the energy management system proposed.

The different powers which flow between the generation sources, the storage systems and the required demand are calculated as:

$$P_g = P_{AC} = P_{L-AC} + P_{gAC} \quad (50)$$

$$P_{rAC} = P_g - P_s \quad (51)$$

$$P_{rg} = P_{L-DC} + P_{rAC} \tag{52}$$

$$P_{Net} = P_r + P_{rg} \tag{53}$$

$$P_{FC} = P_{Net} - P_{Bat} - P_{Elect} \tag{54}$$

where, P_s and P_r are the stator and rotor powers generated by DFIG. The $P_{L\text{-}AC}$ and $P_{L\text{-}DC}$ are the variable AC and DC load powers, P_{rAC} is the power transferred through DC/AC converter II, P_{Elc} and P_{Bat} are the Electrolyzer and BBESS powers, also P_{FC} and P_{pv} are the generated FC and photovoltaic powers, respectively. Whereas P_{AC} and P_g are the powers exchanged between the proposed micro_grid system and the AC grid, P_{gAC} is the power exchanged between the proposed MG associated with its local variable AC load and the external AC grid.

The energy control and management strategies consist that, at any given time, the BBESS provides the averaged power and the FC provides the instantaneous power. When the BBESS SOC is reached its maximum at 80% in charge mode, the electrolyzer is turned on to store the excess power and to begin producing hydrogen, as given in **Figure 11**, which is delivered to the hydrogen storage tanks, as shown in **Figure 1**, but if this limit is reached its minimum at 20% in discharge mode, the FC provides the deficit required power. However, in case of a high wind or solar energies conditions or lower demand ($P_{Net} > 0$), the surplus power at first is supplied to the BBESS until it reaches its upper limit of charge carrying capacity, the Electrolyzer is used to support the BBESS operations to absorb the additional power, as given in **Figure 14**.

In case of low or no power generation from RESs and even high demand ($P_{Net} < 0$), the power requirement will be supplied by the BBESS up to SOC lower limit. At this moment, the FC ensures continuity of service by supplying the power requirement.

The DC-grid voltage control by the Buck-Boost actually means managing the flow of available power at this DC-grid, while respecting the flowchart given in **Figure 14**, imposed by the energy management system proposed.

The local reactive power compensation exchanged between the proposed MG and the bus-bar from the AC side (AC grid), Q_{AC} or Q_g, is provided by the DFIG. Knowing that, the reactive power at the DC/AC converter II output is forced to zero (operating at unity power factor).

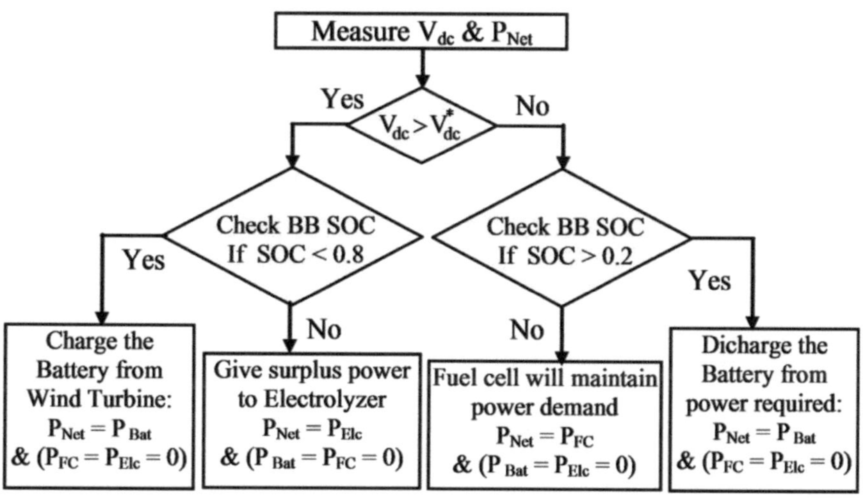

Figure 14.
Flowchart of control coordination and energy management algorithm.

$$Q_{AC} = Q_g = Q_s + Q_{rAC};\ Q_{rAC} = 0 \tag{55}$$

$$Q_{AC} = Q_s = Q_g = Q_{gAC} + Q_{L-AC} \tag{56}$$

$$Q_{L-AC} = P_{L-AC} \cdot tg(\varphi_{L-AC}) \tag{57}$$

where, Q_s is the reactive power flowing between the DFIG stator and the AC grid side. While, $Q_{L\text{-}AC}$ and Q_{AC} or Q_g represent respectively, the reactive powers exchanged with the local AC load and the AC grid together with the proposed system. Q_{rg} is the reactive power transferred through DC/AC converter II. The local load phase shift is given by φ_{L-AC}, where P_{gAC} is the power exchanged between the proposed MG associated with its local variable AC load and the external AC grid.

4. Simulation results and discussion

In this section, the WT-DFIG control simulation and the utility of the proposed MG simulation have been presented.

4.1 WT-DFIG control simulation results and discussion

In this section, the simulation of the Variable Wind speed generator (VSWG) behavior using DFIG in conjunction with an indirect AC-AC converter connected to the grid is developed. So, in order to test the control efficiency and robustness of the WT-DFIG under constraints reflecting an operation that is close to its actual behavior, we opted for the wind speed profile illustrated in **Figure 15a**. In the aim to take advantage and to verify the validity and highlight of the proposed system performances, a model allowing the machine with its control behavior simulation, for different operating zones, is developed, as illustrated in **Figure 4**.

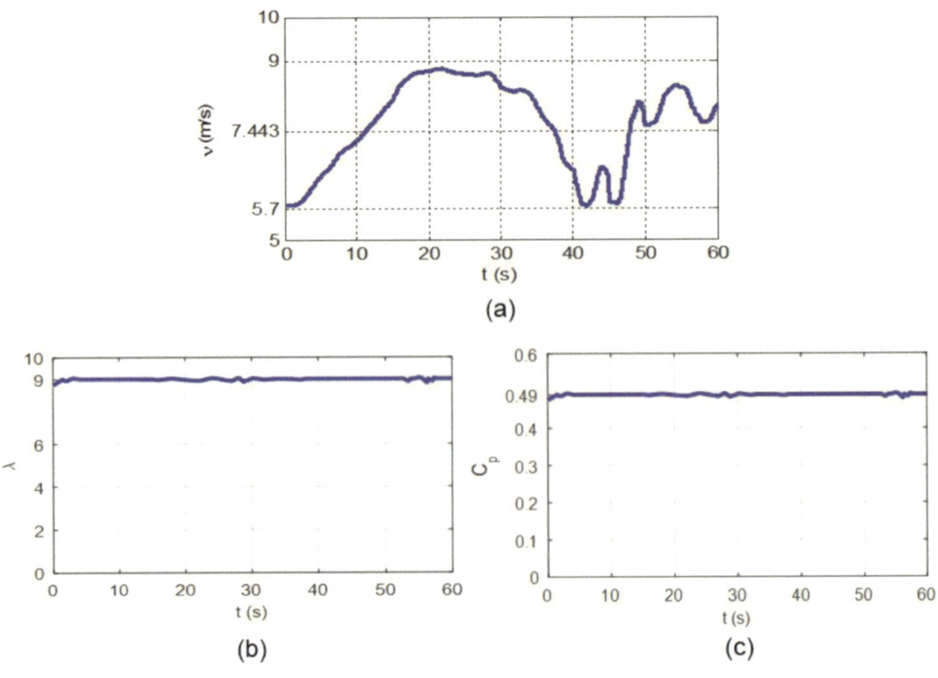

Figure 15.
Waveforms of WT characteristics: (a) wind speed profile. (b) Tip-speed ratio. (c) Power coefficient.

The machine and control parameters details considered in this studied are given in [3, 19], where the simulation results are presented thereafter, as illustrated in **Figures 15–22**.

The MPPT algorithm applied makes it possible to have the ratio of speed equal to its optimum value ($\lambda_{opt} = 9$) with a maximum power coefficient ($C_{p\text{-}max} = 0.49$) as indicated in **Figure 15b** and **c**. These last variables are almost constant despite the fluctuations caused by the wind speed profile variations.

The results illustrated in this section show that the references tracking is very well achieved, which is verified for the two variables, the active power delivered to

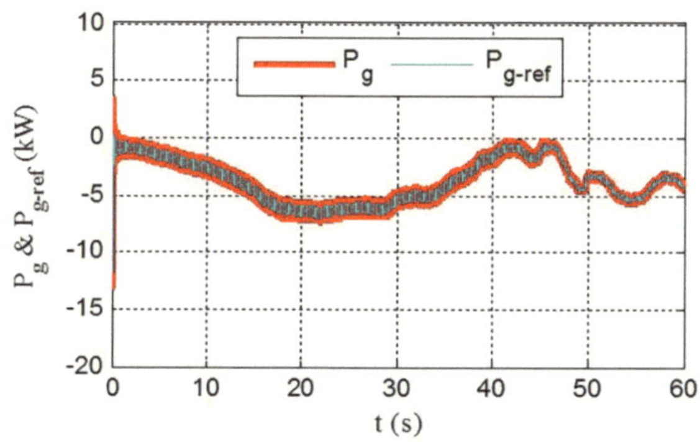

Figure 16.
Total active power generated by the WT-DFIG.

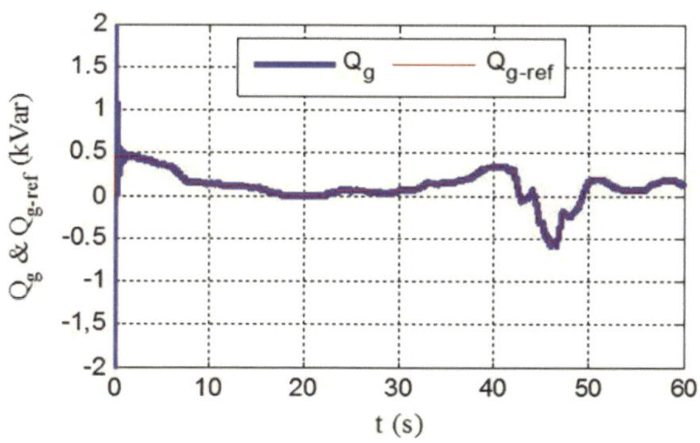

Figure 17.
Total reactive power compensated by the DFIG.

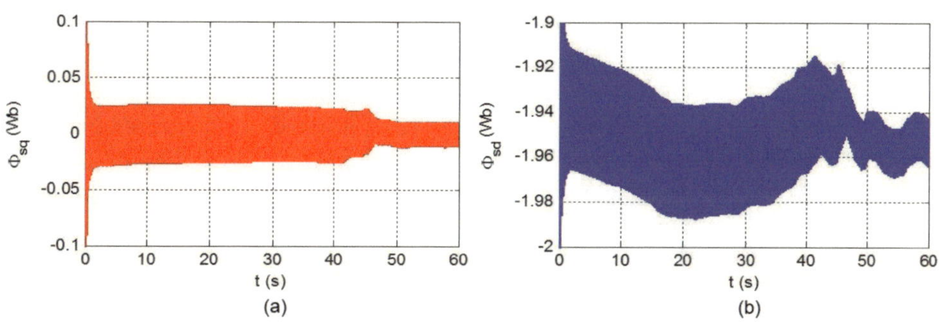

Figure 18.
DFIG direct and quadrature stator flux. (a) Direct stator flux. (b) Quadrature stator flux.

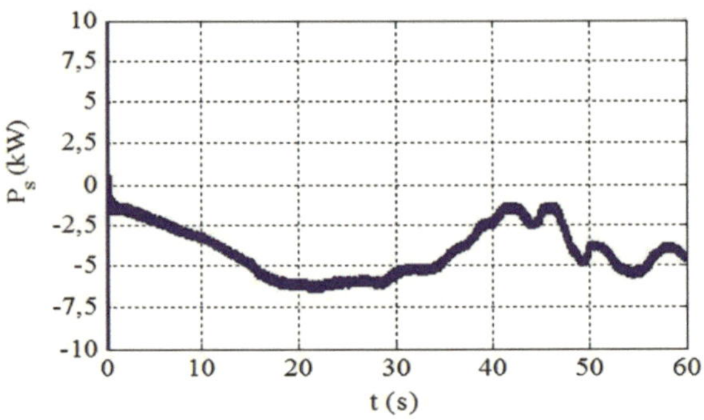

Figure 19.
Active power generated by the DFIG stator.

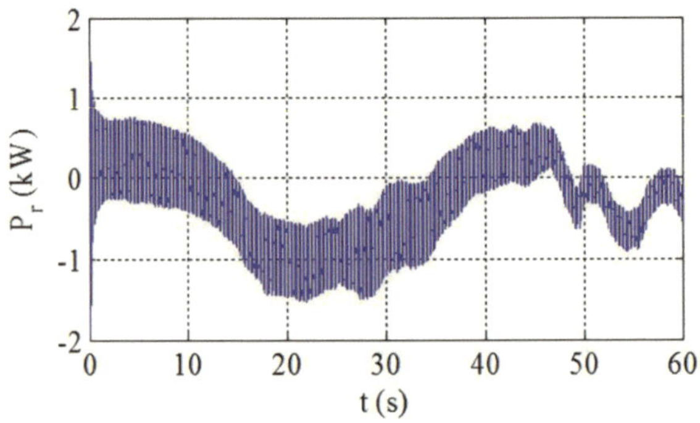

Figure 20.
DFIG rotor active power.

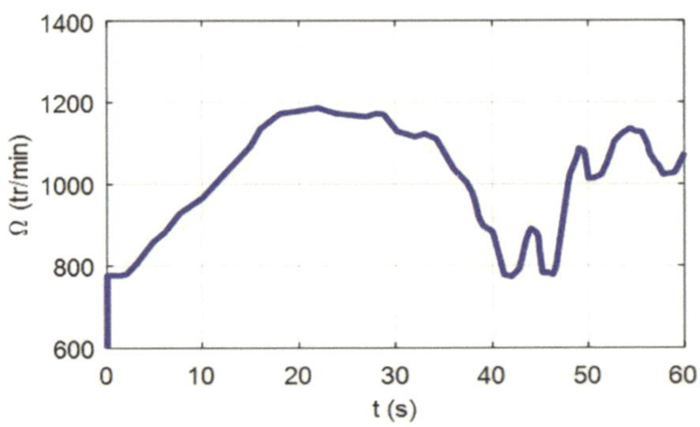

Figure 21.
DFIG mechanical speed.

the AC grid and the local reactive power to be compensated, as illustrated in **Figures 16** and **17**.

In addition, the control decoupling is represented on the stator flux evolution, as given in **Figure 18**; where, the quadrature component is always kept zero (**Figure 18a**), while that of direct axis follows the form of the active power produced, as presented in **Figures 18b** and **19**.

The machine operates in the three operation modes, as shown in **Figures 20–22**. For $s > 0$, the DFIG is run in the sub-synchronous mode, indicating that the rotor power is supplied by the grid. For $s < 0$, the DFIG is operated in the

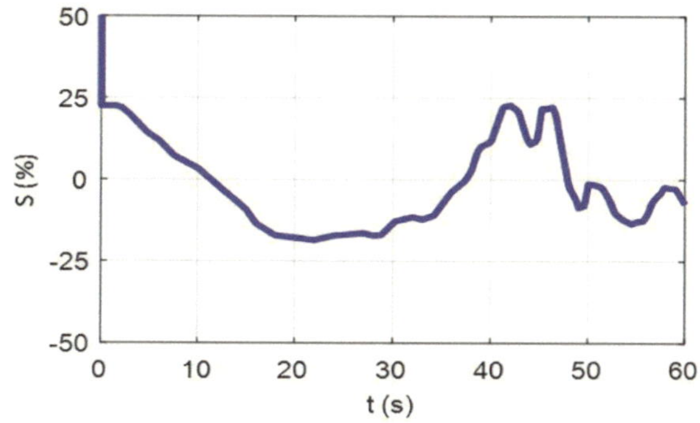

Figure 22.
DFIG slip.

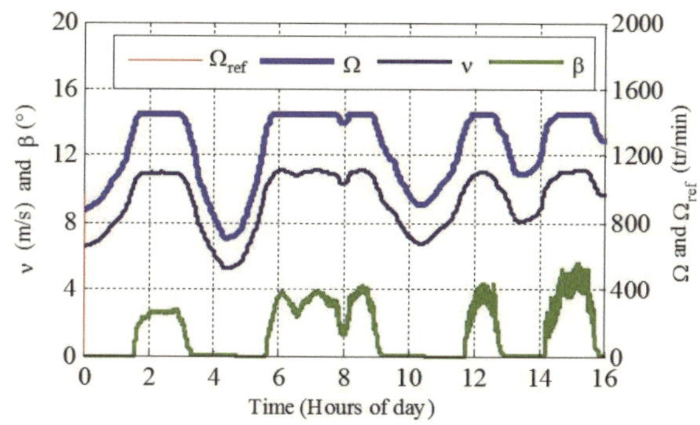

Figure 23.
Wind speed profiles, DFIG mechanical speed with its reference with blade pitch angle evolution.

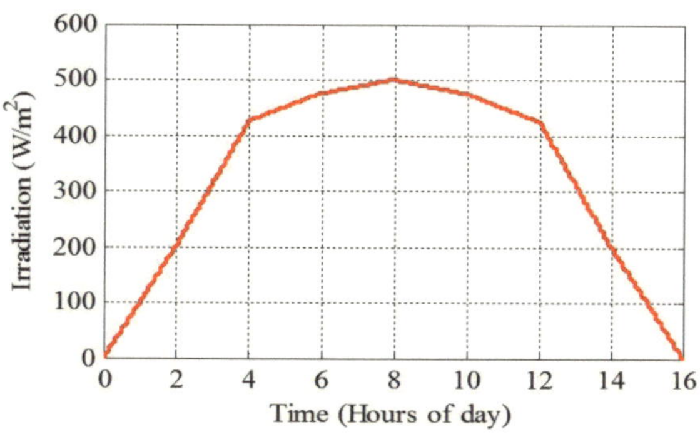

Figure 24.
Irradiation profiles waveform.

super-synchronous mode. It can be seen, from **Figure 21**, that the rotor power flow direction is reversed. For s = 0, the DFIG functions as a synchronized asynchronous generator, the rotor active power is absorbed as rotor Joules losses.

4.2 Global MG control simulation results and discussion

In this part, the analysis of the performances and behavior of the proposed MG, as presented in **Figure 1**, will be studied. The proposed system simulation is realized

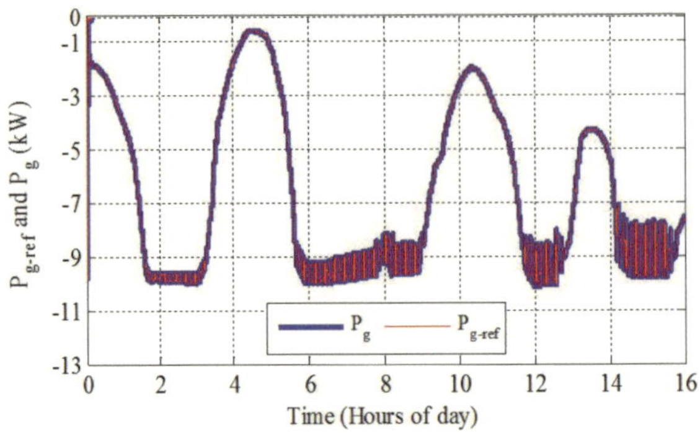

Figure 25.
WT-DFIG generated active power evolution with its reference.

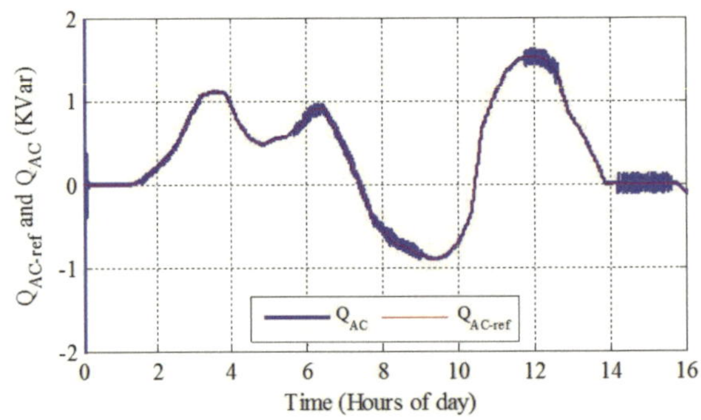

Figure 26.
Evolution of the reactive power compensation with its reference.

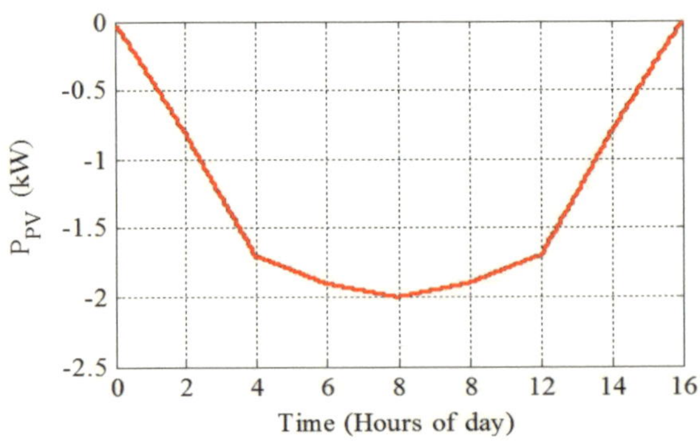

Figure 27.
Evolution of the PV power generation.

under different situations, as variable load demand and random weather conditions data (wind speed, solar irradiance, etc.). The wind speed profile is selected shown the two useful zones operating ranges of the turbine, as given in **Figure 2**. In zones I and II, the maximum power extraction algorithm (MPPT) is applied, but in zone III, defined as over-speed turbine operations, constant power is generated by acting on the pitch angle control, as illustrated in **Figure 23**. Simulation results for this case are given and discussed in the following section, as illustrated in **Figures 23–32**.

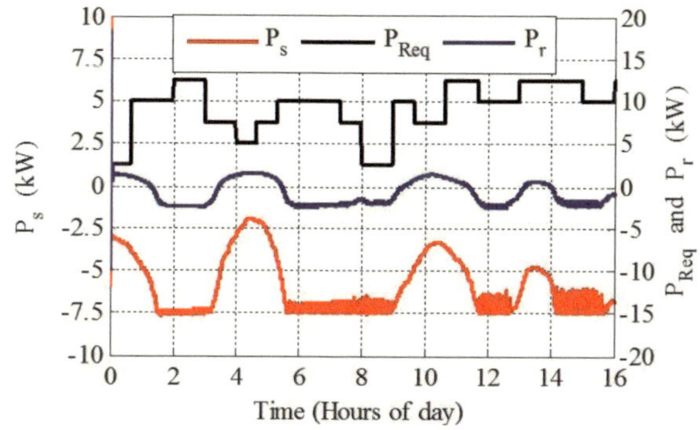

Figure 28.
Average required demand power and the stator active power generated by DFIG with its rotor active power.

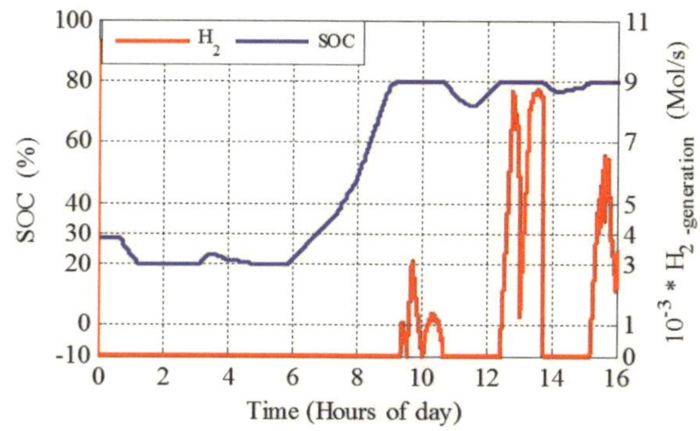

Figure 29.
Waveforms of the hydrogen (H_2) generation rate and the battery SOC evolution.

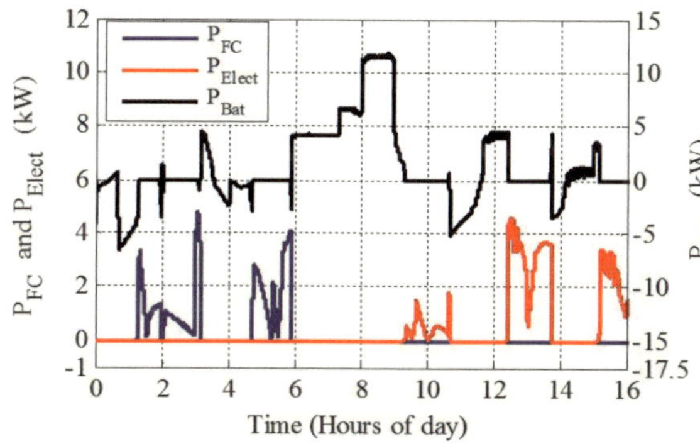

Figure 30.
Evolution of the hybrid storage powers (power available for H_2 generation and BBESS power) and FC generated power.

The simulation parameters of the different elements composing the proposed system, associated controls and proposed energy management are given in [2, 3, 19].

The total hourly averages of the weather data (solar radiation and wind speed profile) and powers required profiles, over 16 hours by day are shown in **Figures 23** and **24**. The wind speed profile is chosen in a manner that the WT-DFIG operates in a different area (MPPT and blade pitch control zone), as presented in **Figure 23**.

Also, this figure illustrates the blade pitch angle. When the generated stator active power reached its maximum value (P_{smax}), the pitch angle control is activated, it generates the variation of the tip speed ratio and the power coefficient. The DFIG mechanical speed is kept constant, at its rated value, as shown in **Figure 23**. However, in the MPPT regions, the produced active power is maximized. Where the blade pitch angle, tip speed ratio and the power coefficient are maintained at constant value ($\beta = 0$, $\lambda = \lambda_{opt} = 9$, $C_p = C_{p\text{-}max} = 0.49$).

The waveform of the generator mechanical speed is depicted in **Figure 23**, where the generated grid active power and the local reactive power compensation are illustrated in **Figures 25** and **26**, respectively. These results affirm that the good performances of the proposed control, where the generated grid active power varies as a result of the wind speed profile to maximize the power coefficient in the MPPT zones. While in the blade pitch control, the generated power (**Figure 25**) is constant to maintain the constant electromagnetic torque, whereas, the compensated reactive power is kept at its reference and varies according to the magnitude of the direct stator flux. Thus, the WT is used in an optimal way and justifying the usefulness of the DFIG in the possibility of management and the active-reactive powers control.

Figures 27 and 28 show that all the same of changes in the power distribution curve of generated solar power (**Figure 27**), stator active power generated by DFIG with its rotor active power and the required power (**Figure 28**).

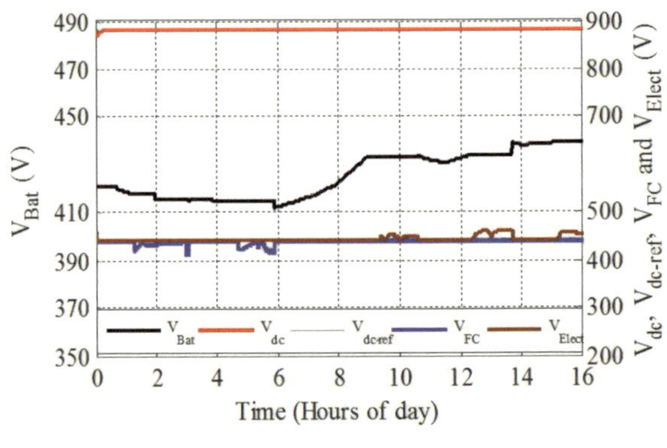

Figure 31.
Evolution of the DC-grid, electrolyzer, battery and FC voltages waveforms.

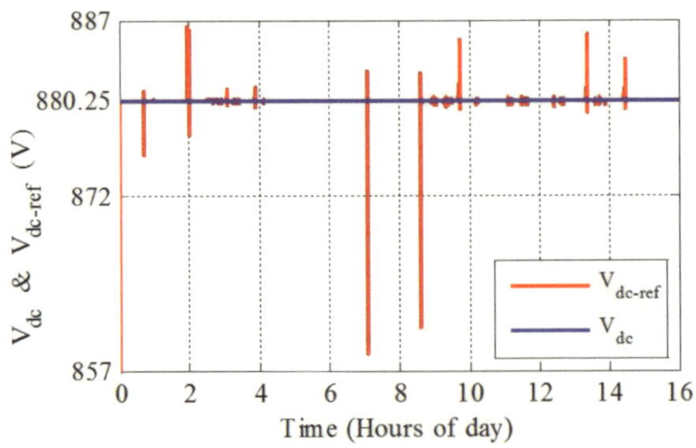

Figure 32.
Zoom of the DC-grid voltage evolution with its reference.

The waveforms of the battery SOC and the H_2 generation rate over are shown in **Figure 29**. The SOC is limited within range (20–80%).

Hence, **Figures 29** and **30** show when the generation powers are insufficient to supply the required power and the SOC of the battery reach its minimum at 20%, the FC supplements the deficit energy to meet the demand requirement. While, when the generated power is greater than the required demand and the battery is fully charged, the SOC reaches its maximum limit at 80%, the excess energy is used to generate the H_2, as illustrated in **Figure 29**.

Figure 31 shows that the FC, the Electrolyzer, the battery bank and the DC-grid output voltage variations are respectively expected. Indeed, the evolution of the different voltages represented in this figure follow the evolution of the power transited by the element that represents. In addition, the reference tracking of the DC-grid voltage is depicted in **Figure 32**. This voltage is constant and follows its reference, despite the variations in hybrid powers generation and required demand. So, it can be confirmed that the performances of the buck-boost DC-DC bidirectional converter controller are quite satisfactory in both transient and steady-state hybrid power and required demand conditions.

5. Conclusions

In this paper, a renewable multi-source microgrid associated with a hybrid energy storage system is studied. The energy sources combined WT-DFIG with a PV generator and a Fuel cell. Whereas, the hybrid energy storage is composed of batteries bank energy storage system (BBESS) associated with an electrolyzer system, which is a hydrogen generator.

The simulation model of the overall hybrid MG system has been developed using MATLAB/Simulink.

The system configuration and the characteristics of the main components in the system are given, and the overall control and power management strategy for the proposed hybrid energy system is presented. The WT-DFIG and PV generation systems are the main power generation devices and the electrolyzer-BB act as a dump load using any excess power available to BB charging and produce H_2. The FC and BB system is the backup generation and supply power to the system when there is a power deficit.

The obtained results confirm also the robustness and the efficiency of the proposed control that allows us to enjoy the benefits of the combination between a stator flux-oriented control (SFOC) with variable structure control (VSC) process, using the nonlinear Sliding Mode control (SMC) technique to control the generated active power to the AC grid and to govern the local reactive energy compensation. This latter justifies the usefulness of the DFIG in the possibility of management and the active-reactive powers control, especially, operating as a local reactive power compensator, where it can work in the synchronized asynchronous mode as a synchronous generator.

Furthermore, the obtained results showed clearly the satisfactory performances and the effectiveness of the proposed micro-grid. Indeed, the main objectives of this contribution, which are: efficient control of the coordination between the various energy sources, a proper operation of the proposed micro-grid (DC and AC grid) providing a better energy quality and an uninterrupted service are satisfied. Then, better energy management is observed, permitting an optimal operation of the whole system with high performances of the micro-grid under unfavorable and critical conditions such as power required demand variations and random weather conditions data (wind speed random profile, variable solar irradiance).

Author details

Salah Tamalouzt[1*], Nabil Benyahia[2], Abdelmounaim Tounzi[3] and Amar Bousbaine[4]

1 Laboratory LTII, Faculty of Technology, University of Bejaia, Algeria

2 Laboratory LATAGE, University M/Mammeri of Tizi-Ouzou, Algeria

3 Laboratory L2EP, University of Lille, Lille, France

4 College of Engineering and Technology, University of Derby, UK

*Address all correspondence to: tamalouztsalah@yahoo.fr

References

[1] Boudries Z, Tamalouzt S. Study on sliding mode control of a small size wind turbine permanent magnet synchronous generator system. Revue Roumaine des Sciences Techniques-Serie Electrotechnique et Energetiqu. 2019;**64, 2**:157-162

[2] Tamalouzt S, Benyahia N, Rekioua T, Rekioua D, Abdessemed R. Performances analysis of WT-DFIG with PV and fuel cell hybrid power sources system associated with hydrogen storage hybrid energy system. International Journal of Hydrogen Energy. 2016;**41**(45):21006-21021. DOI: 10.1016/j.ijhydene.2016.06.163

[3] Tamalouzt S. Contribution to the study, design and control of double fed electromagnetic converters [thesis]. A. MIRA University of Bejaia: Faculty of Technology, Department of Electrical Engineering, LTII Laboratory; 2017

[4] Tamalouzt S, Benyahia N, Rekioua T, Abdesemed R. A doubly fed induction generator wind turbine and fuel cell hybrid power sources system for micro-grid applications. In: Proceeding of 3rd International Conference on Control, Engineering & Information Technology (CEIT'2015); May 25–27 2015; Tlemcen. Algeria: IEEE; 2015. pp. 1-6

[5] Tamalouzt S, Rekioua T, Rekioua D, Hamoudi F. Variable speed wind generator associated with hybrid energy storage system-application for micro-grids. In: Proceeding of 5th International Renewable Sustainable Energy Conference (IRSEC'2017); December 4–7 2017; Marrakech & Ouarzazate. Morocco: IEEE; 2018. pp. 1-6

[6] Tamalouzt S, Benyahia N, Tounzi A, Idjdarene K. Performance analysis of a micro-grid system composed of renewable energy sources with hybrid energy storage system. In: Proceeding of ELECTRIMACS'2019; 21–23 May 2019. Salerno, Italy. pp. 1-6

[7] Tamalouzt S, Rekioua T, Abdesemed R. Direct torque and reactive power control of grid connected doubly fed induction generator for the wind energy conversion. In: Proceeding of International Conference on Electrical Sciences and Technologies in Maghreb (CISTEM); 3–6 Nov 2014. Tunis: IEEE; 2015. pp. 1-7

[8] Sahri Y, Tamalouzt S, Lalouni Belaid S. Direct torque control of DFIG driven by wind turbine system connected to the grid. In: Proceeding of International Conference on Wind Energy and Applications in Algeria (ICWEAA'2018); Nov. 6–7 2018. Algiers, Algeria: IEEE; 2019. pp. 1-6

[9] Tamalouzt S, Idjdarene K, Rekioua T, Abdessamed R. Direct torque control of wind turbine driven doubly fed induction generator. Revue Roumaine des Sciences Techniques-Serie Electrotechnique et Energetiqu. 2016; **61**(3):244-249

[10] Tamalouzt S. Performances of direct reactive power control technique applied to three level-inverter under random behavior of wind speed. Revue Roumaine des Sciences Techniques-Serie Electrotechnique et Energetiqu. 2019;**64**(1):33-33

[11] Tamalouzt S, Rekioua T, Abdessamed R, Idjdarene K. Direct torque control of grid connected doubly fed induction generator for the wind energy conversion. In: Proceeding of International Renewable Energy Congress (IREC'2012), December 19–22 2012. Sousse, Tunisia. pp. 1-6

[12] Abad J, Rodriguez MA, Wanski G, Poza L. Direct power control of doubly

fed induction-generator based wind turbines under unbalanced grid voltage. IEEE Transactions on Power Electronics. 2010;**25**(2):586-588. DOI: 10.1109/TPEL.2009.2027438

[13] Shang L, Hu J. Sliding-mode-based direct power control of grid-connected wind-turbine-driven doubly fed induction generators under unbalanced grid voltage conditions. IEEE Transactions on Energy Conversion. 2012;**27**(2):362-373. DOI: 10.1109/TEC.2011.2180389

[14] Abad J, Rodriguez MA, Slama-Belkhodja I, Charaabi L. Direct virtual torque control for doubly fed induction generator grid connection. IEEE Transactions on Industrial Electronics. 2009;**56**(10):4163-4173. DOI: 10.1109/TIE.2009.2021590

[15] Hua J, Yuan X. VSC-based direct torque and reactive power control of doubly fed induction generator. Renewable Energy. 2012;**40**(1):13-23. DOI: 10.1016/j.renene.2011.08.023

[16] Payam AF, Hashemnia MN, Faiz J. Robust DTC control of doubly-fed induction machines based on input-output feedback linearization using recurrent neural networks. Journal of Power Electronics (JPE). 2011;**11**(5):719-725

[17] Shehata EG. Active and reactive power control of doubly fed induction generators for wind energy generation under unbalanced grid voltage conditions. Electric Power Components & Systems. 2013;**41**(6):619-640. DOI: 10.1080/15325008.2013.763308

[18] Ghedamsi K, Aouzellag D. Improvement of the performances for wind energy conversions systems. Electrical Power and Energy Systems. 2010;**32**(9):936-945. DOI: 10.1016/j.ijepes.2010.02.012

[19] Tamalouzt S, Rekioua T, Abdesemed R. A novel power sliding mode control of DFIG used in wind power generation fed by matrix converter. In: Proceeding of International Conference on Electrical Engineering, Electronics and Automatic (ICEEA'10), November 1–3 2010. Bejaia, Algeria. pp. 1-6

[20] Soares O, Gonçalves H, Martins A, Carvalho A. Nonlinear control of the doubly-fed induction generator in wind power systems. Renewable Energy. 2010;**58**(8):1662-1670. DOI: 10.1016/j.renene.2009.12.008

[21] Abdelfetteh A, Abid M. Hybrid fuzzy sliding mode control of a doubly fed induction generator in wind turbines. Revue Roumaine des Sciences Techniques-Serie Electrotechnique et Energetiqu. 2012;**57**(4):15-24

[22] Abdeddaim S, Betka A, Drid S, Becherif M. Implementation of MRAC controller of a DFIG based variable speed grid connected wind turbine. Energy Conversion and Management. 2014;**79**:281-288. DOI: 10.1016/j.enconman.2013.12.003

[23] Pichan M, Rastegar H, Monfared M. Two fuzzy-based direct power control strategies for doubly-fed induction generators in wind energy conversion systems. Energy. 2013;**51**:154-162. DOI: 10.1016/j.energy.2012.12.047

[24] Bhende CN, Mishra S, Malla SG. Permanent magnet synchronous generator-based standalone wind energy supply system. IEEE Transactions on Sustainable Energy. 2011;**2**(4):361-373. DOI: 10.1109/TSTE.2011.2159253

[25] Wang C, Hashem NM. Power management of a stand-alone wind/photovoltaic/fuel cell energy system. IEEE Transactions on Energy Conversion. 2008;**23**:957-967. DOI: 10.1109/TEC.2007.914200

[26] Onar OC, Uzunoglu M, Alam MS. Modeling, control and simulation of an autonomous wind turbine/photovoltaic/ fuel cell/ultra-capacitor hybrid power system. Journal of Power Sources. 2008; **185**:1273-1283. DOI: 10.1016/j.jpowsour. 2008.08.083

[27] Belfkira R, Zhang L, Barakat G. Optimal sizing study of hybrid wind/ PV/diesel power generation unit. Solar Energy. 2011;**85**:100-110. DOI: 10.1016/ j.solener.2010.10.018

[28] El-Shater TF, Eskander MN, El-Hagry MT. Energy flow and management of a hybrid wind/PV/fuel cell generation system. Energy Conversion and Management. 2006;**47**: 1264-1280. DOI: 10.1080/1478645 0600631483

[29] Gonzàlez I, Ramiro A, Calderòn M, Calderòn AJ, Gonzàlez JF. Estimation of the state-of-charge of gel lead-acid batteries and application to the control of a stand-alone wind-solar test-bed with hydrogen support. International Journal of Hydrogen Energy. 2012;**37**: 11090-11103. DOI: 10.1016/j.ijhydene. 2012.05.001

[30] Maleki A, Askarzadeh A. Comparative study of artificial intelligence techniques for sizing of a hydrogen-based stand-alone photovoltaic/wind hybrid system. International Journal of Hydrogen Energy. 2014;**39**(19):9973-9984. DOI: 10.1016/j.ijhydene.2014.04.147

[31] Malla SG, Bhende CN. Voltage control of stand-alone wind and solar energy system. Electrical Power and Energy Systems. 2014;**56**:361-373. DOI: 10.1016/j.ijepes.2013.11.030

[32] Yang H, Lu L, Zhou W. A novel optimization sizing model for hybrid solar-wind power generation system. Journal of Solar Energy. 2007;**81**: 76-84. DOI: 10.1016/j.solener.2006. 06.010

[33] Benyahia N, Rekioua T, Benamrouche N. Modeling and simulation of a stand-alone wind/ photovoltaic/fuel cell system associated with a hybrid energy storage. In: Proceedings of the Third International Renewablle Energy Congress (IREC'2011); 20–22 December 2011. Hammamet, Tunisia. pp. 1-7

[34] Zhou T, François B. Energy management and power control of a hybrid active wind generator for distributed power generation and grid integration. IEEE Transactions on Industrial Electronics Energy Conversion. 2011;**58**(1):95-104. DOI: 10.1109/TIE.2010.2046580

[35] Ou TC, Hong CM. Dynamic operation and control of microgrid hybrid power systems. Energy. 2014;**66**: 314-323. DOI: 10.1016/j.energy.2014. 01.042

[36] Samrat NH, Ahmad N, Choudhury IA, Taha Z. Modeling, control, and simulation of battery storage photovoltaic-wave energy hybrid renewable power generation systems for island electrification in Malaysia. The Scientific World Journal. 2014;**21**:1-21. DOI: 10.1155/2014/436376

[37] Osman Haruni AM, Negnevitsky M, Haque ME, Gargoom A. A novel operation and control strategy for a standalone hybrid renewable power system. IEEE Transactions on Sustainable Energy. 2013;**4**(2):402-413. DOI: 10.1109/TSTE.2012.2225455

[38] Obara S, Morizane Y, Morel J. Economic efficiency of a renewable energy independent microgrid with energy storage by a sodiume-sulfur battery or organic chemical hydride. International Journal of Hydrogen Energy. 2013;**38**:8888-8902. DOI: 10.1016/j.ijhydene.2013.05.036